Friedhelm Weick

Owls (Strigiformes) · Annotated and Illustrated Checklist

Eulen (Strigiformes) · Kommentierte und illustrierte Artenliste

Friedhelm Weick

Owls (Strigiformes)

Annotated and Illustrated Checklist

With 63 Illustrations and 23 Color Plates

Friedhelm Weick

Pommernstraße 34
76646 Bruchsal-Untergrombach
Germany

About the Author

Friedhelm Weick was born in Karlsruhe, Germany. For 35 years he worked as a technician in the construction industry. Throughout this time he painted birds, and since 1965 he has also been a professional illustrator and wildlife artist. Since 1988 he has been a graphic designer at the Museum for Natural History in Karlsruhe, a job which entailed painting a complete range of animals and plants. He illustrated over 120 books, including most volumes of the *Handbuch der Vögel Mitteleuropas* (1971–1997) and all volumes of *Die Vögel Baden-Württembergs* (1987–2006). His work has appeared in all leading wildlife magazines. In addition, he has published books in his own right, including *Birds of Prey of the World*, in collaboration with L. H. Brown, *Owls, a Guide to the Owls of the World*, in collaboration with C. König and J. H. Becking and *Anmut im Federkleid*, in collaboration with O. Kröher.

Library of Congress Control Number: 2006927285

ISBN-10 3-540-35234-1 **Springer Berlin Heidelberg New York**
ISBN-13 978-3-540-35234-1 **Springer Berlin Heidelberg New York**

Springer is a part of Springer Science+Business Media
springer.com

Printed in Germany

Editor: Dr. Dieter Czeschlik, Heidelberg, Germany
Desk Editor: Dr. Jutta Lindenborn, Heidelberg, Germany
Cover design: *design&production*, Heidelberg, Germany
Typesetting: Stasch, Bayreuth, Germany
Production: LE-TEX Jelonek, Schmidt & Vöckler GbR, Leipzig, Germany
Printing and binding: Stürtz AG, Würzburg, Germany

Printed on acid-free paper 31/3100 YL – 5 4 3 2 1

Dedicated to three splendid owl-experts and friends /
Den drei hervorragenden Eulenkennern und Freunden gewidmet:

Dr. h. c. Siegfried Eck †, Dresden, Germany
Dr. Gerrit Paulus Hekstra, Harich, The Netherlands
Professor Dr. Claus König, Ludwigsburg, Germany

Foreword

The owl is a magical creature: it exudes an ambivalent fascination that can trigger completely antithetical responses. On the one hand, there is something unsettling about the owl's soundless flight, eerie call, and nocturnal activity that can result in its demonization as a messenger of darkness. On the other hand, however, its masklike visage and seemingly penetrating gaze prove enthralling, and this may have contributed to the deification of the owl as the companion spirit of an omniscient mother earth. In addition, the highly specialized captor of prey aroused the competitive envy of hunting parties, even provoking a bit of resentment when they lost out to owls in their hunt for small game. Moreover, the excrement and pellets (hairballs) they deposited in barns and church towers contributed to their reputation as undesirables or even pests. With the overcoming of superstition and suspicion, our relationship with these creatures of the night has now become one of unreserved admiration. They grant us insights into the rich diversity of adaptation to nocturnal hunting as well as into sound inventories, brooding biology, and nestling development.

In a complete transformation of what was once a symbol of fear and hatred, charismatic owls now enjoy an almost singular popularity in large parts of the population. Whether it be the barn owl, eagle owl, or boreal (Tengmalm's) owl, each is well-suited as the flagship species for a biotope and species protection in keeping with our times. A gratifyingly large number of specialists, amateurs, and enthusiasts are involved in owl protection: creating potential nesting hollows, banding birds, collecting pellets, guarding endangered breeding areas, or mitigating the risk of electrocution by power poles. The organization of owl friends on the local, regional, and international level ensures that advisory support and coordination take place at a high level of expertise.

Other friends of owls seek to regenerate owl populations in as natural a way as possible, whether for resettlement or for securing the continued existence of highly endangered species. This effort encompasses not only zoos, bird parks, and special breeding stations, but also those work-intensive rehabilitation centers in which injured or orphaned owls can be given professional care. All of these organizations and facilities serve, directly or indirectly, to inform and enlighten the public and are thus instrumental in ensuring a positive image for the once so detested "wraiths of the night." The spectrum of owl-related hobbies includes nature photography, painting and sculpture, with the lattermost encompassing the most diverse forms of owl representation. Owls it appears are highly reluctant to let go of anyone who has fallen into their clutches.

This certainly applies to the author of this taxonomic list of owl species. Deeply impressed by the grace and patterns of behavior of wild animals, Friedhelm Weick moved from a very different profession to become – about 40 years ago – a painter of animals and an illustrator of textbooks. The beginnings of his collections of illustrations and sources from the literature, as well as his study of museum skins (to obtain an overview of the world of owls) go back at least 25 years. These activities were considerably boosted by his joint efforts with Claus König to create the first compendium of the entire order of Strigiformes (1999). Such tenacious and long-term pursuit of this project undoubtedly reflects a profound enthusiasm for owls.

In a time of ever-new media that bombard our senses and constantly seek to outdo each other in emotional impact, that dazzle us with floods of images, 3-D projections, and virtual reconstructions of long-lost worlds, a systematic list of a specific group of birds may seem anachronistic and antiquated. And, in fact, there is no lack of voices that ridicule the outmoded museum work of classical taxonomists and see no practical value in discussions of the systematic classification of organisms. Since academia too did not remain untinged by this subliminal disdain, the foundations of species information and systematic classification were largely forced out of research and teaching in the competitive struggles among the various academic disciplines. If the university, however, shirks its responsibility, taxonomy threatens to become an endangered discipline. And this despite the fact that sound knowledge of the systematic subdifferentiation of organisms represents the essential precondition for understanding evolutionary processes. In a multifaceted interplay with geography, ecology, ethology, and genetics, systematics makes it possible to trace the natural history of populations. This includes the origin and paths of dissemination of populations, their development through competition and predation, through changes in climate or shifts in available habitat, and their isolation and radiation into distinctive forms of adaptation.

Ever since the environmental protection conference in Rio de Janeiro in 1992 (the "Earth Summit"), the catch-all term of "biodiversity" has enjoyed pervasive use. Even if biodiversity extends the variety of biological phenomena to habitats and landscapes, there is no doubt that diversity of organisms continues to represent its most important foundation. This lends new weight to the genealogies of organisms. The majority of developed nations signed the Biodiversity Treaty, which makes mandatory the protection of the worldwide diversity of lifeforms. Within the European Union, the concept of "Natura 2000" is based on this objective. Aside from the "Red Lists" of endangered species, criteria were also established for Germany for determining responsibility in the protection of species. Even if these are highly commendable developments, they nevertheless remain first and foremost a reflection of a troubling process of extinction, an ever-accelerating disappearance of species caused by a historically unprecedented expansion of human settlement.

Whether projects of research involve questions of evolutionary biology, conservation programs for the preservation of the species, or the breeding of economically relevant livestock, knowledge and correct taxonomy of "species" represent the unconditional precondition for all of them. Even if the naming and categorizing of animate objects is as old as the development of human language itself, the scientifically grounded definition and differentiation of particular groups of organisms as a "species" is nonetheless a highly complex discipline. And here, apart from objectifiable decision-making criteria, subjective aspects such as worldview, system of social values, or even national ambitions and political correctness may have a part to play. Moreover, the state of knowledge regarding global diversity and its diffusion patterns remains full of gaps.

According to present listings we know of 1.7 million recent species on earth from the realms of protozoa, fungi, plants, and animals. Approximately 50 000 of them are vertebrates, of which 9 935 are birds; a good 200 of these, in turn, are owls. However, since many of the tropical forests have not yet been explored and may possibly be destroyed by logging operations prior to a species survey, well-founded speculations reckon with the possibility of up to ten times as many species.

Early museum taxonomy was only concerned with classificatory systems for designating and administering the diversity of organisms, and this was especially true as long as even biologists considered the immutability of creation as incontrovertible dogma. Ever since Darwin, however, the goal has become more ambitious: to classify species as the basis for reconstructing their evolutionary history (phylogenesis). A comparative examination of morphological, physiological, and ethological properties that takes the results of ecology and zoogeography into account expects, on the one hand, that all individuals of a population demonstrating a high level of matching traits can be brought together as of the same species and, on the other hand, that a high level of matching traits between heterogeneous populations correlates with a high degree of kinship between them. Ernst Mayr added, as an essential criterion of differentiation of individual species, the reproductive barrier developing between them in evolution, which minimizes the risk of interspecies hybridization. Nevertheless, all classic species concepts are confronted by an abundance of convergent forms, which is why they have to place great weight on the distinction between homologous and analogous traits. Additionally, a comparison of traits is frequently insufficient for the systematic differentiation of subspecies and even less sufficient for assessing higher-level taxa in terms of their phylogenetic connections.

Ultimately, molecular biology, at the end of the 20th century, has led the way out of this stagnation on the systematic level in an almost revolutionary manner, for it turned out that very fundamental questions could only be approached by incorporating modern genetics into the analysis. Above all the structural analysis of genetic material, of DNA, has revived the discussion of taxonomic classification and given it back the societal recognition it needs. Since DNA analysis looks for similarities in genetic material, it is more immune to irritations arising from convergences in morphology. The goal here goes beyond the grouping of related species; the effort is to reconstruct as realistically as possible the process of speciation with its individual branchings of the phylogenetic tree. It is even possible to locate such splitting off points in time with the help of the "molecular clock," which makes the phylogenetic tree into a multidimensional construct.

Of all the groups of organisms, the phylogenetic classification of birds is the most advanced. Currently, recent forms have been arranged into 23 orders, 146 families, and 2 141 genera of 9 935 species (as of December 2005, according to *zoonomen.net*). Spectacular fossil discoveries of recent years, especially from mesozoic deposits in China, even allow the documentation of the derivation of birds from feathered, bipedal saurians (of the Sauriurae and Ornithurae families). These fossils, of at least 120 million years in age, can already be assigned to the plovers, loons, petrels, and anserines. Owls have existed for 50–60 million years.

Within the class of birds, owls represent a conspicuously uniform morphological group. With their usually large and rigidly forward oriented eyes within the "face" of a large, round, and "neckless" head, with their loose, fleecy plumage usually in camouflaging colors, and with their broad and compact bodies, it is not hard for anyone to recognize them as owls. Insofar, however, as the uniform appearance of owls was selected by a nocturnally active lifeform as predator, this uniformity of morphological traits made systematic subdifferentiation more difficult. Given this unity, there was never any controversy regarding the owls as a uniform group, even though as early as 1840 Nitsch had proposed splitting them up into a group of barn owls and a group of "typical owls." There was, however, a great deal of uncertainty regarding their integration with their next of kin, as Sibley and Ahlquist (1990) showed in great detail in their historical survey. On the basis of their similar beak forms, Linnaeus (1758) grouped owls, as the genus *Strix*, together with falcons and shrikes (in the Accipitres order). Baird (1858) was the first to use Strigidae as a family name for nocturnal birds of prey, in the context of the bird world of North America. Huxley (1867) classified owls with vultures and secretary birds. Fürbringer (1888) was the first to dissociate owls from diurnal raptors, incorporating the Striges lineage with the family Strigidae among the Coraciiformes. Peters (1931) was the first to recognize owls as an independent order *Striges,* ranking them among nightjars, parrots, and cuckoos.

According to Sibley and Ahlquist, the methods of molecular biology made it possible to partially corroborate these kinship relationships. Nonetheless, in this interpretation, owls were separated from the Coraciiformes and at the same time moved closer to the swifts and hummingbirds. DNA analysis requires the dissociation, once and for all, of nocturnal from diurnal birds of prey. These authors add a new superorder, Strigimorphae, which includes the Strigiformes order, with its 319 species in 51 genera. A suborder, Strigi, within the latter, includes the two owl families, Tytonidae, with its 15 species in 2 genera, and Strigidae, with its 184 species in 27 genera. As of late, the frogmouths and the nightjars are also incorporated as suborders within the Strigiformes order. Within the Strigimorphae superorder, the Strigiformes order is joined by the turacos (Musophagiformes order).

The systematic list of the owls of the world provided here encompasses 220 species of owls and with the subspecies classification results in 539 taxa. By providing synonyms, the aim is to clear up the often confusing duplication found in the designations. (Even today, valid species names are sometimes assigned twice: *Glaucidium passerinum,* for instance, designates not only the Eurasian pygmy owl, but also a blue-blossomed plant in Japan.) This checklist has been brought up-to-date and already takes the results of genetic analysis into account. This has led to the splitting up of previous species complexes, as is especially striking in the case of the genera *Otus* and *Glaucidium.* In this context, Weick names considerably more owl species than do all authors before him: Peters (1940) cited 141 species, del Hoyo et al. (1999) cited 195, the *World Birds Taxonomic List* (*zoonomen.net*) cites 199, and König et al. (1999) cited 212. Recently, however, the *Global Owl Project* actually named 225 owl species (*www.globalowlproject.com/species*). The current upheaval in owl systematics is documented by countless individual publications. By bringing together sources that are in part widely dispersed, the succinctly annotated checklist provided here presents a kind of status report.

It goes without saying that no list can ever be final, since the process of bringing the classification into line with the findings of evolutionary biology represent a continuous process of approximation and convergence. On the one hand, animal forms exist in a continuum of superspecies, species, and subspecies, of ecotype and morph, and in this way the construct of the species concept ultimately originates from a convention, which has no equivalent in real population dynamics. Thus, in spite of all scientific differentiation criteria, there remains a large amount of leeway for personal assessments. On the other hand, the temporal and spatial differentiation of the branchings in the phylogenetic tree is much more multifaceted than can be reproduced through the limitations of rigid classificatory categories (Newton 2003). Furthermore, evolutionary biology has demonstrated that in the course of evolutionary history a cooperative fusioning of heterogeneous organisms must have repeatedly occurred, which means that individuals at a higher level of organization actually have emerged from "species conglomerates," which raises a problem of philosophical theory for the definition of species.

This checklist is neither an identification guide, nor a mere picture book, nor an owl monograph. By reflecting a high level of scientific expertise and the most up-to-date research, it provides outstanding support in the handling of scientific collections and in studies of global biodiversity. But it should also address the needs of practitioners of scientific species preservation, since global, national, or regional responsibility for ensuring the continuance of endangered forms frequently first becomes apparent on the basis of taxonomic and geographic differentiation. Not least of all, this list can support scientific conservation and breeding programs, whose contribution to species preservation is largely dependent on knowledge of the systematic subclassification of the animals bred. I too would like to support this intention to the best of my abilities.

St. Oswald, Germany
January 2006

Dr. Wolfgang Scherzinger
Zoologist in Bavarian Forest National Park

References

Baird S (1858) Birds, vol 9. Tucker, Washington, DC (in cooperation with Cassin J and Lawrence G)
del Hoyo J, Elliot A, Sargatal J (1999) Handbook of the birds of the world, vol 5. Lynx Edition, Barcelona (pp 34–65)
Fürbringer M (1888) Untersuchungen zur Morphologie und Systematik der Vögel, vol 1–2. Von Holkema, Amsterdam, 1751 pp
Huxley T (1867) On the classification of birds; and on the taxonomic value of modification of certain of the cranial bones observable in that class. Proc Zool Soc Lond (1867) 415–472
König C, Weick F, Becking J-H (1999) Owls. A guide to the owls of the world. Pica Press, Kent, UK
Mayr E (1975) Grundlagen der zoologischen Systematik. Paul Parey, Hamburg, Berlin
Newton I (2003) The speciation and biogeography of birds. Academic Press, Amsterdam, Boston, London, New York
Nitzsch C (1840) System der Pterylographie. Anton, Halle, 228 pp
Peters J (1931) Check-list of the birds of the world, vol 1. Museum of Comperative Zoology, Cambridge, MA
Peters J (1940) Check-list of the birds of the world, vol 4. Harvard University Press, Cambridge, MA
Sibley Ch, Ahlquist J (1990) Phylogeny and classification of birds. A study in molecular evolution. Yale University Press, New Haven, London
von Linné C (1758) Systema naturae per regna tria naturae. Holmiae

Geleitwort

Eulen sind Zauberwesen; von ihnen geht eine zwiespältige Faszination aus, die beim menschlichen Betrachter völlig gegensätzliche Reaktionen auslösen kann. Zum einen verunsichern ihr lautloser Flug, ihre unheimliche Stimme und die Nachtaktivität, was zur Verteufelung der Eule als Bote der Finsternis geführt hat. Zum anderen aber fesselt ihr maskenhaftes Gesicht mit dem scheinbar durchdringenden Blick, was wiederum zur Vergötterung der Eule als Hilfsgeist einer allwissenden Urmutter Erde beigetragen haben mag. Außerdem schürten die hochspezialisierten Beutefänger den Konkurrenzneid der Jägerschaft und verursachten durch die Verluste beim Niederwild bei diesen zumindest eine gewisse Verärgerung. Weiterhin trugen in Kirchtürmen und Scheunen hinterlassener Kot und Gewölle dazu bei, dass die Eulen zu Lästlingen oder gar Schädlingen gestempelt wurden. Doch nach Überwinden von Aberglaube und Argwohn können wir uns heute eines vorbehaltlosen bewundernden Umgangs mit diesen Geschöpfen der Nacht erfreuen. Sie ermöglichen uns Einblicke in die reiche Vielfalt der Anpassung an nächtlichen Beutefang, in Lautinventare, Brutbiologie und Nestlingsentwicklung.

In regelrechter Umkehr überholter „Feindbilder“ haben die Eulen mit ihrem Charisma heute eine geradezu herausragende Beliebtheit in breiten Bevölkerungskreisen erreicht. Ob Schleiereule, Uhu oder Raufußkauz, sie eignen sich als attraktive *flagship*-Arten eines zeitgemäßen Biotop- und Artenschutzes. Eine erfreulich große Zahl von Fachleuten, Amateuren und Liebhabern engagiert sich heute im Eulenschutz, sei es durch das Bereitstellen von Nisthöhlenangeboten, durch Beringung, Aufsammeln von Gewöllen, beim Bewachen störungsgefährdeter Brutplätze oder durch Entschärfung des Stromschlagrisikos an Lichtmasten. Durch den Zusammenschluss der Eulenfreunde zu lokalen, regionalen und internationalen Organisationen sind Beratung und Koordination heute auf hohem fachlichen Niveau gesichert.

Wieder andere Eulenfreunde bemühen sich um eine naturnahe Nachzucht von Eulen, sei es für die Wiederansiedlung oder für die Existenzsicherung hoch bedrohter Arten. Neben den Zoos, Vogelparks und speziellen Zuchtstationen sind hier auch die arbeitsaufwendigen Pflegestationen einzureihen, in denen verunglückte oder verwaiste Eulen fachgerecht versorgt werden können. Alle diese Verbände und Einrichtungen leisten direkt und indirekt einen volksbildnerischen Beitrag, der den positiven Stellenwert der einst so verabscheuten „Nachtgespenster“ maßgeblich sichert. Zur Palette der Eulenliebhaberei gehört auch die Naturfotografie, Malerei und Bildhauerei, die bis zu den vielgestaltigen Sammlungen von Eulenplastiken reichen kann. Wen die Eulen einmal in ihren Bann gezogen haben, den scheinen sie so leicht nicht mehr los zu lassen.

Das trifft zweifellos auch für den Autor dieser taxonomischen Eulenartenliste zu. Friedhelm Weick wurde vor rund 40 Jahren unter dem prägenden Eindruck der Anmut von Wildtieren und ihren Verhaltensweisen aus einem völlig anders ausgerichteten Beruf zum Tiermaler und Illustrator von Fachbüchern. Die Anfänge seiner Sammlungen von Literaturquellen und Illustrationen sowie sein Studium von Museumsbälgen für eine Übersicht der Eulen der Welt reichen mindestens 25 Jahre zurück. Diese Aktivitäten wurden im Zuge seiner Mitarbeit am ersten Kompendium der gesamten Ordnung Strigiformes (1999) von Claus König nochmals deutlich verstärkt. Solch konsequentes Festhalten an einer langjährigen Projektbearbeitung spiegelt zweifellos eine tiefe Begeisterung für die Eulen wider.

Im Zeitalter immer neuer Medien, die alle unsere Sinne okkupieren und sich in ihrer emotionalen Wirkung laufend überbieten möchten, dabei mit Bilderfluten, 3-D-Projektionen und virtuellen Rekonstruktionen längst vergangener Welten auftrumpfen, mag eine systematische Artenliste einer bestimmten Vogelgruppe auf den ersten Blick anachronistisch und antiquiert erscheinen. Tatsächlich fehlt es nicht an Stimmen, die über die „verstaubte" Museumsarbeit klassischer Taxonomen spötteln und in Diskussionen der systematischen Einordnung von Organismen keinen praktischen Nutzen sehen. Da diese unterschwellige Geringschätzung auch auf die Hochschulen abfärbte, wurden im Konkurrenzkampf der einzelnen Fachdisziplinen die Grundlagen von Artenkenntnis und systematischer Zuordnung weitgehend aus Forschung und Lehre gedrängt. Wenn sich aber der Hochschulunterricht aus dieser Verantwortung stiehlt, dann droht die Taxonomie zur aussterbenden Disziplin zu werden. Dabei stellen gerade fundierte Kenntnisse zur feinsystematischen Gruppierung von Organismen die essentielle Voraussetzung für das Verständnis evolutiver Prozesse dar. In vielfältiger Wechselwirkung mit Geographie, Ökologie, Ethologie und Genetik macht es die Systematik möglich, die Naturgeschichte von Populationen nachzuzeichnen. Dazu gehören die Herkunft und Ausbreitungswege von Populationen, ihre Entwicklung durch Konkurrenz und Predation, durch Klimaänderungen oder wechselndes Lebensraumangebot und ihre Verinselung oder Aufspaltung in unterschiedliche Anpassungsformen.

Seit der internationalen Naturschutzkonferenz von Rio de Janeiro 1992 ist der neue Sammelbegriff „Biodiversität" in aller Munde. Auch wenn Biodiversität die Vielfalt biologischer Phänomene auf Lebensräume und Landschaften ausdehnt, so stellt die Mannigfaltigkeit der Organismen doch zweifellos das wichtigste Fundament dar. Die Stammbäume der Organismen erhalten dadurch neues Gewicht. Der Großteil der entwickelten Staaten hat die Biodiversitäts-Konvention unterzeichnet, die zur Bewahrung der weltweiten Vielfalt an Lebensformen verpflichtet. Innerhalb der Europäischen Gemeinschaft ist das Konzept „Natura 2000" auf dieses Ziel ausgerichtet. Neben den „Roten Listen" existenzgefährdeter Arten wurden für Deutschland auch Kriterien zur Ermittlung der Verantwortlichkeit im Artenschutz aufgestellt. Wenn dies auch sehr positiv zu bewertende Entwicklungen sind, sind sie doch leider vor allem das Spiegelbild eines beunruhigenden Aussterbeprozesses, eines durch die menschliche Expansion verursachten und mit zunehmendem Tempo voranschreitenden Artenschwunds von bisher nicht gekanntem Ausmaß.

Ob Forschungsfragen der Evolutionsbiologie, Schutzprogramme zur Artensicherung oder Züchtung wirtschaftlich relevanter Nutztiere, die Kenntnis und die korrekte Taxation von „Arten" sind für alle Projekte die unbedingte Voraussetzung. Wenn auch das Benennen und Kategorisieren von belebten Objekten so alt ist wie die Sprachentwicklung des Menschen selbst, so ist die wissenschaftlich fundierte Definition und Abgrenzung bestimmter Organismengruppen als „Art" dennoch eine hochkomplexe Disziplin. Dabei können in die objektivierbaren Entscheidungskriterien auch subjektive Aspekte wie Weltanschauung, gesellschaftliches Wertesystem oder auch nationaler Ehrgeiz und *political correctness* hineinspielen. Außerdem ist der Kenntnisstand der globalen Mannigfaltigkeit und ihrer Verbreitungsmuster noch immer sehr lückenhaft.

Nach gegenwärtiger Auflistung kennen wir rund 1,7 Millionen rezente Arten aus den Reichen der Einzeller, Pilze, Pflanzen und Tiere auf der Erde. Etwa 50 000 davon sind Wirbeltiere, 9 935 davon wiederum Vögel und gut 200 von diesen stellen die Eulen. Da aber viele der Tropenwälder noch nicht erforscht sind und womöglich schon vor einer Artenerfassung durch Abholzung vernichtet werden, rechnen begründete Spekulationen auch mit der Möglichkeit eines bis zu zehnfachen Artenreichtums.

Ging es der frühen Museumstaxonomie nur um Ordnungssysteme zur Benennung und Verwaltung der Organismenvielfalt – zumal solange die Unveränderlichkeit der Schöpfung selbst den Biologen als unumstößliches Dogma erschien – so gilt seit Darwin das ehrgeizige Ziel, mit der Klassifikation der Arten auch ihre Stammesgeschichte nachzuzeichnen. Die vergleichende Betrachtung von morphologischen, physiologischen und ethologischen Merkmalen, unter Berücksichtigung der Ergebnisse aus Ökologie und Tiergeographie, erwartet zum einen, dass alle Individuen einer Population mit hoher Merkmalsübereinstimmung als „gleichartig" zusammengefasst werden können, und zum anderen, dass eine hohe Merkmalsübereinstimmung zwischen verschiedenartigen Populationen mit deren naher Verwandtschaft korreliert. Ernst Mayr (1975) fügte als wesentliches Kriterium der Abgrenzung von Einzelarten deren evolutiv entwickelte Fortpflanzungsschranke hinzu, die das Risiko zwischenartlicher Hybridisierung minimiert. Doch alle klassischen Artkonzepte sind mit einer Fülle konvergenter Formen konfrontiert, weshalb sie großes Gewicht auf die Unterscheidung homologer und analoger Merkmale legen müssen. Auch reicht ein Merkmalsvergleich häufig nicht zur feinsystematischen Auftrennung von Unterarten und noch weniger, um höhere Taxa auf ihre stammesgeschichtlichen Verbindungen zu prüfen.

Aus dieser Stagnation der Systematik hat am Ende des 20. Jahrhunderts letztlich die Molekularbiologie in geradezu revolutionärer Weise herausgeführt, denn ganz wesentliche Fragen konnten erst mit der Einbindung der modernen Genetik angegangen werden. Nicht zuletzt hat die Strukturanalyse der Erbsubstanz DNA die taxonomische Klassifikation wieder diskussionsfähig gemacht und ihr die notwendige Anerkennung in der Gesellschaft zurückgebracht. Da die DNA-Analyse nach Ähnlichkeiten in der Erbsubstanz sucht, ist sie gegen Irritationen durch Konvergenzen der Morphologie besser gefeit. Das Ziel geht über die Gruppierung verwandter Arten hinaus; der Artbildungsprozess mit den einzelnen Verzweigungen am Stammbaum soll möglichst realistisch rekonstruiert werden. Mit Hilfe der „molekularen Uhr" lassen sich solche Artabspaltungen sogar zeitlich einordnen, wodurch der Stammbaum zum mehrdimensionalen Konstrukt wird.

Von allen Organismengruppen ist die phylogenetische Klassifikation der Vögel am weitesten fortgeschritten. Aktuell sind die rezenten Formen in 23 Ordnungen, 146 Familien, 2 141 Gattungen mit 9 935 Arten gegliedert (Stand Dezember 2005, nach *zoonomen.net*). Durch spektakuläre Fossilfunde der letzten Jahre, speziell aus mesozoischen Ablagerungen in China, lässt sich heute sogar die Ableitung der Vögel von befiederten, biped laufenden Sauriern belegen (Familien Sauriurae und Ornithurae). Diese wenigstens 120 Millionen Jahre alten Fossilien lassen sich bereits den Regenpfeifern, Seetauchern, Sturmvögeln und Gänseartigen zuordnen. Eulen gibt es seit 50–60 Millionen Jahren.

Die Eulen stellen innerhalb der Klasse der Vögel eine auffällig einheitliche Formengruppe dar. Durch die meist großen, starr nach vorne gerichteten Augen im „Gesicht" eines großen, runden und „halslosen" Kopfes, durch ihr locker-flauschiges Gefieder, in meist tarnenden Farben und einem gedrungenbreiten Körper sind sie von jedermann unschwer den Eulen zuzuordnen. Soweit das einheitliche Erscheinungsbild der Eulen durch ihre dunkelaktive Lebensweise als Beutegreifer selektiert wurde, erschwert diese Uniformität morphologischer Merkmale allerdings die feinsystematische Untergliederung. Durch diese Geschlossenheit waren die Eulen als einheitliche Gruppe nie umstritten, wobei die Aufspaltung in eine Gruppe der Schleiereulen und in „eigentliche Eulen" bereits durch Nitzsch (1840) vorgeschlagen wurde. Doch große Unsicherheit herrschte bei der Eingliederung ihrer nächsten Verwandtschaft, wie der historische Abriss in Sibley and Ahlquist (1990) sehr detailliert ausführt. Auf Grund ähnlicher Schnabelformen gruppierte Linné (1758) die Eulen als Gattung *Strix* zusammen mit den Greifvögeln und Würgern (Ordnung Accipitres). Der Begriff Strigidae als Familiennamen für die Nachtgreifvögel wird erstmals von Baird (1858) am Beispiel der nordamerikanischen Vogelwelt verwendet. Unter dieser Bezeichnung stuft Huxley (1867) die Eulen bei den Geiern und Sekretären ein. Erst Fürbringer (1888) löst die Eulen von den Taggreifvögeln und gliedert das Geschlecht der Striges mit der Familie Strigidae bei den Rackenvögeln (Coraciformes) ein. Der Rang einer eigenen Ordnung *Striges* wird den Eulen erst von Peters (1931) zuerkannt, der sie zwischen Nachtschwalben, Papageien und Kuckucken einreiht.

Diese verwandtschaftlichen Verhältnisse konnten nach Sibley and Ahlquist (1990) durch die molekularbiologischen Methoden teilweise erhärtet werden, doch werden die Eulen in dieser Interpretation von den Rackenvögeln abgetrennt, dabei gleichzeitig näher an die Segler und Kolibris gerückt. Die DNA-Analyse fordert die endgültige Loslösung der Nachtgreifvögel von den Taggreifvögeln. Diese Autoren fügen als neue Überordnung die Strigimorphae ein, der die Ordnung Strigiformes mit 319 Arten aus 51 Gattungen zugeordnet ist, zu der innerhalb der Unterordnung Strigi die beiden Eulenfamilien Tytonidae mit 15 Arten aus 2 Gattungen, und Strigidae mit 184 Arten aus 27 Gattungen zählen. In die Ordnung Strigiformes werden neuerdings auch die Eulenschwalme und Nachtschwalben im Rang von Unterordnungen eingereiht. In der Überordnung Strigimorphae wurden neben die Ordnung Strigiformes die Turakos (Ordnung Musophagiformes) gestellt.

Die vorliegende systematische Liste der Eulen der Welt erfasst 220 Eulenarten, mit der Unterartengliederung sind es 539 Taxa. Durch die Nennung der Synonyma soll die oft verwirrende Duplizität in der Namengebung aufgeklärt werden. (Selbst heute gültige Artnamen wurden mitunter zweimal vergeben, wie *Glaucidium passerinum*, mit dem nicht nur der eurasische Sperlingskauz, sondern auch eine blaublühende Pflanze aus Japan benannt wurden.) Die Artenliste wurde auf den neuesten Stand gebracht und berücksichtigt bereits die Ergebnisse aus der genetischen Analyse. Soweit diese zum *splitting* bisheriger Artkomplexe geführt haben, wie besonders auffällig bei der Gattung *Otus* und *Glaucidium* geschehen, nennt Weick deutlich mehr Eulenarten als alle Autoren zuvor, mit 141 Arten bei Peters (1940), 195 Arten bei del Hoyo et al. (1999), 199 Arten in der *World Birds Taxonomic List* (*zoonomen.net*) und 212 Arten bei König et al. (1999). Die Liste des *Global Owl Project* nennt neuerdings allerdings sogar 225 Eulenarten (*www.globalowlproject.com/species*). Der aktuelle Umbruch in der Eulensystematik ist durch unzählige Einzelpublikationen dokumentiert. In dieser kompakt kommentierten Artenliste werden die zum Teil breit gestreuten Quellen in einer Art Statusbericht zusammengeführt.

Selbstverständlich kann keine Liste endgültig sein, da die Angleichung der Klassifikation an die evolutionsbiologischen Erkenntnisse einem steten Prozess der Annäherung entspricht. Zum einen können Tierformen in einem Kontinuum aus Superspezies, Spezies und Subspezies, aus Ökotyp und Morphe vorkommen und damit entstammt das Konstrukt des Artbegriffs letztlich einer Konvention, die in der realen Populationsdynamik keine Entsprechung hat, so dass trotz aller wissenschaftlichen Abgrenzungskriterien weiterhin ein breiter Spielraum für persönliche Einschätzungen gegeben bleibt. Zum anderen ist die zeitliche und räumliche Differenzierung der Verzweigungen im Stammbaum wesentlich vielfältiger als sie durch die Beschränkung starrer Klassifizierungskategorien wiedergegeben werden könnte (Newton 2003). Des weiteren hat die Evolutionsbiologie aufgezeigt, dass im Laufe der Stammesgeschichte immer wieder eine kooperative Fusionierung verschiedenartiger Organismen erfolgt sein dürfte, weshalb Individuen höherer Organisationsstufe eigentlich aus „Art-Konglomeraten" hervorgegangen sind, was ein philosophisch-theoretisches Problem der Artdefinition aufwirft.

Diese Artenliste ist weder ein Bestimmungsschlüssel noch ein reines Bilderbuch noch eine Eulenmonographie. Bei hohem Anspruch an das fachliche Niveau und die Aktualität bietet sie eine hervorragende Hilfe bei der Bearbeitung wissenschaftlicher Sammlungen und bei Studien zur globalen Biodiversität. Aber auch der Praktiker im wissenschaftlichen Artenschutz soll mit dieser Liste angesprochen werden, da die globale, nationale oder regionale Verantwortlichkeit für die Bestandssicherung gefährdeter Formen häufig erst aus einer taxonomischen und geographischen Differenzierung ersichtlich wird. Nicht zuletzt kann sie wissenschaftliche Erhaltungs- und Zuchtprogramme unterstützen, deren Beitrag zum Artenschutz stark von der Kenntnis der feinsystematischen Zuordnung der Zuchttiere abhängt. Diese Intention möchte auch ich nach Kräften unterstützen.

St. Oswald,
Januar 2006

Dr. Wolfgang Scherzinger
Zoologe im Nationalpark Bayerischer Wald

Acknowledgements

The author is deeply indebted to many institutions and individuals for their kind and most helpful cooperation. My most cordial thanks goes to Claus König, Ludwigsburg, Germany, for his many tips and large amounts of information, and Gerrit Paulus Hekstra, Harich, The Netherlands, for making available a fount of unpublished data.

I would also like to thank the following museums and other scientific institutions: American Museum of Natural History, New York, USA (M. Le Croy), British Museum of Natural History, Tring, UK (Bird section, M. Adams); Brighton Young University, Utah, USA (C. M. White); Forschungsinstitut und Naturmuseum Senckenberg, Frankfurt, Germany (D. Peters, J. Steinbacher †); Institute of Zoology, Taipeh, Taiwan (L. Liu Severinghaus); Institut für Zoologie, Universität Heidelberg, Germany (H. Möller); Museum für Naturkunde, Humboldt-Universität zu Berlin, Germany (F. H. Steinheimer, B. Stephan, K. Wunderlich †); Museum Heineanum, Halberstadt, Germany (B. Nicolai); Muséum National d'Histoire Naturelle, Paris, France (E. Pasquet); Musée Zoologique de l'Université Strasbourg, France (M. Wandhammer); Nationaal Natuurhistorisch Museum Leiden, The Netherlands (R. Dekker); Naturhistorisches Museum Wien, Österreich (E. Bauernfeind, K. Bauer); Naumann Museum Köthen, Germany (W.-D. Busching); Pfalzmuseum für Naturkunde, Bad Dürkheim, Germany (R. Flößer): The Society for Conservation and Research of Owls, Smithers, BC, Canada (R. Krahe) and Sektion Deutschland (R. Steinberg); Staatliches Museum für Naturkunde, Karlsruhe, Germany (S. Rietschel, V. Wirth); Staatliches Museum für Naturkunde, Stuttgart, Germany (C. König, A. Schlüter); Staatliches Museum für Tierkunde, Dresden, Germany (S. Eck †); Tierpark Berlin, Germany (W. Grummt, H. Klös), Übersee Museum, Bremen, Germany (H. Hohmann); Vogelpark Niendorf, Germany; Vogelpark Walsrode, Germany; Vogelwarte Radolfzell, Möggingen, Germany (R. Schlenker); Zoologisches Forschungsinstitut und Museum A. Koenig, Bonn, Germany (R. v. d. Elzen); Zoologisches Museum Hamburg, Germany (C. Bracker); Zoologische Sammlung des Bayerischen Staates, München, Germany (J. H. Reichholf).

Moreover, I am indebted to the following experts for their kind cooperation and support: H. J. Becking, Wageningen, The Netherlands; E. Bezzel, Garmisch-Partenkirchen, for a fruitful collaboration over many years; T. Butynski, Nairobi, Kenya; G. Ehlers, Leipzig, Germany; S. R. and J. M. Goodell, North Logan, Utah, USA; U. N. Glutz von Blotzheim, Schwyz, Switzerland, for successful collaboration and his leadership of thirty years of work on the "Handbuch"; S. Dudley, BOU Administrator,Oxford, UK; J. Haffer, Essen, Germany; J. Hölzinger, Remseck, Germany; G. Kirwan, BOU Editor, Norwich, UK; O. Lakus †, Hambrücken, Germany; D. W. Morton, Emu, CSIRO, Collingwood, Australia; T. Mebs, Castell, Germany; B. U. Meyburg, Berlin, Germany; G. Müller, Rheinstetten-Mörsch, Germany; O. v. d. Rootselaar, Renkum, The Netherlands; W. Scherzinger, St. Oswald, Germany; D. Schmidt, Editor *Gefiederte Welt*, Backnang, Germany; S. Schönn, Oschatz, Germany; W. Thiede, Publisher *Ornithologische Mitteilungen*, Köln, Germany; U. Veith, Erwitte-Eikeloh, Germany; K. H. Voous †, Huizen, The Netherlands, W. Weise †, Claußnitz, Germany, for his lifelong, true friendship and constant support of this work; H. Wuchner, Kleinostheim, Germany, representing all private owl breeders and aviaries.

I especially want to thank my dear wife Christel, for her immeasurable and indefatigable support at home and in visiting museums, institutes, etc., and for her warm hospitality to all visiting ornithological friends and guests from near and far.

I am greatly indebted to Dr. Wolfgang Scherzinger, St. Oswald, Germany, for writing the Preface and for critically reviewing several parts of this systematic check list. His great knowledge of owls was an invaluable help.

Lastly, I gratefully acknowledge Dr. Dieter Czeschlik and Dr. Jutta Lindenborn (Springer, Heidelberg, Germany) for their cooperation and support throughout the entire production of this book. I particularly thank Dr. Jutta Lindenborn for her patience and guidance, and her many good suggestions concerning the layout and realization of this project.

Bruchsal-Untergrombach, Germany
February 2006

Friedhelm Weick

Danksagung

Der Autor steht tief in der Dankesschuld zahlreicher Personen und Institutionen für die hilfreiche Unterstützung dieser Arbeit. Ein herzlicher Dank gilt Claus König, Ludwigsburg, für die vielen Hinweise und Informationen und Gerrit Paulus Hekstra, Harich, Holland, für das Überlassen einer Fülle unveröffentlichter Daten.

Den folgenden Museen und Institutionen möchte ich ebenfalls herzlich danken:

American Museum of Natural History, New York, USA (M. Le Croy); British Museum of Natural History, Tring, UK (Bird section, M. Adams); Brighton Young University, Utah, USA (C. M. White); Forschungsinstitut und Naturmuseum Senckenberg, Frankfurt (D. Peters und J. Steinbacher †); Institute of Zoology, Taipeh, Taiwan (L. Liu Severinghaus); Institut für Zoologie, Universität Heidelberg (H. Möller); Museum für Naturkunde, Humboldt-Universität zu Berlin (F. H. Steinheimer, B. Stephan und K. Wunderlich †); Museum Heineanum, Halberstadt (B. Nicolai); Muséum National d'Histoire Naturelle Paris, France (E. Pasquet); Musée Zoologique de l'Université Strasbourg, France (M. Wandhammer); Nationaal Natuurhistorisch Museum Leiden, The Netherlands (R. Dekker); Naturhistorisches Museum, Wien, Österreich (E. Bauernfeind und K. Bauer); Naumann Museum Köthen (W.-D. Busching); Pfalzmuseum für Naturkunde, Bad Dürkheim (R. Flößer); The Society for Conservation and Research of Owls, Smithers, BC, Canada (R. Krahe) und Sektion Deutschland (R. Steinberg); Staatliches Museum für Naturkunde Karlsruhe (S. Rietschel und V. Wirth); Staatliches Museum für Naturkunde Stuttgart (C. König und A. Schlüter); Staatliches Museum für Tierkunde Dresden (S. Eck †), Tierpark Berlin (W. Grummt und H. Klös); Übersee Museum Bremen (H. Hohmann); Vogelpark Niendorf; Vogelpark Walsrode; Vogelwarte Radolfzell, Möggingen (R. Schlenker); Zoologisches Forschungsinstitut und Museum A. Koenig, Bonn (R. v. d. Elzen); Zoologisches Museum Hamburg (C. Bracker); Zoologische Sammlung des Bayerischen Staates, München (J. H. Reichholf)

Weiterhin gilt mein Dank auch den folgenden Wissenschaftlern und Eulenexperten für Ihre freundliche Unterstützung: H. J. Becking, Wageningen, The Netherlands; E. Bezzel, Garmisch-Partenkirchen, für eine fruchtbare Zusammenarbeit über viele Jahre; T. Butynski, Nairobi, Kenya; G. Ehlers, Leipzig; S. R. und J. M. Goodell, North Logan, USA; U. N. Glutz von Blotzheim, Schwyz, Switzerland, für eine erfolgreiche Zusammenarbeit und die vertrauensvolle Führung über 30 Jahre „Handbuchmitarbeit"; S. Dudley, BOU Administrator, Oxford, UK; J. Haffer, Essen; J.Hölzinger, Remseck; G. Kirwan, BOU Editor, Norwich, UK; O. Lakus †, Hambrücken; D. W. Morton, Emu, CSIRO, Collingwood, Australia; T. Mebs, Castell; B. U. Meyburg, Berlin; G. Müller, Rheinstetten-Mörsch; O. v. d. Rootselaar, Renkum, The Netherlands; W. Scherzinger, St. Oswald; D. Schmidt, Editor *Gefiederte Welt*, Backnang; S. Schönn, Oschatz; W. Thiede, Herausgeber *Ornithologische Mitteilungen*, Köln; U. Veith, Erwitte-Eikeloh; K. H. Voous †, Huizen, The Netherlands; W. Weise †, Claußnitz, für eine lebenslange, treue Freundschaft und stetige Ermutigung zu dieser Arbeit; H. Wuchner, Kleinostheim, stellvertretend für alle privaten Eulenzüchter.

Mein ganz besonderer Dank aber gilt meiner lieben Frau Christel für Ihre unermessliche und unverdrossene Unterstützung bei meiner Arbeit und bei den Besuchen von Museen, etc., außerdem für Ihre Gastfreundschaft für alle Ornithologen-Freunde aus nah und fern.

Wärmster Dank geht auch an Herrn Dr. Wolfgang Scherzinger, St. Oswald, für das Schreiben des Vorworts und für eine kritische Durchsicht der wichtigsten Teile der Artenliste. Sein guten Eulenkenntnisse waren eine wertvolle Hilfe.

Letztendlich gilt mein aufrichtiger Dank dem Springer-Verlag Heidelberg, insbesondere Herrn Dr. Dieter Czeschlik und Frau Dr. Jutta Lindenborn für das freundliche Entgegenkommen und die Unterstützung in vielfältiger Weise während der gesamten Produktion dieses Buches. Frau Lindenborn darüber hinaus meinen herzlichen Dank für den großen Zeitaufwand, Ihre Beharrlichkeit und Geduld sowie eine Menge wertvoller Verbesserungsvorschläge zur Übersichtlichkeit und Gestaltung dieses Projektes.

Bruchsal-Untergrombach
Februar 2006

Friedhelm Weick

Contents / Inhalt

Part I · Introduction/Einführung ... 1

Introduction ... 3
Einführung ... 5

Owls: A Brief Overview ... 7
Eulen: Eine kurze Übersicht ... 9

Overview of the Order Strigiformes/
Überblick über die Ordnung Strigiformes ... 11

Part II · Order Strigiformes/Ordnung Strigiformes ... 13

Familia/Family Tytonidae ... 15
Subfamilia/Subfamily Tytoninae ... 15
Genus *Tyto* ... 15
Tyto alba ... 15
Tyto glaucops ... 21
Tyto insularis ... 21
Tyto punctatissima ... 22
Tyto detorta ... 22
Tyto thomensis ... 22
Tyto soumagnei ... 23
Tyto deroepstorffi ... 23
Tyto crassirostris ... 23
Tyto delicatula ... 24
Tyto aurantia ... 25
Tyto nigrobrunnea ... 25
Tyto inexspectata ... 26
Tyto sororcula ... 26
Tyto manusi ... 27
Tyto rosenbergii ... 27
Tyto novaehollandiae ... 28
Tyto castanops ... 29
Tyto capensis ... 29
Tyto longimembris ... 30
Tyto tenebricosa ... 31
Tyto multipunctata ... 32
Tyto prigoginei ... 33

Subfamilia/Subfamily Phodilinae 34
Genus *Phodilus* 34
Phodilus badius 34
Familia/Family Strigidae 36
Subfamilia/Subfamily Striginae 36
Tribus/Tribe Otini 36
Genus *Otus* 36
Otus sagittatus 36
Otus rufescens 36
Otus thilohoffmanni 37
Otus icterorhynchus 37
Otus ireneae 38
Otus balli 38
Otus alfredi 39
Otus stresemanni 39
Otus spilocephalus 39
Otus angelinae 42
Otus mirus 42
Otus longicornis 42
Otus mindorensis 43
Otus hartlaubi 43
Otus rutilus 43
Otus madagascariensis 44
Otus mayottensis 44
Otus moheliensis 44
Otus capnodes 45
Otus pauliani 45
Otus pembaensis 46
Otus flammeolus 46
Otus scops 47
Otus brucei 49
Otus senegalensis 51
Otus sunia 52
Otus elegans 55
Otus magicus 55
Otus beccarii 58
Otus mantananensis 58
Otus manadensis 59
Otus (manadensis) siaoensis 59
Otus (manadensis) kalidupae 60
Otus collari 60
Otus insularis 61
Otus alius 61
Otus umbra 61
Otus enganensis 62
Otus mentawi 62
Otus brookii 62
Otus lempiji 63
Otus lettia 64
Otus bakkamoena 65

Otus semitorques 67
Otus megalotis 68
Otus fuliginosus 69
Otus silvicola 69
Subgenus *Megascops* 70
Otus kennicotti 70
Otus seductus 72
Otus cooperi 73
Otus lambi 73
Otus asio 74
Otus trichopsis 76
Otus choliba 77
Otus koepckeae 79
Otus roboratus 80
Otus pacificus 80
Otus clarkii 80
Otus barbarus 81
Otus ingens 82
Otus colombianus 82
Otus petersoni 83
Otus marshalli 83
Otus watsonii 83
Otus atricapillus 84
Otus sanctae-catarinae 85
Otus hoyi 85
Otus guatemalae 86
Otus vermiculatus 87
Otus napensis 88
Otus roraimae 88
Otus nudipes 89
Subgenus *Macabra* 90
Otus albogularis 90
Genus *Pyrroglaux* 92
Pyrroglaux podarginus 92
Genus *Gymnoglaux* 93
Gymnoglaux lawrencii 93
Genus *Ptilopsis* 94
Ptilopsis leucotis 94
Ptilopsis granti 94
Genus *Mimizuku* 96
Mimizuku gurneyi 96
Tribus/Tribe Bubonini 97
Genus *Nyctea* 97
Nyctea scandiaca 97
Genus *Bubo* 99
Bubo virginianus 99
Bubo magellanicus 103
Bubo bubo 104
Bubo bengalensis 109
Bubo ascalaphus 109

Bubo capensis ... 110
Bubo africanus ... 111
Bubo (africanus) milesi ... 112
Bubo cinerascens ... 112
Bubo poensis ... 113
Bubo vosseleri ... 114
Bubo nipalensis ... 114
Bubo sumatranus ... 115
Bubo shelleyi ... 116
Bubo lacteus ... 117
Bubo coromandus ... 117
Bubo leucostictus ... 118
Bubo philippensis ... 119
Bubo blakistoni ... 120
Genus *Ketupa* ... 121
Ketupa zeylonensis ... 121
Ketupa flavipes ... 122
Ketupa ketupu ... 123
Genus *Scotopelia* ... 125
Scotopelia peli ... 125
Scotopelia ussheri ... 126
Scotopelia bouvieri ... 126
Tribus/Tribe Strigini ... 127
Genus *Strix* ... 127
Strix seloputo ... 127
Strix ocellata ... 128
Strix leptogrammica ... 129
Strix bartelsi ... 131
Strix newarensis ... 131
Strix aluco ... 133
Strix butleri ... 136
Strix woodfordi ... 137
Strix virgata ... 138
Strix albitarsis ... 141
Strix chacoensis ... 141
Strix rufipes ... 141
Strix hylophila ... 142
Strix nigrolineata ... 142
Strix huhula ... 143
Strix fulvescens ... 144
Strix occidentalis ... 144
Strix varia ... 146
Strix uralensis ... 148
Strix (uralensis) davidi ... 150
Strix nebulosa ... 151
Genus *Jubula* ... 153
Jubula lettii ... 153
Genus *Lophostrix* ... 154
Lophostrix cristata ... 154

Genus *Pulsatrix* ... 156
Pulsatrix perspicillata ... 156
Pulsatrix pulsatrix ... 157
Pulsatrix koeniswaldiana ... 158
Pulsatrix melanota ... 158
Subfamilia/Subfamily Surniinae ... 159
Tribus/Tribe Surniini ... 159
Genus *Surnia* ... 159
Surnia ulula ... 159
Genus *Glaucidium* ... 161
Glaucidium passerinum ... 161
Glaucidium perlatum ... 162
Glaucidium californicum ... 163
Glaucidium hoskinsii ... 164
Glaucidium gnoma ... 165
Glaucidium nubicola ... 166
Glaucidium jardinii ... 166
Glaucidium bolivianum ... 166
Glaucidium peruanum ... 167
Glaucidium nanum ... 167
Glaucidium siju ... 168
Glaucidium ridgwayi ... 169
Glaucidium brasilianum ... 169
Glaucidium tucumanum ... 171
Glaucidium palmarum ... 172
Glaucidium sanchezi ... 172
Glaucidium griseiceps ... 173
Glaucidium parkeri ... 173
Glaucidium hardyi ... 174
Glaucidium minutissimum ... 174
Glaucidium sicki ... 174
Glaucidium tephronotum ... 175
Glaucidium brodiei ... 176
Subgenus *Taenioglaux* ... 178
Glaucidium radiatum ... 178
Glaucidium castanonotum ... 179
Glaucidium cuculoides ... 179
Glaucidium castanopterum ... 181
Glaucidium sjoestedti ... 181
Glaucidium castaneum ... 182
Glaucidium etchecopari ... 182
Glaucidium capense ... 183
Glaucidium albertinum ... 184
Genus *Xenoglaux* ... 185
Xenoglaux loweryi ... 185
Genus *Micrathene* ... 186
Micrathene whitneyi ... 186
Genus *Athene* ... 188
Athene noctua ... 188

Athene brama ... 192
Athene blewitti ... 193
Athene cunicularia ... 194
Tribus/Tribe Aegolini ... 200
Genus *Aegolius* ... 200
Aegolius funereus ... 200
Aegolius acadicus ... 202
Aegolius ridgwayi ... 203
Aegolius harrisii ... 204
Tribus/Tribe Ninoxini ... 206
Genus *Ninox* ... 206
Ninox rufa ... 206
Ninox strenua ... 207
Ninox connivens ... 208
Ninox rudolfi ... 210
Ninox boobook ... 210
Ninox leucopsis ... 214
Ninox novaeseelandiae ... 214
Ninox scutulata ... 215
Ninox affinis ... 219
Ninox superciliaris ... 219
Ninox philippinensis ... 220
Ninox mindorensis ... 222
Ninox sumbaensis ... 222
Ninox dubiosa sp. nov. ... 223
Ninox burhani ... 223
Ninox ochracea ... 223
Ninox ios ... 224
Ninox squamipila ... 224
Ninox natalis ... 225
Ninox meeki ... 226
Ninox theomacha ... 226
Ninox punctulata ... 227
Ninox odiosa ... 228
Ninox variegata ... 228
Ninox jacquinoti ... 229
Genus *Uroglaux* ... 231
Uroglaux dimorpha ... 231
Genus *Sceloglaux* ... 232
Sceloglaux albifacies ... 232
Subfamilia/Subfamily Asioninae ... 233
Genus *Pseudoscops* ... 233
Pseudoscops grammicus ... 233
Genus *Asio* ... 234
Asio clamator ... 234
Asio stygius ... 235
Asio otus ... 237
Asio abyssinicus ... 239
Asio madagascariensis ... 240

Asio flammeus 240
Asio capensis 244
Genus *Nesasio* 246
Nesasio solomonensis 246

Part III · Owls in Flight/Eulen im Flug 247

Plate 1 *Tyto, Phodilus* 249
Plate 2 *Otus* spp. 250
Plate 3 *Otus (Megascops), Ptilopsis, Aegolius, Athene* 251
Plate 4 *Bubo* spp. 252
Plate 5 *Bubo, Scotopelia, Nyctea* 253
Plate 6 *Strix, Pulsatrix* 254
Plate 7 *Glaucidium (Taenioglaux), Micrathene* 255
Plate 8 *Ninox, Surnia* 256
Plate 9 *Asio, Jubula, Lophostrix* 257

Part IV · Wing Formula and Topography/ Schwingenformel und Topographie 259

Table 1 Shortfall of Primary Tips/ Projektionen zur Handschwingenspitze 261
Plate 10 Topography of an Owl/Topographie einer Eule 267

Part V · Owls Described or Rediscovered in the Last 20 Years/ Neu beschriebene und wiederentdeckte Eulen der vergangenen 20 Jahre 269

Figure 1 *Tyto soumagnei* and *Tyto alba hypermetra* 271
Figure 2 *Tyto prigoginei* and *Phodilus badius* 272
Figure 3 *Otus alfredi, O. stresemanni* and *O. magicus albiventris* 272
Figure 4 *Otus thilihoffmanni, O. sunia leggei* and *O. rufescens malayensis* 273
Figure 5 *Otus capnodes* 274
Figure 6 *Otus rutilus* and *Otus madagascariensis* 274
Figure 7 *Otus alius, Otus umbra* and *Otus enganensis* 275
Figure 8 *Otus collari, Otus manadensis* and *Otus siaoensis* 275
Figure 9 *Otus moheliensis* and *Otus mayottensis* 276
Figure 10 *Otus cnephaeus, Otus lempiji* and *Otus lettia* 276
Figure 11 *Otus pacificus, O. roboratus* and *O. koepckeae* 277
Figure 12 *Otus hoyi, O. atricapillus, O. petersoni* and *O. marshalli* 278
Figure 13 *Glaucidium bolivianum, G. peruanum* and *G. tucumanum* 279
Figure 14 *Glaucidium nubicola, G. costaricanum* and *G. jardinii* 279
Figure 15 *Glaucidium hardyi, G. minutissimum, G. sicki* and *G. griseiceps* 280
Figure 16 *Glaucidium parkeri, G. sanchezi* and *G. palmarum* 280
Figure 17 *Xenoglaux loweryi, Micrathene whitneyi, Glaucidium albertinum* and *Glaucidium c. capense* 281
Figure 18 *Athene blewitti* and *Athene brama* 282

Figure 19 *Ninox ios* and *Ninox ochracea* 282
Figure 20 *Ninox sumbaensis, Ninox rudolfi* and *Ninox dubiosa* sp. novae 283
Figure 21 *Ninox burhani* and *Ninox punctulata* 283
Figure 22 Taxonomy of American Great Horned Owls (*Bubo* spp.) 284
Figure 23 Taxonomy of American Great Horned Owls (*Bubo* spp.) 285

Part VI · References/Literaturverzeichnis 287

Part VII · Indices 317

Index of Scientific Owl Names/
Index der wissenschaftlichen Eulennamen 319
Index of Vernacular English Owl Names/
Index der englischen Eulennamen 332
Index of Vernacular German Owl Names/
Index der deutschen Eulennamen 334
Index of Vernacular French Owl Names/
Index der französischen Eulennamen 336
Index of Vernacular Spanish (Portuguese) Owl Names/
Index der spanischen (portugiesischen) Eulennamen 338
Index of Geographical Terms/
Index der geographischen Namen 340

Terms, Abbreviations and Acronyms / Begriffe und Abkürzungen

Terms and abbreviations used in the checklist /
In der Artenliste verwendete Begriffe und Abkürzungen

Term / Abbreviation	Explanation / Definition	Erläuterung / Definition
Terra typica	Known area of type locality	Erster benannter Fundort des Typus
Synonym	Synonymous name	Gleichbedeutende Doppelbenennung
Distribution		Verbreitungsgebiet
Habitat		Habitat, Lebensraum
Museum	Museum collection	Museumssammlung
Length	Total length	Gesamtlänge
Body mass	Weight	Gewicht
Wing length	From bend (carpal joint) to wingtip (see also Plate 10)	Flügellänge vom Bug bis zur Spitze (s. auch Tafel 10)
Tail length	From base of the central tail feathers to tip (see also Plate 10)	Schwanzlänge vom Ansatz der zentralen Schwanzfeder bis zur Spitze (s. auch Tafel 10)
Tarsus length	Length of tarsometetarsus (see also Plate 10)	Tarsal- oder Lauflänge (s. auch Tafel 10)
Length of bill	From tip to frontal feathering (see also Plate 10)	Schnabellänge von der Spitze bis zur Stirnbefiederung (s. auch Tafel 10)
Length of bill (cere)	From tip to cere (see also Plate 10)	Schnabellänge von der Spitze bis zur Wachshaut (s. auch Tafel 10)
Illustration	Lithograph, etching etc. or selection of colour plates, drawings in recent literature	Lithographie, Stich etc. oder Auswahl von Farbtafeln, Zeichnungen in neuerer Literatur
Photograph	Photograph of living owl	Fotographie einer lebenden Eule
Literature	References in journals, field guides, etc.	Literatur wie Zeitschriften, Feldführer, Handbüchern, etc.
Remarks	To systematic or morphological aspects	Bemerkungen zur Systematik oder zu morphologischen Aspekten
spp	Species	Spezies, Art
ssp	Subspecies	Subspezies, Unterart
adult		Ausgefärbter Altvogel
juvenile		Jungvogel
immature		Unausgefärbt, nicht erwachsen
a.s.l.	Above sea-level	Über Normalnull
∅	Average	Durchschnittlich

List of acronyms used for museums and institutions/
Verzeichnis der in der Artenliste verwendeten Abkürzungen für Museen und Institute

AMS	Australian Museum Sydney, Australia
AMNH	American Museum of Natural History, New York, USA
ANSP(h)	Academy of Natural Sciences, Philadelphia, Pennsylvania, USA
ANWC = CSIRO	Australian National Wildlife Collection, Camberra, Australia
BBM	Bernice P. Bishop Museum, Honolulu, Hawaii, USA
BGSU	Bowling Green State University, Bowling Green, Ohio, USA
BMNH	British Museum of Natural History, Tring, London, UK
BNHS	Bombay Naturhistory Society, Bombay, India
BMMF = MFM	Biol. Museum Mun. Funchal, Madeira
BYUU	Brighton Young University, Utah, USA
C(am)UM	Cambridge University Museum, Cambridge, UK
CAS	California Academy of Sciences, California, USA
ChiMNH	Chicago Museum of Natural History, Chicago, Illinois, USA
CIT	Californian Institute of Technology, USA
CMC	Canterbury Museum Christchurch, New Zealand
CMNH	Cincinnati Museum of Natural History, Cincinnati, Ohio, USA
CM(P)	Carnegie Museum, Pittsburgh, Pennsylvania, USA
CPV	Colección Phelps de Venezuela
CSIRO = ANWC	Australian National Wildlife Collection, Camberra, Australia
CULO	Cornell University, Laboratory of Ornithology, Ithaca, N.Y., USA
D(el)MNH	Delaware Museum of Natural History, Greenville, Delaware, USA
D(en)MNH	Denver Museum of Natural History, Denver, Colorado, USA
DML	Derby Museum, Liverpool, UK
DMW	Dominian Museum Wellington, New Zealand
FCUNAM	Facultad de Ciencas, University National Autonoma de México
FM(NH)	Field Museum of Natural History, Chicago, Illinois, USA
FMV	Fairbanks Museum, St. Johnsburg, Vermont, USA
FNSF	Forschungsinstitut und Naturkundemuseum Senckenberg, Frankfurt/M., Germany
HLMD	Hessisches Landesmuseum, Darmstadt, Germany
HLW	H. L. White Collection in NMV
HNMBud	Hungarian Natural Museum Budapest, Hungary
IHN	Instituto de Historia Natural, Mexico
IML	Instituto Miguel Lillo, Tucumán, Argentina
IRSNB = ISB	Institut Royal de Scienses Naturelles de Belgique, Brussels, Belgium
KCP	Koepcke Collection, Lima, Peru
LACM	Los Angeles County Museum, Los Angeles, California, USA
LivCM	Liverpool Museum, Liverpool, UK
LRUR	Liceul Real – Umanist, Reghin, Rumania
LSUMZ	Louisiana State University, Museum of Zoology, Baton Rouge, Louisiana, USA
MACN	Museo Argentino de Ciencias Naturale, Buenos Aires, Argentina
MBL	Museu Bocage, Lisbon, Portugal
MCNUS	Museo de Ciencias Naturales Univesity of Salta, Argentina
MCZ	Museum of Comparative Zoology, Harvard University, Cambridge, Mass., USA
MESN	Museo Ecuatoriana de Siencias Naturales, Ecuador
MFM = BMMF	Museo Funchal, Madeira
MGD	Museo Civico di Storia Naturale "Giacomo Dorie", Genova, Italy

Continued / Fortsetzung

MHH	Museum Heineanum Halberstadt, Germany
MHNG	Muséum d'Histoire Naturelle, Genève, Switzerland
MHNParaguay	Museo de Historia Natural de Paraguay
MHNP(aris)	Muséum National d'Histoire Naturelle, Paris, France
MHNUC	Museo de Historia Natural, Universidad de Cauca, Popayán, Colombia
MHNUSM	Museo de Historia Natural de la Universidad de San Marcos, Lima, Peru
M(in)SUNHM	Mindanao State Univ. Natural History Museum, Mindanao, Philippines
MJPL	Museo de Historia "Javier Prado" Lima, Peru
MLA	Bureau of Science Manila, Philippines (destroyed 1945)
MLP	Museo de la Plata, Argentina
MLUH	Martin Luther Universität, Halle, Germany
M(oore)LZ	Moore Laboratory of Zoology, Occidental College, Pasadena, California, USA
MM	Malmö Museum, University of Malmö, Sweden
MMNH	James Ford Belle Museum, Minneapolis (formerly Minnesota Mus. Nat. Hist.), Minnesota, USA
MNBHU = ZMB	Museum für Naturkunde Berlin, Humboldt Universität = Zoologisches Museum Berlin, Germany
MNCNM	Museo Nacional de Ciencas Naturales, Madrid, Spain
MNHB	Museum of Natural History, Broom, Western Australia
MNHNLP	Museo Nacional de Historia Natural, La Paz, Bolovia
MNP = PNM	Philippine National Museum, Manila, Philippines
MNRJ	Museo Nacional de Rio de Janeiro, Brazil
MPEG	Museu Paranese Emilio Goeldi, Brazil
MSBUNM	Museum of Southwestern Biology, Dep. of Biology, University of New Mexico, USA
MVZ	Museum of Vertebrate Zoology, University of California, Berkely, California, USA
MZSUT(or)	Museo et Istituto Zoologia Sistematica, Università di Torino, Turin, Italy
MZUMich	Museum of Zoology, University Michigan, USA
MZUS	Musée Zoologique de la Université Strasbourge, France
MZUSP	Museu Zoologia da Universidade de Sao Paulo, Brazil
NCSM	North Carolina State Museum of Natural History, Raleigh, North Carolina, USA
NHMB(as)	Naturhistorisches Museum Basel, Switzerland
NHMB(ern)	Naturhistorisches Museum Bern, Switzerland
NHMW(ien)	Naturhistorisches Museum Wien, Vienna, Austria
NHRMSt	Naturhistoriska Riksmuseet, Stockholm, Sweden
NMC	National Museum Colombo, Sri Lanka
NMRC	National Museum Costa Rica
NMKN	National Museum of Kenya, Nairobi, Kenya
NMP	Natal Museum Pietermaritzburg, Natal, South Africa
NMPraha	Narodni Muzeum Praha, Csechia
NMV	National Museum Victoria, Victoria, Australia
NMSZ	National Museums of Scotland, Edinburgh, UK
NMZB	National Museum of Zimbabwe, Bulawayo, Zimbabwe (formerly Nat. Mus. of Rhodesia)
NNML = RMNH	Nationaal Natuurhistorisch Museum Leiden, Leiden, The Netherlands
NSMT	National Sciences Museum, Tokyo, Japan
OSUMZ	Ohio State University Museum of Zoology, Columbus, Ohio, USA
PMV(ict)	Provincial Museum, Victoria, British Columbia, Canada
PNM = MNP	Philippine National Museum, Manila, Philippines
QM	Queensland Museum, Queensland, Australia
RAMM	Royal Albert Museum, Exeter, UK

Continued / Fortsetzung

RMNH = NNML	Rijksmuseum van Natuurlijke Historie, Leiden, The Netherlands (older name)
RMTB	Royal Museum Tervuren, Brussels (Koninklijk Museum voor Midden Africa, Tervuren), Belg.
ROM	Royal Ontario Museum of Zoology, Toronto, Canada
RNA	Rijksinstituut vor Natuurbeheer, The Netherlands
RSM	Royal Scottish Museum, Edinburgh, UK
SAM	South Australian Museum, Adelaide, Australia
SAMC	South African Museum Capetown, South Africa
SBCM	San Bernardino County Museum, California, USA
SDNHM	San Diego Natural History Museum, San Diego, California, USA
SMNKa	Staatliches Museum für Naturkunde Karlsruhe, Germany
SMNSt	Staatliches Museum für Naturkunde Stuttgart, Germany
SMTD	Staatliches Museum für Tierkunde Dresden, Germany
SNMB	Staatliches Naturhistorisches Museum Braunschweig, Germany
SU	Silliman University, Dumaguente City, Negros Oriental, Philippines
TMBD	Transvaal museum, Bird Departement, Transvaal, South Africa
TUMZ	Turin University Museum of Zoology, Turin, Italy
ÜMB	Übersee Museum Bremen, Germany
UA(riz)DZ	University of Arizona, Depr. of Zoology, Tucson, Arizona, USA
UBCMZ	University of British Columbia, Museum of Zoology, Vancouver, British Columbia, Canada
UC(al)MVZ	University of California, Museum of Vertebrate Zoology, Berkely, California, USA = MVZ
UCLA	University of California at Los Angeles, California, USA
UFPE	Universidade Federal de Pernambuco (Bird Coll.), Pernambuco, Brazil
UK(ans)MNH	University of Kansas, Museum of Natural History, Lawrence, Kansas, USA
UM(ich)MZ	University of Michigan, Museum of Zoology, Ann Arbor, Michigan, USA
UMZC = ZMK	University Museum of Zoology, Copenhagen, Danmark
UPLB	University of the Philippines, Los Banos, Philippines
UPQC	University of the Philippines, Quezon City, Metro Manila, Philippines
USNM	United States National Museum, Washington D.C., USA
UW(isc)ZM	University of Wisconsin, Zoological Museum, Madison, Wisconsin, USA
V(anc)MBS	Vertebrate Museum of Biological Sciences, Vancouver, British Columbia, Canada
WAM	Western Australian Museum, Perth, Australia
W(ash)SM	Washington State Museum, Seattle, Washington, USA
WFVZ	Western Foundation for Vertebrate Zoology, Los Angeles, California, USA
Y(ale)PM	Yale Peabody Museum, New Haven, Connecticut, USA
ZFMK	Zoologisches Forschungsinstitut und Museum, A. Koenig, Bonn, Germany
ZILE	Zoological Institute, St. Petersburg, Russia
ZIUT	Zoological Institute, University Turku, Finland
ZMA(ms)	Zoological Museum Amsterdam, The Netherlands
ZMB = MNBHU	Zoologisches Museum Berlin, Humboldt Universität, Berlin, Germany
ZMH(amb)	Zoologisches Museum Hamburg, Germany
ZMK = UMCZ	Zoologigk Museet Kobenhavn, Danmark
ZMM	Zoological Museum Moscow, Russia
ZMO	Zoologisgk Museum, Universitet Oslo, Norway
ZMTAU	Zoological Museum Tel Aviv University, Tel Aviv, Israel
ZMUZ	Zoologisches Museum der Universität Zürich, Switzerland
ZSBS	Zoologische Sammlung des Bayerischen Staates, München, Germany
ZUEC	Zool. Universidade Estadual de campinas, Brazil

Part I Introduction

Teil I Einführung

Introduction

Not since the publication in 1949 of James Lee Peters' *Checklist of the Birds of the World*, which included the order Strigiformes (owls), has any detailed checklist of owls been compiled. Later checklists, such as those of Amadon and Bull (1988) and Sibley and Monroe (1990), listed primarily only species names, and did not include important data such as subspecies, synonyms, terra typica (type locality), or details of first description of type.

The primary objective in preparing the present list has been to provide clear, concise and up-to-date information on all living or recently extinct owl taxons. Hence, the taxonomy of all species and subspecies is described. In addition to the scientific names, the common names of each owl species in English, German, French and Spanish are listed. For each taxon, the name of the author who first described it and the year of publication are given, followed by aids to identification such as the first known location (type locality/terra typica), and scientific synonyms. Geographical distribution and preferred habitat are then briefly described. For each species, a detailed list of hitherto unpublished measurements is provided, including body length and mass, wing and tail length, as well as tarsus and bill length, the latter particularly in the case of the tribe Otini. These measurements were collected by visiting museums, aviaries, and zoos, as well as from the available literature. This is the first taxonomic checklist to include information on the location of museums housing collections of skins and mounted specimens.

To further address the interests of the ever increasing number of professional and amateur ornithologists and owl enthusiasts, references to figures, paintings, early lithographs, etchings and woodcuts from the period of 1700–1920 as well as a critical selection of recent journals and books containing paintings, line drawings, prints or photographs of live owls have been included. Last but not least a list of publications – journals, field guides, monographs and handbooks – complements each description.

Naturally, gaps are inevitable when compiling such a checklist. Such omissions mainly concern the measurements of body masses of numerous taxons, since a great number of collected and labelled skins lack data on the weight of the freshly killed bird. Because several species are extinct or known only from a few skins, data on the body masses of such owls are probably lost for all time.

While superspecies are generally not considered, comments on taxa are given in cases where the species or subspecies status is unclear, or where the synonym or taxon has notable morphological aspects.

The shortfall of each primary tip, measured from the wingpoint of the longest primary, is significant in determining the wing formula, i.e. whether the wing is pointed or rounded in shape. Furthermore, such measurements can be helpful in establishing the degree of relationship of species within a genus. An extensive list of measurements of such shortfalls, mainly of the complex tribe Otini but with some examples of other genera and species, is given in Table 1. The table is complemented by around 100 line drawings of flying owls with upward wingstrokes, showing both the wingshape (to scale) and the underwing pattern.

Besides the line drawings of owls in flight and numerous drawings of different owl species, the book includes several colour plates illustrating owl species that have been newly described within the last 20 years or have been recently rediscovered. In most cases, these owls are depicted together with species with similar plumage, or which have overlapping territories, and which therefore can easily be confused. Two colour paintings showing the subspecies and colour variations of the American Great Horned Owls complete the collection of watercolours.

The motivation to create this list was first inspired by the excellent booklet *Eulen* by Eck and Busse (1973), a real treasury of information. The checklist is not intended to be used as an identification guide, but rather offers information not usually published in monographs or handbooks. For the owl enthusiast seeking further information, I can recommend for further reading the handbooks of König, Weick and Becking (1999) and Del Hoyo, Elliott and Sargatal (1999). The statement by Edward C. Dickinson (1991) that "*Every new checklist stimulates the work that later outdates it*" will surely also be the fate of this checklist.

Einführung

Seit James Lee Peters im Jahre 1940 seine *Checklist of the Birds of the World* mit der Ordnung der Eulen Strigiformes veröffentlicht hatte, gab es in dieser ausführlichen Fassung keine neue Artenliste mehr. Später erschienene systematische Eulenlisten wie z. B. die von Amadon and Bull (1988) oder von Sibley and Monroe (1990) führten in erster Linie nur die Spezies, nicht aber die Subspezies auf. Auch fehlten Angaben zur Terra typica, dem ersten benannten Fundort, zu Synonymen oder zur Literatur der Erstbeschreibung (Typus).

Mit der vorliegenden Artenliste wird erstmals wieder eine aktuelle Eulenübersicht mit bündigen Informationen zu allen lebenden und zu den in jüngster Zeit ausgestorbenen Eulentaxa gegeben. Sämtliche Spezies und Subspezies werden in ihrer taxonomischen Einordnung beschrieben. Neben den wissenschaftlichen Namen sind die englischen, deutschen, französischen und spanischen Trivialnamen der einzelnen Eulenarten genannt. Zu jedem Taxon ist der Name des erstbeschreibenden Autors und das Publikationsjahr angegeben, es folgen Hinweise zur Typusbeschreibung, zum ersten benannten Fundort, sowie zu den Synonymen. Weiter werden die geografische Verbreitung und der bevorzugte Lebensraum kurz beschrieben. Für jede Eulenart sind zahlreiche bisher nicht publizierte Maßangaben, wie Körperlänge und -gewicht sowie Flügel-, Schwanz-, Lauf- und Schnabellänge aufgeführt, letztere besonders zum Tribus Otini. Diese Daten wurden nach Messungen an Bälgen in Museen und an lebenden Vögeln sowie aus der Literatur zusammengestellt. Erstmalig in einer Artenliste finden sich Angaben zu den Standorten von Bälgen und aufgestellten Präparaten in Museen.

Besonders mit Blick auf die wachsende Zahl von Ornithologen und Eulenliebhabern wurden Quellenverweise sowohl auf Abbildungen, Lithografien, Holzschnitte und Radierungen aus der Zeit von 1700–1920 als auch eine kritische Auswahl an rezenten Zeitschriften und Büchern mit Aquarellen, Federzeichnungen oder Drucken mit Fotos von lebenden Eulen aufgenommen. Literaturangaben zu Zeitschriften, Feldführern, Monografien oder Handbüchern zu jedem Taxon ergänzen die Beschreibungen.

Ein solch umfangreicher Datenbestand muss naturgemäß lückenhaft bleiben; so fehlen beispielsweise oft die Angaben zum Körpergewicht, da die Etikette vieler älterer Bälge keine Hinweise zum Gewicht frisch-toter Vögel enthalten. Zahlreiche Arten sind aber bereits ausgestorben und nur durch wenige Balgbelege bekannt; deshalb ist zu befürchten, dass solche Daten für immer verloren sind.

Angaben zu Superspezies werden im allgemeinen nicht gemacht, jedoch finden sich Bemerkungen zu Taxa, deren Art- oder Unterartstatus zweifelhaft ist, außerdem zu Synonymen und zu Taxa mit bemerkenswerten morphologischen Aspekten.

Handschwingenprojektionen, die den Überstand der längsten Handschwinge, der Flügelspitze, zu den anderen Handschwingenspitzen aufzeigen, lassen Rückschlüsse auf die Flügelform wie spitz- oder rundflügelig zu. Außerdem können die Schwingenformeln hilfreich sein, um die Verwandtschaft der Arten innerhalb einer Gattung aufzuzeigen. Eine Tabelle bietet umfangreiche Messdaten; diese betreffen in erster Linie den unübersichtlichen Tribus Otini sowie einige Beispielen anderer Gattungen und Arten. Die Tabelle wird ergänzt durch Tafeln mit ca. 100 Eulenflugbildern. Die aufwärts gerichtete Flügelhaltung zeigt dabei die maßstäbliche Flügelform und die Unterseitenzeichnung.

Neben diesen Eulenflugbildern und den zahlreichen Federzeichnungen verschiedener Eulenarten finden sich am Ende des Buches Farbabbildungen von Eulen, die in den letzten 20 Jahren neu beschrieben oder wiederentdeckt wurden. Meist sind diese Eulen im Vergleich mit ihnen ähnlichen Arten, mit denen sie verwechselt wurden, dargestellt oder mit Arten, die in ihrer unmittelbarer Nachbarschaft vorkommen. Zwei Farbzeichnungen mit den Unterarten und Farbmorphen der amerikanischen Uhus ergänzen die Aquarelle.

Einen ersten Anstoß zur Konzeption dieser Artenliste gab das schöne Büchlein *Eulen* von Eck and Busse (1973), eine echte Fundgrube an Informationen. Die Artenliste soll kein Bestimmungsbuch sein, enthält dafür aber ergänzende Angaben, die sonst in keiner Monografie oder einem Handbuch zu finden sind. Dem an Eulen Interessierten seien für weitere Informationen die Handbücher von König, Weick and Becking (1999) oder Del Hoyo, Elliott and Sargatal (1999) empfohlen. Sicher wird auch diese Artenliste das Schicksal teilen, das Eduard C. Dickinson (1991) voraussah: „*Every new checklist stimulates the work, that later outdates it.*“

Owls: A Brief Overview

The owls form a natural, easily defined group of chiefly nocturnal birds, classified as the order Strigiformes. All owls have a rather large, rounded head, with forward facing eyes and soft, often cryptically coloured plumage. The curved bill with its pointed tip, and powerful talons with sharp claws, are similar to those of birds of prey. Owls have a caeca but no crop. When perched, owls exhibit an upright posture. Many species have so-called "ear-tufts" – elongated feathers on the sides of the forehead. These play a behavioural role, primarily for camouflage, but have nothing to do with hearing. The true ears are openings situated at the sides of the head behind the rim of the facial disc.

The order Strigiformes comprises two families: Tytonidae (Barn- and Bayowls) and Strigidae (Typical Owls). Both families have distinctive morphological and anatomical features.

The family Tytonidae is devided in two subfamilies: Barn Owls (Tytoninae) and Bay Owls (Phodilinae). The subfamily Barn Owls (Tytoninae) includes only one genus with about 22 species. In this checklist, the Prigogine or Itombwe Owl is included in the Tytoninae, as it possesses typical barn owl features, e.g. a heart-shaped facial disc and relatively small eyes. Some barn owls normally considered as subspecies of *Tyto alba* have been accorded full species status in this list (BSC, Mayer 1975). It is very likely that the American barn owls, with the exception of the small owl *Tyto bargei*, should also be regarded as a distinct species: *Tyto furcata*. The Bay Owls (Phodilinae) consist of only one genus and one species, and occur only in south-east Asia.

The family Typical Owls (Strigidae) is subdivided into three subfamilies: the subfamily Striginae, i.e. Scops and Screech Owls, Eagle Owls and Wood Owls; the subfamily Hawk Owls (Surniinae), and the subfamily Ear Owls and relatives (Asioninae). These subfamilies are further subdivided into various tribes (tribus) and into 25 genera. Most owl species belong to the Strigidae family of Typical Owls.

The American screech owls, assigned to the genus *Otus*, are here treated in the subgenera *Megascops* and *Macabra*. The Flammulated Owl (*Otus flammeolus*) holds a questionable position in the genus *Otus*. Recent molecular biological studies suggest this owl should be assigned to a separate genus, *Psiloscops* Coues (personal communication, König 2005). The Blakiston's fish owl is here included within the genus *Bubo*, while the remaining fish owl species (*Ketupa*), and the snowy owl (*Nyctea*) retain their position within their accepted genera.

The Pernambuco pygmy owl in the genus *Glaucidium* was named *Glaucidium mooreorum* by da Silva et al. (2002). However, as early as 1830, Wied had already named these same birds from north-eastern Brazil as *Glaucidium minutissimum*, thus the name *G. mooreorum* became a synonym. It was therefore necessary to find a new name for the pygmy owls of south-eastern Brazil, and the adjacent Paraguay, since the name *G. minutissimum* could no longer been used. König and Weick (2005) described them as a new species, *Glaucidium sicki*. The untypical, much larger Owlets from Africa and Asia, hitherto included in the genus *Glaucidium*, are now defined as a subgenus *Taenioglaux*, a name already defined in 1848 by Kaup.

In the hawk owls, an indeterminate skin, located in the Senckenberg Museum, labelled only with the inscription "*Ninox* spec.", with neither the location of the find nor the collector indicated, is now named as *Ninox dubiosa*, in the hope that further information concerning its distribution may be obtained. The Striped Owl, *Asio clamator*, is not included in the genus *Pseudoscops* together with the Jamaican Owl, as was suggested by Olsen (1995); molecular biological analysis has indicated that *Asia clamator* as well as *Asio stygius* are distinct species within the genus Asio (Wink and Heidrich 1999).

Owls do not exhibit distinct regional "dialects"; all vocalisations are inherited. Bioacoustic indications are thus among the most important taxonomic criteria used to distinguish species groups that are otherwise difficult to separate. Species status can be substantiated by DNA evidence and field studies, complemented by morphological and anatomical data (König 1999).

Eulen: Eine kurze Übersicht

Eulen bilden eine natürliche, einheitliche und deutlich abgegrenzte Gruppe hauptsächlich nächtlich lebender Vögel. Taxonomisch werden sie in der Ordnung Strigiformes zusammengefasst. Alle Eulen haben einen relativ großen, runden Kopf mit nach vorne gerichteten Augen und ein weiches, meist kryptisch gefärbtes Gefieder. Der krumme, spitze Schnabel und die kräftigen, mit scharfen Krallen versehenen Fänge sind denen der Greifvögel sehr ähnlich. Eulen besitzen einen Blinddarm, aber keinen Kropf. Sitzend nehmen sie eine aufrechte Haltung ein. Viele Eulenarten besitzen sogenannte Federohren, bewegliche verlängerte Federn an den Stirnseiten. Diese Federohren spielen nur eine Rolle im Verhalten, in erster Linie sollen sie als Unterstützung der Tarnstellung von Eulen dienen; mit dem Gehör haben sie nichts zu tun. Die eigentlichen Ohröffnungen sitzen an den Kopfseiten, verborgen hinter dichtstehenden Federn des Schleierrandes.

Die Ordnung Strigiformes wird zunächst in zwei Gruppen unterteilt, in die Familie der Schleier- und Maskeneulen (Tytonidae) und die Familie der eigentlichen Eulen (Strigidae). Beide Familien zeigen deutliche morphologische und anatomische Unterschiede.

Die Familie Tytonidae teilt sich in die Unterfamilien Schleiereulen (Tytoninae) und Maskeneulen (Phodilinae). Die Unterfamilie der Schleiereulen (Tytoninae) besteht aus nur einer Gattung mit etwa 22 Arten. In der vorliegenden Artenliste wird die Prigogine oder Itombwe-Eule den Tytoninae zugerechnet. Sie hat typische Schleiereulenmerkmale, wie z. B. einen herzförmigen Schleier und relativ kleine Augen. Einige Schleiereulen, die sonst meist als Unterarten der Species *Tyto alba* zugerechnet werden, erhielten in der vorliegenden Liste Artstatus (Biol. Artkonzept, BSC, Mayer 1975). Mit großer Wahrscheinlichkeit stellt auch die Gruppe der großen amerikanischen Schleiereulen eine eigenständige Art dar, die Species *Tyto furcata*, allerdings ohne die kleine Eule *Tyto bargei*. Die Maskeneulen (Phodilinae) sind mit nur einer Gattung und nur einer Art ausschließlich in Süd-Ost-Asien verbreitet.

Die Familie der eigentlichen Eulen (Strigidae) umfasst drei Unterfamilien: die Unterfamilie Striginae, das sind Zwergohr- und Kreischeulen, Uhus und echte Käuze, die Unterfamilie der Falkenkäuze (Surniinae) und die Unterfamilie der Ohreulen und Verwandte (Asioninae). Die Unterfamilien gliedern sich weiterhin in zahlreiche Stämme (Tribus) sowie in 25 Gattungen. Die Familie der eigentlichen Eulen Strigidae umfasst den Großteil der Eulenarten.

In der vorliegenden Eulensystematik wurden für die amerikanischen Kreischeulen in der Gattung *Otus* die Untergattungen *Megascops* und *Macabra* verwendet. Eine etwas rätselhafte Stellung nimmt *Otus flammeolus* ein, die nach neuesten molekularbiologischen Analysen einer eigenen Gattung *Psiloscops Coues* zugeordnet werden sollte (persönliche Mitteilung, König 2005). Der Riesenfischuhu wird nun zur Gattung *Bubo*, den eigentlichen Uhus, gerechnet, die restlichen Fischuhus (*Ketupa*) sowie die Schnee-Eule (*Nyctea*) behalten ihre bislang benutzten Gattungsnamen.

Der Pernambuco Zwergkauz in der Gattung *Glaucidium* wurde von da Silva et al. (2002) als *Glaucidium mooreorum* benannt. Da aber bereits 1830 von Wied diese Vögel aus Nordostbrasilien *Glaucidium minutissimum* nannte, wurde *G. mooreorum* zum Synonym. Es war daher nötig, Zwergkäuze aus Südostbrasilien und dem angrenzenden Paraguay neu zu benennen, da für diese der Name *G. minutissimum* nicht mehr verwendet werden kann. König und Weick (2005) gaben den Zwergkäuzen dieses Gebietes nun den Namen *Glaucidium sicki*. Die untypischen, viel größeren „Sperlingskäuze" Afrikas und Asiens wurden zur Untergattung *Taenioglaux* zusammengefasst, ein Name, der bereits von Kaup im Jahr 1848 geprägt wurde.

Bei den Falkenkäuzen wurde ein rätselhafter Balg aus dem Senckenberg-Museum, der nur mit dem Etikett „*Ninox* spec." versehen war, also weder Fundort noch den Namen des Sammlers enthielt, in unserer Liste als „*Ninox dubiosa*" benannt und abgebildet, um damit eventuell zu weiteren Informationen zu gelangen. Die Streifenohreule *Asio clamator* wird hier nicht, wie von Olson (1995) vorgeschlagen, mit der Jamaika-Eule in einer Gattung *Pseudoscops* vereint. Durch molekularbiologische Analysen wurde belegt, dass es sich bei *Asio clamator* genauso wie bei *Asio stygius* um echte Arten der Gattung *Asio* handelt (Wink und Heidrich 1999).

Alle Lautäußerungen und Gesänge sind bei Eulen angeboren; Eulen weisen daher keine regionalen Dialekte auf. Diese bioakustischen Merkmale sind damit die wichtigsten Kriterien zur taxonomischen Abgrenzung von sonst schwer zu unterscheidenden Artengruppen. Ihr Artstatus lässt sich durch DNA-Untersuchungen und Feldstudien, die durch morphologische und anatomische Ergebnisse ergänzt werden, bestätigen (König 1999).

Overview of the Order Strigiformes
Überblick über die Ordnung Strigiformes

Ordo / Order Strigiformes

Familia / Family	Subfamilia / Subfamily	Tribus / Tribe	Genus	Subgenus
Tytonidae	Tytoninae		*Tyto* Billberg 1828	
	Phodilinae		*Phodilus* Geoffroy Saint Hilaire 1830	
Strigidae	Striginae	Otini	*Otus* Pennent 1769	*Megascops* Kaup 1848
				Macabra Bonaparte 1854
			Pyrroglaux Yamashina 1938	
			Gymnoglaux Cabanis 1855	
			Ptilopsis Kaup 1851	
			Mimizuku Hachisuka 1934	
		Bubonini	*Nyctea* Stephens 1826	
			Bubo Duméril 1806	
			Ketupa Lesson 1830	
			Scotopelia Bonaparte 1850	
		Strigini	*Strix* Linné 1758	
			Jubula Bates 1929	
			Lophostrix Lesson 1836	
			Pulsatrix Kaup 1848	
	Surniinae	Surniini	*Surnia* Duméril 1806	
			Glaucidium Boie 1826	*Taenioglaux* Kaup 1848
			Xenoglaux O'Neill and Graves 1977	
			Micrathene Coues 1866	
			Athene Boie 1822	
		Aegolini	*Aegolius* Kaup 1829	
		Ninoxini	*Ninox* Hodgson 1837	
			Uroglaux Mayr 1937	
			Sceloglaux Kaup 1848	
	Asioninae		*Pseudoscops* Kaup 1848	
			Asio Brisson 1760	
			Nesasio J. L. Peters 1937	

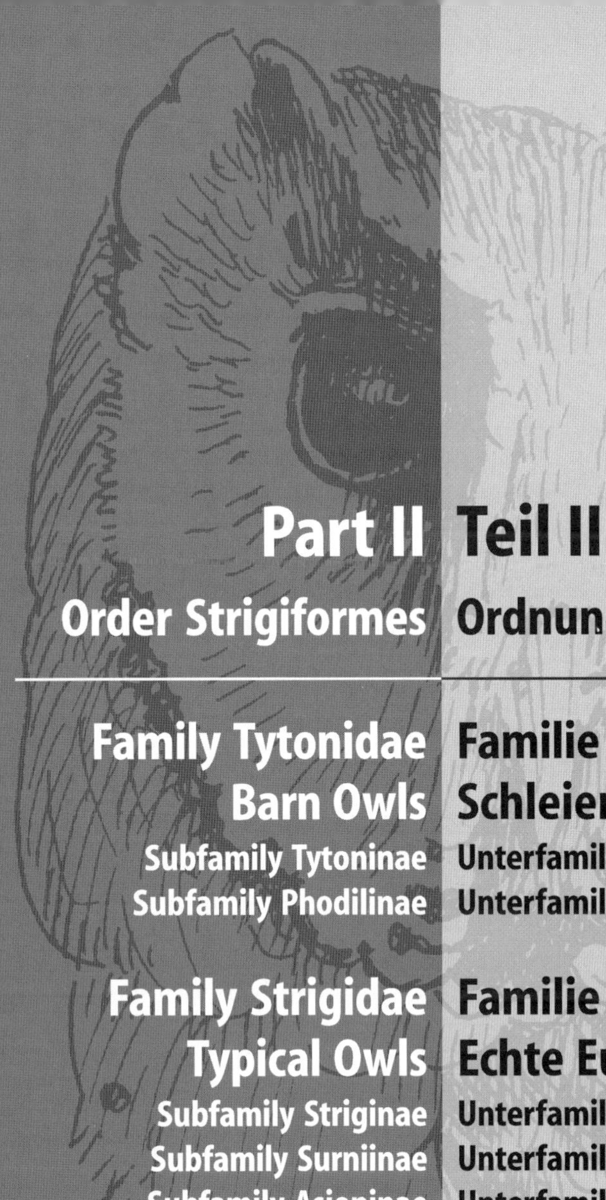

Part II Teil II

Order Strigiformes | Ordnung Strigiformes

Family Tytonidae | Familie Tytonidae
Barn Owls | Schleiereulen
Subfamily Tytoninae | Unterfamilie Tytoninae
Subfamily Phodilinae | Unterfamilie Phodilinae

Family Strigidae | Familie Strigidae
Typical Owls | Echte Eulen
Subfamily Striginae | Unterfamilie Striginae
Subfamily Surniinae | Unterfamilie Surniinae
Subfamily Asioninae | Unterfamilie Asioninae

Familia / Family
Tytonidae · Barn Owls · Schleiereulen

Subfamilia / Subfamily
Tytoninae · Barn Owls, Grass Owls, Masked Owls · Schleiereulen, Graseulen

Genus
Tyto **Billberg 1828**
Tyto Billberg 1828, Syn Fauna Scand. 1(2): Table 1. Type by *Strix flammea* auct. = *Strix alba* Scop

Tyto alba **(Scopoli) 1769**
Common Barn Owl · Schleiereule · Effraie de clochers · Lechuza Común
(see Plate 1)

Length: 330–430 mm
Body mass: 320–480 g

Distribution: Widely distributed in warm and temperate regions, including many islands. Partial migrant in north. Avoids cold areas and high altitudes. Several insular forms are very distinct and may be true species
Habitat: Occurs in a great variety of habitats in mixed or open country, also in human settlements

- ***Tyto alba schmitzi*** **(Hartert) 1900**
  Strix flammea schmitzi Hartert 1900, Novit. Zool. 7: 534; Terra typica: Funchal, Madeira

Distribution: Madeira and Porto Santo
Museum: AMNH, MFM = BMMF, ZFMK, BMNH
Remarks: Similar to light morphs of *T. a. alba*, but larger and coarser spots below

Wing length: 268–286 mm
Tail length: 107–116 mm
Tarsus length: ?
Length of bill: 33.5 mm
Body mass: ♂ 304 g, ♀ 323 g

Illustration: M. Hulme in del Hoyo et al. 1999: Pl. 2 (head)
Photograph: ?
Literature: Hartert 1900: 534; Bunn et al. 1982: 27; Hartert 1912–1921: 1036, 1037; Vaurie 1965: 631; Schneider and Eck 1995; Hazevoet 1995; Cramp et al. 1985: 447, 448; del Hoyo et al. 1999: 71, 72; König et al. 1999: 195

- ***Tyto alba gracilirostris*** **Hartert 1905**
  Strix flammea gracilirostris (Hartert) 1905, Bull. Brit. Ornith. Cl.: 16; Terra typica: Fuertoventura, Canary Islands

Distribution: Eastern Canary Islands: Fuertoventura and Lanzarote
Museum: BMNH, AMNH (Rothsch.), ZFMK, ZMA, ZMH

Wing length: 235–270 mm
Tail length: 94–105 mm
Tarsus length: 55–61 mm
Length of bill: 27.2 mm
Body mass: ♂ 250–400 g, ♀ 252–330 g

Illustration: H. Delin in Cramp et al. 1985: Pl. 43
Photograph: ?
Literature: Hartert 1912–1921: 1037; Vaurie 1965: 630; Cramp et al. 1985: 447, 448; Jaume et al. 1993; Schneider and Eck 1995: 92; König et al. 1999: 195; del Hoyo et al. 1999: 71, 72

- ***Tyto alba alba*** **(Scopoli) 1769**
  Strix alba Scopoli 1769, Annus I, Hist.-Nat. 21; Terra typica: Ex Foro Juli = Friuli (Friaul), Italy

Synonym:
- *Strix kirchhoffi* A.E. Brehm 1857, Allg. Deutsche Naturhist. Zeit.: 440; Terra typica: South Spain
- *Strix hostilis* Kleinschmidt 1915, Falco 11: 18; Terra typica: England (holotype ZFMK)
- *Tyto alba kleinschmidti* von Jordans 1924, J. Orn. 72: 409; Terra typica: Alcudia, Mallorca

Tyto alba alba
Barn Owl
Schleiereule

Distribution: British Isles, Channel Islands, western France, Iberian Peninsula, Italy south of the Alps. Countries bordering the Mediterranean basin and islands, with the exception of Corsica and Sardinia, Mauretania, Tunisia and Egypt. From eastern France eastwards, sympatric with subspecies *guttata* in western Germany
Museum: BMNH, FNSF, NNML, SMNKa, SMNSt, ZFMK, ZMA

Wing length:	♂ 280–302 mm, ♀ 278–296 mm (1 × 310 mm)
Tail length:	110–125 mm
Tarsus length:	56–63 mm
Length of bill:	30–33 mm
Body mass:	♂ 240–313 g, ♀ 245–360 g

Illustration: T. Bewick in Bewick 1809: 89 (wood engraving); J. G. Keulemans in Dresser 1871–1896: Pl. 302; C. Tunnicliffe in Cusa 1984: 4; C. Tunnicliffe in Niall 1985: Frontcover and 43; H. Delin in Cramp et al. 1985: Pl. 43; A. Cameron in Voous 1988: 11; A. Thorburn in Thorburn 1990: Pl. facing p. 30; F. Weick in Brandt and Seebaß 1994: Abb. 2.5 col.; F. Weick in König et al. 1999: Pl. 1; M. Hulme in del Hoyo et al. 1999: Pl. 2; F. Weick in Weick 2004: Oktober
Photograph: E. Hosking in Hosking and Flegg 1982: 112–129; M. Rogl in Epple and Rogl 1989; Danegger in Mebs and Scherzinger 2000: 115–117; Nill in Mebs and Scherzinger 2000: 115, 123; M. Read in del Hoyo et al. 1999: 45, 46; A. Rouse in del Hoyo et al. 1999: 53, 54
Literature: Hartert 1912–1921: 1031–1036; Hartert 1929a: 94–96; Vaurie 1965: 631; Bunn et al. 1982; Glutz von Blotzheim and Bauer 1980: 234–276; Cramp et al. 1985: 432–449; Voous 1988: 9–22; Brandt and Seebaß 1994; Schneider and Eck 1995: 89, 90; König et al. 1999: 193–195; del Hoyo et al. 1999: 71, 72; Mebs and Scherzinger 2000: 114–131

- ***Tyto alba ernesti*** (Kleinschmidt) 1901
  Strix ernesti Kleinschmidt 1901, Orn. Monatsb. 9: 168;
  Terra typica: Loceri, Sardinia

Distribution: Corsica and Sardinia
Museum: FNSF, SMNSt, SMTD, ZFMK
Remarks: Above and on wings very light coloured, below clear atlas-white, often without spots

Wing length:	281–309 mm
Tail length:	110–126 mm
Tarsus length:	57–62 mm
Length of bill:	19.2 mm
Body mass:	?

Illustration: H. Quintscher in Schneider and Eck 1995: Abb. 34a; F. Weick in Brandt and Seebaß 1994: Abb. 2.5; F. Weick in König et al. 1999: Pl. 1
Photograph: ?
Literature: Hartert 1912–1921: 1036; Hartert 1929a: 95–96; Vaurie 1965: 632; Cramp et al. 1985: 447, 448; Schneider and Eck 1995: 90; König et al. 1999: 194; del Hoyo et al. 1999: 71, 72

- ***Tyto alba guttata*** (C.L. Brehm) 1831
  Strix guttata C.L. Brehm 1831, Handb. Naturg. Vögel Deutschl.: 106; Terra typica: Rügen, Germany

Synonym:
- *Strix splendens* C.L. Brehm 1855, Vogelfang: 40; Terra typica: Strasbourg, France
- *Strix adspersa* C.L. Brehm 1858, Vogelfang: 215; Terra typica: Eisenberg, Germany
- *Strix flammea obscura* C.L. Brehm 1858, Naumannia: 214; **Terra typica:** Altenburg, Germany
- *Strix Flammea rhenana* Kleinschmidt 1906, Berajah: *Strix Flammea*: 20; **Terra typica:** Darmstadt, Germany

Distribution: Southern Sweden, Germany (sympatric in the west with *alba*), east through the Baltic countries and Poland to western Russia; south

to the Alps, Austria, Hungary, Bulgaria and the Crimea
Museum: FNSF, SMNKa, SMNSt, SMTD, ZFMK, ZMH

Wing length:	270–297 (1 × 310) mm
Tail length:	122–137 mm
Tarsus length:	55–59 mm
Length of bill:	29–33 mm
Body mass:	♂ 290–345 g, ♀ 335–380 g (405 g)

Illustration: J. F. Naumann in Naumann 1822: Pl. 47; O. Kleinschmidt in Kleinschmidt 1958: Pl. 17, 40; F. Weick in König et al. 1999: Pl. 1; H. Delin in Cramp et al. 1985: Pl. 43; M. Hulme in del Hoyo et al. 1999: Pl. 2; F. Weick in Brandt and Seebaß 1994: Abb. 2.5; A. Cameron in Voous 1988: 11; Mullarney and Zetterström in Svensson et al. 1999: 213
Photograph: G. W. Robinson in Everett 1977: 20; A. Limbrunner in Pforr and Limbrunner 1980: 44; A. Limbrunner in Mebs and Scherzinger 2000: 126; Schmidt in Mebs and Scherzinger 2000: 116
Literature: Naumann 1822: 483–492; Kleinschmidt 1906; Kleinschmidt 1958: 40; Hartert 1912–1921: 1029–1031; Hartert 1929a: 97; Dementiev and Gladkov 1951: 474–477; Vaurie 1965: 632; Cramp et al. 1985: 432–448; Mikkola 1983: 37–57; Schneider and Eck 1995: 87–89; König et al. 1999: 193–196; del Hoyo et al. 1999: 71, 72

- ***Tyto alba affinis*** **(Blyth) 1862**
Strix affinis Blyth 1862, Ibis: 388; Terra typica: Capetown, South Africa

Synonym:
- *Strix poensis* Fraser 1843, Proc. Zool. Soc. Lond. 10, 1842: 189; Terra typica: Fernando Póo (Bioko) Island
Remarks: *Poensis* is possibly the oldest name for an African *Tyto* species! The correct name of the Afrotropical mainland subspecies of Barn Owl is *Tyto alba poensis*! (Eck and Busse 1973: 60; Bruce and Dowsett 2004: 184)
- *Strix maculata* Brehm 1855, Vogelfang: 40; Terra typica: Northeast Africa

Distribution: Tropical Africa, from Gambia, south of the Sahara and the Sudan to southern South Africa, as well as the Bioko, Zanzibar and Pemba Islands
Museum: BMNH, SMNKa, SMNSt, TMBD, SMTD etc.
Remarks: Stronger feet than in Mediterranean subspecies; distinctly larger size than nominate *alba*

Wing length:	270–312 mm
Tail length:	110–123 mm
Tarsus length:	63–70 mm (1 × 57 mm)
Length of bill:	31–36 mm
Body mass:	266–470 g

Illustration: F. Neubaur in Koenig 1936: Pl. 55; M. Woodcock in Fry et al. 1988: Pl. 7; H. Quintscher in Schneider and Eck 1995: Abb. 33g; P. Hayman in Kemp and Kemp 1998: 283; M. Hulme in del Hoyo et al. 1999: Pl. 2; N. Borrow in Borrow and Demey 2001: Pl. 98; J. Gala in Stevenson and Fanshawe 2002: Pl. 99
Photograph: J. L. Vieljoen and P. Barichievy in Ginn et al. 1989: 331
Literature: Hartert 1912–1921: 1038, 1039; Bannerman 1953: 528, 529; Vaurie 1965: 633; Fry et al. 1988: 108–110; Schneider and Eck 1995: 91; Kemp and Kemp 1998: 282, 283; König et al. 1999: 194; del Hoyo et al. 1999: 71, 72; Bruce and Dowsett 2004: 184–187

- ***Tyto alba hypermetra*** **Grote 1928**
Tyto alba hypermetra Grote 1928, Orn. Monatsb. 36: 79; Terra typica: Central Madagascar
(see Figure 1)

Distribution: Comoro Islands and Madagascar
Museum: FNSF, MHNP, ZFMK (Kleinschmidt)
Remarks: Probably only a synonym of *affinis*, but with distinctly larger and more powerful feet than the African species

Wing length:	300–320 mm
Tail length:	120–140 mm
Tarsus length:	63–71 mm
Length of bill:	34–37 mm
Body mass:	?

Illustration: F. Weick in König 1999a: Pl. 1; F. Weick in *Gefiederte Welt* 2003(7): 213
Photograph: B. Wright in Morris and Hawkins 1998: 166
Literature: Hartert 1912–1921: 1038; Hartert 1929a: 97; Langrand 1990: 226; Schneider and Eck 1995: 91, 92; Morris and Hawkins 1998: 166; König et al. 1999: 194

- ***Tyto alba erlangeri*** (W.L. Sclater) 1921
  Tyto alba erlangeri W.L. Sclater 1921, Bull. Brit. Ornith. Cl. 42: 24; Terra typica: Lajeh, Arabia

Synonym:
- *Strix pusilla* Blyth 1850a, J. As. Soc. Bombay 18: 801; Terra typica: Egypt
- *Strix parva* Blyth 1850b, J. As. Soc. Bombay 18: 802; Terra typica: North Africa
- *Tyto alba microsticta* Koelz 1950, Proc. Biol. Soc. Wash. 1452: 3; Terra typica: Jahrum, Fars, southern Iran

Distribution: Near East, Iran, Iraq and Arabia. With North African birds of this species (Mauretania, Algeria and the Nile Valley); the older name *pusillus* Blyth (1850a) has priority?!
Museum: AMNH, SMTD, ZFMK, ZMTAU etc.
Remarks: Similar to *ernesti*, but more golden and less grey above, wings and tail more darkly and broadly banded, more spotted below

Wing length:	280–315 mm
Tail length:	107–126 mm
Tarsus length:	59–66 mm
Length of bill:	30–33 mm
Body mass:	?

Illustration: J. G. Keulemans in Koenig 1936: Pl. 54; H. Delin in Cramp et al. 1985: Pl. 43
Photograph: R. Pinna in Shirihai 1996: Pl. 66
Literature: Vaurie 1965: 632, 633; Cramp et al. 1985: 447–449; Schneider and Eck 1995: 90, 91; Shirihai 1996: 308, 309

- ***Tyto alba stertens*** Hartert 1929
  Tyto alba stertens Hartert 1929, Novit. Zool. 35: 98; Terra typica: Cachar

Synonym:
- *Tyto alba crypta* Koelz 1939, Proc. Biol. Soc. Wash. 52: 80; Terra typica: Londa, Bombay Presid

Distribution: India, Sri Lanka, west and east Pakistan, Assam and Myanmar, east of southern central China (Yunnan), Vietnam and southern Thailand
Museum: BNHS, AMNH, FNSF, USNM etc.
Remarks: Lighter coloured above and below than *javanica*, especially the grey pattern

Wing length:	262–322 mm
Tail length:	119–129 (134) mm
Tarsus length:	60–77 mm
Length of bill:	30–32 mm
Body mass:	?

Illustration: A. M. Hughes in Ali and Ripley 1969: Pl. 46; D. Cole in Grimmett et al. 1998: Pl. 28; F. Weick in König et al. 1999: Pl. 1; T. Worfolk in Robson 2000: Pl. 22
Photograph: ?
Literature: Hartert 1929a: 98; Deignan 1945: 171, 172; Ali and Ripley 1969: 250, 251; Schneider and Eck 1995: 93; König et al. 1999: 194; del Hoyo et al. 1999: 71; Robson 2000: 289; Rasmussen and Anderton 2005: 233

- ***Tyto alba javanica*** (Gmelin) 1788
  Strix javanica Gmelin 1788, Syst. Nat. 1(1): 295; Terra typica: Java

Distribution: Central and southern Myanmar, south-west China, Thailand, Cambodia, northern and central Laos, South Vietnam, the Malay Peninsula, Indonesia, southern Borneo, Sumatra, Java (including Kangean Island), Flores, Lombok, Kalao and Kalaotoa (as straggler?)
Museum: NNML, SMTD, ZFMK (Kleinschmidt)

Wing length:	275–323 mm
Tail length:	119–127 mm
Tarsus length:	68–77 mm
Length of bill:	30–32 mm
Body mass:	♂ 555 g, ♀ 612 g

Illustration: P. Burton in *The Raptor* 1989, Front-cover; K. Phillipps in McKinnon and Phillipps 1993: Pl. 39; H. Quintscher in Schneider and Eck 1995: Abb. 33f; D. Gardner in Coates et al. 1997: Pl. 33; H. Quintscher in Eck and Busse 1973: Abb. 29
Photograph: ?
Literature: Baker 1927: 385; Delacour and Jabouille 1931: 148, 149; Burton et al. 1984: 50; White and Bruce 1986: 246; McKinnon and Phillipps 1993: 188; Schneider and Eck 1995: 93, 94; Coates et al. 1997: 357; Rasmussen and Anderton 2005: 233

- ***Tyto alba pratincola*** (Bonaparte) 1838
  Strix pratincola Bonaparte 1838, Geogr. and Comp. List: 7;
  Terra typica: Pennsylvania

Synonym:
- *Strix flammea* var. *Guatemalae* Ridgway 1873, Bull. Essex Inst. 5: 200; Terra typica: Chimandega, Nicaragua
- *Tyto perlatus lucayanus* Riley 1913, Proc. Biol. Soc. Wash. 26: 153; Terra typica: New Providence, The Bahamas
- *Tyto alba bondi* Parkes and Phillips 1978, Ann. Carnegie Mus. 47: 479–492; Terra typica: Bay Island, Honduras

Distribution: North America (British Columbia), south Central America (Panama and possibly western Colombia), The Bahamas, Bermuda and Hispaniola. Bay Island off North Honduras
Museum: AMNH, BMNH, CMP (holotype *bondi)*, FNSF, SMTD, UKMNH, USNM (holotype *lucayanus)*, ZFMK, ZMK

Wing length: ♂ 314–346 mm, ♀ 320–360 (370) mm
Tail length: 125–158 mm
Tarsus length: 68–79 mm
Length of bill (cere): 22–30 mm
Body mass: ♂ 311–507 g, ♀ 383–573 g

Illustration: R. Ridgway in Merriam and Fisher 1893: Pl. 19; L. A. Fuertes in McCracken-Peck 1952: 93; A. Brooks in Sprunt 1955: 169; L. Malick in Breeden et al. 1983: 239; Landsdowne in Landsdowne and Livingston 1968: Pl. 30; K. E. Karalus in Karalus and Eckert 1974: Pl. 1; A. Wilson in Wilson 1975 (repr.): 50/2 (wood engraving); D. Gardner in Stiles and Skutch 1991: Pl. 20; D. Sibley in Sibley 2000: 272; F. Weick in König et al. 1999: Pl. 2; M. Hulme in del Hoyo et al. 1999: Pl. 2; L. A. Fuertes in Johnsgard 2002: Pl. 1
Photograph: H. Harrison and R. Robinson in Murphy and Amadon 1953: 34 and 35; T. Martin in del Hoyo et al. 1999: 47; F. Gohier in del Hoyo et al. 1999: 39; R. G. Krahe in *SCRO Mag.* 1999(2): 19
Literature: Ridgway 1914: 604–611; Sprunt 1955: 165–168; Bent 1961: 140–153; Karalus and Eckert 1974: 3–19; Schneider and Eck 1995: 96, 97; König et al. 1999: 194; del Hoyo et al. 1999: 71, 72; Johnsgard 2002: 65–73

- ***Tyto alba furcata*** (Temminck) 1827
  Strix furcata Temminck 1827, Pl. col. livr. 73: Pl. 432 and text;
  Terra typica: Cuba

Synonym:
- *Strix Cubae* C.L. Brehm 1831, Ges. Naturgesch. Vögel Deutschl.: 104
- *Tyto alba niveicauda* Parkes and Phillips 1978, Ann. Carnegie Mus. 47: 479–492; Terra typica: Isle of Pines, Cuba

Distribution: Cuba and Isle of Pines, Grand Cayman Islands, Cayman Brac and Jamaica
Museum: AMNH, CMP (holotype *niveicauda*), FNSF, MLUH, USNM, SMTD, ZFMK
Remarks: Secondaries and tail feathers often uniformly white. Long legs and powerful feet. Much larger than *pratincola*

Wing length: ♂ 316–349 mm, ♀ 328–358 mm
Tail length: ♂ 114–138 mm, ♀ 123–152 mm
Tarsus length: 67–77 mm
Length of bill: ?
Body mass: ?

Illustration: H. Quintscher in Schneider and Eck 1995: Abb. 33b; F. Weick in König et al. 1999: Pl. 2
Photograph: ?
Literature: Ridgway 1914: 602–604; Hartert 1929a: 101; Burton et al. 1984: 51; Parkes and Phillips 1978: 479–492

- ***Tyto alba bargei*** (Hartert) 1892
  Strix flammea bargei Hartert 1892, Bull. Brit. Ornith. Cl. 1: 13;
  Terra typica: Curacao

Distribution: Island of Curacao. Possibly also an endemic island species like *T. glaucops* or *T. insularis*
Museum: AMNH(Rothsch.), NNML, FMNH
Remarks: Very small size, endemic insular form, resembling some Mediterranean and Egypt specimens. Taxa *pratincola* (*guatemalae*), *furcata*, *contempta*, *hellmayri* and *tuidara* are sometimes considered as subspecies of the American Barn Owl *Tyto furcata* (oldest name). Thus *bargei* may be a separate species!

Wing length: 246–258 mm
Tail length: 109 mm
Tarsus length: 56 mm
Length of bill: ?
Body mass: ?

Illustration: M. Hulme in del Hoyo et al. 1999: Pl. 2
Photograph: ?
Literature: Hartert 1892: 13; Rothschild and Hartert 1902: 495; Hartert 1912–1921: 1039; Burton et al. 1984: 51; Schneider and Eck 1995: 99

- ***Tyto alba contempta*** (Hartert) 1898
  Strix flammea contempta Hartert 1898, Novit. Zool. V: 500; Terra typica: Cayambe, Ecuador

Synonym:
- *Strix stictica* von Madarász 1904, Ann. Mus. Nat. Hungar. 2: 115; Terra typica: Venezuela
- *Tyto alba subandeana* L. Kelso 1938, Biol. Leafl. 9: unpaged; Terra typica: Bogotá, Colombia

Distribution: Colombia, Ecuador, Peru and Venezuela
Museum: AMNH, FNSF, SMTD, ZFMK, ZMH
Remarks: Great variation in plumage, similar to *Tyto alba guttata*. Requires further studies (DNA sequencing)

Wing length: 293 and 300 mm
Tail length: ?
Tarsus length: ?
Length of bill: ?
Body mass: ?

Illustration: J. Fjeldsa in Fjeldsa and Krabbe 1990: Pl. 26; F. Weick in König et al. 1999: Pl. 2; P.J. Greenfield in Ridgely and Greenfield 2001: II: Pl. 36(1)
Photograph: ?
Literature: Hartert 1898b: 500; Hartert 1929a: 101; Burton et al. 1984: 51; Fjeldsa and Krabbe 1990: 221, 222; Schneider and Eck 1995: 99; König et al. 1999: 195; Ridgely and Greenfield 2001: I: 302, II: 211; Hilty 2003: 358

- ***Tyto alba hellmayri*** Griscom and Greenway 1937
  Tyto alba hellmayri Griscom and Greenway 1937, Bull. Mus. Comp. Zool. 81: 421; Terra typica: Paramaribo, Surinam

Distribution: Eastern Venezuela (including Margarita Island), the Guianas south to the Amazon Valley, west to Surinam, northern Brazil, Trinidad and Tobago
Museum: AMNH, CMC, MCZ, NNML
Remarks: The separation of *hellmayri* from *tuidara*, based on size characteristics, is questionable!

Wing length: 315–335 mm
Tail length: ?
Tarsus length: ?
Length of bill: ?
Body mass: ♂ 460–485 g, ♀ 446–558 g

Illustration: P. Barruel in Haverschmidt 1968: Pl. 13; S. Webb in Hilty 2003: Pl. 24
Photograph: ?
Literature: Haverschmidt 1962: 236–342; Haverschmidt 1968: 156–158; Schneider and Eck 1990: 99–101; del Hoyo et al. 1999: 71; Hilty 2003: 358

- ***Tyto alba tuidara*** (J.E. Gray) 1829
  Strix Tuidara J.E. Gray 1829, in Griffith of Cuvier's Anim. Kingdom 6: 75; Terra typica: Brazil

Synonym:
- *Strix perlata* (invalid name) Lichtenstein 1823, Verz. Doubl. Mus. Berlin: 59; Terra typica: Brazil
- *Tyto alba zottae* L. Kelso 1938, Biol. Leafl. 9: unpaged; Terra typica: Cordillera de Rio Chico, Patag., Argentina
- *Tyto alba hauchecorni* Kleinschmidt 1940, Falco 36: 60 (footnote); Terra typica: Chile

Distribution: From Brazil south of the Amazon Valley, to Chile and Argentina (Tierra del Fuego)
Museum: AMNH, MNBHU, SMTD, ZFMK, ZMH

Wing length: 290–338 mm
Tail length: 113–143 mm
Tarsus length: 79.5 mm
Length of bill: ?
Body mass: 387–560 g

Illustration: H. Quintscher in Schneider and Eck 1995: Pl. facing p. 48; M. Hulme in del Hoyo et al. 1999: Pl. 2; F. Weick in König et al. 1999: Pl. 2 (light and dark morph)
Photograph: ?
Literature: Kleinschmidt 1940: 60; Eck 1971: 177, 178; Fjeldsa and Krabbe 1990: 221; Schneider and Eck 1995: 99; König et al. 1999: 194; del Hoyo et al. 1999: 71

Tyto glaucops **(Kaup) 1852**
Ashy-faced Barn Owl · Hispaniolaschleiereule · Effraie d'Hispaniola · Lechuza de la Espaniola
Strix glaucops **Kaup 1852, Contrib. Orn. [Jardine]: 118;**
Terra typica: **Jamaica error = Hispaniola**
(see Plate 1)

Length: 330–350 mm
Body mass: 260–500 g

Distribution: Islands of Hispaniola and Tortuga
Habitat: Open country with trees and bushes, often near human settlements. Also in scrubland, caves and open forest. Sympatric with *Tyto alba*, endangered due to civilization
Museum: AMNH, BMNH (holotype)
Remarks: Separation as distinct species, widely accepted by sympatric breeding with *T. alba* and distinct vocalisations

Wing length: ♂ 240–250 mm, ♀ 260–280 mm
Tail length: 127–160 mm
Tarsus length: 60–77 mm
Length of bill: ?
Body mass: ♂ 260–346 g, ♀ 465–535 g

Illustration: J. G. Keulemans in Sharpe 1875a: Pl. 14; M. Hulme in del Hoyo et al. 1999: Pl. 2; F. Weick in König et al. 1999: Pl. 2; K. Williams in Raffaele et al. 2003: Pl. 43
Photograph: H. Hecht in Schneider and Eck 1995: 153; del Hoyo in del Hoyo et al. 1999: 36; R. Steinberg 1999: 18
Literature: Ridgway 1914: 612, 613; Schneider and Eck 1995: 152–154; König et al. 1999: 196; del Hoyo et al. 1999: 72; Raffaele et al. 2003: 100

Tyto insularis **(Pelzeln) 1872**
Lesser Antilles Barn Owl (St. Vincent and Dominica Barn Owll) · Kleine Antillen-Schleiereule

Length: 270–330 mm
Body mass: 260 g

Distribution: Lesser Antilles (St. Vincent, Grenada, Carriacou, Union and Bequia, Dominica)
Habitat: Open woodland, scrubland, bushes and caves
Remarks: Geographically isolated distribution suggests separation as a distinct species!

- ***Tyto insularis insularis*** (Pelzeln) 1872
  Strix insularis von Pelzeln 1872, J. Orn. 20: 23;
  Terra typica: St. Vincent, Lesser Antilles

Synonym:
- *Hybris nigrescens noctividus* Barbour 1912, Proc. Biol. Soc. Wash. 24: 57; Terra typica: Grenada

Distribution: Islands of St. Vincent, Grenada, Carriacou, Union and Bequia, Lesser Antilles
Museum: FNSF, NHMWien, NNML, USNM, ZFMK

Wing length: ♂ 226–236 mm, ♀ 241–243 mm
Tail length: 101–108 mm
Tarsus length: 48–55 mm
Length of bill: ?
Body mass: 264 g

Illustration: F. Weick in König et al. 1999: Pl. 2; M. Hulme in del Hoyo et al. 1999: Pl. 2; K. Williams in Raffaele et al. 2003: Pl. 43
Photograph: ?
Literature: Ridgway 1914: 613–615; Schneider and Eck 1995: 97, 98; del Hoyo et al. 1999: 72; König et al. 1999: 195

- ***Tyto insularis nigrescens*** (Lawrence) 1878
  Strix flammea var. *nigrescens* Lawrence 1878, Proc. US Nat. Mus. 1: 64; Terra typica: Dominica, Lesser Antilles

Distribution: Island of Dominica, Lesser Antilles
Museum: USNM (holotype), AMNH, FNSF, ZFMK

Wing length: ♂ 230–235 mm, ♀ 240–247 mm
Tail length: 103–113 mm
Tarsus length: 50–53 mm
Length of bill: ?
Body mass: ?

Illustration: M. Hulme in del Hoyo et al. 1999: Pl. 2
Photograph: ?
Literature: Ridgway 1914: 615; Schneider and Eck 1995: 98; del Hoyo et al. 1999: 71, 72

Tyto punctatissima **(G.R. Gray) 1838**
Galapagos Barn Owl · Galapagos-Schleiereule · Effraie des Galapagos
Strix punctatissima **G.R. Gray 1838, Gould's Zool. Voyage *"Beagle"* 3(3): Pl. 4 and (9): 34;** Terra typica: **James Island, Galapagos Archipelago**

Length:	330 mm
Body mass:	?

Distribution: Galapagos Archipelago
Habitat: Scrubland, bushes, caves in dry creek beds, humid highlands up to 500–600 m
Museum: BMNH, AMNH
Remarks: Well marked, endemic island bird, suggest separation as species!

Wing length:	229–234 mm
Tail length:	104–117 mm
Tarsus length:	55–68 mm
Length of bill:	43 mm
Body mass:	?

Illustration: G. R. Gray in Gray 1838: Pl. 4; M. Hulme in del Hoyo et al. 1999: Pl. 2
Photograph: R. Behrstock in del Hoyo et al. 1999: 36; T. de Roy in del Hoyo et al. 1999: 59
Literature: Sharpe 1875a: 297, 298; Ridgway 1914: 616, 617; Burton et al. 1984: 52; Schneider and Eck 1995: 98; del Hoyo et al. 1999: 71; König et al. 1999: 193–196

❖

Tyto detorta **Hartert**
Cape Verde Barn Owl · Kap Verde-Schleiereule
Tyto alba detorta **Hartert 1913, Bull. Brit. Ornith. Cl. 31: 38;** Terra typica: **Sao Tiago, Cape Verde Islands**

Length:	350 mm
Body mass:	?

Distribution: Cape Verde Islands; Santiago and Sao Vincente
Habitat: Open country, with trees, bushes, rocky canyons and ravines with steep walls, also near human settlements
Museum: AMNH, BMNH, SMTD
Remarks: Mostly included as a subspecies of *Tyto alba*, but is in fact a distinct endemic Island species. (see Hazevoet 1995)

Wing length:	272–297 mm
Tail length:	116–124 mm
Tarsus length:	60–66 mm
Length of bill:	28–33 mm
Body mass:	?

Illustration: J. G. Keulemans in Sharpe 1875a: Pl. 14; D. M. Reid Henry in Bannerman and Bannerman 1968: Pl. 44; H. Delin in Cramp et al. 1985: Pl. 43; M. Hulme in del Hoyo et al. 1999: Pl. 2
Photograph: ?
Literature: Hartert 1912–1921: 1037; Burton et al. 1984: 49; Cramp et al. 1985: 447–449
Schneider and Eck 1995: 92, 93 and 152–156; Hazevoet 1995: AOU Checklist

❖

Tyto thomensis **(Hartlaub) 1852**
Sâo Tomé Barn Owl · Sâo Tomé-Schleiereule · Effraie de Sâo Tomé
Strix thomensis **Hartlaub 1852, Rev. Mag. Zool. II(4)4: 3;** Terra typica: **Sâo Tomé and Principe Islands**

Length:	330 mm
Body mass:	380 g

Distribution: Sâo Tomé and Principe Islands
Habitat: Occurs in a great variety of habitats, common
Museum: AMNH, ZFMK, ZMH
Remarks: Often included as a subspecies of *Tyto alba*

Wing length:	241–264 mm
Tail length:	97–114 mm
Tarsus length:	60–66 mm
Length of bill:	?
Body mass:	380 g (1 ♂)

Illustration: ? in von Müller 1853/1854: Pl. 15; M. Woodcock in Fry et al. 1988: Pl. 7; F. Weick in König et al. 1999: Pl. 1; M. Hulme in del Hoyo et al. 1999: Pl. 2; N. Borrow in Borrow and Demey 2001: Pl. 50
Photograph: F. Höhler in Günther and Feiler 1950: Pl. 7 (skin)
Literature: Hartert 1929a: 98; Fry et al. 1988: 108; Burton et al. 1984: 49, 50; Schneider and Eck 1995: 93; König et al. 1999: 194; del Hoyo et al. 1999: 71

Tyto soumagnei **(Milne-Edwards) 1878**
Madagascar Red Owl · Malegasseneule · Effraie de Madagascar · Lechuza Malgache
(see Plate 1 and Figure 1)

Synonym:
- *Heliodilus Soumagnei* Milne-Edwards 1878, Compt. Rend. Acad. Sci. Paris 85(1877): 66; Terra typica: Madagascar

Length: 280–300 mm
Body mass: 320–430 g

Distribution: Northeast and east Madagascar
Habitat: Forests and secondary growth
Museum: BMNH, MHNP
Remarks: Sometimes in seperate genus *Heliodilus*. Rare and endangered

Wing length: 190, 212, 215 and 222 mm
Tail length: 93–100 mm (120 mm?)
Tarsus length: 54–59 mm
Length of bill: 27 mm
Body mass: 323 and 435 g

Illustration: J. G. Keulemans in Grandidier 1879: Atlas 1, Pl. 36 T. Boyer in Boyer and Hume 1991: 18; P. Hayman in Kemp and Kemp 1998: 281; V. Bretagnolle in Langrand 1990: Pl. 26; F. Weick in Brandt and Seebaß 1994: 2.4; M. Hulme in del Hoyo et al. 1999: Pl. 3; F. Weick in König et al. 1999: Pl. 2; F. Weick in Weick 2003: 217
Photograph: R. Thorstrom in Thorstrom et al. 1995: 22; R. Thorstrom in Morris and Hawkins 1998: 165; R. Thorstrom in del Hoyo et al. 1999: 40 and 64; R. Toft in Duncan 2003: 202
Literature: Eck and Busse 1973: 66, 67; Langrand 1990: 225, 226; Boyer and Hume 1991: 18; Schneider and Eck 1995: 102; Thorstrom et al. 1995; Thorstrom et al. 1997: 477–481; Morris and Hawkins 1998: 165; Kemp and Kemp 1998: 280; König et al. 1999: 197 del Hoyo et al. 1999: 73; Weick 2003: 217

❖

Tyto deroepstorffi **(Hume) 1875**
Andaman Barn Owl · Andamanen-Schleiereule · Effraie des Andamanes
Strix De-Roepstorffi **Anonymus = Hume 1875, Str. Feath. 3: 390;** Terra typica: **Aberdeen, South Andaman Islands**

Length: 330–360 mm
Body mass: ?

Distribution: South Andaman Islands
Habitat: Possibly similar to *Tyto alba*
Museum: AMNH (Rothsch.), BMNH (holotype), only two skins exist
Remarks: Plumage very different to that of *T. alba*, a typical endemic island species!

Wing length: 250–264 mm
Tail length: 110–113 mm
Tarsus length: 61 mm
Length of bill: 32 mm
Body mass: ?

Illustration: D. Cole in Grimmett et al. 1998: Pl. 28; F. Weick in König et al. 1999: Pl. 2; M. Hulme in del Hoyo et al. 1999: Pl. 2; McQueen in Rasmussen and Anderton 2005: Pl. 75
Photograph: R. Saldino in internet: Owl Pages, photo gallery 2002
Literature: Hume 1875: 390, 391; Baker 1927: 386, 387; Burton et al. 1984: 50; König et al. 1999: 196; Rasmussen and Anderton 2005: 233

❖

Tyto crassirostris **Mayr 1935**
Boang Barn Owl · Boang-Schleiereule · Effraie de Boang
Tyto alba crassirostris **Mayr 1935, Am. Mus. Novit. 820: 3 (Whitney South Sea Exp.);** Terra typica: **Boang Island**

Length: 330 mm
Body mass: ?

Distribution: Tanga Islands, East Bismarck Archipelago
Habitat: Grassland, farmland, woodland
Museum: AMNH (holotype)
Remarks: Differs from *T. delicatula* by stronger bill and feet and much darker plumage. Needs further studies!

Wing length: 285–290 mm
Tail length: ?
Tarsus length: ?
Length of bill: ?
Body mass: ?

Illustration: ?
Photograph: ?
Literature: Burton et al. 1984

Tyto delicatula **(Gould) 1837**
Australian Barn Owl · Australien-Schleiereule

Length: 330–380 mm
Body mass: 230–470 g

Distribution: Australia and offshore islands, Sawu, Roti, Timor, Jaco, Wetar, Kisar and Tanimbar Islands, Long Island, possibly northern New Britain and New Ireland, Nissan, Buka, Solomon Islands, Vanuatu, New Caledonia, Loyalty Islands, Fiji, north of Rotuma, Tonga, Wallis and Futuna Islands, Nive and Samoa, Southeast New Guinea
Habitat: Open country, farmland, particularly with cereal crops, heath, moorland, desert and semi-desert, open woodland and offshore islands

- ***Tyto delicatula meeki*** **(Rothschild and Hartert) 1907**
  Strix flammea meeki Rothschild and Hartert 1907, Novit. Zool. 14: 446; **Terra typica:** Collingwood Bay, New Guinea

Distribution: Southeast New Guinea, from Collingwood Bay in the north and Port Moresby in the south, eastward. Also Sepik River, Karkas and Manam Islands
Museum: NNML (holotype), SMTD, ZFMK
Remarks: Melanesian Barn Owl population requires further study; *crassirostris, meeki, sumbaensis* and *interposita* may merit further splits. However, most subspecies in this area are to merge with the nominate *delicatula*

Wing length: 282–300 mm
Tail length: 110–114 mm
Tarsus length: 70 mm
Length of bill: 37 mm
Body mass: ?

Illustration: T. Medland in Iredale 1956: Pl. 7
Photograph: ?
Literature: Rothschild and Hartert 1907: 446; T. Iredale 1956: 134, 135; Rand and Gilliard 1967: 250, 251; Beehler et al. 1986: 130; Burton et al. 1984: 50

- ***Tyto delicatula sumbaensis*** **(Hartert) 1897**
  Strix flammea sumbaensis Hartert 1897, Novit. Zool. 4: 270; **Terra typica:** Waingapo, Sumba Island

Distribution: Sumba Island
Museum: AMNH (holotype)
Remarks: Plumage very similar to that of *meeki* and *delicatula*

Wing length: no data!
Tail length: ?
Tarsus length: ?
Length of bill: ?
Body mass: ?

Illustration: ?
Photograph: ?
Literature: Hartert 1897a: 446; Hartert 1929a: 99, 100!

- ***Tyto delicatula interposita*** **Mayr 1935**
  Tyto alba interposita Mayr 1935, Am. Mus. Novit. 820: 3; **Terra typica:** Santa Cruz Island, Banks Islands

Distribution: Santa Cruz Island, Banks Islands, northern Vanuatu
Museum: AMNH (holotype)
Remarks: Distinctly different plumage from nominate *delicatula*

Wing length: 268–279 mm
Tail length: ?
Tarsus length: ?
Length of bill: ?
Body mass: ?
no other data!

Literature: Hartert 1929a: 100; Mayr 1935: 3; Burton et al. 1984: 51; Schneider and Eck 1995: 96

- ***Tyto delicatula delicatula*** **(Gould) 1837**
  Strix delicatula Gould 1837, Proc. Zool. Soc. Lond. 1836: 140; **Terra typica:** New south Wales, Australia

Synonym:
- *Strix lulu* Peale 1848, U.S. Expl. Exp. 8: 74; **Terra typica:** Upolu Island, Samoa Group; **Museum:** AMNH; **Wing length:** 265–280 mm
- *Tyto alba alexandrae* Mathews 1912, Novit. Zool. 18: 99; **Terra typica:** Alexandra, Northern Territory
- *Tyto alba lifuensis* Brasil 1915, Rev. France Orn. 4: 202; **Terra typica:** Lifu, Loyalty Island; **Wing length:** 279 mm
- *Tyto alba everetti* Hartert 1929, Novit. Zool. 35: 99; **Terra typica:** Sawu Island; **Museum:** AMNH; **Wing length:** 247–283 mm
- *Tyto alba kuehni* Hartert 1929, Novit. Zool. 35: 99; **Terra typica:** Kisar Island; **Museum:** AMNH; **Wing length:** 288–300 mm
- *Tyto alba bellonae* Bradley 1962, Nat. Hist. Rennell Is. 4: 12; **Terra typica:** Bellona Island, Solomon Island

Distribution: Australia and offshore islands, Lesser Sundas, south-west Pacific Islands, (Solomon, Loyalty, New Caledonia, Fiji, Tonga, Samoa)
Museum: AMNH, AMS, ANWC, FNSF, MNHB, NNML, QM, WAM, ZFMK
Remarks: Taxa *lifuensis*, *everetti*, *kuehni* and *bellonae* are inseparable from *delicatula*!

Wing length:	247–300 mm (Australia: 273–294 mm)
Tail length:	110–119 mm
Tarsus length:	60 mm
Length of bill:	
Body mass:	230–470 g

Illustration: B. Fremlin in Fremlin 1986: 29; P. Slater in Slater et al. 1989: 192; N. Day in Simpson and Day 1988: 155; H. Quintscher in Schneider and Eck 1995: Abb. 33e; M. Hulme in del Hoyo et al. 1999: Pl. 2; F. Weick in König et al. 1999: Pl. 2; S. Elsner in Olson 2001: 203
Photograph: D. Hollands in Hollands 1991: 70–77; J. Calaby and K. Ireland in Frith et al. 1983: 305; J. P. Ferrero in del Hoyo et al. 1999: 43; E. Hannecart in Hannecart and Letocart 1983: 84 = *T. d. lifuensis*
Literature: Hartert 1929a: 99, 100; Simpson and Day 1988: 154; Mees 1964a: 37–39; Hollands 1991: 71–79 and 210, 211; Schneider and Eck 1995: 95, 96

❖

Tyto aurantia (Salvadori) 1881

Golden Masked Owl · Goldeule · Effraie dorée · Lechuza Dorada
***Strix aurantia* Salvadori 1881, Atti R. Accad. Sci. Torino 16: 619; Terra typica: New Britain**

Length:	270–330 mm
Body mass:	?

Distribution: New Britain, Bismarck Archipelago
Habitat: Tropical rainforest, from lowland up to 1 830 m
Museum: BMNH, MNBHU, ÜMB, ZMH

Wing length:	220–230 mm
Tail length:	99 mm
Tarsus length:	62 mm
Length of bill:	?
Body mass:	?

Illustration: J. G. Keulemans in *Ibis* 1882: Pl. 2; T. Boyer in Boyer and Hume 1991: 24; F. Weick in Brandt and Seebaß 1994: 2.6; H. Quintscher in Schneider and Eck 1995: Abb. 34g; F. Weick in König et al. 1999: Pl. 3; M. Hulme in del Hoyo et al. 1999: Pl. 1
Photograph: ?
Literature: Meyer 1934: 575; Gurney 1882: 132; Eck and Busse 1973: 66; Boyer and Hume 1991: 24; König et al. 1999: 198; del Hoyo et al. 1999: 68

❖

Tyto nigrobrunnea Neumann 1939

Taliabu Masked Owl · Taliabu-Eule · Effraie de Taliabu · Lechuza de la Taliabu
***Tyto nigrobrunnea* Neumann 1939, Bull. Brit. Ornith. Cl. 59, 421: 89; Terra typica: Taliabu, Sula Islands**
(see Plate 1)

Length:	310 mm
Body mass:	?

Distribution: Taliabu and Sula Islands
Habitat: Rainforest and secondary forest
Museum: SMTD (holotype)

Tyto nigrobrunnea · Taliabu Masked Owl · Taliabu-Eule

Remarks: Probably a subspecies of the Minahassa Owl *Tyto inexspectata*. Known only from a single female skin. Recently rediscovered. Vulnerable

Wing length:	283 mm
Tail length:	125 mm
Tarsus length:	?
Length of bill:	?
Body mass:	?

Illustration: H. Quintscher in Eck and Busse 1973: Pl. 1; T. Boyer in Boyer and Hume 1991: 22; F. Weick in Brandt and Seebaß 1994: 2.6; D. Gardner in Coates and Bishop 1997: Pl. 33; F. Weick in König et al. 1999: Pl. 3; M. Hulme in del Hoyo et al. 1999: Pl. 1

Photograph: ?

Literature: Neumann 1939: 89; Eck and Busse 1973: 64, 65; White and Bruce 1986: 248, 249; Boyer and Hume 1991: 22; Schneider and Eck 1995: 105, 106; Stones et al. 1997: 58, 59; Coates and Bishop 1997: 358; König et al. 1999: 199; del Hoyo et al. 1999: 69

Tyto inexspectata **(Schlegel) 1879**
Minahassa Masked Owl · Minahassaeule · Effraie de Minahassa · Lechuza de Minahassa
Strix inexspectata **Schlegel 1879, Not. Leyden Mus. 1: 50–52;**
Terra typica: **Minahassa, Sulawesi**

Length:	270–310 mm
Body mass:	?

Distribution: Minahassa Peninsula, north and north-central Sulawesi

Habitat: Primary hill forest and lower montane forest, forest edges and riverside forest, from 250 up to 1 500 m

Museum: BMNH, MNBHU, NNML

Remarks: Known from only a few specimens and sightings. Widely considered to include *Tyto nigrobrunnea* as a subspecies

Wing length:	239–272 mm
Tail length:	102–122 mm
Tarsus length:	58–63.5 mm
Length of bill:	?
Body mass:	?

Illustration: H. Quintscher in Eck and Busse 1973: Abb. 31; T. Boyer in Boyer and Hume 1991: 22; F. Weick in Brandt and Seebaß 1994: 2; H. Quintscher in Schneider and Eck 1995: Abb. 34f; D. Gardner in Coates and Bishop 1997: Pl. 33; M. Hulme in del Hoyo et al. 1999: Pl. 1; F. Weick in König et al. 1999: Pl. 5

Photograph: ?

Literature: Stresemann 1931: 105; Stresemann 1940: 432, 433; Eck and Busse 1973: 65, 66; White and Bruce 1986: 248; Boyer and Hume 1991: 22; Schneider and Eck 1995: 102, 103; Coates and Bishop 1997, 133 and 358; König et al. 1999: 205; del Hoyo et al. 1999: 69

Tyto sororcula **(P.L. Sclater) 1883**
Lesser Masked Owl · Tanimbareule · Effraie de Tanimbar · Lechuza de la Tanimbar

Length:	290–310 mm
Body mass:	?

Distribution: Tanimbar Islands (Larat, Yamdena) Moluccas (Buru, Seram)

Habitat: Rainforest, primary and secondary monsoon forest

- *Tyto sororcula cayelii* (Hartert) 1900
  Strix cayelii Hartert 1900, Novit. Zool. 7: 228;
  Terra typica: Buru

Distribution: Buru, Seram, Moluccas

Museum: AMNH (Rothsch.)

Remarks: *Cayelii* possibly represents a separate species. The Seram population may represent an undescribed subspecies!

Wing length:	♀ 251 mm
Tail length:	?
Tarsus length:	?
Length of bill:	?
Body mass:	?

Illustration: M. Hulme in del Hoyo et al. 1999: Pl. 1

Photograph: ?

Literature: Hartert 1900: 7; del Hoyo et al. 1999: 68, 69

- *Tyto sororcula sororcula* (P.L. Sclater) 1883
  Strix sororcula P.L. Sclater 1883, Proc. Zool. Soc. Lond.: 52; Terra typica: Larat

Distribution: Larat and Yamdena, Tanimbar Islands
Museum: NNML

Wing length:	227–235 mm
Tail length:	?
Tarsus length:	?
Length of bill:	?
Body mass:	?

Illustration: T. Boyer in Boyer and Hume 1991: 22; D. Gardner in Coates and Bishop 1997: Pl. 33; F. Weick in König et al. 1999. Pl. 4; M. Hulme in del Hoyo et al. 1999. Pl. 1
Photograph: ?
Literature: Stresemann 1934: 177; White and Bruce 1986: 247, 248; Coates and Bishop 1997: 358; del Hoyo et al. 1999: 68, 69; König et al. 1999: 201, 202

❖

Tyto manusi **Rothschild and Hartert 1914**
Manus Masked Owl · Manus-Schleiereule · Effraie de Manus · Lechuza de la Manus
Tyto manusi **Rothschild and Hartert 1914, Novit. Zool. 21: 291; Terra typica: Manus Island, Admirality Islands**

Length:	330 mm
Body mass:	?

Distribution: Manus Island, Admirality Islands
Habitat: Rainforest, woodland in hilly elevations
Museum: AMNH (Rothsch.), MNBHU
Remarks: Sometimes considered as conspecific with *T. novaehollandiae*. Monotypic. No recent records, vulnerable

Wing length:	275–301 mm
Tail length:	122–133 mm
Tarsus length:	78 and 79 mm
Length of bill:	41 mm
Body mass:	?

Illustration: H. Quintscher in Schneider and Eck 1995: Abb. 34c; F. Weick in König et al. 1999: Pl. 4; M. Hulme in del Hoyo et al. 1999: Pl. 1
Photograph: ?
Literature: Rothschild and Hartert 1914a: 291; Schneider and Eck 1995: 104; Buckingham et al. 1995; Steadman and Kirch 1998: 12–22; del Hoyo et al. 1999: 68; König et al. 1999: 202

Tyto rosenbergii **(Schlegel 1866)**
Sulawesi Masked Owl · Sulawesi-Schleiereule, Rosenberg-eule · Effraie de Rosenberg · Lechuza de Célebes

Length:	410–460 mm
Body mass:	?

Distribution: Sulawesi, Sangihe and adjacent islands, Peleng (Banggai Island)
Habitat: Wooded cultivations, forest edges, plantations, grassland with few trees, from sea-level up to 1 100 m
Remarks: Sometimes considered as conspecific with *T. novaehollandiae*, but possibly most closely related to *Tyto alba*, this species requires further study, particularly DNA sequence analysis

- *Tyto rosenbergii rosenbergii* (Schlegel) 1866
  Strix Rosenbergii Schlegel 1866, Nederl. Tijdschr. Dierk. 3: 99, 181; Terra typica: Monelido, Boni and Gorontalo, Sulawesi

Distribution: Sulawesi, Sangihe and adjacent islands
Museum: BMNH, MNBHU, NNML, SMTD, ZFMK

Wing length:	331–360 mm
Tail length:	141–165 mm
Tarsus length:	72–74 mm
Length of bill:	?
Body mass:	?

Illustration: T. Boyer in Boyer and Hume 1991: 21; D. Gardner in Coates and Bishop 1997: Pl. 33; F. Weick in Brandt and Seebaß 1994: 2.6; M. Hulme in del Hoyo et al. 1999: Pl. 1; F. Weick in König et al. 1999: Pl. 4
Photograph: ?
Literature: Schlegel 1866: 99 and 181; Sharpe 1875a: 298, 299; White and Bruce 1986: 246, 247; Boyer and Hume 1991: 21; Schneider and Eck 1995: 103, 104; Coates and Bishop 1997: 133 and 357, 358; del Hoyo et al. 1999: 69; König et al. 1999: 203, 204

- ***Tyto rosenbergii pelengensis*** Neumann 1939
  Tyto rosenbergii pelengensis Neumann 1939, Bull. Brit. Ornith. Cl. 59, 421: 92; Terra typica: Peleng = Banggai Islands

Distribution: Banggai Islands
Museum: SMTD

Wing length:	296 mm
Tail length:	152 mm
Tarsus length:	?
Length of bill:	?
Body mass:	?

Illustration: ?
Photograph: ?
Literature: Neumann 1939: 92; White and Bruce 1986: 247; Schneider and Eck 1995: 104; Coates and Bishop 1997: 357, 358; del Hoyo et al. 1999: 69; König et al. 1999: 204

Tyto novaehollandiae (Stephens) 1826
Australian Masked Owl · Neuhollandeule · Effraie masquée · Lechuza Australiana
(see Plate 1)

Length:	330–470 mm
Body mass:	420–670 g

Distribution: Lowlands of southern New Guinea and Daru Islands, Australia (except the arid interior)
Habitat: Tall open forest with large trees, open woodland with clearings and dry, open understorey

- ***Tyto novaehollandiae calabyi*** Mason 1983
  Tyto novaehollandiae calabyi Mason 1983, Bull. Brit. Ornith. Cl. 103: 122–128; Terra typica: Southern New Guinea: Merauke to Tarara

Distribution: Southern New Guinea, southern flight region: Merauke to Tarara and Daru Islands
Museum: AMNH (cotype), NNML (holotype), MNBHU

Wing length:	317–328 mm
Tail length:	127–137 mm
Tarsus length:	71.5–74 mm
Length of bill (cere):	25.5–28 mm
Body mass:	?

Illustration: F. Weick in König et al. 1999: Pl. 4; M. Hulme in del Hoyo et al. 1999: Pl. 1
Photograph: I. J. Mason in Mason 1983: Pl. 2 (skins)
Literature: Mason 1983: 122–128; Beehler et al. 1986: 130; König et al. 1999: 203; del Hoyo et al. 1999: 67, 68

- ***Tyto novaehollandiae kimberli*** Mathews 1912
  Tyto novaehollandiae kimberli Mathews 1912, Aust. Avian Rec. 1: 35; Terra typica: East Kimberley, Western Australia

Synonym:
- *Tyto novaehollandiae mackayi* Mathews 1912, Aust. Avian Rec. 1: 34; Terra typica: Mackay, Queensland (holotype AMNH)
- *Tyto novaehollandiae melvillensis* Mathews 1912, Aust. Avian Rec. 1: 35; Terra typica: Melville Island, Northern Territory (holotype AMNH, only 2 skins)
- *Tyto novaehollandiae perplexa* Mathews 1912, Novit. Zool. 18: 257; Terra typica: East Beverly, Western Australia (holotype AMNH)
- *Tyto novaehollandiae galei* Mathews 1914, S. Austral. Orn. 1(2): 12; Terra typica: Cape York Peninsula (MNBHU)

Distribution: Melville Island, Western Australia, Northern Territory and North Queensland
Museum: AMNH, HLW, MNBHU, NNML, WAM, ZFMK

Wing length:	293–332 mm
Tail length:	123–144 mm
Tarsus length:	67–74 mm
Length of bill (cere):	25.5–28 mm
Body mass:	?

Illustration: N. Day in Simpson and Day 1986: 155; F. Weick in König et al. 1999: Pl. 4
Photograph: D. Hollands in Hollands 1991: 80
Literature: Rothschild and Hartert 1913/1914: 280–284; Mees 1964a: 1–62; Simpson and Day 1986: 154; Schneider and Eck 1995: 105; del Hoyo et al. 1999: 68; König et al. 1999: 202, 203

- ***Tyto novaehollandiae novaehollandiae*** (Stephens) 1826
  Strix (?) Novae Hollandiae Stephens 1826, in Shaw's Gen. Zool. 13(2): 61; Terra typica: New South Wales

Synonym:
- *Tyto novaehollandiae whitei* Mathews 1912, Aust. Avian Rec. 1: 34; Terra typica: Adelaide, South Australia, AMNH

- *Tyto novaehollandiae riordani* Mathews 1912, Aust. Avian Rec. 1: 35; Terra typica: Warnambool, Victoria, AMNH (Rothsch. holotype)
- *Tyto longimembris dombraini* Mathews 1914, Aust. Avian Rec. 2: 91; Terra typica: Victoria
- *Tyto novaehollandiae troughtoni* Cayley 1931, What bird is that?: 32, Pl. 5; Terra typica: South Queensland, New South Wales, Victoria and South Australia

Distribution: Southwest and western Australia, east of Victoria, north to northeast Queensland, mainly in coastal areas, rare and scattered inland
Museum: AMNH, AMS, BMNH, CSIRO, HLW, MNBHU, MNV, NNML, SMNSt, SMTD, ZFMK

Wing length: ♂ 290–318 mm, ♀ 299–358 mm
Tail length: 119–150 mm
Tarsus length: 64–79 mm
Length of bill (cere): 24–28 mm
Body mass: 545–673 g

Illustration: T. Boyer in Boyer and Hume 1991: 28; N. Day in Simpson and Day 1986: 155; M. Hulme in del Hoyo et al. 1999: Pl. 1; F. Weick in König et al. 1999: Pl. 4
Photograph: D. Hollands in Hollands 1991: 93; Lindgren in Burton et al. 1992: 45; E. and D. Hoskings in del Hoyo et al. 1999: 48
Literature: Rothschild and Hartert 1913/1914: 280–284; Mees 1964a; Eck and Busse 1973: 61–63; Frith et al. 1983: 306; Simpson and Day 1986: 154; Boyer and Hume 1991: 23; Burton et al. 1992: 45, 46; Schneider and Eck 1995: 104; del Hoyo et al. 1999: 67, 68; König et al. 1999: 202, 203

❖

Tyto castanops (Gould) 1837
Tasmanian Masked Owl · Tasmanien-Schleiereule · Effraie de Tasmanie
***Strix castanops* Gould 1837, Proc. Zool. Soc. Lond. 1836: 140;**
Terra typica: **Tasmania**

Length: 470–550 mm
Body mass: 600–1 260 g

Distribution: Tasmania and Maria Island, introduces Lord Howe Island
Habitat: Semi-open forest, woodland with clearings and open understorey
Museum: AMNH, AMS, BMNH, MNBHU, NNML, NMV, SMNSt, SMTD, WAM
Remarks: Often included as a subspecies of *novaehollandiae*, but size and geographical isolation support distinctness

Wing length: ♂ 310–347 mm, ♀ 344–387 mm
Tail length: ♂ 140–163 mm, ♀ 150–178 mm
Tarsus length: 64–82 mm
Length of bill (cere): 25–27.5 mm
Body mass: 600–1 260 g

Illustration: N. Day in Simpson and Day 1986: 155; F. Weick in Brandt and Seebaß 1994: 2.6; N. Hulme in del Hoyo et al. 1999: Pl. 1; F. Weick in König et al. 1999: Pl. 4
Photograph: Calaby in Frith et al. 1983: 306; E. Hosking in Hosking and Flegg 1982: 162, 163; D. Hollands in Hollands 1991: 83; Watts in *Australian Nature* 1995(Spring): 265
Literature: Rothschild and Hartert 1913/1914: 283; Fleay 1955: 203–210; Skemp 1955: 210–211; Mees 1964a: 1–62; Schneider and Eck 1995: 105; del Hoyo et al. 1999: 67, 68; König et al. 1999: 204, 205

❖

Tyto capensis (A. Smith) 1834
African Grass Owl · Kapgraseule · Effraie du Cap · Lechuza de El Cabo
***Strix Capensis* A. Smith 1834, S. Afr. Q. J.: 317;**
Terra typica: **South Africa near Cape Town**

Length: 340–420 mm
Body mass: 355–520 g

Synonym:
- *Strix punctata* G.R. Gray 1869, Hand List Bds. 1: 53; Terra typica: near Cape Town
- *Strix cabrae* Dubois 1902, Syn. Av. 2: 900, note 1; Terra typica: Lower Congo
- *Tyto Capensis Damarensis* Roberts 1922, Ann. Transvaal Mus. 8: 212; Terra typica: Damaraland, Southern Angola
- *Tyto capensis libratus* Peters and Loveridge 1935, Proc. Biol. Soc. Wash. 48: 77; Terra typica: Kenya

Distribution: East Africa (single record from Ethiopia), Highlands of Cameroon (three records), Congo and northern Angola, to Uganda in the east and

Kenya in the west, south to Tanzania and Zambia, western Mozambique and eastern South Africa to west of Cape Town
Habitat: Moist and dry open grassland, marshes, moors up to 3 200 m above sea-level
Museum: BMNH, NNML, SMNKa, TMBD. ZFMK

Wing length:	283–345 mm
Tail length:	115–125 mm
Tarsus length:	77–83 mm
Length of bill:	?
Body mass:	355–520 g

Illustration: T. I. Ford in Smith 1839: Pl. 45; A. Cameron in Voous 1988: 27; G. Arnott in Steyn 1982: Pl. 24; T. Boyer in Boyer and Hume 1991: 26; F. Weick in Brandt and Seebaß 1994: 2.5; P. Hayman in Kemp and Kemp 1998: 285; M. Hulme in del Hoyo et al. 1999: Pl. 3; F. Weick in König et al. 1999: Pl. 3; N. Borrow in Borrow and Demey 2001: Pl. 98; J. Gale in Stevenson and Fanshawe 2002: Pl. 99
Photograph: P. Steyn in Steyn 1982: 248, 249; Lindgren in Burton et al. 1984: 52; P. J. Ginn in Ginn et al. 1989: 332; P. J. Ginn in Kemp and Kemp 1998: 284; J. Carlyon in Kemp and Kemp 1998: 285; J. Carlyon in del Hoyo et al. 1999: 60; P. J. Ginn in del Hoyo et al. 1999: 61
Literature: Steyn 1982: 247–251; Fry et al. 1988: 106–108; Voous 1988: 23–28; Boyer and Hume 1991: 25–27; Burton et al. 1992: 40–42; Zimmerman et al. 1996: 440; Kemp and Kemp 1998: 284, 285; del Hoyo et al. 1999: 74; König et al. 1999: 200, 201; Borrow and Demey 2001: 492; Stevenson and Fanshawe 2002: 198

Tyto longimembris (Jerdon) 1839

Eastern Grass Owl · Graseule · Effraie de prairie · Lechuza de Patilarga
(see Plate 1)

Length:	320–400 mm
Body mass:	250–450 g

Distribution: From India to Vietnam and southeast China, Taiwan, Philippines, Sulawesi, Flores, southeast New Guinea, Australia, New Caledonia and the Fiji Islands
Habitat: Tall and open grassland, savannas, marshes, heathland, from coastal marshy heath up to grassy hillsides and grassland at 2 500 m

- *Tyto longimembris longimembris* (Jerdon) 1839
  Strix longimembris Jerdon 1839, Madras J. Lit. Sci. 10: 86; Terra typica: Neilgherri Mountains, India

Synonym:
- *Strix walleri* Diggles 1866, Orn. Austral. 7: 1, Pl. 14; Terra typica: Brisbane, Queensland
- *Strix pithecops* Swinhoe 1866, Ibis: 396; Terra typica: Taiwan. Does not resemble the subspecies from China but rather the Indian birds! (See Hartert 1929a: 103) (ZMH)
- *Strix oustaleti* Hartlaub 1879, Proc. Zool. Soc. Lond.: 295; Terra typica: Viti Levu, Fiji Islands (ZMH)
- *Tyto longimembris georgiae* Mathews 1912, Aust. Avian Rec. 1: 75; Terra typica: Victoria River, Northern Territory (holotype AMNH)
- *Tyto maculosa* Glauert 1945, Emu 44: 229, 230; Terra typica: Western Australia

Distribution: India, southern Nepal, Bangladesh, Myanmar, East Tonkin, central and south Annam, Cochinchina, Taiwan, Tukangbesi Island (Kalidupa), Flores, Sumba, Fiji Islands, northern, central and eastern Australia
Museum: AMNH, AMS, BMNH, HLW, NNML, SMNSt, ZFMK, ZMH

Wing length:	♂ 273–322 mm, ♀ 321–348 mm
Tail length:	114–139 mm
Tarsus length:	79–96 mm
Length of bill:	36 mm
Body mass:	265–450 g

Illustration: Richter in Gould 1831–1832: Pl. 28 (adult and pull.); N. Day in Simpson and Day 1984: 155; T. Boyer in Boyer and Hume 1991: 27; F. Weick in Brandt and Seebaß 1994: 2.5; D. Gardner in Coates and Bishop 1997: Pl. 33; D. Cole in Grimmett et al. 1998: Pl. 28; M. Hulme in del Hoyo et al. 1999: Pl. 3; F. Weick in König et al. 1999: Pl. 3 (adult and pull.); H. Burn in Robson 2000: Pl. 22; McQueen in Rasmussen and Anderton 2005: Pl. 75
Photograph: Calaby/Chaffer in Frith et al. 1983: 307; D. Hollands in Hollands 1991: 42; Pizzey in Burton et al. 1992: 41; D. Hollands in Burton et al. 1992: 42; R. Seitre in del Hoyo et al. 1999: 40; E. Mc Nabb in internet: Owl Pages, photo gallery 2002

Literature: Hartert 1929a: 102–104; Delacour and Jabouille 1931: 147, 148; Ali and Ripley 1969: 252, 253; Eck and Busse 1973: 56, 57; Simpson and Day 1984: 155; White and Bruce 1986: 249; Hollands 1991: 96–107 and 211, 212; Coates and Bishop 1997: 358; Grimmett et al. 1998: 429; del Hoyo et al. 1999: 74, 75; König et al. 1999: 109, 110; Rasmussen and Anderton 2005: 233, 234

- ***Tyto longimembris chinensis*** Hartert 1929
  Tyto longimembris chinensis Hartert 1929, Novit. Zool. 35(2): 104; Terra typica: Shuikow, Fukien

Synonym:
- *Tyto longimembris albifrons* Caldwell and Caldwell 1931, South China Birds: 232; Terra typica: Futsing, Fukien
- *Tyto longimembris melli* Yen 1932, Ois. Rev. France Orn. 3: 242; Terra typica: Yao-Shan, Kwangtung

Distribution: Southeast China (southeast Yunnan to Jiangsu), Vietnam
Museum: AMNH, MNBHU (holotype), ZFMK

Wing length:	340 mm
Tail length:	125 mm
Tarsus length:	?
Length of bill:	?
Body mass:	?

Illustration: F. Berille in Etchécopar and Hüe 1978: Pl. 18; F. Weick in König et al. 1999: Pl. 3; M. Hulme in del Hoyo et al. 1999: Pl. 3 (head)
Photograph: ?
Literature: Hartert 1929a: 102–104; Etchécopar and Hüe 1978: 491; König et al. 1999: 199, 200; del Hoyo et al. 1999: 74, 75

- ***Tyto longimembris papuensis*** Hartert 1929
  Tyto longimembris papuensis Hartert 1929, Novit. Zool. 35: 103; Terra typica: Ogwarra, 1 800 m, New Guinea

Synonym:
- *Tyto longimembris baliem* Ripley 1964, Peabody Mus. Nat. Hist. Bull. 19: 37, 38; Terra typica: Baliem Valley Netherlands, New Guinea (YPM)

Distribution: Eastern and western New Guinea
Museum: AMNH (holotype), YPM

Wing length:	♂ 330 mm, ♀ 350 mm
Tail length:	120, 122 mm
Tarsus length:	70–90 mm
Length of bill (cere):	22 mm
Body mass:	?

Illustration: ?
Photograph: ? in Mayr and Gilliard 1954: Pl. 17
Literature: Hartert 1929a: 103; Mayr and Gilliard 1954: 341; Ripley 1964: 37, 38; Rand and Gilliard 1967: 252, 253

- ***Tyto longimembris amauronota*** (Cabanis) 1872
  Strix amauronota Cabanis 1872, J. Orn. 20: 316; Terra typica: Luzon, Philippines

Distribution: Philippines
Museum: MNBHU, (holotype not located)

Wing length:	330–360 mm
Tail length:	120–135 mm
Tarsus length:	83–89 mm
Length of bill:	?
Body mass:	582 g

Illustration: G. Sandström in du Pont 1971: Pl. 39; F. Weick in König et al. 1999: Pl. 3; A. P. Sutherland in Kennedy et al. 2000: Pl. 34
Photograph: ?
Literature: Cabanis 1872: 316; Hartert 1929a: 104; du Pont 1971: 166; Dickinson et al. 1991: 222, 223; del Hoyo et al. 1999: 74, 75; König et al. 1999: 200; Kennedy et al. 2000: 173

❖

Tyto tenebricosa **(Gould) 1845**
Greater Sooty Owl · Rußeule · Effraie ombrée · Lechuza Tenebrosa

Length:	370–430 mm
Body mass:	875–1 160 g

Distribution: New Guinea and southeast Australia
Habitat: Lowland and montane rainforest with clearings, up to 4 000 m. In Australia, wet forest with tall eucalyptus trees, rainforest trees and tree-ferns

Tyto t. tenebricosa
Greater Sooty Owl (perched and in flight)
Große Rußeule (sitzend und im Flug)

- ***Tyto tenebricosa tenebricosa*** (Gould) 1845
  Strix tenebricosus Gould 1845, Proc. Zool. Soc. Lond.: 80; Terra typica: Clarence River, New South Wales

Synonym:
- *Tyto tenebricosa magna* Mathews 1912, Novit. Zool. 18: 258; Terra typica: Victoria (holotype AMNH)

Distribution: Southeast Australia, Southeast Queensland to eastern Victoria
Museum: AMNH, AMS, FNSF, NMV, NNML, SMNSt, SMTD

Wing length:	♂ 286–300 mm, ♀ 327–343 mm
Tail length:	145 mm
Tarsus length:	66 mm
Length of bill:	?
Body mass:	875–1 160 g

Illustration: N. Day in Simpson and Day 1986: 155; T. Boyer in Boyer and Hume 1991: 25; F. Weick in Brandt and Seebaß 1994: 2.6; M. Hulme in del Hoyo et al. 1999: Pl. 1; F. Weick in König et al. 1999: Pl. 5
Photograph: Calaby and Lindgren in Frith et al. 1983: 307; D. Hollands in Hollands 1991: 108–118; D. Hollands in Burton et al. 1992: 44; C. Huber in internet: Owl Pages, photo gallery 2002
Literature: Rothschild and Hartert 1913/1914: 280–284; Mees 1964a: 1–62; Eck and Busse 1973: 63, 64; Frith et al. 1983: 307; Hollands 1991: 108–123 and 214; Boyer and Hume 1991: 24, 25; del Hoyo et al. 1999: 67; König et al. 1999: 206, 207

- ***Tyto tenebricosa arfaki*** (Schlegel) 1879
  Strix tenebricosa arfaki Schlegel 1879, Notes Leyden Mus. 1: 101; Terra typica: Hattam, Mount Arfak, New Guinea

Synonym:
- *Megastrix tenebricosa perconfusa* Mathews 1916, Birds Austr. 5: 108; Terra typica: Aroa River, British New Guinea

Distribution: New Guinea, all parts of the mainland and Japen Island
Museum: NNML (holotype), AMNH, FNSF, SMTD, ZFMK, YPM
Remarks: Plumage of some specimens very similar to that of *Tyto multipunctata*

Wing length:	♂ 243–260 mm, ♀ 253–305 mm
Tail length:	115–132 mm
Tarsus length:	60–70 mm
Length of bill (cere):	24–26 mm
Body mass:	500–750 g

Illustration: T. Medland in Iredale 1956: Pl. 7; A. E. Gilbert in Rand and Gilliard 1967: Pl. 34; H. Quintscher in Schneider and Eck 1995: Abb. 34d; F. Weick in König et al. 1999: Pl. 5
Photograph: L. Robinson in del Hoyo et al. 1999: 37
Literature: Rothschild and Hartert 1913/1914: 280–284; T. Iredale 1956: 136; Mees 1964a: 1–62; Rand and Gilliard 1967: 251, 252; Beehler et al. 1986: 130; Schneider and Eck 1995: 106; del Hoyo et al. 1999: 67; König et al. 1999: 206, 207

Tyto multipunctata **Mathews 1912**

Lesser Sooty Owl · Fleckenrußeule · Effraie piquetée · Lechuza Moteada
Tyto tenebricosa multipunctata **Mathews 1912, Novit. Zool. 18: 257; Terra typica: Johnston River, Queensland (see Plate 1)**

Length:	310–380 mm
Body mass:	430–540 g

Distribution: Northeast Australia (Northeast Queensland)
Habitat: Rainforest, mainly above 300 m, with tall eucalyptus trees and clearings to hunt
Museum: AMNH, HLW, MNBHU

Wing length: ♂ 237–253 mm, ♀ 243–266 mm
Tail length: ?
Tarsus length: ?
Length of bill: ?
Body mass: ♂ 430 g, ♀ 540 g

Illustration: H. Quintscher in Eck and Busse 1973: Abb. 30; N. Day in Simpson and Day 1986: 155; F. Weick in König et al. 1999: Pl. 5; M. Hulme in del Hoyo et al. 1999: Pl. 1
Photograph: D. Hollands in Hollands 1991: 124–140; D. Hollands in Burton et al. 1992: 43; D. Hollands 1995: cover and 39–44; C. and D. Frith in del Hoyo et al. 1999: 42, 46; M. Sacchi in del Hoyo et al. 1999: 55; J. Young in del Hoyo et al. 1999: 57; D. Hollands in internet: Owl Pages, photo gallery 2003
Literature: Rothschild and Hartert 1913/1914: 280–284; Mees 1964a: 1–62; Simpson and Day 1986: 155; Hollands 1991: 124–141 and 215; Burton et al. 1992: 44; del Hoyo et al. 1999: 67; König et al. 1999: 205, 206

Tyto prigoginei **Schouteden 1952**
Itombwe or Prigogine Owl, Congo Bay Owl · Prigogine-Eule · Effraie de Prigogine, Phodile de Prigogine · Lechuza del Congo
Phodilus Prigoginei **Schouteden 1952 Rev. Zool. Bot. Afr. 46: 423–428; Terra typica: Muusi, 2 432 m**
(see Plate 1 and Figure 2)

Length: 230–250 mm
Body mass: 195 g

Distribution: Itombwe (Mitumba) Massive, extreme east of Zaire, probably southwest Rwanda (Nyungwe forest) and northwest Burundi
Habitat: Montane forest, interspersed with bamboo thickets and grassland, at 1 830–2 430 m. Rare and elusive. Vulnerable
Museum: RMTB
Remarks: Occasionally treated as a species in the genus *Phodilus*, but the relatively small eyes and heart-shaped facial disc are more typical of the genus *Tyto* or genus novae *Pusilltyto*. Requires further study (vocalisations and DNA sequencing). Rare and elusive. Vulnerable

Wing length: 192 mm
Tail length: 93 mm
Tarsus length: 37–41 mm
Length of bill: 27 mm
Body mass: 195 g

Illustration: S. Berger in Prigogine 1973: 176; F. Weick in König et al. 1999: Pl. 5; M. Hulme in del Hoyo et al. 1999: Pl. 3; F. Weick in *Gefiederte Welt* 2003(10): 303
Photograph: T. Butynski in Butynski et al. 1996: 2–4, covers; T. Butynski in Butynski et al. 1997: 35; T. Butynski in 1998 *SCRO Mag.* 2: 23; T. Butynski in del Hoyo et al. 1999: 65
Literature: Schouteden 1952: 423–428; Prigogine 1973: 176–185; Fry et al. 1988: 105, 106; Butynski et al. 1996: 2–4; Butynski et al. 1997: 32–35; Steinberg 1998: 28–31; König et al. 1999: 207, 208; del Hoyo et al. 1999: 75

Subfamilia / Subfamily
Phodilinae · Bay Owls · Maskeneulen

Phodilus

Genus
Phodilus **Geoffroy Saint Hilaire**
Phodilus Geoffroy Saint Hilaire 1830, Ann. Sci. Nat. Zool. 21: 199; Type by *Strix badia* Horsfield

Phodilus badius (Horsfield) 1821
Bay Owl · Maskeneule · Phodile calong · Lechuza cornuta
(see Plate 1 and Figure 2)

Length:	225–290 mm
Body mass:	220–300 g

Distribution: Southwest India, Sri Lanka, Nepal, Sikkim, Assam, Nagaland, Manipur, Myanmar, Thailand, eastern to southern China, south through Malay Peninsula to the Greater Sunda Islands and the Philippines (Samar)
Habitat: Evergreen and mixed deciduous forest, plantations, mangroves. Wet lowland up to foothills at 2 200 m

- ***Phodilus badius saturatus*** Robinson 1927
  Phodilus badius saturatus Robinson 1927, Bull. Brit. Ornith. Cl. 47: 121; Terra typica: Sikkim

Distribution: Sikkim, northeast India, Myanmar, Thailand, east to Vietnam and southeast China
Museum: AMNH, BMNH, BNHS

Wing length:	214–237 mm
Tail length:	89–97 mm
Tarsus length:	45–48 mm
Length of bill:	30–32 mm
Body mass:	?

Illustration: H. Grönvold in Delacour and Jabouille 1931: Pl. 20; P. Barruel in Ali and Ripley 1968: Pl. 17; D. Cole in Grimmett et al. 1998: Pl. 28; M. Hulme in del Hoyo et al. 1999: Pl. 3; F. Weick in König et al. 1999: Pl. 5; McQueen in Rasmussen and Anderton 2005: Pl. 75
Photograph: ?
Literature: Robinson 1927b: 121; Baker 1927: 390, 391; Delacour and Jabouille 1931: 143, 144; Ali and Ripley 1969: 253, 254; Grimmett et al. 1998: 429, 430; del Hoyo et al. 1999: 75; König et al. 1999: 208–210; Robson 2000: 290; Rasmussen and Anderton 2005: 235

- ***Phodilus (badius) ripleyi*** Hussain and Reza Khan 1978
  Phodilus badius ripleyi Hussain and Reza Khan 1978, J. Bombay Nat. Hist. Soc. 74: 335; Terra typica: Anaimalai-Nelliampathy Hills, Kerala, Southwest India

Distribution: Anaimalai-Nelliampathy Hills, Kerala, southwest India
Museum: BNHS

Wing length:	208 mm
Tail length:	81 mm
Tarsus length:	51 mm
Length of bill:	33 mm
Body mass:	?

Illustration: ?
Photograph: Reza Khan in Hussain and Reza Khan 1978: 334
Literature: Hussain and Reza Khan 1978: 333–335; del Hoyo et al. 1999: 75; König et al. 1999: 208, 209; Rasmussen and Anderton 2005: 234

Phodilus badius · Bay Owl · Maskeneule

- ***Phodilus (badius) assimilis*** Hume 1877
  Phodilus assimilis Hume 1877, Str. Feath. 5: 138;
  Terra typica: Ceylon

Distribution: Central and southern Sri Lanka. Low and wet zones, up to 1 200 m above sea-level
Museum: BMNH, BNHS, AMNH (1)

Wing length:	197–203 mm
Tail length:	81–89 mm
Tarsus length:	44–50 mm
Length of bill:	30 mm
Body mass:	?

Illustration: J. G. Keulemans in Legge 1878a: Pl. 6; G. M. Henry in Henry 1978: Pl. 16; F. Weick in König et al. 1999: Pl. 5; McQueen in Rasmussen and Anderton 2005: Pl. 75
Photograph: Lindgren in Burton et al. 1992: 49
Literature: Hume 1877: 138; Legge 1878a: Pt. 1; Baker 1927: 391, 392; Ali and Ripley 1969: 255; del Hoyo et al. 1999: 75; König et al. 1999: 208, 209; Rasmussen and Anderton 2005: 235

- ***Phodilus badius badius*** (Horsfield) 1821
  Strix badia Horsfield 1821, Trans. Linn. Soc. London 13(1): 139; Terra typica: Java

Synonym:
- *Phodilus badius abbotti* Oberholser 1924, Journ. Wash. Acad. Sci. 14: 302; Terra typica: Prov. Wellesley, Fed. Malay States
- *Phodilus riverae* McGregor 1927, Philip. J. Sci. 32(4): 302; Terra typica: Loquilocom, Samar

Distribution: Malay Peninsula, Greater Sundas including Nias, Samar (Philippines, 1 specimen)
Museum: BMNH, FNSF, MNBHU, NNML, SMNSt, ÜMB, USNM (holotype *riverae* destroyed in World War 2, 1945)

Wing length:	207–215 mm
Tail length:	79–90 mm
Tarsus length:	48–56 mm
Length of bill:	?
Body mass:	300 g

Illustration: Gould in Gould 1831–1832: 113 (Faksim.); M. Hulme in del Hoyo et al. 1999: Pl. 3; F. Weick in König et al. 1999: Pl. 5; H. Burn in Robson 2000: Pl. 21; A. R. Sutherland in Kennedy et al. 2000: Pl. 34; F. Weick in *Gefiederte Welt* 2003(10): 308
Photograph: H. Busse in *Der Falke* 1977(6): 215; K. W. Fink in MacKenzie 1986: 141; W. Mayr in *ZEITmagazin* 1992(46): 13; R. Steinberg in *SCRO Mag.* 1998(2): 22; D. Farrow in internet: Owl Pages, photo gallery 2002; M. Lenton in del Hoyo et al. 1999: 38
Literature: Inglis 1945: 93–96; Miller 1965: 536–538; Busse 1977, *Der Falke* 1977(6): 214; Dickinson et al. 1991: 223; Burton et al. 1992: 49, 50; del Hoyo et al. 1999: 75; R. Steinberg in SCRO Mag 1998: 21–28; König et al. 1999: 208, 209; Robson 2000: 290

- ***Phodilus badius arixuthus*** Oberholser 1932
  Phodilus arixuthus Oberholser 1932, US Nat. Mus. Bull. 159: 40; Terra typica: Bunguran Island (North Natuna Islands)

Distribution: Bunguran Island, North Natuna Islands, off northwest Borneo, known only from the holotype!
Museum: USNM (holotype)

Wing length:	185 mm
Tail length:	71 mm
Tarsus length:	?
Length of bill:	?
Body mass:	?

Illustration: A. Hughes in Smythies 1960: Pl. 15
Photograph: ?
Literature: Oberholser 1932: 40; Smythies 1960: 276; del Hoyo et al. 1999: 75; König et al. 1999: 208, 209

- ***Phodilus badius parvus*** Chasen 1937
  Phodilus badius parvus Chasen 1937, Treubia 16: 216;
  Terra typica: Biliton Island = Belitung Island

Distribution: Belitung Island, off southeast Sumatra
Museum: BMNH, NNML (holotype)

Wing length:	172–180 mm
Tail length:	?
Tarsus length:	?
Length of bill:	?
Body mass:	?

Illustration: F. Weick in König et al. 1999: Pl. 5
Photograph: A. Mercicca in MacKenzie 1986: 141
Literature: Chasen 1937: 216; del Hoyo et al. 1999: 75; König et al. 1999: 208, 209

Familia / Family
Strigidae · Typical Owls · Echte Eulen

Subfamilia / Subfamily
Striginae · Scops and Screech Owls, Eagle Owls, Wood Owls · Zwergohr- und Kreischeulen, Uhus, Käuze

Tribus / Tribe
Otini · Scops and Screech Owls · Zwergohr- und Kreischeulen

Otus

Genus
Otus **Pennant**
Otus Pennant 1769, Indian Zool.: 3. Type by *Otus bakkamoena* Pennant

Synonym:
- *Psiloscops* Coues 1899, Osporey 3: 144. Type by *Otus flammeolus* Kaup

Otus sagittatus (Cassin) 1849

White-fronted Scops Owl · Weißstirneule · Petit-duc à front blanc · Autillo Frontiblanco
Ephialtes sagittatus Cassin 1849, Proc. Acad. Nat. Sci. Phila. 4 (1848): 121; Terra typica: Malacca (see Plate 2)

Length: 250–280 mm
Body mass: 110–140 g

Distribution: Southern Myanmar, southern Thailand, Malay Peninsula and possibly northern Sumatra (Aceh)
Habitat: Tropical primary forest, lowland evergreen forest, swampy forest, tall secondary forest, from lowland up to 700 m
Museum: AMNH, ASNPh, BMNH, FNSF, IRSNB, MHNP, MNBHU, MZUSUSNM, SMNSt, ZMK

Wing length: 173–192 mm
Tail length: 108–125 mm
Tarsus length: 29–32 mm
Length of bill: 23–26 mm
Body mass: 109–139 g

Illustration: H. Quintscher in Eck and Busse 1973: Abb. 32; T. Boyer in Boyer and Hume 1991: 32; K. Phillipps in McKinnon and Phillipps 1993: Pl. 40; F. Weick in König et al. 1999: Pl. 6; H. Burn in Robson 2000: Pl. 21
Photograph: ?

Literature: Cassin 1849: 121; Sharpe 1875a: 98–100; Oates 1877: 247, 248; Baker 1927: 430, 431; Chasen 1939: 59–61; Eck and Busse 1973: 69, 70; Boyer and Hume 1991: 32; McKinnon and Phillipps 1993: 189; del Hoyo et al. 1999: 153; König et al. 1999: 211, 212; Robson 2000: 290; Hekstra unpublished

Otus rufescens (Horsfield) 1821

Reddish Scops Owl · Röteleule · Petit-duc roussâtre · Autillo Rojizo

Length: 150–180 mm
Body mass: 77 g

Distribution: South of peninsular Thailand, Malay Peninsula, Sumatra, Java, Borneo
Habitat: From lowlands to submontane forest, rainforest, swampforest and tall secondary forest, up to 600 m (1 000 m?)

- ***Otus rufescens rufescens*** (Horsfield) 1821
  Strix rufescens Horsfield 1821, Trans. Linn. Soc. London 13(1): 140; Terra typica: Java

Synonym:
- *Strix mantis* Temminck and Schlegel 1850, Siebold's Fauna Japonica, Aves: 25; Terra typica: Sumatra
- *Otus rufescens burbidgei* Hachisuka 1934, Bds. Philippine Is. 3: 51; Terra typica: Jolo Island (Sulu Island, The Philippines), doubtful, record not accepted, plumage not outside normal variation of nominate subspecies (see Dickinson et al. 1991: 227)

Distribution: Sumatra, Banka, Java, Borneo
Museum: BMNH, MCZ, MHNP, MZUS, NNML, ZSBS (*rufescens*), IRSNB, MCZ, MZUS, NNML, ZMA (*mantis*), BMNH (holotype *burbidgei*)

Wing length: 127–137 mm
Tail length: 62–68 mm
Tarsus length: 23–25 mm
Length of bill: 19–21.5 mm
Body mass: 77 g

Illustration: A. Hughes in Smythies 1960: Pl. 15; G. Sandström in du Pont 1971: Pl. 38 (*burbidgei*); T. Boyer in Boyer and Hume 1991: 33; K. Phillipps in McKinnon and Phillipps 1993: Pl. 40; F. Weick in König et al. 1999: Pl. 6
Photograph: P. Fodgen in Burton et al. 1992: 96; P. Fodgen in del Hoyo et al. 1999: 27
Literature: Horsfield 1821: 140; Hachisuka 1934: 51; Smythies 1960: 276; du Pont 1971: 168; Boyer and Hume 1991: 33; McKinnon and Phillipps 1993: 189; del Hoyo et al. 1999: 153; König et al. 1999: 212, 213

- ***Otus rufescens malayensis*** Hachisuka 1934
  Otus rufescens malayensis Hachisuka 1934, Bds. Philipp. Is. 3: 52; Terra typica: Mount Ophir, Malacca (see Figure 4)

Distribution: Southern peninsular Thailand, Malay Peninsula
Museum: AMNH, BMNH, NNML, USNM, ZMA, ZSBZ

Wing length: 121–132 mm
Tail length: 64–66 mm
Tarsus length: 21–24 mm
Length of bill: 19.5–21.5 mm
Body mass: ?

Illustration: H. Burn in Robson 2000: Pl. 21
Photograph: G. Ehlers, Leipzig in collection Weick (in letter)
Literature: Hachisuka 1934: 52; del Hoyo et al. 1999: 153; König et al. 1999: 212, 213; Robson 2000: 290; Hekstra unpublished

❖

Otus thilohoffmanni Warakagoda and Rasmussen 2004
Serendib Scops Owl · Serendib-Zwergohreule
***Otus thilohoffmanni*, Warakagoda and Rasmussen 2004, Bull. Brit. Ornith. Cl. 124(2): 87; Terra typica: Southwest Sri Lanka (see Plate 2 and Figure 4)**

Length: 165–170 mm
Body mass: ?

Distribution: Endemic to southwest Sri Lanka
Habitat: Lowland rainforest, dense secondary growth with bamboo, creepers and tree ferns, at 30–530 m asl.
Museum: NMC (only the holotype)
Remarks: Nearest related species seems to be *O. rufescens*, but DNA evidence is lacking?

Wing length: 128–140 mm
Tail length: 63–66 mm
Tarsus length: 27.5–33.5 mm
Length of bill (cere): 13 mm
Body mass: ?

Illustration: P. Samaraweera in Warakagoda and Rasmussen 2004: 86; McQueen in Rasmussen and Anderton 2005: Pl. 75
Photograph: C. Kahandawala in Warakagoda and Rasmussen 2004: 89, Fig. 1a–e
Literature: Warakagoda and Rasmussen 2004: 85–105; Rasmussen and Anderton 2005: 235

❖

Otus icterorhynchus (Shelley) 1873
Cinnamon or Sandy Scops Owl · Gelbschnabelzwergohreule · Petit-duc à bec jaune · Autillo Piquigualdo

Length: 180–200 mm
Body mass: 61–80 g

Distribution: Liberia, Ivory Coast, Ghana, Cameroon, northern Congo, north and east Zaire, northern Gabon
Habitat: Lowland, evergreen forest, forest-shrub-grassland mosaic, open canopy forest. From sea-level up to 1 000 m

- ***Otus icterorhynchus icterorhynchus*** (Shelley) 1873
  Scops icterorhynchus Shelley 1873, Ibis: 138; Terra typica: Fanti, Gold Coast

Synonym:
- *Scops spurrelli* Ogilvie-Grant 1912, Bull. Brit. Ornith. Cl. 29: 116; Terra typica: Ashanti, Gold Coast

Distribution: Liberia, Ivory Coast, Ghana
Museum: BMNH, SMNSt

Wing length: 117–134 mm
Tail length: 64–71 mm
Tarsus length: 20 and 21 mm
Length of bill: 15.2–16.5 mm
Body mass: 61–80 g

Illustration: M. Woodcock in Fry et al. 1988: Pl. 9; T. Boyer in Boyer and Hume 1991: 33; P. Hayman in Kemp and Kemp 1998: 315; F. Weick in König et al. 1999: Pl. 6; N. Borrow in Borrow and Demey 2001: Pl. 59
Photograph: ?
Literature: Shelley 1873: 138; Reichenow 1901: 667, 668; Bannerman 1953: 535, 536; Eck and Busse 1973: 71, 72; Fry et al. 1988: 111; Boyer and Hume 1991: 33, 34; Kemp and Kemp 1998: 314; del Hoyo et al. 1999: 153; König et al. 1999: 213; Hekstra unpublished

- ***Otus icterorhynchus holerythrus*** (Sharpe) 1901
Scops holerythra Sharpe 1901, Bull. Brit. Ornith. Cl. 12: 3;
Terra typica: Efulen, Cameroon
(see Plate 2)

Synonym:
- *Pisorhina badia* Reichenow 1903, Orn. Monatsb. 11: 40; Terra typica: Bipindi, Cameroon

Distribution: Southern Cameroon, northern Congo, northern and eastern Zaire, northern Gabon
Museum: AMNH, BMNH, ChiMNH, CMP, FNSF, MNBHU, RMTB, SMNSt

Wing length:	129–144 mm
Tail length:	67–85 mm
Tarsus length:	22–26 mm
Length of bill:	15.5–18.5 mm
Body mass:	♂ 69–80 g, ♀ 61–80 g

Illustration: Reichenow 1902, Atlas; J. G. Keulemans, *Ibis* 1904: Pl. 2; F. Weick in König et al. 1999: Pl. 6; N. Borrow in Borrow and Demey 2001: Pl. 59
Photograph: ?
Literature: Sharpe 1901: 3; Eck and Busse 1973: 70, 71; Fry et al. 1988: 111; del Hoyo et al. 1999: 153; König et al. 1999: 213

❖

Otus ireneae Ripley 1966

Sokoke Scops Owl · Sokoke-Zwergohreule · Petit-duc d'Irene · Autillo de Sokoke
Otus ireneae **Ripley 1966, Ibis 108: 136;**
Terra typica: **Sokoke-Arabuku Forest, Kenya**
(see Plate 2)

Length:	160–180 mm
Body mass:	♂ 45–55 g

Distribution: Sokoke-Arabuku Forest Kenya and NE Tanzania (lowlands, northern and eastern Usambara Mts.)
Habitat: Cynometra-Manilkara forest, also Brachystegia woodland, 200–400 m
Museum: BMNH, NMKN, USNM, YPM (holotype)

Wing length:	112–124 mm
Tail length:	56–66 mm
Tarsus length:	17–23 mm
Length of bill:	?
Body mass:	45–55 g

Illustration: Fennessy in *Ibis* 1966: Pl. 1 (grey morph); M. Woodcock in Fry et al. 1988: Pl. 9; P. Hayman in Kemp and Kemp 1998: 309; F. Weick in König et al. 1999: Pl. 6; J. Gale in Stevenson and Fanshawe 2002: Pl. 100; Hekstra unpublished
Photograph: A. Weaving in Kemp and Kemp 1998: 308; A. Pauw in Kemp and Kemp 1998: 309; L.C. Marigo in del Hoyo et al. 1999: 145; A. Weaving in internet: Owl Pages, photo gallery 2002; R.A. Behrstock in Duncan 2003: 205
Literature: Ripley 1966: 136; Fry et al. 1988: 114; Boyer and Hume 1991: 34; Zimmerman et al. 1996: 441; Kemp and Kemp 1998: 308, 309; del Hoyo et al. 1999: 153, 154; König et al. 1999: 214

❖

Otus balli (Hume) 1873

Andaman Scops Owl · Andamanen-Zwergohreule · Petit-duc des Andamanes · Autillo de Andamán
Ephialtes Balli **Hume 1873, Str. Feath. 1: 407;**
Terra typica: **South Andaman Islands**

Length:	180–190 mm
Body mass:	?

Distribution: Andaman Islands (Gulf of Bengal)
Habitat: Semi-open areas, cultivated countrysides with gardens, bushes, trees, also near buildings and settlements
Museum: BMNH

Wing length:	133–143 mm
Tail length:	75–80 mm
Tarsus length:	26–28 mm
Length of bill:	19.5–20 mm
Body mass:	?

Illustration: J. G. Keulemans in Sharpe 1891: Pl. 20; T. Boyer in Boyer and Hume 1991: 35; F. Weick in

König et al. 1999: Pl. 8; McQueen in Rasmussen and Anderton 2005:Pl. 78
Photograph: ?
Literature: Hume 1873: 407; Sharpe 1875a: 100–102; Baker 1927: 429, 430; Abdulali 1965: 534: Ali and Ripley 1969: 613, 614; Eck and Busse 1973: 73; Boyer and Hume 1991: 35; del Hoyo et al. 1999: 154; König et al. 1999: 222, 223; Hekstra unpubl.; Rasmussen and Anderton 2005: 235

❖

Otus alfredi (Hartert) 1897
Flores Scops Owl · Flores-Zwergohreule · Petit-duc de Florès · Autillo de Flores
***Pisorhina alfredi* Hartert 1897, Novit. Zool. 4: 527; Terra typica: Repok Mountains, Flores Islands. (See Figure 3)**

Length: 190–210 mm
Body mass: ?

Distribution: Flores Islands, (Ruteng and Todo Mountains), Lesser Sundas
Habitat: Humid forests, above 1 000 m
Museum: AMNH (Rothsch.), MZB

Wing length: 137–160 mm
Tail length: (69) 77–82 mm
Tarsus length: 23–27 mm
Length of bill: 19–23 mm
Body mass: ?

Illustration: J. G. Keulemans in *Novit. Zool.* 1898(5): Pl. 1; J. Marshall jr. in Amadon and Bull 1988: col. Pl.; D. Gardner in Coates and Bishop 1997: Pl. 34; F. Weick in König et al. 1999: Pl. 8; F. Weick in *Gefiederte Welt* 2004(1): 19
Photograph: Y. de Fretes in Monk et al. 1997; Y. de Fretes in Widodo et al. 1999: 16
Literature: Hartert 1897b: 527; Coates and Bishop 1997: 360; Widodo et al. 1999: 15–23; del Hoyo et al. 1999: 154; König et al. 1999: 220, 221; Weick 2003–2005, *Gefiederte Welt* 2004(1): 19, 20

❖

Otus stresemanni (Robinson) 1927
Stresemann's Mountain Scops Owl · Stresemann-Zwergohreule · Petit-duc de Stresemann
***Athenoptera spilocephalus stresemanni* Robinson 1927, Bull. Brit. Ornith. Cl. 47: 126; Terra typica: Scolah Dras, 3 000 feet (900 m), Sumatra. (See Figure 3)**

Length: 180 mm
Body mass: ?

Distribution: Scolah Dras, Korinchi, Sumatra
Habitat: Dense evergreen forests, foothills, 900 m
Museum: BMNH (spiritus coll.), NNML (holotype) (collected March 1914!)
Remarks: Status uncertain, often regarded as conspecific with *O. balli* or *Otus spilocephalus* or colour morph of subspecies *vandewateri*, but very different in plumage-pattern and so regarded as full species

Wing length: 141–150 mm
Tail length: 63 and 86 mm
Tarsus length: 23 and 25.5 mm
Length of bill: 17.2 mm
Body mass: ?

Illustration: K. Phillipps in McKinnon and Phillipps 1993: Pl. 40; F. Weick in König et al. 1999: Pl. 7; F. Weick in *Gefiederte Welt* 2004(1): 19
Photograph: ? (only photograph from skin exist)
Literature: Stresemann 1925: 179–195; Robinson 1927a: 126; Chasen 1935: 86; Eck and Busse 1973: 72, 73; del Hoyo et al. 1999: 154 (Taxonomy); König et al. 1999: 216

❖

Otus spilocephalus Blyth 1846
Mountain Scops Owl · Fuchseule · Petit-duc de Montagne (ou tacheté) · Autillo Montano

Length: 170–205 mm
Body mass: 50–110 g

Distribution: Pakistan, Nepal, Himalayas in northern and eastern India, to Sikkim and Myanmar, southeast China, Taiwan. South to southeast Asia. Malay Peninsula, Sumatra, Borneo
Habitat: Dense, evergreen forest, mainly chestnut, oak, rhododendron and pine. Montane forest with gullies and ravines. From foothills up to 2 600 m

- ***Otus spilocephalus spilocephalus* (Blyth) 1846**
  Ephialtes spilocephalus Blyth 1846, J. As. Soc. Beng. 15: 8; Terra typica: Darjeeling

Synonym:
- *Otus spilocephalus rupchandi* Koelz 1952, J. Zool. Soc. India 4: 45; Terra typica: Kohima, Nagaland

Distribution: Central Nepal, east to Aruncha Pradesh and Myanmar, 1 500–2 750 m
Museum: AMNH, BMNH, ChiMNH, MNBHU, SMNSt, UMMZ

Wing length:	136–152 mm
Tail length:	73–87 mm
Tarsus length:	25–32 mm
Length of bill:	17–18 mm
Body mass:	60–77 g

Illustration: T. Boyer in Boyer and Hume 1991: 35; D. Cole in Grimmett et al. 1998: Pl. 28; F. Weick in König et al. 1999: Pl. 7; H. Burn in Robson 2000: Pl. 21; McQueen in Rasmussen and Anderton 2005: Pl. 78
Photograph: ?
Literature: Stresemann 1925: 179–195; Baker 1927: 428; Ali and Ripley 1969: 257, 258; Eck and Busse 1973: 71, 72; Voous 1988: 29–32; Boyer and Hume 1991: 35; Grimmett et al. 1998: 430; del Hoyo et al. 1999: 154, 155; König et al. 1999: 214–216; Robson 2000: 290; Rasmussen and Anderton 2005: 236; Hekstra unpublished

- ***Otus spilocephalus huttoni*** (Hume) 1870
Ephialtes Huttoni Hume 1870, Rough Notes 1(2): 393; Terra typica: Jerripani/Mussorie

Distribution: Northern Pakistan, east to central Nepal (Himalayas), up to 2 600 m
Museum: AMNH (Rothsch.), BMNH, ChiMNH, ZMH

Wing length:	135–149 mm
Tail length:	71–85 mm
Tarsus length:	25–32 mm
Length of bill:	17 mm
Body mass:	?

Illustration: G. M. Henry in Ali 1949: Pl. 53; D. Cole in Grimmett et al. 1998: Pl. 28; F. Weick in König et al. 1999: Pl. 7; McQueen in Rasmussen and Anderton 2005: Pl. 78
Photograph: ?
Literature: Baker 1927: 429; Ali 1949: 156, 157; del Hoyo et al. 1999: 154, 155; König et al. 1999: 214–216; Rasmussen and Anderton 2005: 236; Hekstra unpublished

- ***Otus spilocephalus latouchi*** (Rickett) 1900
Scops latouchi Rickett 1900, Bull. Brit. Ornith. Cl. 10: 56; Terra typica: Ah Chung, Fukien

Distribution: Northern Thailand and Laos to southeast China and Hainan
Museum: AMNH, BMNH, MCZ, MNBHU, NHRMSt
Remarks: Sometimes regarded as identical to subspecies *siamensis*

Wing length:	140–152 mm
Tail length:	82–89 mm
Tarsus length:	28–31 mm
Length of bill:	14–17 mm
Body mass:	?

Illustration: H. Grönvold in Delacour and Jabouille 1931: Pl. 21
Photograph: ?
Literature: Rickett 1900: 56; Delacour and Jabouille 1931: 128; Etchécopar and Hüe 1978: 477, 478; del Hoyo et al. 1999: 154, 155; König et al. 1999: 214–216; Hekstra unpublished

- ***Otus spilocephalus hambroecki*** Swinhoe 1870
Ephialtes Hambroecki Swinhoe 1870, Ann. Mag. Nat. Hist. 4(6): 153; Terra typica: Formosa = Taiwan

Distribution: Taiwan
Museum: BMNH, MNBHU, NHRMSt, NNML, SMTD, USNM
Remarks: Hekstra regarded *hambroecki* as subspecies of *Otus longicornis* (personal communication)

Wing length:	142–152 mm
Tail length:	82–89 mm
Tarsus length:	28–31 mm
Length of bill:	18–20 mm
Body mass:	53–112 g

Illustration: F. Weick in König et al. 1999: Pl. 7
Photograph: ?

Literature: Eck and Busse 1973: 71; Etchécopar and Hüe 1978: 477, 478; Meyer de Schauensee 1984: 265, 266; del Hoyo et al. 1999. 154, 155; König et al. 1999: 214–216; Hekstra unpublished

- ***Otus spilocephalus siamensis*** Robinson and Kloss 1922
  Otus luciae siamensis Robinson and Kloss 1922, J. Fed. Malay St. Mus. 10: 261; Terra typica: Kai Nong, Siam (Thailand)

Distribution: Southern Thailand to southern Vietnam
Museum: AMNH, BMNH, MNHP, USNM

Wing length:	138–142 mm
Tail length:	65–78 mm
Tarsus length:	23–27 mm
Length of bill (cere):	12 mm
Body mass:	?

Illustration: H. Grönvold in Delacour and Jabouille 1931: Pl. 21; W. Monkol in Lekagul and Round 1991: Pl. 60; H. Burn in Robson 2000: Pl. 21
Photograph: ?
Literature: Robinson and Kloss 1922: 261; Delacour and Jabouille 1931: 128, 129; Deignan 1945: 173; Lekagul and Round 1991: 177; del Hoyo et al. 1999: 154, 155; König et al. 1999: 214–216; Robson 2000: 293; Hekstra unpublished

- ***Otus spilocephalus vulpes*** (Ogilvie-Grant) 1906
  Heteroscops vulpes Ogilvie-Grant 1906, Bull. Brit. Ornith. Cl. 19: 11; Terra typica: Gunong Tahan, 5 300 feet, Malay Peninsula

Distribution: Mountains of the Malay Peninsula
Museum: AMNH, BMNH, ZFMK

Wing length:	133–139 mm
Tail length:	65–68 mm
Tarsus length:	24–25 mm
Length of bill:	17.5–20 mm
Body mass:	?

Illustration: F. Weick in König et al. 1999: Pl. 7; H. Burn in Robson 2000: Pl. 21
Photograph: ?
Literature: Ogilvie-Grant 1906b: 11; König et al. 1999: 214–216; del Hoyo et al. 1999: 154, 155; Robson 2000: 290; Hekstra unpublished

- ***Otus spilocephalus vandewateri*** (Robinson and Kloss) 1916
  Pisorhina vandewateri Robinson and Kloss 1916, J. Str. Branch R. As. Soc. 73: 275; Terra typica: Korinchi Peak, 7 300 feet, Sumatra

Distribution: Mountains of Sumatra
Museum: AMNH, ANSP, BMNH, FNSF, MCZ, NNML, ZMA (de Bussy)
Remarks: Hekstra regards this taxon as a subspecies of *Otus angelinae* (personal communication)

Wing length:	136–149 mm
Tail length:	71–79 mm
Tarsus length:	25–27 mm
Length of bill:	17–19.5 mm
Body mass:	85 g

Illustration: K. Phillipps in McKinnon and Phillipps 1993: Pl. 40; F. Weick in König et al. 1999: Pl. 7
Photograph: ?
Literature: Robinson and Kloss 1916: 275; Chasen 1935: 86; Eck and Busse 1973: 72, 73; McKinnon and Phillipps 1993: 189, 190; del Hoyo et al. 1999: 154, 155; König et al. 1999: 214–216

- ***Otus spilocephalus luciae*** (Sharpe) 1888
  Scops luciae Sharpe 1888, Ibis: 478; Terra typica: Mount Kina Balu, Borneo

Distribution: Mountains of Borneo
Museum: AMNH, BMNH (holotype), CUM, NNML, USNM

Wing length:	129–141 mm
Tail length:	63–72 mm
Tarsus length:	24–27 mm
Length of bill:	17.5–20 mm
Body mass:	?

Illustration: J. G. Keulemans in *Ibis* 1889: Pl. 3; A. Hughes in Smythies 1960: Pl. 15; K. Phillipps in McKinnon and Phillipps 1993: Pl. 40; F. Weick in König et al. 1999: Pl. 7
Photograph: ?
Literature: Sharpe 1888b: 478; Smythies 1960: 276, 277; McKinnon and Phillipps 1993: 189, 190; del Hoyo et al. 1999: 154, 155; König et al. 1999: 214–216; Hekstra unpublished

Otus angelinae · *Javan Scops Owl* · *Angelina-Zwergohreule*

Otus angelinae (Finsch) 1912
Javan Scops Owl · Angelina-Zwergohreule, Java-Zwergohreule · Petit-duc de Java · Autillo de Java
***Pisorhina angelinae* Finsch 1912, Orn. Monatsb. 20: 156; Terra typica: Pangerango Mountains, 6 000 feet, Java**

Length: 160–180 mm
Body mass: 75–90 g

Distribution: Mountains of western Java
Habitat: Montane primary rainforest with understorey, between 900 and 2 500 m
Museum: MZB, NNML
Remarks: Sometimes treated as a subspecies of *Otus brookii* or *Otus spilocephalus*, but differs in morphology far less vocal habits. Monotypic

Wing length: 135–149 mm
Tail length: 63–69 mm
Tarsus length: 25–27 mm
Length of bill: 19.5–21.5 mm
Body mass: 75–90 g

Illustration: J. Marshall jr. in Amadon and Bull 1988: col. Pl.; K. Phillipps in McKinnon and Phillipps 1993: Pl. 40; F. Weick in König et al. 1999: Pl. 7
Photograph: M. Ruedi in Becking 1994, Frontispiece; M. Ruedi in del Hoyo et al. 1999: 148
Literature: Finsch 1912: 156–159; McKinnon 1990: 179; McKinnon and Phillipps 1993: 190; Becking 1994: 211–224; del Hoyo et al. 1999: 157; König et al. 1999: 216–218; Hekstra unpubl.

Otus mirus Ripley and Rabor 1968
Mindanao Scops Owl · Mindanao-Zwergohreule · Petit-duc de Mindanao · Autillo de Mindanao
***Otus scops mirus* Ripley and Rabor 1968, Proc. Biol. Soc. Wash. 81: 31–36; Terra typica: Hilong, Hilong Peak, Province Agusau, Mindanao**

Length: 170–200 mm
Body mass: 65 g

Distribution: Mindanao Island, southern Philippines
Habitat: Montane rainforest, between 1 000 and 1 500 m
Museum: BMNH, FMNH, FNSF, NMP / CMNH (Mt. Apo), SU, USNM (holotype)

Wing length: 127–132 mm
Tail length: 57–58.5 mm
Tarsus length: 21–25 mm
Length of bill: ?
Body mass: 65 g

Illustration: J. Marshall jr. in Amadon and Bull 1988: col. Pl.; F. Weick in König et al. 1999: Pl. 8; A. Sutherland in Kennedy et al. 2000: Pl. 34
Photograph: Wechsler in del Hoyo et al. 1999: 87
Literature: Ripley and Rabor 1968: 31–36; Dickinson et al. 1991: 223; del Hoyo et al. 1999: 159, 160; König et al. 1999: 218; Kennedy et al. 2000: 175

Otus longicornis Ogilvie-Grant 1894
Luzon Scops Owl · Luzon-Zwergohreule · Petit-duc longicorn · Autillo de Luzón
***Scops longicornis* Ogilvie-Grant 1894, Bull. Brit. Ornith. Cl. 3: 51; Terra typica: Mountains of North Luzon**

Length: 180–190 mm
Body mass: ?

Distribution: Luzon, northern Philippines
Habitat: Humid forests and pine woodland, from 1 000 m up to about 2 000 m
Museum: BMNH (holotype), CMNH, Del MNH, FMNH

Wing length: 136–152 mm
Tail length: 63–74 mm
Tarsus length: 28–30 mm
Length of bill: 18.5–20.5 mm
Body mass: ?

Illustration: J. Marshall jr. in Amadon and Bull 1988: col. Pl.; F. Weick in König et al. 1999: Pl. 8; A. Sutherland in Kennedy et al. 2000: Pl. 34
Photograph: G. Ehlers, Leipzig in letter (freshly killed specimen); D. Wechsler in del Hoyo et al. 1999: 87
Literature: Dickinson et al. 1991: 224; del Hoyo et al. 1999: 160; König et al. 1999: 219; Kennedy et al. 2000: 176

❖

Otus mindorensis (Whitehead) 1899
Mindoro Scops Owl · Mindoro-Zwergohreule · Petit-duc de Mindoro · Autillo de Mindoro
***Scops mindorensis* Whitehead 1899, Ibis: 98;**
Terra typica: **Highlands of Mindoro**

Length: 180–190 mm
Body mass: ?

Distribution: Mindoro Island, Philippines (Mount Hálcon, Mount Baco and Mount Dulaugan)
Habitat: Montane mossy forest, above 1 000 m
Museum: BMNH (holotype), CMNH

Wing length: 133–136 mm
Tail length: 63 mm
Tarsus length: 25–28 mm
Length of bill: ?
Body mass: ?

Illustration: J. Marshall jr. in Amadon and Bull 1988: col. Pl.; F. Weick in König et al. 1999: Pl. 8; A. Sutherland in Kennedy et al. 2000: Pl. 34
Photograph: ?
Literature: Whitehead 1899: 98, 99; Dickinson et al. 1991: 224; del Hoyo et al. 1999: 160; König et al. 1999: 219, 220; Kennedy et al. 2000: 175, 176

Otus hartlaubi (Giebel) 1872
São Tomé Scops Owl · Hartlaub-Zwergohreule · Petit-duc de São Tomé · Autillo de Santo Tomé
***Athene leucopsis* Hartlaub 1849, Rev. Mag. Zool. II, 1: 496;**
Terra typica: **São Tomé (invalid name)**
***Noctua hartlaubi* Giebel 1872, Thes. Ornith. 1: 448;**
Terra typica: **São Tomé**

Length: 170–190 mm
Body mass: 79 g

Distribution: São Tomé Island (Gulf of Guinea)
Habitat: Dense primary and secondary forest and plantations, from sea-level up to 1 300 m
Museum: AMNH, BMNH, BÜM, FNSF, SMTD, ZFMK, ZMB, ZMH

Wing length: 123–139 mm
Tail length: 60–72 mm
Tarsus length: 20–27 mm
Length of bill: ?
Body mass: 79 g

Illustration: J. Marshall jr. in Amadon and Bull 1988: col. Pl.; M. Woodcock in Fry et al. 1988: Pl. 9; P. Hayman in Kemp and Kemp 1998: 329; F. Weick in König et al. 1999: Pl. 8
Photograph: D. Amadon 1953: Pl. 4 (b/w)
Literature: Fry et al. 1988: 117, 118; Kemp and Kemp 1998: 328, 329; del Hoyo et al. 1999: 170; König et al. 1999, 221, 222

❖

Otus rutilus (Pucheran) 1849
Madagascar Scops Owl · Madagaskar-Zwergohreule · Petit-duc de malagache · Autillo Malagache
***Scops rutilus* Pucheran 1849, Rev. Mag. Zool. II, 1: 17–29;**
Terra typica: **Madagascar**
(see Plate 2 and Figure 6)

Length: 190–220 mm
Body mass: 85–120 g

Distribution: Eastern and northern Madagascar
Habitat: Humid primary and secondary forest, thickets, urban parks
Museum: BMNH, BÜM, MHNP, NHMBas, NHMWien, NHRMSt, NNML, ROM, ZMH, ZSBS

Wing length: 151–161 mm
Tail length: 74–82 mm
Tarsus length: 28.5–31.5 mm
Length of bill: 20–22 mm
Body mass: ♂ 85–107 g, ♀ 112–116 g

Illustration: V. Bretagnolle in Langrand 1990: Pl. 26; J. Lewington in Kemp and Kemp 1998: 317; F. Weick in König et al. 1999: Pl. 9; J. Lewington in Rasmussen et al. 2000: 76; F. Weick in *Gefiederte Welt* 2004(1): 21
Photograph: O. Langrand in Kemp and Kemp 1998: 317; P. Morris in Morris and Hawkins 1998: 167 b
Literature: Pucheran 1849: 17–29; Grandidier 1867: 84–88; Langrand 1990: 227; Kemp and Kemp 1998: 316, 317; Morris and Hawkins 1998: 167; del Hoyo et al. 1999: 170; König et al. 1999: 223; Rasmussen et al. 2000: 75–102; Weick 2003–2005, *Gefiederte Welt* 2004(1): 20, 21

❖

Otus madagascariensis **(Grandidier) 1867**
Torotoroka Scops Owl · Torotoroka-Zwergohreule · Petit-duc Torotoroka · Autillo Torotoroka
Scops madagascariensis **Grandidier 1867, Rev. Mag. Zool. II, 19: 84–88;** Terra typica: **Madagascar**
(see Figure 6)

Length: 200–220 mm
Body mass: similar to *O. rutilus*

Distribution: Western and southwestern Madagascar
Habitat: Deciduous and dry forests, sometimes in trees near human settlements
Museum: BMNH, FNSF, MNHP, MNHN?, NHRMSt, NNML

Wing length: 152–161 mm
Tail length: 82–88 mm
Tarsus length: 29–31.5 mm
Length of bill: 20, 21 mm
Body mass: similar to *rutilus*

Illustration: J. G. Keulemans in Milne-Edwards and Grandidier 1876: Pl. 40; V. Bretagnolle in Langrand 1990: Pl. 26; J. Lewington in Kemp and Kemp 1998: Pl. 317; F. Weick in König et al. 1999: Pl. 9; F. Weick in *Gefiederte Welt* 2004(1): 21
Photograph: P. Morris in Morris and Hawkins 1998: 167a; N. Dennis in Kemp and Kemp 1998: 316; T. Quested in internet: Owl Pages, photo gallery 2002
Literature: Grandidier 1867: 84–88; Marshall 1978: 1–58; Goodman et al. 1998; Kemp and Kemp 1998: 316, 317; Morris and Hawkins 1998: 167; del Hoyo et al. 1999: 170; König et al. 1999: 223; Rasmussen et al. 2000: 75–102; Weick 2003–2005, *Gefiederte Welt* 2004(1): 20, 21

❖

Otus mayottensis **Benson 1960**
Mayotte Scops Owl · Mayotte-Zwergohreule · Petit-duc de Mayotte · Autillo de Mayotte
Otus rutilus mayottensis **Benson 1960, Ibis 103b: 60–61;** Terra typica: **Mayotte Island, Comoro Island**
(see Figure 9)

Length: 240 mm
Body mass: 120 g

Synonym:
- *Scops humbloti* Oustalet 1888, means Milne-Edwards and Oustalet, Nouv. Arch. Mus. Hist. Nat. Ser. 2

Distribution: Mayotte Island = Maore Island, Comoro Islands
Habitat: Evergreen forest
Museum: BMNH, MHNP
Remarks: Mostly regarded as a subspecies of *O. rutilus*, but is possibly a distinct species!

Wing length: 166–178 mm
Tail length: 80–86.5 mm
Tarsus length: 35–37.5 mm
Length of bill (cere): 24.5–26 mm
Body mass: ?

Illustration: F. Weick in König et al. 1999: Pl. 9; F. Weick in *Gefiederte Welt* 2004(10): 312
Photograph: ?
Literature: Benson 1960: 60, 61; Louette 1988: 82, 83; König et al. 1999: 223; del Hoyo et al. 1999: 70; Rasmussen et al. 2000: 82–93

❖

Otus moheliensis **Lafontaine and Moulaert 1999**
Mohéli Scops Owl · Moheli-Zwergohreule · Petit-duc de Moheli · Autillo de Moheli
Otus moheliensis **Lafontaine and Moulaert 1999, Bull. Afric. Bird Cl. 6: 61–65;** Terra typica: **Mohéli Island = Mwali Island**
(see Plate 2 and Figure 9)

Length: 220 mm
Body mass: 95–119 g

Distribution: Mohéli = Mwali Island, Comoro Islands
Habitat: Dense humid forest on hillsides, 450–790 m
Museum: IRSNB (holotype)

Wing length:	155–164 mm
Tail length:	71 mm
Tarsus length:	34–36 mm
Length of bill:	23–24 mm
Body mass:	95–119 g

Illustration: F. Weick in König et al. 1999: Pl. 9; F. Weick in *Gefiederte Welt* 2004(10): 312
Photograph: R. M. Lafontaine in Lafontaine and Moulaert 1999: backcover
Literature: Lafontaine and Moulaert 1999: 61–65; del Hoyo et al. 1999: 170; König et al. 1999: 225, 226; Rasmussen et al. 2000: 87, 95, 96; Weick 2003–2005, *Gefiederte Welt* 2004(10): 312, 313

❖

Otus capnodes **(Gurney) 1889**
Anjouan Scops Owl · Anjouan-Zwergohreule · Petit-duc d'Anjouan · Autillo de Anjouan
Scops capnodes **Gurney 1889, Ibis: 104;**
Terra typica: Anjouan Island

Length: 220 mm
Body mass: ♂ 119 g ($n = 1$)

Distribution: Anjouan = Ndzuani Island, Comoro Islands
Habitat: Montane primary forest, above 550 m, mostly above 800 m
Museum: AMNH, BMNH (holotype), CUM, FNSF, LivCM, MCZ, MHN, MNHN, MNBHU, MZUS, ROM, USNM, ZMUZ
Remarks: Sometimes regarded as a subspecies of *O. rutilus*, but with very different vocalisation!

Wing length:	153–167 mm
Tail length:	80–86 mm
Tarsus length:	33.5–36 mm
Length of bill:	20–24.5 mm
Body mass:	♂ 119 g ($n = 1$)

Illustration: J. Lewington in Kemp and Kemp 1998: 323; F. Weick in König et al. 1999: Pl. 9; F. Weick in *Gefiederte Welt* 2004(1): 20
Photograph: R. Safford (in letter), collection Weick; R. Safford in Kemp and Kemp 1998: 322; A. Lewis in Kemp and Kemp 1998: 322; A. Lewis in internet: Owl Pages, photo gallery 2002
Literature: Gurney 1889: 104–107; Benson 1960: 5–106; Safford 1993: 57–74; Lewis 1996: 131–133; Safford 1996 in letter; del Hoyo et al. 1999: 169; König et al. 1999: 224, 225; Lafontaine and Moulaert 1999: 62–64; Rasmussen et al. 2000: 82, 94; Weick 2003–2005, *Gefiederte Welt* 2004(1): 20

❖

Otus pauliani **Benson 1960**
Grand Comoro Scops Owl · Comoren-Zwergohreule · Petit-duc de Karthala · Autillo de las Comores
Otus rutilus pauliani **Benson 1960, Ibis: 62–63;**
Terra typica: Mount Karthala, Grand Comoro Island
(see Plate 2)

Length: 180–200 mm
Body mass: 69.5 g

Distribution: Grand Comoro = Ngadzidja Island, Comoro Islands
Habitat: Primary and degraded mountain forest, forest edges, tree heath. From 400 m up to 1 000 m
Museum: BMNH (holotype), LivCM, MNHN?

Wing length:	138–144 mm
Tail length:	70–73 mm
Tarsus length:	25–27 mm
Length of bill:	21–23 mm
Body mass:	69.5 g

Illustration: T. Boyer in Boyer and Hume 1991: 48; J. Lewington in Kemp and Kemp 1998: 321; F. Weick in König et al. 1999: Pl. 9
Photograph: A. Lewis in Kemp and Kemp 1998: 321; A. Lewis in del Hoyo et al. 1999: 149; A. Lewis in Duncan 2003: 150
Literature: Benson 1960: 62, 63; Louette 1988: 83; Herremans et al. 1991: 123–133; Boyer and Hume 1991: 48; Kemp and Kemp 1998: 320, 321; del Hoyo et al. 1999: 169; König et al. 1999: 224; Rasmussen et al. 2000: 82, 95

Otus pembaensis **Pakenham 1937**
Pemba Scops Owl · Pembaeule · Petit-duc de Pemba · Autillo de Pemba
Otus pembaensis **Pakenham 1937, Bull. Brit. Ornith. Cl. 37: 112; Terra typica: Pemba Island**
(see Plate 2)

Length: 170–180 mm
Body mass: ?

Distribution: Pemba Island, off northern Tanzania
Habitat: Wooded and semi-open areas, plantations
Museum: BMNH (holotype)

Wing length: 146–152.5 mm
Tail length: 73–79 mm
Tarsus length: 22–27 mm
Length of bill: 22–23.5 mm
Body mass: ?

Illustration: M. Woodcock in Fry et al. 1988: Pl. 9; P. Hayman in Kemp and Kemp 1998: 319; F. Weick in König et al. 1999: Pl. 9; J. Gale in Stevenson and Fanshawe 2002: Pl. 100
Photograph: ?
Literature: Pakenham 1937: 112; Fry et al. 1988: 118; Kemp and Kemp 1998: 318, 319; del Hoyo et al. 1999: 169; König et al. 1999: 226; Lafontaine and Moulaert 1999: 62–64; Rasmussen et al. 2000: 82, 96, 97; Stevenson and Fanshawe 2002: 200

Otus flammeolus **(Kaup) 1853**
Flammulated Owl · Ponderosa-Zwergohreule · Petit-duc nain · Tecolote Flameado, Autillo Flameado
Scops (Megascops) flammeola **"Licht." Kaup 1853, Contrib. Orn. [Jardine] 1852: 111; Terra typica: Mexico**
(see Plate 2)

Length: 160–180 mm
Body mass: 45–63 g

Synonym:
- *Megascops flammeola idahoensis* Merriam 1891: North Am. Fauna 5: 96; Terra typica: Ketchum, Idaho
- *Otus flammeolus guatemalae* Griscom 1935, Ibis: 549; Terra typica: Duenas, Guatemala
- *Otus flammeolus rarus* Griscom 1937, Auk 54: 391; Terra typica: Duenas, Guatemala, new name for *guatemalae*

Otus flammeolus · Flammulated Owl · Ponderosa-Zwergohreule

- *Otus flammeolus meridionalis* Hekstra 1982, Bull. Zool. Mus. Univ. Amsterdam 9(7): 55, 56; Terra typica: Guapongo, Guerrero, Mexico
- *Otus flammeolus frontalis* Hekstra 1982, ibid.: 56; Terra typica: Front range, Colorado, USA
- *Otus flammeolus borealis* Hekstra 1982, ibid.: 56; Terra typica: Okanagan Valley, British Columbia, Canada

Remarks: Individual plumage variation appears to be con-tinuous, biometric differences possibly clinal. Subspecies names are problematic so treated as monotypic

Distribution: Mountains of southwest Canada, United States south to Mexico (winters to central and southern Mexico), Guatemala and El Salvador
Habitat: Open coniferous montane forest with bushy undergrowth. Mixed forest (pine, oaks, aspen or pine/douglas fir)
Museum: AMNH, BMNH, ChiMNH, DMNH, LUSMZ, MNBHU (holotype), MCZ, ROM, UC(al)MVZ, USNM
Remarks: Differs from subgenus *Megascops* by lacking typical trilled song, but probably differs also

from genus *Otus*! This species possibly belongs to its own genus [C. König 2004, personal communication. Confirmed M. Wink. The generic name is then *Psiloscops* (Coues 1899)]

Wing length: 126–148 mm
Tail length: 59–67 mm
Tarsus length: 20–23 mm
Length of bill: 15–16.5 mm
Body mass: 45–63 g

Illustration: R. Ridgway in *Auk* 1892: Pl. 2 (*idahoensis*); R. Ridgway in Merriam and Fisher 1893: Pl. 26; A. Brooks in Grosvenor and Wetmore 1937: 18; K. Karalus in Karalus and Eckert 1974: Pl. 35; L. Malick in Breeden et al. 1983: 244; A. Cameron in Voous 1988: 55; T. Boyer in Boyer and Hume 1991: 43; F. Weick in König et al. 1999: Pl. 10; D. Sibley in Sibley 2000: 279

Photograph: A. Nelson in Johnsgard 2002: Pl. 11; B. Marcot in internet: Owl Pages, photo gallery 2003; R. and N. Bowers in internet: Owl Pages, photo gallery 2004; R. A. Behrstock in Duncan 2003: 214

Literature: Sharpe 1875a: 105, 106; Ridgway 1914: 728–732; Marshall 1939: 71–78; Johnson 1963: 174–178; Hekstra 1982b: 55, 56; Voous 1988: 53–58; del Hoyo et al. 1999: 164, 165; König et al. 1999: 227

Otus scops · Common Scops Owl · Zwergohreule (left active, right camouflaged)

Otus scops **(Linnaeus) 1758**
Eurasian Scops Owl · Zwergohreule · Petit-duc scops, Hibou Petit-duc · Autillo Europeo
(see Plate 2)

Length: 160–200 mm
Body mass: 60–135 g

Distribution: Southern Europe (especially Mediterranean region), locally in central, eastern and western Europe, Africa north of the Sahara, from Morocco to Tunisia, Asia Minor and east to central Asia

Habitat: Open and semi-open woodland, parks, orchards, plantations, etc. Locally in scrub and garrigue, open coniferous forest, also in dry or rocky country. Winters in wooded savanna and bushy country. From sea-level up to about 2 000 m, locally to 3 000 m

- ***Otus scops scops*** (Linnaeus) 1758
  Strix Scops Linnaeus 1758, Syst. Nat. ed. 10, 1: 92;
  Terra typica: Europe, (Italy)

Synonym:
- *Strix giu* Scopoli 1769, Annus I Hist.-Nat.: 19; Terra typica: Krain
- *Strix zorca* Gmelin 1788, Syst. Nat. 1(1): 289; Terra typica: Sardinia
- *Pisorhina scops erlangeri* Tschusi zu Schmidhoffen 1904, Orn. Jahrb. 15: 101; Terra typica: Tallah, Tunisia
- *Pisorhina scops graeca* Tschusi zu Schmidhoffen 1904, ibid.: 102; Terra typica: Lamia, Greece
- *Pisorhina scops tuneti* Tschusi zu Schmidhoffen 1904, ibid.: 103; Terra typica: Tunis, Tunisia
- *Scops scops tschusii* Schiebel 1910, Orn. Jahrb. 21: 102; Terra typica: Ajaccio, Corsica

Distribution: France, Italy, and Mediterranean region, east to Volga River, south to northern Greece, northern Turkey and Transcaucasia

Museum: AMNH, BMNH, FNSF, MNBHU, NHMWien, NNML, SMNKa, SMNSt, SMTD, ZFMK, ZMA, ZMH, ZSBS

Wing length: 145–168 mm
Tail length: 65–80 mm
Tarsus length: 24–32 mm
Length of bill: 17–20 mm
Body mass: ♂ 66–113 g, ♀ 65–135 g

Illustration: N. Robert 1614–1685 in *Catal. Sotheby's* 1996: 33; J. F. Naumann in Naumann 1822: vol. 1, Pl. 43; H. C. Richter in Gould 1832–1837: Faksim.; J. G. Keulemans in Dresser 1871–1896: Pl. 314; J. G. Keulemans in Hennicke 1905: Faksim.; P. A. Robert in Géroudet 1979: 359; P. Barruel in Sutter and Barruel 1958: Pl. 28; L. Binder in Wüst 1970: 232; H. Delin in Cramp et al. 1985: Pl. 44 and 54; A. Cameron in Voous 1988: 43; A. Thorburn in Thorburn 1990: Pl. 26; L. Jonsson in Jonsson 1992: 325; P. Hayman in Kemp and Kemp 1998: 311; F. Weick in König et al. 1999: Pl. 10; D. Zetterström and K. Mullarney in Svensson et al. 1999: 215; J. Gale in Stevenson and Fanshaw 2002: Pl. 100; F. Weick in Weick 2004: Juli

Photograph: K. Paysan in Everett 1977: 21; Diemer in *Gefiederte Welt* 1993(1): 22, 23; numerous photographs in Mebs and Scherzinger 2000: 133 (Hofmann), 134 (Brandl), 14, 134, 137 (Nill), 139 (Diemer)

Literature: Naumann 1822: 466–472; Hartert 1912–1921: 978–980; Delacour 1941: 133–142; Dementiev and Gladkov 1951: 414, 417; Glutz von Blotzheim and Bauer 1980: 278–302; Bezzel 1985: 636–638; Cramp et al. 1985: 454–465; Voous 1988: 41–46; del Hoyo et al. 1999: 163, 164; König et al. 1999: 228, 229; Mebs and Scherzinger 2000: 133–146

- ***Otus scops pulchellus*** (Pallas) 1771
  Stryx (sic) *pulchella* Pallas 1771, Reise versch. Prov. Russ. Reichs 1: 456; **Terra typica:** ad Volgam, Samaram, Taicum

Synonym:
- *Pisorhina scops zarudnyi* Tschusi zu Schmidhoffen 1903, Orn. Jahrb. 14: 139; **Terra typica:** Sarepta
- *Pisorhina scops bascanica* Johansen 1907, ibid. 18: 202; **Terra typica:** Baskan, northeast Turkestan
- *Scops scops sibirica* Buturlin 1910, Mess. Orn. 1: 260; **Terra typica:** Krasnoyarsk and Minusinsk, upper Yennisei
- *Scops scops ferghanensis* Buturlin 1912, Nascha Ochota: 45; **Terra typica:** Osh district, 5 000 feet, Ala: Ferghana Mountains
- *Scops scops irtyshensis* Buturlin 1912, Nascha Ochota: 46; **Terra typica:** Tara on the upper Irtysh, western Siberia

Distribution: Volga River, east to Lake Baikal, south to Altai and Tien Shan. Migratory to Sindh, India

Museum: BMNH, FNSF, MHNP, NHMWien, NNML, SMTD, ZMH, ZMM, ZSBS

Wing length:	145–167 mm
Tail length:	66–74 mm
Tarsus length:	24–26 mm
Length of bill:	16 and 17 mm
Body mass:	78–82.5 g

Illustration: McQueen in Rasmussen and Anderton 2005: Pl. 78

Photograph: ?

Literature: Hartert 1912–1921: 980, 981; Baker 1927: 433; Dementiev and Gladkov 1951: 417, 418; Vaurie 1965: 599, 600; Ali and Ripley 1969: 261, 262; Cramp et al. 1985: 454–465; Voous 1988: 42; del Hoyo et al. 1999: 163, 164; König et al. 1999: 228, 229; Rasmussen and Anderton 2005: 236

- ***Otus scops mallorcae*** Jordans 1924
  Otus scops mallorcae von Jordans 1924, J. Orn. 72: 407; **Terra typica:** Alcudia, Mallorca

Distribution: Balearic Islands, northwestern Africa (north and central Morocco to Tunisia), northern and central Iberian Peninsula

Museum: BMNH, BÜM, MNBHU, NHMWien, SMNSt, ZFMK

Wing length:	151–161 mm
Tail length:	65–73 mm
Tarsus length:	24–27 mm
Length of bill:	16–19 mm
Body mass:	?

Illustration: H. Dehlin in Cramp et al. 1985: Pl. 44

Photograph: ?

Literature: Vaurie 1965: 598; Cramp et al. 1985: 454–465; del Hoyo et al. 1999: 163, 164; König et al. 1999: 228, 229

- ***Otus scops cycladum*** (Tschusi) 1904
  Pisorhina scops cycladum Tschusi zu Schmidhoffen 1904, Orn. Jahrb. 15: 104; **Terra typica:** Naxos Island, Cyclades

Synonym:
- *Otus scops powelli* Meinertzhagen 1920, Bull. Brit. Ornith. Cl. 41: 21; **Terra typica:** Candia district, Crete

Distribution: Crete, Cyclades (Naxos and Paros), southern Greece, and South Asia Minor, to central Israel and Jordan

Museum: AMNH, BMNH, FNSF, SMTD, ZFMK

Wing length: 153–166 mm
Tail length: ?
Tarsus length: 27–29 mm
Length of bill: 17.5–19.5 mm
Body mass: ?

Illustration: ?
Photograph: ?
Literature: Vaurie 1965: 598; Cramp et al. 1985: 454–465; del Hoyo et al. 1999: 163, 164

- ***Otus scops cyprius*** (Madarász) 1901
  Scops cypria von Madarász 1901, Termes. Fuzet. 24: 272;
  Terra typica: Livadia, Cyprus

Distribution: Cyprus
Museum: BMNH, MNBHU, NNML, Hung. NMBud (holotype), ZFMK, ZMH

Wing length: 152–167 mm
Tail length: 62–76 mm
Tarsus length: 27–29 mm
Length of bill: 17.5–19.5 mm
Body mass: ?

Illustration: ? in *Ann. Mus. Natl. Hungar.* 1904(2): Pl. 15; F. Weick in König et al. 1999: Pl. 10
Photograph: ?
Literature: Hartert 1912–1921: 982, 983; Vaurie 1965: 598; Cramp et al. 1985: 454–465; del Hoyo et al. 1999: 163, 164; König et al. 1999: 228, 229; Hekstra unpublished

- ***Otus scops turanicus*** (Loudon) 1905
  Pisorhina scops turanicus von Loudon 1905, Orn. Monatsb. 13: 129;
  Terra typica: Desert of Kara Kum, Transcaspia

Distribution: Iraq, Iran, (southeast Turkey?), east to northwestern Pakistan
Museum: BMNH, ZMM
Remarks: Separated from *O. sunia* and *O. senegalensis* by ecological and bioacoustic patterns, further reinforced by DNA evidence

Wing length: 149–165 mm
Tail length: 60–70 mm
Tarsus length: ?
Length of bill: ?
Body mass: ?

Illustration: D. Cole in Grimmett et al. 1998: Pl. 28; F. Weick in König et al. 1999: Pl. 10
Photograph: A. Roberts in Shirihai 1996: Pl. 64; H. Shirihai in Shirihai 1996: Pl. 64
Literature: Vaurie 1965: 600; Cramp et al. 1985: 454–465; Shirihai 1996: 311–313; Grimmett et al. 1998: 431; del Hoyo et al. 1999: 163, 164; König et al. 1999: 228, 229

Otus brucei **(Hume) 1873**
Pallid Scops Owl, Striated Scops Owl · Streifen-Zwergohreule · Petit-duc de Bruce · Autillo Persa
(see Plate 2)

Length: 180–210 mm
Body mass: 110 g

Distribution: Middle East (southern and central Turkey, northern Syrie, Iraq, Iran, eastern Arabia), to West and Central Asia, south to Afghanistan, Pakistan and Northwest India (formery bred in Israel). Socotra Island and southern Saudi Arabia. Remarks: Birds from Socotra Island *socotranus* and southern Saudi Arabia *pamelae*, are sometimes regarded as subspecies of *O. senegalensis*
Habitat: Semi-open landscape with trees and bushes. Cultivation, palmgroves, orchards, parks; riverine woodland. Sometimes in rocky and arid country. From lowland up to 1 800 m

- ***Otus brucei brucei*** (Hume) 1873
  Ephialtes Brucei Hume 1873, Str. Feath. 1: 8; Terra typica: Rahuri, Ahmedanggar = Ahmadnagar, Bombay

Distribution: Eastern Aral Sea to Kirgizia and Tadjikistan
Museum: AMNH, BMNH (holotype), FNSF, NHMW, ZMH

Wing length: 158–170 mm
Tail length: 72–83 mm
Tarsus length: 28–35 mm
Length of bill: 18–19.5 mm
Body mass: 110 g (?)

Illustration: J. G. Keulemans in Dresser 1871–1896: Pl. 691; J. G. Keulemans in Sharpe 1891: Pl. 1; H. Delin in Cramp et al. 1985: Pl. 44; A. Cameron in

Voous 1988: 34; T. Boyer in Boyer and Hume 1991: 36; F. Weick in König et al. 1999: Pl. 11; McQueen in Rasmussen and Anderton 2005: Pl. 78

Photograph: J. P. Smith in internet: Owl Pages, photo gallery 2003

Literature: Hume 1873: 8; Sharpe 1891; Hartert 1912–1921: 977, 978; Baker 1927: 431, 432; Demen-tiev and Gladkov 1951: 411–413; Vaurie 1965: 601, 602; Cramp et al. 1985: 450–454; Voous 1988: 37–40; Boyer and Hume 1991: 36, 37; Shirihai 1996: 310, 311; del Hoyo et al. 1999: 163; König et al. 1999: 229–231; Rasmussen and Anderton 2005: 236

- ***Otus brucei obsoletus*** (Cabanis) 1875
  Scops obsoleta Cabanis 1875, J. Orn. 23: 26;
  Terra typica: Buchara, Syria

Distribution: Southern Turkey, northern Syria, northern Iraq, Turkmenia, Uzbekistan, northern Afghanistan (formerly bred in Israel, but possibly the subspecies *exiguus*?)

Museum: AMNH, BMNH, MNBHU, NHMWien

Wing length:	150–168 mm
Tail length:	68–75 mm
Tarsus length:	29–31 mm
Length of bill:	?
Body mass:	110 g

Illustration: H. Delin in Cramp et al. 1985: Pl. 44; F. Weick in König et al. 1999: Pl. 11

Photograph: Y. Eshbol in Shirihai 1996: Pl. 64?; H. Shirihai in Shirihai 1996: Pl. 64?

Literature: Cramp et al. 1985: 450–454; Shirihai 1996: 310, 311; del Hoyo et al. 1999: 163; König et al. 1999: 229–231

- ***Otus brucei semenowi*** (Sarudny and Härms) 1902
  Scops Semenowi Sarudny and Härms 1902, Orn. Jahrb. 13: 49;
  Terra typica: Pers. Baluchistan

Distribution: Southern Tadjikistan, western China, east to the central Tarim Basin, south to eastern Afghanistan and northern Pakistan

Museum: NHMWien, ZMM

Wing length:	160–169 mm
Tail length:	75–79 mm
Tarsus length:	28–30 mm
Length of bill:	19–20 mm
Body mass:	?

Illustration: D. Cole in Grimmett et al. 1998: Pl. 28; McQueen in Rasmussen and Anderton 2005: Pl. 78

Photograph: ?

Literature: Cramp et al. 1985: 450–454; Grimmett et al. 1998: 430, 431; del Hoyo et al. 1999: 163; Rasmussen and Anderton 2005: 236

- ***Otus brucei exiguus*** Mukherjee 1958
  Otus brucei exiguus Mukherjee 1958, Rec. Indian Mus. 53(1–2) (1955): 301; Terra typica: Baghdad, Iraq

Distribution: Central and eastern Iraq, southern Iran, Oman, southern Afghanistan, western Pakistan, northern Egypt(?). Possibly *exiguus* and not *obsoletus*; bred formerly in Israel, but subspecies status unclear!

Museum: AMNH, BMNH, ChiMNH, MNBHU, ZMTAU

Wing length:	145–163 mm
Tail length:	69–78 mm
Tarsus length:	26–30 mm
Length of bill:	?
Body mass:	100 g

Illustration: P. Hayman in Kemp and Kemp 1998: 317; F. Weick in König et al. 1999: Pl. 11

Photograph: Y. Eshbol in Shirihai 1996: Pl. 64 (see *obsoletus*); H. Shirihai in Shirihai 1996: Pl. 64

Literature: Cramp et al. 1985: 450–454; Shirihai 1996: 310, 311; del Hoyo et al. 1999: 163; König et al. 1999: 229–231

- ***Otus (brucei) pamelae*** Bates 1937
  Otus senegalensis pamelae Bates 1937, Bull. Brit. Ornith. Cl. 57: 150; Terra typica: Dailami, Wadi Bisha, Arabia

Distribution: Southern Saudi Arabia

Remarks: Sometimes placed with *Otus senegalensis* or may even be a distinct species

Museum: AMNH, BMNH, ZMH, ZMK

Wing length:	134–148 mm
Tail length:	60–68 mm
Tarsus length:	25–29 mm
Length of bill:	?
Body mass:	62–71 g

Illustration: ? (painting in del Hoyo not identical with skin!)

Photograph: ? (possibly photo P. Bison in *Birds of Dhofar* 1980: Pl. 15)

Literature: Gallagher and Roger 1980: 369, 370; Cramp et al. 1985: 450–454; König et al. 1999: 231, 232 (*senegal. pamelae*!); del Hoyo et al. 1999: 163 (*senegal. pamelae*!); Hekstra unpublished

- ***Otus (brucei) socotranus*** (Ogilvie-Grant and Forbes) 1899
  Scops socotranus Ogilvie-Grant and Forbes 1899, Bull. Liverp. Mus. 2: 2, 3; Terra typica: Socotra Island

Distribution: Socotra Island
Remarks: Sometimes placed with *Otus senegalensis* or as a distinct species!
Museum: LivCM (holotype), BMNH, USNM

Wing length:	124–135 mm
Tail length:	58–63 mm
Tarsus length:	22–25 mm
Length of bill:	?
Body mass:	64–85 g

Illustration: M. Woodcock in Fry et al. 1988: Pl. 9; F. Weick in König et al. 1999: Pl. 11
Photograph: ?
Literature: Ogilvie-Grant and Forbes 1899: 2, 3; Ripley and Bond 1966: 23, 24; del Hoyo et al. 1999: 163 (*senegalensis socotranus*!); König et al. 1999: 229–231; Hekstra unpublished

Otus senegalensis (Swainson) 1837
African Scops Owl · Afrika-Zwergohreule · Petit-duc Africain · Autillo Africano

Length:	160–190 mm
Body mass:	45–100 g

Distribution: Africa south of the Sahara, including Annobon Island (off coast of Gabon), absent from the forested areas of Mali, Equatorial Guinea, Gabon and Congo, also from the deserts of Namibia
Habitat: Wooded savanna, dry open woodland, park-like habitats, bushveld, acacia scrub and gardens with tall trees. From sea-level up to 2 000 m

- ***Otus senegalensis senegalensis*** (Swainson) 1837
  Scops Capensis A. Smith 1834, S. Afr. Q. J.: 314; Terra typica: South Africa, name valid!
  Scops Senegalensis Swainson 1837, Birds W. Afr. 1: 127; Terra typica: Senegal, but holotype from Gambia!

Otus senegalensis · African Scops Owl · Afrika-Zwergohreule

Synonym:
- *Scops latipennis* Kaup 1853, Contrib. Orn. [Jardine] 1852: 110; Terra typica: Caffaria, Cape Province
- *Ephialtes hendersoni* Cassin 1853, Proc. Acad. Nat. Sci. Phila. 1852: 186; Terra typica: Novo Redondo, Angola
- *Scops pygmea* C.L. Brehm 1855, Vogelfang: 43; Terra typica: Sennar
- *Pisorhina ugandae* Neumann 1899, J. Orn. 47: 56; Terra typica: Kwa Mtessa, Uganda
- *Pisorhina capensis intermedia* Gunning and Roberts 1911, Ann. Transvaal Mus. 3: 111; Terra typica: Pretoria
- *Pisorhina capensis grisea* Gunning and Roberts 1911, ibid.: 111; Terra typica: Orange Free State, South Africa
- *Pisorhina capensis pusilla* Gunning and Roberts 1911, ibid.: 111; Terra typica: Namableda, Boror, East Africa
- *Otus senegalensis caecus* Friedmann 1929, Auk 46: 521; Terra typica: Sadi Malka, Ethiopia
- *Otus senegalensis graueri* Chapin 1930, Am. Mus. Novit. 412: 4; Terra typica: Tanganjika Sea

Distribution: Senegal, Sierra Leone, east to northwestern Ethiopia and Somaliland (except southeast Kenya), south to southeastern South Africa
Museum: AMNH, BMNH, CMNH, FNSF, MHNP, MNBHU, NHMBern, NMKN, RMTB, SMNSt, USNM, YPM, ZMH, ZSBS
Remarks: The synonyms from some subspecies, described on the basis of differences in plumage and/or size, are probably only colour morphs

Wing length:	126–144 mm
Tail length:	54–70 mm
Tarsus length:	20–25 mm
Length of bill:	?
Body mass:	♂ 46–65 g, ♀ 58–100 g

Illustration: J. G. Keulemans in Sharpe 1875a: Pl. 2; M. Woodcock in Fry et al. 1988: Pl. 9; T. Boyer in Boyer and Hume 1991: 41; P. Hayman in Kemp and Kemp 1998: 307; F. Weick in König et al. 1999: Pl. 11; N. Borrow in Borrow and Demey 2001: Pl. 59; J. Gale in Stevenson and Fanshawe 2002: Pl. 100
Photograph: Mc Illeron in Ginn et al. 1989: 335; T. Dressler in Kemp and Kemp 1998: 306; N. P. H. Photo in Kemp and Kemp 1998: 307; B. Marcot in internet: Owl Pages, photo gallery 2002; S. Porter in internet: Owl Pages, photo gallery 2002
Literature: Chapin 1930: 1–11; Fry et al. 1988: 115–117; Ginn et al. 1989: 335; Boyer and Hume 1991: 41, 42; Kemp and Kemp 1998: 306, 307; del Hoyo et al. 1999: 163; König et al. 1999: 231, 232; Borrow and Demey 2001: 493; Stevenson and Fanshawe 2002: 200; Hekstra unpublished

- ***Otus senegalensis feae*** (Salvadori) 1903
Scops feae Salvadori 1903, Mem. R. Accad. Sci. Torino (2), 53: 95; Terra typica: Annobon Island

Distribution: Annobon Island (= Pagalu, in the Gulf of Guinea)
Museum: BMNH, FNSF, MNBHU

Wing length:	120–135 mm
Tail length:	60–65 mm
Tarsus length:	20–23 mm
Length of bill:	?
Body mass:	?

Illustration: F. Weick in König et al. 1999: Pl. 11
Photograph: ?
Literature: Salvadori 1903: 95; Chapin 1930: 1–11; Fry et al. 1988: 115–117; del Hoyo et al. 1999: 163; König et al. 1999: 231, 232

- ***Otus senegalensis nivosus*** Keith and Twomey 1968
Otus senegalensis nivosus Keith and Twomey 1968, Ibis 110: 538; Terra typica: Lali Mountains, southeast Kenya

Distribution: Southeast Kenya (along Tana River and Lali Mountains)
Museum: ANSPh, BMNH, ChiMNH, ZFMK

Wing length:	117–121 mm
Tail length:	53–58 mm
Tarsus length:	19–23 mm
Length of bill:	?
Body mass:	?

Illustration: F. Weick in König et al. 1999: Pl. 11
Photograph: ?
Literature: Keith and Twomey 1968: 538; Fry et al. 1988: 115–117; Zimmerman et al. 1996: 442, 443; del Hoyo et al. 1999: 163; König et al. 1999: 231, 232; Hekstra unpublished

❖

Otus sunia (Hodgson) 1836
Oriental Scops Owl · Orient-Zwergohreule · Petit-duc d'Orient · Autillo Oriental

Length:	170–210 mm
Body mass:	75–95 g

Distribution: Northern Pakistan (Punjab), India, Nepal, east to Bangladesh and Assam, Sri Lanka, East Asia, from Japan (Hokkaido and Kijushu), eastern Siberia, Manchuria, Taiwan, eastern China to Malay Peninsula (Malacca). Andaman Island. Nicobar Island and vagrant to Hong Kong and Aleutian Island

Habitat: Deciduous, mixed and open evergreen forest, riparian woodland, parks, orchards and cultivated land, also near settlements. From sea-level up to 1 500 m, in the Himalayas up to 2 300 m

- ***Otus sunia sunia*** (Hodgson) 1836
  Scops sunia Hodgson 1836, As. Res. 19: 175; Terra typica: Nepal

Remarks: Replaces *Scops pennatus* Hodgson 1837 of Sharpe's Handlist from 1899!

Distribution: Northern Pakistan, east to Bangladesh and northern India
Museum: AMNH, BMNH, FNSF, MHNP, MNBHU, NHMWien, ZMH, ZSBS

Wing length:	135–154 mm
Tail length:	61–71 mm
Tarsus length:	20–24 mm
Length of bill:	16–18 mm
Body mass:	75–95 g

Illustration: D. Watson in Ali and Ripley 1969: Pl. 43; T. Boyer in Boyer and Hume 1991: 39; D. Cole in Grimmett et al. 1998: Pl. 28; F. Weick in König et al. 1999: Pl. 12
Photograph: ?
Literature: Sharpe 1875a: 53, 54; Baker 1927: 435, 436; Dementiev and Gladkov 1951: 419–421; Ali and Ripley 1969: 262, 263; Voous 1988: 33–36; Boyer and Hume 1991: 39, 40; Grimmett et al. 1998: 431, 432; del Hoyo et al. 1999: 164; König et al. 1999: 232, 233; Hekstra unpublished

- ***Otus sunia rufipennis*** (Sharpe) 1875
  Scops rufipennis Sharpe 1875, Cat. Birds Brit. Mus. 2: 60; Terra typica: Eastern Ghats, Madras

Distribution: Southern India
Museum: BMNH (holotype), BNHS, MHNP, MZUS,

Wing length:	122–140 mm
Tail length:	52–68 mm
Tarsus length:	21–23 mm
Length of bill:	17 and 18 mm
Body mass:	?

Illustration: J. G. Keulemans in Sharpe 1875a: Pl. 4 (*Scops gymnopodes*); D. Watson in Ali and Ripley 1969: Pl. 43; F. Weick in König et al. 1999: Pl. 12; McQueen in Rasmussen and Anderton 2005: Pl. 78
Photograph: ?
Literature: Sharpe 1875a: 60 and 65; Baker 1927: 434; Ali and Ripley 1969: 263; Voous 1988: 33–36; Grimmett et al. 1998: 431; del Hoyo et al. 1999: 164; König et al. 1999: 232, 233; Hekstra unpublished; Rasmussen and Anderton 2005: 236, 237

- ***Otus sunia leggei*** Ticehurst 1923
  Scops minutus Legge 1878, Ann. Mag. Nat. Hist. 5(1): 175; Terra typica: Kotmalie, Ceylon (invalid name)
  Otus sunia leggei Ticehurst 1923, Ibis: 242; Terra typica: Ceylon (new name for *Scops minutus*!) (see Figure 4)

Distribution: Sri Lanka
Museum: BMNH, MZUS

Wing length:	119–127 mm
Tail length:	49–54 mm
Tarsus length:	20–21 mm
Length of bill:	17 and 18 mm
Body mass:	?

Illustration: J. G. Keulemans in Legge 1878a: 1, Pl. 4; G. M. Henry in Henry 1955: Pl. 16; F. Weick in König et al. 1999: Pl. 12; McQueen in Rasmussen and Anderton 2005: Pl. 78
Photograph: ?
Literature: Legge 1878b: 175; Baker 1927: 434, 435; Henry 1955: 197, 198; Ali and Ripley 1969: 264, 265; del Hoyo et al. 1999: 164; König et al. 1999: 232, 233; Rasmussen and Anderton 2005: 237; Hekstra unpublished

- ***Otus sunia modestus*** (Walden) 1874
  Scops modestus Walden 1874, Ann. Mag. Nat. Hist. (4), 13: 123

Synonym:
- *Ephialtes nicobaricus* Hume 1876, Str. Feath. 4: 283; Terra typica: Camorta Island, Nicobars
- *Otus sunia distans* Friedmann and Deignan 1941, Journ. Wash. Acad. Sci 29: 287; Terra typica: Chiang Mai, northern Thailand
- *Otus sunia khasiensis* Koelz 1954, Contrib. Inst. Reg. Expl. 1: 22; Terra typica: Khasi Mountains, Assam

Distribution: Assam, south to Brahmaputra River, Myanmar, northwestern Thailand, Indochina and Andaman Island, also central Nicobar Island (Camorta)
Museum: AMNH, BMNH, UM(ich)MZ, USNM (holotype *distans*)

Wing length:	137–145 mm
Tail length:	50–68 mm
Tarsus length:	22–25 mm
Length of bill:	17–19.5 mm
Body mass:	?

Illustration: W. Monkol in Lekagul and Round 1991: Pl. 60; D. Cole in Grimmett et al. 1998: Pl. 28; H. Burn in Robson 2000: Pl. 21; McQueen in Rasmussen and Anderton 2005: Pl. 78
Photograph: N. Sivasothi in internet: Owl Pages, photo gallery 2003
Literature: Walden 1874: 123; Baker 1927: 437; Delacour and Jabouille 1931: 129, 130; Smythies and Hughes 1984: 318, 319; Friedmann and Deignan 1941: 287; Ali and Ripley 1969: 265, 266; Lekagul and Round 1991: 178; Grimmett et al. 1998: 431, 432; del Hoyo et al. 1999: 164; König et al. 1999: 232, 233; Rasmussen and Anderton 2005: 237; Hekstra unpublished

- ***Otus sunia malayanus*** (Hay) 1847
  Scops malayanus Hay 1847, Madras J. Lit. Sci. 13(2): 147;
  Terra typica: Malacca

Distribution: Southern China (western Yunnan and Guangdong), south to Malay Peninsula
Museum: AMNH, BMNH, MNBHU, NNML, USNM, ZFMK

Wing length: (134) 140–140 mm
Tail length: (56) 61–66 mm
Tarsus length: 20–23 mm
Length of bill: 17 and 18 mm
Body mass: ?

Illustration: J. G. Keulemans in Sharpe 1875a: Pl. 4; K. Phillipps in McKinnon and Phillipps 1993: Pl. 21; F. Weick in König et al. 1999: Pl. 12; H. Burn in Robson 2000: Pl. 21
Photograph: H. H. Chew in *SCRO Mag.* 1999: 54; N. Sivasothi in internet: Owl Pages, photo gallery 2003
Literature: Hay (= of Tweeddale) 1847: 147; Sharpe 1875a: 60 and 65–67; Baker 1927: 437, 438; del Hoyo et al. 1999: 164; König et al. 1999: 232, 233

- ***Otus sunia stictonotus*** (Sharpe) 1875
  Scops stictonotus Sharpe 1875, Cat. Birds Brit. Mus. 2: 54;
  Terra typica: China

Distribution: Southeast Siberia, Sachalin, northeast China, northern Korea. Winter in southeast China and Taiwan
Museum: AMNH, BMNH, MNBHU, SMNKa, SMNSt, SMTD, USNM, YPM, ZFMK, ZMH

Wing length: 138–158 mm
Tail length: 65–72 mm
Tarsus length: 22–24 mm
Length of bill: 17–21.5 mm
Body mass: ♂ 75–86 g, ♀ 78–95 g

Illustration: J. G. Keulemans in Sharpe 1875a: Pl. 3; A. Cameron in Voous 1988: 34; F. Weick in König et al. 1999: Pl. 12
Remarks: More coarsely and boldly patterned below than in *japonicus*
Photograph: Pukinski in Pukinski 1975: 51, 52 (b/w)
Literature: Sharpe 1875a: 54; Hartert 1912–1921: 983; Hartert 1923: 387, 388; Dementiev and Gladkov 1951: 419–421; Vaurie 1965: 600; Voous 1988: 33–36; del Hoyo et al. 1999: 164; König et al. 1999: 232, 233

- ***Otus sunia japonicus*** Temminck and Schlegel 1850 (1844?)
  Otus sunia japonicus Temminck and Schlegel 1850 (1844?), in Siebold's Fauna Japonica, Aves: 27, Pl. 9;
  Terra typica: Japan

Distribution: Japan
Museum: AMNH, BMNH, MNBHU, NNML, SMNKa, ÜMB, ZFMK
Remarks: *O. s. stictonotus*, somtimes included with *japonicus*

Wing length: 140–148 mm
Tail length: 65–75 mm
Tarsus length: 21–24 mm
Length of bill: 16.5–21 mm
Body mass: ?

Illustration: J. Wolf in Temminck and Schlegel 1850 (1844?): Pl. 9; S. Takano in Wild Bird Soc. Japan 1983: 193; F. Weick in König et al. 1999: Pl. 12
Photograph: ? in Takano et al. 1991 (4 colour photographs)
Literature: Temminck and Schlegel 1850 (1844?): 27; Hartert 1912–1921: 983; Hartert 1923: 387; Vaurie 1965: 601; Wild Bird Soc. Japan 1983: 192; del Hoyo et al. 1999: 164; König et al. 1999: 232, 233

Otus elegans (Cassin) 1852
Ryukyu Scops Owl · Schmuck-Zwergohreule · Petit-duc élégant · Autillo Elegante

Length: 200 mm
Body mass: 100–107 g

Distribution: Ryukyu Island, southern Japan, Lanyu Island (off southeast Taiwan), also Batan, Calayan and probably other small Islands, north of Luzon, Philippines
Habitat: Subtropical, dense evergreen forests, also near or in urban settlements. From sea-level up to 550 m

- ***Otus elegans elegans*** (Cassin) 1852
  Ephialtes elegans Cassin 1852, Proc. Acad. Nat. Sci. Phila. 6: 185; Terra typica: At sea off the coast of Japan

Synonym:
- *Ephialtes japonicus interpositus* Kuroda 1923, Bull. Brit. Ornith. Cl. 43: 122; Terra typica: Minami-Daito-jima = Daito Island

Distribution: Ryukyu Islands (Nasei, Shoto), southern Japan
Museum: AMNH, BMNH, FNSF, LSUMZ, MNBHU, NHRMSt, SMNSt,

Wing length: 165–178 mm
Tail length: 75–81 mm
Tarsus length: 27–30 mm
Length of bill: 20.5–23.5 mm
Body mass: 100–107 g

Illustration: F. Weick in König et al. 1999: Pl. 12
Photograph: ?
Literature: Cassin 1852: 186; Kuroda 1823: 122; del Hoyo et al. 1999: 167; König et al. 1999: 233, 234; Hekstra unpublished

- ***Otus elegans botelensis*** Kuroda 1928
  Otus sunia botelensis Kuroda 1928, Tori 5: 25, 26; Terra typica: Botal Tobago

Distribution: Lanyu Island, off southeastern Taiwan
Museum: NSMT?
Remarks: No data! Tori 5 also not seen!

Wing length: ?
Tail length: ?
Tarsus length: ?
Length of bill: ?
Body mass: ?

Illustration: Not painted in König et al. because plumage identical to nominate in both morphs!
Photograph: L. Liu Severinghaus in letter (2 colour photograph); Wayne Hsu in internet: Owl Pages, photo gallery 2002
Literature: Kuroda 1928: 26; Severinghaus 1986: 143–196; Severinghaus 1989: 423–429; del Hoyo et al. 1999: 233, 234

- ***Otus elegans calayensis*** McGregor 1904
  Otus calayensis McGregor 1904, Bull. Philip. Mus. 4: 18; Terra typica: Calayan Island

Synonym:
- *Otus bakkamoena batanensis* Manuel and Gilliard 1952, Am. Mus. Novit. 1545: 4; Terra typica: Basco, Batan Island

Distribution: Batan Island, Sabtang Island and Calayan Island, off northern Philippines
Museum: ChiMNH, FMNH, (syntype = MLA), AMNH, ChiMNH, PNM (holotype *batanensis*)

Wing length: 167–174 mm
Tail length: 81–84 mm
Tarsus length: 28–29 mm
Length of bill: 21 and 22 mm
Body mass: ?

Illustration: A. Sutherland in Kennedy et al. 2000: Pl. 34
Photograph: ?
Literature: McGregor 1904: 18; Dickinson et al. 1991: 256, 257; del Hoyo et al. 1999: 167; König et al. 1999: 233, 234; Kennedy et al. 2000: 176, 177; Hekstra unpublished

Otus magicus (S. Müller) 1841
Moluccan Scops Owl · Molukken-Zwergohreule · Petit-duc mystérieux · Autillo Moluqueno

Length: 230–250 mm
Body mass: 114–165 g

Distribution: Lesser Sunda Islands (Lombok, Sumbawa, Flores, Lomblen, Wetar), Moluccas (Morotai, Halmahera, Ternate, Bacan, Obi, Buru, Ambon, Seram, Kasiruta)
Habitat: Lowlands, near coast, primary and secondary forest, also coastal swamp forest. Plantations, farming areas, up to 900 m, sometimes to 1 500 m above sea-level

- ***Otus magicus morotensis*** (Sharpe) 1875
Scops morotensis Sharpe 1875, Cat. Birds Brit. Mus. 2: 75 and Pl. 7; **Terra typica:** Morotai Island, Moluccas

Distribution: Morotai Island, Ternate Island, northern Moluccas
Museum: AMNH, BMNH, FNSF, NNML, SMTD, ZMK
Remarks: Plumage very similar to that of *O. m. leucospilus* thus possibly only a synonym

Wing length: 167–188 mm
Tail length: 82–96 mm
Tarsus length: 27–31 mm
Length of bill: 23–25 mm
Body mass: ?

Illustration: J. G. Keulemans in Sharpe 1975: Pl. 7
Photograph: ?
Literature: Sharpe 1875a: 75; Finsch 1898/1899: 163 and 184, Pl. 9 and 10; White and Bruce 1986: 250–253; Coates and Bishop 1997: 360, 361; del Hoyo et al. 1999: 165; König et al. 1999: 236, 237; Hekstra unpublished

- ***Otus magicus leucospilus*** (G.R. Gray) 1861
Ephialtes leucospila G.R. Gray 1861, Proc. Zool. Soc. Lond. 1860: 344; **Terra typica:** Batjan and Halmahera, Moluccas

Distribution: Halmahera, Kasiruta and Bacan
Museum: BMNH, MNBHU, NNML, SMTD
Remarks: Together with *morotensis* probably synonyms of the nominate *magicus*

Wing length: 160–186 mm
Tail length: 78–93 mm
Tarsus length: 27–34 mm
Length of bill: 23–25 mm
Body mass: ?

Illustration: J. G. Keulemans in Sharpe 1875a: Pl. 6
Photograph: F. Lambert in del Hoyo et al. 1999: 87
Literature: Sharpe 1875a: 72, 73; Finsch 1898/1899: 163–184, Pl. 9 and 10; White and Bruce 1986: 250–253; Coates and Bishop 1997: 360, 361; del Hoyo et al. 1999: 165; König et al. 1999: 236, 237; Hekstra unpublished

- ***Otus magicus obira*** Jany 1955
Otus obira Jany 1955, J. Orn. 96: 106;
Terra typica: Madopolo auf Bisa (Kep Obi)

Distribution: Obi Islands, Moluccas
Museum: MNBHU (holotype), NNML (paratype)
Remarks: Possibly synonym; included in nominate *magicus*

Wing length: 168–174 mm
Tail length: 82 mm
Tarsus length: 32 mm
Length of bill: ?
Body mass: 128–148 g

Illustration: ?
Photograph: ?
Literature: Jany 1955: 106; White and Bruce 1986: 250–253; Coates and Bishop 1997: 360, 361; del Hoyo et al. 1999: 165; König et al. 1999: 236, 237

- ***Otus magicus magicus*** (S. Müller) 1841
Strix magica S. Müller 1841, Verh. Nat. Gesch. Nederl., Land- en Volkenk. 4: 110; **Terra typica:** Amboina

Distribution: Seram, Ambon, Moluccas, Aru Island?
Museum: NNML (holotype), BMNH, MNBHU, MZUS, SMNSt, ÜMB, ZMA(ms)
Remarks: Two specimens collected by Rosenberg, NNML, probably belong to *O. m. magicus*, also one in the SMNSt?

Wing length: 172–186 mm (Seram), 175–197 mm (Ambon)
Tail length: 81–91 mm
Tarsus length: 30–33 mm
Length of bill: 23–25 mm
Body mass: 165 g

Illustration: J. G. Keulemans in Sharpe 1875a: Pl. 5; D. Gardner in Coates and Bishop 1997: Pl. 34; F. Weick in König et al. 1999: Pl. 13

Photograph: ?
Literature: Sharpe 1875a: 70, 71; Finsch 1898/1899: 163–185, Pl. 9, 10; White and Bruce 1986: 250–253; Coates and Bishop 1997: 360, 361; del Hoyo et al. 1999: 165; König et al. 1999: 236, 237

- ***Otus magicus bouruensis*** (Sharpe) 1875
  Scops bouruensis Sharpe 1875, Cat. Birds Brit. Mus. 2: 73; Terra typica: Buru Island, Moluccas

Distribution: Buru Island, South Moluccas
Museum: BMNH (holotype), NNML
Remarks: Possibly synonym to nominate *magicus*!

Wing length:	172–192 mm
Tail length:	87–95 mm
Tarsus length:	31 and 32 mm
Length of bill:	24–26 mm
Body mass:	?

Illustration: J. G. Keulemans in Sharpe 1875a: Pl. 7; F. Weick in König et al. 1999: Pl. 13
Photograph: ?
Literature: Sharpe 1875a: 73; Finsch 1898/1899: 163–184, Pl. 9, 10; White and Bruce 1986: 250–253; Coates and Bishop 1997: 360, 361; del Hoyo et al. 1999: 165; König et al. 1999: 236, 237

- ***Otus magicus albiventris*** (Sharpe) 1875
  Scops albiventris Sharpe 1875, Cat. Birds Brit. Mus. 2: 78; Terra typica: Flores, Lesser Sundas (see Figure 3)

Distribution: Lombok, Sumbawa, Flores, Lomblen, Lesser Sunda Islands
Museum: AMNH, BMNH, FNSF, MNBHU, NNML, SMTD

Wing length:	153–166 mm
Tail length:	71–84 mm
Tarsus length:	23–27 mm
Length of bill:	20.5–23.5 mm
Body mass:	?

Illustration: J. G. Keulemans in Sharpe 1875a: Pl. 8; D. Gardner in Coates and Bishop 1997: Pl. 34; F. Weick in König et al. 1999: Pl. 13; F. Weick in *Gefiederte Welt* 2004(1): 19
Photograph: ?
Literature: Sharpe 1875a: 78; Finsch 1898/1899: 163–184, Pl. 9 and 10; White and Bruce 1986: 250–253; Coates and Bishop 1997: 360, 361; del Hoyo et al. 1999: 165; König et al. 1999: 236, 237

- ***Otus (magicus) tempestatis*** (Hartert) 1904
  Pisorhina manadensis tempestatis Hartert 1904, Novit. Zool. 11: 190; Terra typica: Wetar Island, Lesser Sundas

Distribution: Wetar Island, Lesser Sunda Islands
Remarks: Often confused, and placed together as a subspecies, with *O. manadensis*!
Museum: BMNH, MNBHU, NNML

Wing length:	150–171 mm
Tail length:	70–75 (82) mm
Tarsus length:	23–27 mm
Length of bill:	19.7–22.3 mm
Body mass:	?

Illustration: S. McQueen in Rasmussen 1998: Pl. 3; F. Weick in unpublished (2nd edn. Owls): Pl. 67
Photograph: ?
Literature: Hartert 1904: 190; White and Bruce 1986: 250–253; Rasmussen 1998: 141–152; del Hoyo et al. 1999: 165; König et al. 1999: 236, 237

- ***Otus (magicus) sulaensis*** (Hartert) 1898
  Pisorhina sulaensis Hartert 1898, Novit. Zool. 5: 126; Terra typica: Sula Mangoli

Distribution: Sula Islands (Taliabu, Scho, Mangole, Sanana)
Remarks: Sometimes, like *tempestatis*, as a subspecies included in *Otus manadensis*!
Museum: AMNH (Rothsch.), NNML

Wing length:	161–175 mm
Tail length:	73–77 mm
Tarsus length:	27–30 mm
Length of bill:	23–26 mm
Body mass:	?

Illustration: S. Mc Queen in Rasmussen 1998: Pl. 3; F. Weick in König et al. 1999: Pl. 13
Photograph: ?
Literature: Hartert 1898a: 126; White and Bruce 1986: 250–253; Rasmussen 1998: 141–152; del Hoyo et al. 1999: 167; König et al. 1999: 236, 237

Otus beccarii **(Salvadori) 1876**
Biak Scops Owl · Beccari-Zwergohreule · Petit-duc de Beccari · Autillo de Biak
Scops beccarii **Salvadori 1876, Ann. Mus. Civ. Genova 7(1875): 904; Terra typica: Biak Island (= Misori Island)**

Length: 230 mm
Body mass: ?

Distribution: Biak Island (Geelvink Bay, off northwest New Guinea)
Habitat: Densely wooded areas, forest and forest edges, locally near urban settlements
Museum: ANSPh, MGD (fide AMNH C. Vidani)
Remarks: Often treated as conspecific with either *O. manadensis* or *O. magicus*, but significantly different from both in plumage

Wing length: 170–172 mm
Tail length: (79) 81–84 mm
Tarsus length: 33–35 mm
Length of bill: 22.5 mm
Body mass: ?

Illustration: F. Weick in König et al. 1999: Pl. 13
Photograph: Only from skin (MGD), collection Weick
Literature: Salvadori 1876: 904; Mayr and Meyer de Schauensee 1939: 24, 25; Iredale 1956: 140; Rand and Gilliard 1967: 253, 254; del Hoyo et al. 1999: 167, 168; König et al. 1999: 237, 238; Hekstra unpublished

Otus mantananensis **(Sharpe) 1892**
Mantanani Scops Owl · Philippinen-Zwergohreule · Petit-duc de Mantanani · Autillo de la Mantanani

Length: 180–200 mm
Body mass: 106–110 g

Distribution: Mantanani Island off northern Borneo, west-central and southwest Philippines
Habitat: Lowlands and foothills, forest, woodland, coconut groves. Hunting along forest edges and clearings
Remarks: Sometimes treated as conspecific with *O. scops* or *O. sunia*, but different in vocalisations

▪ ***Otus mantananensis romblonis*** McGregor 1905
Otus romblonis McGregor 1905, Bureau Govt. Labs. 25:12; Terra typica: Romblon Island

Distribution: Banton, Sibuyan, Romblon, Tablas, Tres Reyes, Semirara in west-central Philippines
Museum: PNM, MLA (destroyed?), DMNH?

Wing length: 158 mm
Tail length: 76 mm
Tarsus length: 29 mm
Length of bill: ?
Body mass: ?

Illustration: ?
Photograph: ?
Literature: McGregor 1905: 17; Dickinson et al. 1991: 225; del Hoyo et al. 1999: 165; König et al. 1999: 235

▪ ***Otus mantananensis cuyensis*** McGregor 1904
Otus cuyensis McGregor 1904, Bull. Philip. Mus. 4: 17; Terra typica: Cuyo Island

Distribution: Cuyo Island (South Calamian Island) in west-central Philippines
Museum: ChiMNH, FMNH (formery in MLA), USNM

Wing length: 175–180 mm
Tail length: 84–88 mm
Tarsus length: 32–35 mm
Length of bill: 23–25 mm
Body mass: ?

Illustration: A. Sutherland in Kennedy et al. 2000: Pl. 34
Photograph: ?
Literature: McGregor 1904: 17; Dickinson et al. 1991: 225; del Hoyo et al. 1999: 165; König et al. 1999: 235; Kennedy et al. 2000: 177; Hekstra unpublished

▪ ***Otus mantananensis mantananensis*** (Sharpe) 1892
Scops mantananensis Sharpe 1892, Bull. Brit. Ornith. Cl. 1: 4; Terra typica: Mantanani Island

Distribution: Mantanani Island off Borneo and Ursula Island off southern Palawan
Museum: BMNH, NNML

Wing length: 161–166 mm
Tail length: 81 mm
Tarsus length: 30 mm
Length of bill: ?
Body mass: 106 g

Illustration: K. Phillipps in McKinnon and Phillipps 1993: Pl. 40; F. Weick in König et al. 1999: Pl. 13
Photograph: ?
Literature: Sharpe 1892: 4; Dickinson et al. 1991: 225; König et al. 1999: 235; Kennedy et al. 2000: 177

- ***Otus mantananensis sibutuensis*** (Sharpe 1893)
Scops sibutuensis Sharpe 1893, Bull. Brit. Ornith. Cl. 3: 9; Terra typica: Sibutu Island, Philippines

Synonym:
- *Otus steerei* Mearns 1909, Proc. US Nat. Mus. 36: 437; Terra typica: Tumindao Island

Distribution: Sibutu and Tumindao Islands, southwestern Sulu Islands
Museum: AMNH (Rothsch.), BMNH (holotype *sibutuensis*), DelMNH (Rabor), USNM (holotype *steerei*)

Wing length: 152–156 mm
Tail length: 76–83 mm
Tarsus length: 27–32 mm
Length of bill: 22.5–23.5 mm
Body mass: ?

Illustration: F. Weick in König et al. 1999: Pl. 13
Photograph: ?
Literature: Sharpe 1893: 9; Dickinson et al. 1991: 225; del Hoyo et al. 1999: 165; König et al. 1999: 235; Kennedy et al. 2000: 177; Hekstra unpublished

❖

Otus manadensis (Quoy and Gaimard) 1830
Sulawesi Scops Owl · Manado-Zwergohreule · Petit-duc de Manado · Autillo de Célebes

Length: 190–220 mm
Body mass: 83–93 g

Distribution: Sulawesi, Siau Island, Banggai Islands (Peleng and Labobo), Tukangbesi Islands (Kalidupa)
Habitat: Humid forest (clearings and edges) with high annual rainfall and temperatures. From lowland up to 2 500 m
Remarks: Formerly sometimes treated as conspecific with *O. magicus*

- ***Otus (manadensis) siaoensis*** (Schlegel) 1873
Scops siaoensis Schlegel 1873, Mus. Pays-Bas, 2, Noctuae Rev.: 13; Terra typica: Siao Island
(see Figure 8)

Distribution: Siau Island, north of Sulawesi
Museum: NNML = RMNH (holotype)
Remarks: Known from one skin, perhaps specifically distinct! Measurements of Hekstra and of Lambert and Rasmussen are different!

Wing length: 125 mm (127 mm)
Tail length: 55 mm (57 mm)
Tarsus length: 23 mm (26.8 mm)
Length of bill: 19.9 mm
Body mass: ?

Illustration: C. Anderton in Lambert and Rasmussen 1998: Pl. 4; F. Weick in König et al. 1999: Pl. 14; F. Weick in *Gefiederte Welt* 2004(4): 119
Photograph: ?
Literature: Schlegel 1873: 13; Sharpe 1875a: 78; White and Bruce 1986: 250–253; del Hoyo et al. 1999: 167; König et al. 1999: 239; Weick 2003–2005, *Gefiederte Welt* 2004: 119, 120; Hekstra unpublished

- ***Otus manadensis manadensis*** (Quoy and Gaimard) 1830
Scops manadensis Quoy and Gaimard 1830, Voyage *"Astrolabe"* Zool. 1: 170; Terra typica: Mandano, Celebes
(see Figure 8)

Synonym:
- *Otus manadensis obsti* Eck 1973, Zool. Abh. SMTD 32: 158; Terra typica: Java?, probably a mislabelled *manadensis*?

Distribution: Sulawesi
Museum: AMNH, BMNH, MHNP, MNBHU, MZUS, NHMBas, NHMWien, NNML, SMNSt, SMTD, USNM, ZMH, ZMAms
Remarks: All measurements in brackets are from Minahassa skins!

Wing length: 140–161 mm
(166–190 mm Minahassa)
Tail length: 66–76 mm (–86 mm)
Tarsus length: 24–26 mm (28 mm)
Length of bill: 18–25 mm
Body mass: 86 g

Illustration: ? in Quoy and Gaimard 1830: Atlas Ois. Pl. 2 (b/w); J. G. Keulemans in Sharpe 1875a: Pl. 8; H. Quintscher in Eck 1973: Abb. 1 (b/w); J. Marshall jr. in Amadon and Bull 1988: col. Pl.; T. Boyer in Boyer and Hume 1991: 45; D. Gardner in Coates and Bishop 1997: Pl. 37; J. C. Anderton in Lambert and Rasmussen 1998: Pl. 4; F. Weick in König et al. 1999: Pl. 14; F. Weick in *Gefiederte Welt* 2004(4): 119
Photograph: F. Lambert in del Hoyo et al. 1999: 87
Literature: Sharpe 1875a: 76, 77; Eck 1973: 158; White and Bruce 1986: 250–253; Coates and Bishop 1997: 359, 360; Boyer and Hume 1991: 45; del Hoyo et al. 1999: 167; König et al. 1999: 239; Weick 2003–2005, *Gefiederte Welt* 2004: 119, 120; Hekstra unpublished

- ***Otus manadensis mendeni*** Neumann 1939
Otus mendeni Neumann 1939, Bull. Brit. Ornith. Cl. 59: 106; Terra typica: Peleng Island

Distribution: Banggai Islands, Peleng Island and possibly Labobo
Museum: MCZ (Menden, grey morph), SMTD (red morph)

Wing length: 142–150 mm
Tail length: 64–66 mm
Tarsus length: 25.5–28 mm
Length of bill: 19.5–20.5 mm
Body mass: ?

Illustration: H. Quintscher in *Der Falke* 1975(10): Frontcover; F. Weick in König et al. 1999: Pl. 14
Photograph: ?
Literature: Neumann 1939: 106; White and Bruce 1986: 250–253; del Hoyo et al. 1999: 167; König et al. 1999: 239; Hekstra unpublished

- ***Otus (manadensis) kalidupae*** (Hartert) 1903
Pisorhina manadensis kalidupae Hartert 1903, Novit. Zool. 10: 21; Terra typica: Kalidupa Island

Distribution: Kalidupa Island, Tukangbesi Islands
Museum: AMNH (Rothsch.)
Remarks: Perhaps specifically distinct, but little known

Wing length: 168–176 mm
Tail length: (82) 84–89 mm
Tarsus length: 28–30 (33) mm
Length of bill: 23.5–24 mm
Body mass: ?

Illustration: S. McQueen in Rasmussen 1998: Pl. 3
Photograph: ?
Literature: Hartert 1903: 21; White and Bruce 1986: 250–253; Coates and Bishop 1997: 360, 361; Lambert and Rasmussen 1998: 204–217; del Hoyo et al. 1999: 167; König et al. 1999: 239

❖

Otus collari **Lambert and Rasmussen 1998**
Sangihe Scops Owl · Sangihe-Zwergohreule · Petit-duc de Sangihe · Autillo de la Sangihe
Otus collari **Lambert and Rasmussen 1998, Bull. Brit. Ornith. Cl. 118(4): 204–217; Terra typica: Sangihe Island (see Figure 8)**

Length: 190–200 mm
Body mass: 76 g

Distribution: Sangihe Island, north of Sulawesi
Habitat: Forest, mixed plantations, secondary growth, agricultural country with trees and bushes. Up to 315 m
Museum: SNMB (holotype), NNML, SMTD?
Remarks: Museum specimens collected 1866 and 1867. Often regarded as *O. manadensis* or *O. magicus*, but morphological and vocal differences support treatment as a distinct species

Wing length: 158–166 mm
Tail length: 72–79 mm
Tarsus length: 26–28 mm
Length of bill: 19–21 mm
Body mass: 76 g

Illustration: J. C. Anderton in Lambert and Rasmussen 1998: Pl. 4; F. Weick in König et al. 1999: Pl. 14; F. Weick in *Gefiederte Welt* 2004(4): 119
Photograph: F. Lambert in Lambert and Rasmussen 1998: Pl. 5; F. Lambert in del Hoyo et al. 1999: 85
Literature: Lambert and Rasmussen 1998: 204–217; del Hoyo et al. 1999: 167; König et al. 1999: 240; Weick 2003–2005, *Gefiederte Welt* 2004(4): 119, 120

Otus insularis (Tristram) 1880
Seychelles Scops Owl · Seychelleneule · Petit-duc scieur · Autillo de Seychelles
Gymnoscops insularis Tristram 1880, Ibis: 458;
Terra typica: Mahé Island, Seychelles
(see Plate 2)

Length: 200 mm
Body mass: ?

Distribution: Mahé Island, unconfirmed reports from Prasli north Island and Félicité Island, Seychelles
Habitat: Secondary forest, usually close to water, at 250–600 m from sea-level. First nesting record by Fanchett et al. 2000, *Ibis*: 485, 486
Museum: BMNH, CUM, MHNG, MNBHU

Wing length: 162–173 mm
Tail length: 69–82 mm
Tarsus length: 32.5–35 mm
Length of bill: 21.5–25 mm
Body mass: ?

Illustration: J. G. Keulemans in *Ibis* 1880: Pl. 14; T. Boyer in Boyer and Hume 1991: 49; I. Lewington in Kemp and Kemp 1998: 325; P. Hayman in Kemp and Kemp 1998: 325; F. Weick in König et al. 1999: Pl. 12
Photograph: E. Hosking in Hosking and Flegg 1982: 152; ? in Bullock 1990: Birds of the Republic of the Seychelles: 31; Jones and Watson in Kemp and Kemp 1998: 325; E. and D. Hosking in del Hoyo et al. 1999: 146
Literature: Penny 1974; Safford 1993: 57–74; Boyer and Hume 1991: 49; Collar et al. 1994; Kemp and Kemp 1998: 324, 325; del Hoyo et al. 1999: 168; König et al. 1999: 238; Fanchett et al. 2000, *Ibis*: 485, 486

Otus alius Rasmussen 1998
Nicobar Scops Owl · Nicobaren-Zwergohreule · Petit-duc de Grande Nicobar · Autillo de Nicobar
Otus alius Rasmussen 1998, Bull. Brit. Ornith. Cl. 118(3): 141–152; Terra typica: Cambell Bay, Great Nicobar
(see Figure 7)

Length: 190–200 mm
Body mass: ?

Distribution: Great Nicobar Island – southernmost Nicobar Island
Habitat: Wooded areas near sea-level
Museum: BNHS (holotype and paratype)

Wing length: 160–167 mm
Tail length: 74–78 mm
Tarsus length: 28–30 mm
Length of bill: 21.5–22 mm
Body mass: ?

Illustration: S. McQueen in Rasmussen 1998: Pl. 3; F. Weick in König et al. 1999: Pl. 7; F. Weick in *Gefiederte Welt* 2004(4): 120; McQueen in Rasmussen and Anderton 2005: Pl. 78)
Photograph: ?
Literature: Abdulali 1967: 139–190; Abdulali 1978: 749–772; Rasmussen 1998: 141–152; del Hoyo et al. 1999: 168, 169; König et al. 1999: 241; Weick 2003–2005, *Gefiederte Welt* 2004(4): 120; Rasmussen and Anderton 2005: 237

Otus umbra Richmond 1903
Simeulue Scops Owl · Simeulue-Zwergohreule · Petit-duc de Simalur · Autillo de la Simeulue
Pisorhina umbra Richmond 1903, Proc. US Nat. Mus. 26: 494;
Terra typica: Simalur Island

Length: 160–180 mm
Body mass: 95 g

Distribution: Simalur Island, off northwestern Sumatra
Habitat: Coastal forest and forest edges, plantations
Museum: USNM (holotype), AMNH

Wing length: 142.5–145.5 mm
Tail length: 59.5–61 mm
Tarsus length: 24–26 mm
Length of bill: 20–21.5 mm
Body mass: 95 g

Illustration: J. Marshall jr. in Amadon and Bull 1988: col. Pl.; T. Boyer in Boyer and Hume 1991: 40; K. Phillipps in McKinnon and Phillipps 1993: Pl. 40; F. Weick in König et al. 1999: Pl. 14; F. Weick in *Gefiederte Welt* 2004(4): 120
Photograph: ?

Literature: Richmond 1903: 494; Boyer and Hume 1991: 40; McKinnon and Phillipps 1993: 191; Rasmussen 1998: 141–152; del Hoyo et al. 1999: 168; König et al. 1999: 241, 242; Weick 2003–2005, *Gefiederte Welt* 2004(4): 120; Hekstra unpublished

❖

Otus enganensis Riley 1927
Enggano Scops Owl · Enggano-Zwergohreule · Petit-duc d'Enggano · Autillo de la Enggano
Otus umbra enganensis Riley 1927, Proc. Biol. Soc. Wash. 40: 93;
Terra typica: Enggano
(see Figure 7)

Length: 180–200 mm
Body mass: ?

Distribution: Enggano Island, off southwestern Sumatra
Habitat: Wooded areas, forest edges
Museum: NNML, USNM
Remarks: Often regarded as a subspecies of *Otus umbra* (e.g. Dickinson et al. 2003)

Wing length: (158) 160–165 mm
Tail length: 74–82 mm
Tarsus length: 25–30 mm
Length of bill: 20.5–25.5 mm
Body mass: ?

Illustration: K. Phillipps in McKinnon and Phillipps 1993: Pl. 40; F. Weick in König et al. 1999: Pl. 14; F. Weick in *Gefiederte Welt* 2004(4): 120
Photograph: ?
Literature: Riley 1927: 93; McKinnon and Phillipps 1993: 192; Rasmussen 1998: 141–152; del Hoyo et al. 1999: 168; König et al. 1999: 242, 243; Weick 2003–2005, *Gefiederte Welt* 2004(4): 120

❖

Otus mentawi Chasen and Kloss 1926
Mentawai Scops Owl · Mentawai-Zwergohreule · Petit-duc de Mentawai · Autillo de las Mentawai
Otus bakkamoena mentawi Chosen and Kloss 1926, Ibis: 279;
Terra typica: Sipora Island
(see Plate 2)

Length: 200 mm
Body mass: ?

Distribution: Mentawai Island (Siberut, Sipora and north and south Pagai Islands) off western Sumatra
Habitat: Lowland rainforest and secondary forest, sometimes near human settlements
Museum: BMNH (Raffles), MCZ (Menden), SMTD, USNM

Wing length: 157–166 mm
Tail length: 66–71 (78) mm
Tarsus length: 31–33 mm
Length of bill: 22–23.5 mm
Body mass: ?

Illustration: K. Phillipps in McKinnon and Phillipps 1993: Pl. 40; F. Weick in König et al. 1999: Pl. 16
Photograph: C. König in collection Weick (from Lakus aviary)
Literature: Chasen and Kloss 1926: 279; McKinnon and Phillipps 1993; del Hoyo et al. 1999: 157; König et al. 1999: 249, 250

❖

Otus brookii (Sharpe) 1892
Rajah Scops Owl · Radscha-Zwergohreule · Petit-duc radjah · Autillo Rajá

Length: 215–250 mm
Body mass: ?

Distribution: Borneo and Sumatra
Habitat: Montane rainforest, 1 200–2 400 m above sea-level

- ***Otus brookii solokensis*** (Hartert) 1893
  Pisorhina solokensis Hartert 1893, Bull. Brit. Ornith. Cl. 1: 39;
  Terra typica: Solok Mountains, Sumatra

Distribution: Highlands of Sumatra
Museum: AMNH, BMNH, NNML, SMNSt (holotype), ZNA

Wing length: 163–187 mm
Tail length: 70–90 mm
Tarsus length: 27–32 mm
Length of bill: 23.5–27.5 mm
Body mass: ?

Illustration: F. Weick in König et al. 1999: Pl. 14
Photograph: ?

Literature: Hartert 1893: 39; McKinnon and Phillipps 1993: 192; del Hoyo et al. 1999: 157; König et al. 1999: 243, 244

- ***Otus brookii brookii*** (Sharpe) 1892
  Scops brookii Sharpe 1892, Bull. Brit. Ornith. Cl. 1: 39;
  Terra typica: Mount Dulit, Sarawak, Borneo

Distribution: Mountains of Northwest Borneo
Museum: AMNH (Rothsch.), BMNH (Hose)

Wing length:	162–171 mm
Tail length:	76–78 mm
Tarsus length:	30–32 mm
Length of bill:	26–27 mm
Body mass:	?

Illustration: J.G. Keulemans in *Ibis* 1893: Pl. 11; A. Hughes in Smythies 1968: Pl. 15; T. Boyer in Boyer and Hume 1991: 44; K. Phillipps in McKinnon and Phillipps 1993: Pl. 40; F. Weick in König et al. 1999: Pl. 14
Photograph: ?
Literature: Sharpe 1892: 4; Smythies 1968: 278; Boyer and Hume 1991: 44; McKinnon and Phillipps 1993: 192; del Hoyo et al. 1999: 157; König et al. 1999: 243, 244; Hekstra unpublished

❖

Otus lempiji (Horsfield) 1821
Sunda Scops Owl, Collared Scops Owl · Sunda-Zwergohreule · Petit-duc de Sunda · Autillo de la Sonda

Length:	190–210 mm
Body mass:	90–140 g

Distribution: Malay Peninsula, southern Thai Peninsula, southern Sumatra, Banka, Belitung, Java, Bali, North Natuna Islands, Borneo, northern and central Sumatra, Kangean Islands
Habitat: Secondary evergreen and deciduous forest, plantations, parks, gardens with tall trees, urban and suburban settlements with trees. From sea-level to about 2 000 m
Remarks: Formerly often considered as race of *O. bakkamoena* or *O. lettia*, but with different vocalisations

- ***Otus lempiji lempiji*** (Horsfield) 1821
  Srix (sic) *Lempiji* Horsfield 1821, Trans. Linn. Soc. London 13(1): 140; Terra typica: Java
  (see Plate 2 and Figure 10)

Distribution: Malay Peninsula except southern Thailand, southern Sumatra, Banka, Belitung, Java, Bali, North Natuna Islands, Borneo (except the north)
Museum: AMNH, BMNH (Rothsch.), FNSF, MNHP, MNBHU, MZUS, NHMBas, NNML, SMNSt, SMTD, USNM, ZFMK, ZMH

Wing length:	136–153 mm
Tail length:	69–78 mm
Tarsus length:	27 and 28 mm
Length of bill:	17.5–23.5 mm
Body mass:	90–140 g

Illustration: K. Phillipps in McKinnon and Phillipps 1993: Pl. 40; F. Weick in König et al. 1999: Pl. 15
Photograph: A. Suryadi in internet: Owl Pages, photo gallery 2003; C. König in collection Weick (from Lakus aviary); G. Ehlers, Leipzig in collection Weick (Kualalumpur)
Literature: Horsfield 1821: 140; Sharpe 1875a: 91–93; Deignan 1950: 189–201; McKinnon and Phillipps 1993: 330; del Hoyo et al. 1999: 158; König et al. 1999: 244, 245; Hekstra unpublished

- ***Otus (lempiji) cnephaeus*** Deignan 1950
  Otus bakkamoena cnephaeus Deignan 1950, Auk 67: 195;
  Terra typica: Rumpin River, Pahang State, Malaysia
  (see Figure 10)

Distribution: Southern Malay Peninsula
Museum: AMNH, BMNH, NNML, USNM, ZMA
Remarks: Possibly a distinct species; differs from *lempiji* vocally?

Wing length:	143–157 mm
Tail length:	68–79 mm
Tarsus length:	28–31 mm
Length of bill:	?
Body mass:	?

Illustration: F. Weick unpublished (new edn. Owls)
Photograph: M. Strange in del Hoyo et al. 1999: 120; C. König in collection Weick (from Lakus aviary)
Literature: Deignan 1950: 189–201; del Hoyo et al. 1999: 158; König et al. 1999: 244, 245

- ***Otus lempiji hypnodes*** Deignan 1950
  Otus bakkamoena hypnodes Deignan 1950, Auk 67: 196;
  Terra typica: Pulau Padang Island, mouth of Siak River, Sumatra

Distribution: Eastern, northern and central Sumatra
Museum: AMNH, MNBHU, NHMBas, SMNSt, USNM, ZMA

Wing length: 140–157 mm
Tail length: 65–75 mm
Tarsus length: 26–29 mm
Length of bill: 21.5–24 mm
Body mass: ?

Illustration: F. Weick in König et al. 1999: Pl. 15
Photograph: C. König in collection Weick (from Lakus aviary)
Literature: Deignan 1950: 189–201; McKinnon and Phillipps 1993: 330; del Hoyo et al. 1999: 158; König et al. 1999: 244, 245; Hekstra unpublished

- ***Otus lempiji lemurum*** Deignan 1957
  Otus bakkamoena lemurum Deignan 1957, Proc. Biol. Soc. Wash. 70: 43; Terra typica: Kanowit, Sarawak, northern Borneo

Distribution: Northern Borneo
Museum: AMNH, ChiMNH, FNSF, MNBHU, SMTD, UK(ans)MNH, USNM

Wing length: 140–157 mm
Tail length: 64–75 mm
Tarsus length: 26–30 mm
Length of bill: 20–24 mm
Body mass: ?

Illustration: A. Hughes in Smythies 1968: Pl. 15
Photograph: ?
Literature: Deignan 1957: 43; Smythies 1968: 277, 278; del Hoyo et al. 1999: 158; König et al. 1999: 244, 245; Hekstra unpublished

- ***Otus lempiji kangeana*** Mayr 1938
  Otus bakkamoena kangeana Mayr 1938, Bull. Raffles Mus. 14: 14; Terra typica: Kangean Island

Distribution: Kangean Island, north of Bali
Museum: AMNH (Rothsch.), SMTD

Wing length: 144–147 mm
Tail length: 71–74 mm
Tarsus length: 26 and 27 mm
Length of bill: 20–21.5 mm
Body mass: ?

Illustration: ?
Photograph: ?
Literature: Mayr 1938: 14; del Hoyo et al. 1999: 158; König et al. 1999: 244, 245; Hekstra unpublished

Otus lettia **(Hodgson) 1836**
Collared Scops Owl · Halsband-Zwergohreule · Petit-duc à collier · Autillo Chino

Length: 230–250 mm
Body mass: 100–170 g

Distribution: Eastern Nepal, eastern India, Bangladesh, east to Assam, Myanmar, Thailand and Indochina. Ussuriland, Sakhalin, China, Taiwan, and Hainan Island
Habitat: Deciduous and evergreen forest, secondary forest, groves of trees, bamboo thickets, Gardens with tall trees, also near or in human settlements. From lowlands up to about 2 400 m
Remarks: Specific status due to different vocalisation compared to *O. bakkamoena* and *O. lempiji*

- ***Otus lettia lettia*** (Hodgson) 1836
  Scops lettia Hodgson 1836, As. Res. 19: 176;
  Terra typica: Nepal
  (see Figure 10)

Synonym:
- *Otus bakkamoena condorensis* Kloss 1930, Journ. Siam Soc. Nat. Hist. Suppl. 8: 81; Terra typica: Pulo, Condor
- *Otus bakkamoena manipurensis* Roowal and Nath 1849, Rec. Ind. Mus. 46: 162; Terra typica: Manipur, Assam
- *Otus bakkamoena alboniger* Koelz 1952, J. Zool. Soc. India 4: 45; Terra typica: Lushai Hills, E of Bangladesh

Distribution: Eastern Nepal, eastern India, Bangladesh, east to Assam, Myanmar, Thailand and Indochina
Museum: MNBHU, SMTD, USNM, ZMH

Wing length: 158–180 mm
Tail length: 75–91 mm
Tarsus length: 30–33 mm
Length of bill: 20–23 mm
Body mass: ♂108 g, ♀ 142 g

Illustration: W. Monkol in Lekagul and Round 1991: Pl. 60; F. Weick in König et al. 1999: Pl. 15; H. Burn

in Robson 2000: Pl. 21; McQueen in Rasmussen and Anderton 2005: Pl. 78
Photograph: W. Hsu in internet: Owl Pages, photo gallery 2003
Literature: Hodgson 1836a: 176; Sharpe 1875a: 85, 87; Baker 1927: 437; Delacour and Jabouille 1931: 126, 127; Smythies and Hughes 1984: 317; Deignan 1950: 189–201; Ali and Ripley 1969: 270, 271; del Hoyo et al. 1999: 158; König et al. 1999: 245, 246; Robson 2000: 291; Rasmussen and Anderton 2005: 239; Hekstra unpublished

- *Otus lettia erythrocampe* (Swinhoe) 1874
  Lempijius erythrocampe Swinhoe 1874, Ibis (Ser. 3), 4: 269; Terra typica: Canton, Kwangtung Province, China

Distribution: Southeast China
Museum: AMNH, BMNH, MHNP, MNBHU, NHRMSt, SMNSt, ZFMK

Wing length: 165–180 mm
Tail length: 83–94 mm
Tarsus length: 31–33 mm
Length of bill: 22.5–28 mm
Body mass: ?

Illustration: F. Weick in König et al. 1999: Pl. 15
Photograph: L. Schlawe in Eck and Busse 1973, Frontcover; C. König in collection Weick (from Lakus aviary)
Literature: Swinhoe 1874: 269; Sharpe 1875a: 89, 90; Deignan 1950: 189–201; Etchécopar and Hüe 1978: 476, 477; del Hoyo et al. 1999: 158; König et al. 1999: 245, 246; Hekstra unpublished

- *Otus lettia glabripes* (Swinhoe) 1870
  Ephialtes glabripes Swinhoe 1870, Ann. Mag. Nat. Hist. Ser. 4, 6: 152; Terra typica: Formosa

Distribution: Taiwan
Museum: AMNH, BMNH, ChiMNH, FNSF, MCZ, MNBHU, NNML, SMNSt, SMTD, USNM, ZFMK, ZMK, ZMH

Wing length: 178–188 mm
Tail length: 94–102 mm
Tarsus length: 30–38 mm
Length of bill: 25.5–27.5 mm
Body mass: ?

Illustration: ?
Photograph: ?
Literature: Swinhoe 1870c: 152; Sharpe 1875a: 87, 88; Delacour and Jabouille 1931: 127; Deignan 1950: 189–201; Etchécopar and Hüe 1978: 477; del Hoyo et al. 1999: 158; König et al. 1999: 245, 246; Hekstra unpublished

- *Otus lettia umbratilis* (Swinhoe) 1870
  Ephialtes umbratilis Swinhoe 1870, Ibis (new ser.) 6: 342, footnote; Terra typica: Hainan Island

Distribution: Hainan Island, off southern China
Museum: AMNH (Rothsch.)

Wing length: 161–183 mm
Tail length: 84–91 mm (95 mm)
Tarsus length: 32 and 33 mm
Length of bill: 22.5–26 mm
Body mass: ?

Illustration: ?
Photograph: ?
Literature: Swinhoe 1870a: 342; Sharpe 1875a: 93, 94; Deignan 1950: 189–201; del Hoyo et al. 1999: 158; König et al. 1999: 245, 246; Hekstra unpublished

Otus bakkamoena Pennant 1769
Indian (collared) Scops Owl · Indien-Halsbandeule · Petit-duc indien (des Indes) · Autillo Indio

Length: 200–220 mm
Body mass: 125–150 g

Distribution: Southeast Arabia and southeast Iran to southern Pakistan, northwest Himalayas, India east to western Bengal including the Himalayas from Kashmir east to central Nepal and south to Sri Lanka
Habitat: Forest, secondary woodland, gardens and orchards, plantations, open country with trees, riverine forests in drier areas, often near villages and cultivated areas. From lowlands up to 2 400 m
Remarks: This treatment based only on vocalisations!

- *Otus bakkamoena plumipes* (Hume) 1870
  Ephialtes Plumipes Hume 1870, Rough Notes 1(2): 397; Terra typica: Murree, Kotegurh and Garlawal

Distribution: Western Himalayas from northern Pakistan east to the border of Nepal
Museum: AMNH (Koelz), BMNH, SMNSt

Wing length: 173–185 mm
Tail length: 80–94 mm
Tarsus length: 32–35 mm
Length of bill: 22–24 mm
Body mass: ?

Illustration: D. Cole in Grimmett et al. 1998: Pl. 28; F. Weick in König et al. 1999: Pl. 15
Photograph: ?
Literature: Sharpe 1875a: 85; Baker 1927: 425, 426; Deignan 1950: 189–201; Ali and Ripley 1969: 266, 267; Grimmett et al. 1998: 432; del Hoyo et al. 1999: 157, 158; König et al. 1999: 246, 247; Hekstra unpublished

- ***Otus bakkamoena deserticolor*** Ticehurst 1922
  Otus bakkamoena deserticolor Ticehurst 1922, Bull. Brit. Ornith. Cl. 42: 57, 122; Terra typica: Haidarabad, Sind

Distribution: Southern Pakistan, Southeast Iran(?), old record from Oman
Museum: BMNH (holotype), BNHS, UM(ich)MZ

Wing length: 153–167 (175) mm
Tail length: 77–89 mm
Tarsus length: 31–33 mm
Length of bill: 20–22 mm
Body mass: ?

Illustration: F. Weick in König et al. 1999: Pl. 15; McQueen in Rasmussen and Anderton 2005: Pl. 78
Photograph: ?
Literature: Ticehurst 1922: 57; Baker 1927: 426; Deignan 1950: 189–201; Vaurie 1965: 603, 604; Ali and Ripley 1969: 267; del Hoyo et al. 1999: 157, 158; König et al. 1999: 246, 247; Hekstra unpublished

- ***Otus bakkamoena gangeticus*** Ticehurst 1922
  Otus bakkamoena gangeticus Ticehurst 1922, Bull. Brit. Ornith. Cl. 42: 122; Terra typica: Fatehgarh, Unit. Prov.

Synonym:
- *Otus bakkamoena stewarti* Koelz 1939, Proc. Biol. Soc. Wash. 52: 80; Terra typica: Punjab

Distribution: Northwest India to lowlands of Nepal
Museum: AMNH, BMNH, ChiMNH, MHNP, MNBHU, USNM (Ripley), ZMH

Wing length: (146) 152–167 mm
Tail length: 72–81 mm
Tarsus length: 31–34 mm
Length of bill: 19–21 mm
Body mass: 121 g ($n = 1$)

Illustration: D. Cole in Grimmett et al. 1998: Pl. 28; F. Weick in König et al. 1999: Pl. 15
Photograph: E. Hosking in Hosking and Flegg 1982: 64, 151
Literature: Ticehurst 1922: 122; Baker 1927: 425; Deignan 1950: 189–201; Whistler 1963 (repr.): 345, 346; Ali and Ripley 1969: 268; del Hoyo et al. 1999: 157, 158; König et al. 1999: 246, 247; Hekstra unpublished

- ***Otus bakkamoena marathae*** Ticehurst 1922
  Otus bakkamoena marathae Ticehurst 1922, Bull. Brit. Ornith. Cl. 42: 122; Terra typica: Raipur, Central Prov.

Distribution: Central India east to SW Bengal
Museum: AMNH, BMNH, FNSF, NHMWien, UM(ich)MZ, ZFMK

Wing length: 143–148 mm (Abdulali), 152–165 mm
Tail length: 66–71 mm (Abdulali), 70–81 mm
Tarsus length: 29–33 mm
Length of bill: 20.5–24.5 mm
Body mass: ?

Illustration: McQueen in Rasmussen and Anderton 2005: Pl. 78
Photograph: ?
Literature: Ticehurst 1922: 122; Baker 1927: 424, 425; Ali and Ripley 1969: 268, 269; del Hoyo et al. 1999: 157, 158; König et al. 1999: 246, 247; Hekstra unpublished

- ***Otus bakkamoena bakkamoena*** Pennant 1769
  Otus bakkamoena Pennant 1769, Indian Zool. 3: Pl. 3; Terra typica: Ceylon

Synonym:
- *Scops malabaricus* Jerdon 1844, Madras J. Lit. Sci. 13: 119; Terra typica: Malabar Coast
- *Scops griseus* Jerdon 1844, ibid.: 119; Terra typica: Eastern Ghats

Distribution: Southwest and southeast India, Sri Lanka
Museum: AMNH, BMNH, FNSF, MNBHU, MZUS, NNML, SMNSt, SMTD, ZFMK

Wing length:	149–154 mm
Tail length:	65–74 mm
Tarsus length:	29–31 mm
Length of bill:	20–22 mm
Body mass:	125–150 g

Illustration: G. M. Henry in Henry 1955: Pl. 16; P. Barruel in Ali and Ripley 1968: Pl. 17; T. Boyer in Boyer and Hume 1991: 47; F. Weick in König et al. 1999: Pl. 15

Photograph: Reinhard in Burton et al. 1992: 95 (nominate?); G. Wilson in internet: Owl Pages, photo gallery 2003

Literature: Sharpe 1875a: 94–96; Baker 1927: 422–424; Deignan 1950: 189–201; Dementiev and Gladkov 1951: 408–411; Ali and Ripley 1969: 269, 270; Roberts and King 1986: 299–305; Voous 1988: 48–52; del Hoyo et al. 1999: 157, 158; König et al. 1999: 246, 247; Hekstra unpublished; P. de Bevere in Rice 2004: 78; Rasmussen and Anderton 2005: 239

Otus semitorques Temminck and Schlegel 1850
Japanese Scops Owl · Japan-Halsbandeule · Petit-duc du Japon · Autillo Japonés
(see Plate 2)

Length:	210–255 mm
Body mass:	130 g

Distribution: Southeast Siberia (Ussuriland and Sakhalin), South Kuril Islands, Japan from Hokkaido to Kyushu, also Sado, Shikoku, Tsushima, Quelpart, Goto and Yaku-shima, Izu Islands, Ryukyu Islands (Okinawa and Yagachi-shima)

Habitat: Lowland forests, forested plains, wooded hillsides, up to 900 m above sea-level. Winters near or in human settlements, with parks or gardens, etc.

Remarks: Formerly often regarded as conspecific with the highly variable *O. bakkamoena*, but differs in vocalisations and colour of irides

- ***Otus semitorques ussuriensis*** (Buturlin) 1910
Scops semitorques ussuriensis Buturlin 1910, Mess. Orn. 1: 119; Terra typica: Khanka Lake, Ussuriland

Synonym:
- *Otus bakkamoena aurorae* Allison 1946, Notes d'Ornith. Musée Heude, Shanghai 1(2)2: 2; Terra typica: Vessels of the North China coast

Distribution: South and southeast Manchuria east to Ussuriland, south to Northeast China, Korea and the island of Sakhalin

Museum: BMNH, SMNSt, ZMH, ZMK, ZMM, ZSBS

Remarks: Often included as a subspecies of *O. lettia* or *O. bakkamoena*, but is closer in voice and plumage to *O. semitorques*

Wing length:	(159) 170–180 mm
Tail length:	84–92 mm
Tarsus length:	32–36 mm
Length of bill:	22–25.5 mm
Body mass:	?

Illustration: F. Weick in König et al. 1999: Pl. 16

Photograph: A. Knystaustas and J. Sibnev in Knystaustas and Sibnev 1987: 105

Literature: Buturlin 1910a: 119; Dementiev and Gladkov 1951: 409–411; Vaurie 1965: 602, 603; Polivanov et al. 1971: 85–91; Etchécopar and Hüe 1978: 476, 477; del Hoyo et al. 1999: 158, 159; König et al. 1999: 247, 248; Hekstra unpublished

- ***Otus semitorques semitorques*** Temminck and Schlegel 1850
Otus semitorques Temminck and Schlegel 1850, in Siebold's Fauna Japonica, Aves: 24 and Pl. 8; Terra typica: Japan

Synonym:
- *Otus bakkamoena linae* Floericke 1921, Mitt. Vogelw.: 103; Terra typica: Northern Japan

Distribution: Kuril Islands and Japan (Hokkaido south to Yaku-shima), in winter Izu and Ryukyu Islands

Museum: BMNH, FNSF, MCZ, MZUS, SMNSt, SMTD, ÜMB, USNM, ZMH

Wing length:	166–196 mm
Tail length:	77–96 mm
Tarsus length:	30–40 mm
Length of bill:	21.5–25.5 mm
Body mass:	130 g

Illustration: J. Wolf in Temminck and Schlegel 1850: Pl. 8; T. Myamoto in Kobayashi 1965: Pl. 34; S. Takano in Sonobe et al. 1983: 193; A. Cameron in Voous 1988: 49 (error colour irides); F. Weick in König et al. 1999: Pl. 16

Photograph: ? in Takano et al. 1991 (two colour morphs); F. Weick in collection Weick (from Klee aviary)

Literature: Temminck and Schlegel 1850, *Aves*: 24; Sharpe 1875a: 83–85; Deignan 1950: 189–201; Vaurie 1965: 602, 603; Voous 1988: 48–52; del Hoyo et al. 1999: 158, 159; König et al. 1999: 247, 248; Hekstra unpublished

- ***Otus semitorques pryeri*** (Gurney) 1889
  Scops pryeri Gurney 1889, Ibis: 302; Terra typica: Okinawa, Ryukyu Islands

Synonym:
- *Otus bakkamoena hatchizionis* Momiyama 1923, Dobutsu. Zasshi 35: 400; Terra typica: Hachijo and Okinawa Islands

Distribution: South Izu Islands and south Ryukyu Islands
Museum: AMNH, BMNH, USNM

Wing length: (151) 153–185 mm
Tail length: 80–91 mm
Tarsus length: 32–35 mm
Length of bill: 24–26 mm
Body mass: ?

Illustration: F. Weick in König et al. 1999: Pl. 16
Photograph: ?
Literature: Gurney 1889: 302–305; Dementiev and Gladkov 1951: 411; Vaurie 1965: 602, 603; del Hoyo et al. 1999: 158, 159; König et al. 1999: 247, 248

Otus megalotis (Walden) 1875

Philippine Scops Owl · Philippinen-Halsbandeule · Petit-duc des Philippines (Luzon) · Autillo Filipino

Length: 230–280 mm
Body mass: about 200 g

Distribution: Philippine Islands (excluding Palawan)
Habitat: Tropical forest and secondary woodland, from 300 up to 1 550 m above sea-level

- ***Otus megalotis megalotis*** (Walden) 1875
  Lempijius megalotis Walden 1875, Trans. Zool. Soc. London 9: 145; Terra typica: Manila
  (see Plate 2)

Synonym:
- *Scops whiteheadi* Ogilvie-Grant 1895, Bull. Brit. Ornith. Cl. 4: 40; Terra typica: Lepanto, North Luzon

Distribution: Luzon, Mariduque and Catanduanes
Museum: AMNH (Gilliard), BMNH (holotype), DelMNH, FNSF, MHNP, USNM (Mearns), ZFMK

Wing length: 185–205 mm
Tail length: 89–108 mm
Tarsus length: 35–44 mm
Length of bill: 26–29.5 mm
Body mass: 180 and 310 g

Illustration: J. Smit in Walden 1875: Pl. 25; T. Boyer in Boyer and Hume 1991: 46; F. Weick in König et al. 1999: Pl. 16; A. Sutherland in Kennedy et al. 2000: Pl. 35
Photograph: D. Allen in internet: Owl Pages, photo gallery 2002; G. Ehlers, Leipzig in collection Weick (Luzon)
Literature: Walden 1875: 145; Rand 1950: 1–5; du Pont 1971: 168, 169; Dickinson et al. 1991: 226, 227; Boyer and Hume 1991: 46; del Hoyo et al. 1999: 159; König et al. 1999: 248, 249; Hekstra unpublished

- ***Otus megalotis everetti*** (Tweeddale) 1879
  Scops everetti of Tweeddale 1879, Proc. Zool. Soc. Lond. 1878: 942; Terra typica: Zamboanga, Mindanao

Synonym:
- *Otus boholensis* McGregor 1907, Philip. J. Sci. 2: 323; Terra typica: Sevilla, Bohol (holotype MLA, destroyed)

Distribution: Samar, Leyte, Dinagat, Bohol, Mindanao, Basilan
Museum: AMNH, BMNH, ChiMNH, DelMNH, FNSF, SMTD, YPM, ZFMK

Otus megalotis
Philippine Scops Owl
Phillippinen-Halsbandeule

Wing length: 158–171 mm
Tail length: 71–92 mm
Tarsus length: 29–34 mm
Length of bill: 23–25.5 mm
Body mass: ♂ ∅125 ($n = 4$), ♀ ∅152 ($n = 3$)

Illustration: F. Weick in König et al. 1999: Pl. 16; A. Sutherland in Kennedy et al. 2000: Pl. 35
Photograph: P. Morris in del Hoyo et al. 1999: 87
Literature: of Tweeddale 1878: 942; Rand 1950: 1–5; du Pont 1971: 169; Dickinson et al. 1991: 226; del Hoyo et al. 1999: 159; König et al. 1999: 248, 249; Hekstra unpublished

- ***Otus megalotis nigrorum*** Rand 1950
  Otus bakkamoena nigrorum Rand 1950, Nat. Hist. Misc. Chicago Acad. Sci: 72: 1–5; Terra typica: Negros Island

Distribution: Negros Island, Philippines
Museum: FMNH (holotype), ChiMNH, DelMNH (Rabor), YPM

Wing length: 142–148 mm
Tail length: 66–76 mm
Tarsus length: 27–29 mm
Length of bill: 20.5–22 mm
Body mass: ?

Illustration: F. Weick in König et al. 1999: Pl. 16; A. Sutherland in Kennedy et al. 2000: Pl. 35
Photograph: ?
Literature: Rand 1950: 1–5; du Pont 1971: 169; Dickinson et al. 1991: 227; del Hoyo et al. 1999: 159; König et al. 1999: 248, 249; Hekstra unpublished

❖

Otus fuliginosus (Sharpe) 1888
Palawan Scops Owl · Palawan-Halsbandeule · Petit-duc de Palawan · Autillo de Palawn
Scops fuliginosa **Sharpe 1888, Ibis: 197;**
Terra typica: **Vincinity of Puerto Princesa, Palawan**

Length: 190–200 mm
Body mass: ?

Distribution: Palawan Island, southwest Philippines
Habitat: Lowland forest, mixed cultivations with trees, secondary woodland
Museum: AMNH (Rabor), BMNH (holotype), DelMNH (Rabor), MHNP, USNM, YPM, ZFMK

Wing length: 139–147 mm
Tail length: 72.5–81 mm
Tarsus length: 26 and 27 mm
Length of bill: 22 and 23 mm
Body mass: ?

Illustration: F. Weick in König et al. 1999: Pl. 16; A. Sutherland in Kennedy et al. 2000: Pl. 34
Photograph: ?
Literature: Sharpe 1888b: 197; du Pont 1971: 169; Dickinson et al. 1991: 226; del Hoyo et al. 1999: 159; König et al. 1999: 250; Kennedy et al. 2000: 177, 178; Hekstra unpublished

❖

Otus silvicola (Wallace) 1864
Wallace's Scops Owl · Wallace-Zwergohreule · Petit-duc de Wallace · Autillo de Wallace
Scops silvicola **(sic?) Wallace 1864, Proc. Zool. Soc. Lond. 1863: 487 (from a juvenile bird, corrected by Hartert)**

Length: 230–270 mm
Body mass: 212 g

Distribution: Sumbawa and Flores Island (Lesser Sundas)
Habitat: Tropical, semi-evergreen lowland forest, submontane woodland and bamboo thickets, secondary woodland, also near farms and villages. From 350 up to 1 600 m
Museum: AMNH (Rothsch. and Everett), BMNH (holotype), MCZ, MNBHU (Rensch), NNML

Wing length: 202–231 mm
Tail length: 98–124 mm
Tarsus length: 32–39 mm
Length of bill: 30–34 mm
Body mass: 212 g ($n = 1$)

Illustration: J. G. Keulemans in Hartert 1897b: Pl. 1; T. Boyer in Boyer and Hume 1991: 46 (juvenile); D. Gardner in Coates and Bishop 1997: Pl. 34; F. Weick in König et al. 1999: Pl. 16
Photograph: ?
Literature: Sharpe 1875a: 82, 83; Hartert 1897b: 527; White and Bruce 1986: 253; Boyer and Hume 1991: 46; Coates and Bishop 1997: 361; del Hoyo et al. 1999: 159; König et al. 1999: 251; Hekstra unpublished

Subgenus
Megascops **Kaup 1848**
Megascops Kaup 1848. Type by *Strix asio* Linnaeus

Otus kennicotti (Elliot) 1867
Western Screech Owl · West-Kreischeule · Petit-duc des montagnes, Scops d'Elliot · Tecolote Occidental

Length: 215–240 mm
Body mass: 90–215 g

Distribution: West of the Rocky Mountains from southern Alaska, northwest Canada to central Mexico. Eastern limits of distribution uncertain, overlaps locally with *Otus asio* in southern British Columbia, northern Idaho, western Montana, Colorado, western Oklahoma and western Texas. South to southern Baja California, northern Sinaloa and across the Mexican highlands through Chihuahua and Coahuila (Districto Federal)
Habitat: Arid and semi-open woodland, especially pine-oak forest. Semi-open areas with scattered trees, shrubs. Riparian forest, semi-desert with large cacti. Suburban areas with gardens and parks. Roosts in nestboxes and cavities

▪ ***Otus kennicotti kennicotti*** (Elliot) 1867
Scops Kennicotti Elliot 1867, Proc. Acad. Nat. Sci. Phila.: 99; Terra typica: Sitka, Alaska
(see Plate 3)

Synonym:
- *Megascops asio saturatus* Brewster 1891, Auk 8: 141; Terra typica: Victoria, Vancouver Island
- *Otus asio brewsteri* Ridgway 1914, US Nat. Mus. Bull. 50(6): 685, 700; Terra typica: Salem, Oregon

Distribution: Southern Alaska and northwestern Canada to coastal northern California
Museum: AMNH, BMNH, ChiMNH, MCZ, MNBHU, SDNHM (holotype *saturatus*), USNM (holotype *brewsteri*), YPM

Wing length: 170–190 mm
Tail length: 82–98 mm
Tarsus length: 32–37 mm
Length of bill: 20–24 mm
Body mass: ♂ 131–210 g, ♀ 157–250 g

Illustration: K. Karalus in Karalus and Eckert 1974: Pl. 19 and 25; L. Malick in Breeden et al. 1983: 242; T. Boyer in Boyer and Hume 1991: 51; F. Weick in König et al. 1999: Pl. 17; D. Sibley in Sibley 2000: 280
Photograph: Broder in Marshall 1967: Fig. 12, 13 (skins, b/w); K. Fink in Johnsgard 2002: Pl. 13; B. Marcot in internet: Owl Pages, photo gallery 2002; J. Hobbs in internet: Owl Pages, photo gallery 2003
Literature: Ridgway 1914: 678–700; Bent 1961: 267–270; Marshall 1967: 72 pp; Hekstra 1982a: 131 pp; Voous 1988: 59–62; Boyer and Hume 1991: 51; Burton et al. 1992: 92, 93; König et al. 1999: 251, 252; Johnsgard 2002: 92–97

▪ ***Otus kennicotti bendirei*** (Brewster) 1882
Scops asio bendirei Brewster 1882, Bull. Nuttall Ornith. Clmannnnnn. 7: 31; Terra typica: Nicasio, California

Synonym:
- *Megascops asio macfarlanei* Bendire 1891, Auk 8: 140; Terra typica: Walla Walla, Washington
- *Otus asio quercinus* Grinnell 1915, Auk 32: 60; Terra typica: Pasadena, California

Museum: AMNH, MVZ (holotype *quercinus*), PMV (holotype *macfarlanei*), SDNHM, USNM, YPM

Wing length: 165–188 mm
Tail length: 77–87 mm
Tarsus length: 31.5–34.5 mm
Length of bill: ?
Body mass: ♂ 100–173 g, ♀ 100–223 g

Illustration: K. Karalus in Karalus and Eckert 1974: Pl. 18, 26 and 29; L. Malick in Breeden et al. 1983: 243
Photograph: Sumner jr. in Bent 1961: Pl. 67 (b/w ph); B. Broder in Marshall 1967: Fig. 12, 13 (skins); A. G. Nelson in Johnsgard 2002: Pl. 15
Literature: Ridgway 1914: 697–701; Bent 1961: 266, 267; Marshall 1967: 72 pp; Hekstra 1982a: 131 pp; del Hoyo et al. 1999: 173; König et al. 1999: 251–253; Johnsgard 2002: 92–97

▪ ***Otus kennicotti aikeni*** (Brewster) 1891
Megascops asio aikeni (Brewster) 1891, Auk 8: 139; Terra typica: El Paso County, Colorado

Synonym:
- *Megascops asio cineraceus* Ridgway 1895, Auk 12: 390 – new name for *M. a. trichopsis* Wagler 1887;

Terra typica: Northwestern Mexico and border of the United States (holotype from Fort Huachuca, Arizona)

- *Otus asio inyoensis* Grinnell 1928, Auk 45: 213; Terra typica: Indepedence, Inyo County, California
- *Otus asio mychophilus* Oberholser 1937, Journ. Wash. Acad. Sci. 27: 356; Terra typica: Grand Canyon Village, Arizona

Distribution: Southwestern United States (California east to western Oklahoma, south to northern Mexico (north-central Sonora)

Museum: LSUMZ (Marshall), MVZ (holotype *inyoensis*), USNM (holotype *cineraceus*)

Wing length:	145–185 mm
Tail length:	73–83 mm
Tarsus length:	32–34 mm
Length of bill:	20–22.5 mm
Body mass:	88–190 g

Illustration: K. Karalus in Karalus and Eckert 1974: Pl. 20 and 24; F. Weick in König et al. 1999: Pl. 17; D. Sibley in Sibley 2000: 280

Photograph: J. Marshall jr. in Marshall 1967: Frontispiece; B. Broder in Marshall 1967: Fig. 16, 17 (b/w); G. Lasley in internet: Owl Pages, photo gallery 2003; B. Mathews in internet: Owl Pages, photo gallery 2003

Literature: Ridgway 1914: 695–702; Bent 1961: 274–277; Marshall 1967: 72 pp; Hekstra 1982a: 131 pp; del Hoyo et al. 1999: 173; König et al. 1999: 251–253; Johnsgard 2002: 92–97

- ***Otus kennicotti cardonensis*** Huey 1926
Otus asio cardonensis Huey 1926, Auk 43: 360;
Terra typica: 10 miles east of El Rosario, Lower California

Synonym:

- *Otus asio gilmani* Swarth 1910, Univ. Calif. Publ. Zool. 7: 1; Terra typica: Blackwater, Pinal County, Arizona (superfluous name, see Marshall 1967)
- *Otus asio clazus* Oberholser 1937, Journ. Wash. Acad. Sci. 27: 337; Terra typica: San Jacinto Mountains, California

Distribution: Southern California to Baja California, southern Arizona

Museum: LSUMZ (Marshall), MCZ, MVZ (holotype *gilmani*), SDNHM (holotype *cardonensis*)

Wing length:	145–173 mm
Tail length:	72–90 mm
Tarsus length:	34–36 mm
Length of bill:	?
Body mass:	158.5 g (*O. a. gilmani*)

Illustration: K. Karalus in Karalus and Eckert 1974: Pl. 22

Photograph: B. Broder in Marshall 1967: Fig. 14, 15 (skins, b/w)

Literature: Kelso 1934d: 51; Miller and Miller 1951: 161–177; Bent 1961: 282; Marshall 1967: 72 pp; Hekstra 1982a: 131 pp; del Hoyo et al. 1999: 173; König et al. 1999: 251–253; Johnsgard 2002: 92–97

- ***Otus kennicotti xantusi*** (Brewster) 1902
Megascops asio xantusi Brewster 1902, Bull. Mus. Comp. Zool. 41: 13; Terra typica: Santa Anita, Lower California

Distribution: Southern Baja California

Remarks: Possibly closer to *O. cooperi* based on vocal similitaries (P. Johnsgard 2002)

Museum: MCZ (holotype)

Wing length:	144–149 (152) mm
Tail length:	69.5–77 mm
Tarsus length:	28.5–31.5 mm
Length of bill:	18.5–21 mm
Body mass:	?

Illustration: J. Willis in del Hoyo et al. 1999. Pl. 8 (not useful for identification!)

Photograph: B. Broder in Marshall 1967: Fig. 14, 15 (b/w photograph of skins in two views)

Literature: Ridgway 1914: 703; Bent 1961: 284; Marshall 1967: 72 pp; Hekstra 1982a: 131 pp; del Hoyo et al. 1999: 173; König et al. 1999: 251–253

- ***Otus kennicotti yumanensis*** Miller and Miller 1951
Otus asio yumanensis Miller and Miller 1951, Condor 53: 172;
Terra typica: One mile from Mexican border, Baja California

Distribution: Southeast California and southwest Arizona to northern Mexico (northwest Sonora)

Museum: LSUMZ (Marshall), MVZ

Wing length:	148–160 mm
Tail length:	73–82 mm
Tarsus length:	30–33 mm
Length of bill:	?
Body mass:	164.7 g

Illustration: K. Karalus in Karalus and Eckert 1974: Pl. 35; D. Sibley in Sibley 2000: 280
Photograph: B. Broder in Marshall 1967: Fig. 16, 17 (b/w photograph of skins)
Literature: Miller and Miller 1951: 161–177; Marshall 1967: 72 pp; Hekstra 1982a: 131 pp; del Hoyo et al. 1999: 173; König et al. 1999: 251–253

- ***Otus kennicotti suttoni*** Moore 1941
Otus asio suttoni Moore 1941, Proc. Biol. Soc. Wash. 54: 154;
Terra typica: Portezuelo, Hidalgo, Mexico, 5 800 feet

Synonym:
- *Otus asio sortilegus* Moore 1941, Proc. Biol. Soc. Wash. 54: 155; Terra typica: Jalisco, Mexico, 4 200 feet

Distribution: Southwestern Texas, south to the Mexican Plateau
Museum: LSUMZ (Marshall), M(oore) LZ (holotype)

Wing length:	142–175 mm
Tail length:	75–85 mm
Tarsus length:	31–33 mm
Length of bill:	21 and 22 mm
Body mass:	♂ 87–129 g, ♀ 108–154 g

Illustration: K. Karalus in Karalus and Eckert 1974: Pl. 30; F. Weick in König et al. 1999: Pl. 17; D. Sibley in Sibley 2000: 280
Photograph: B. Broder in Marshall 1967: Fig. 14, 15 (skin b/w photograph)
Literature: Moore 1941: 154, 155; Marshall 1967: 72 pp; Hekstra 1982a: 131 pp; del Hoyo et al. 1999: 173; König et al. 1999: 251–253

- ***Otus kennicotti vinaceus*** (Brewster) 1888
Megascops vinaceus Brewster 1888, Auk 5: 88;
Terra typica: Durasno, Chihuahua, Mexico

Synonym:
- *Otus asio sinaloensis* Moore 1937, Proc. Biol. Soc. Wash. 50: 64; Terra typica: Guamuchil, Sinaloa, Mexico

Distribution: Mexico, southern Sonora and western Chihuahua to northern Sinaloa
Museum: LSUMZ (Marshall), MCZ (holotype), M(oore) LZ (holotype *sinaloensis)*

Wing length:	143–158 mm
Tail length:	71–79 mm
Tarsus length:	30–33 mm
Length of bill:	19.5–21 mm
Body mass:	100 g ($n = 1$)

Illustration: ? in *Auk* 1891(8): Pl. 3; F. Weick in König et al. 1999: Pl. 17
Photograph: W. Fink in Johnsgard 2002: Pl. 16; B. Broder in Marshall 1967: Fig. 18, 19 (skin b/w)
Literature: Ridgway 1914: 708; Marshall 1967: 72 pp; Hekstra 1982a: 131 pp; del Hoyo et al. 1999: 173; König et al. 1999: 251–253; Johnsgard 2002: 92–97

Otus seductus Moore 1941

Balsas Screech Owl · Balsas-Kreischeule · Scops (Petit-duc) du Balsas · Tecolote del Balsas
Otus vinaceus seductus **Moore 1941, Proc. Biol. Soc. Wash. 54: 156; Terra typica: 5 miles northeast of Apatzingan, 1 000 feet, Michoacan, Mexico**

Length:	240–265 mm
Body mass:	150–174 g

Synonym:
- *Otus seductus colimensis* Hekstra 1982, Bull. Zool. Mus. Univ. Amsterdam 9(7): 61; Terra typica: 7 miles south of Colima, southwest Mexico

Distribution: Southwest Mexico from Jalisco and Colima to western Guerrero
Habitat: Arid open and semi-open areas, deciduous woodland, areas with scattered trees and shrubs, also thorny woodland from 600 m up to 1 500 m above sea-level
Museum: LSUMZ (Marshall), (Schaldach holotype *colimensis*), M(oore) LZ

Wing length:	169.5–185 mm
Tail length:	88–99 mm
Tarsus length:	36–38.5 mm
Length of bill:	21.5–23.5 mm
Body mass:	152–174 g

Illustration: S. Webb in Howell and Webb 1995: Pl. 25; F. Weick in König et al. 1999: Pl. 18
Photograph: B. Broder in Marshall 1967: Fig. 18, 19 (skin, b/w photograph); R. and N. Bowers in del

Hoyo et al. 1999: 86; R. and N. Bowers in internet: Owl Pages, photo gallery 2002; D. Lockshaw in internet: Owl Pages, photo gallery 2002
Literature: Marshall 1967: 72 pp; Hekstra 1982a: 131 pp; Howell and Webb 1995: 356; del Hoyo et al. 1999: 173; König et al. 1999: 257, 258; Johnsgard 2002: 98–103

Otus cooperi **(Ridgway) 1878**
Pacific Screech Owl · Mangroven-(Cooper-)Kreischeule · Petit-duc de Cooper · Tecolote de Cooper
Scops cooperi **Ridgway 1878, Proc. US Nat. Mus. 1: 116;**
Terra typica: **Santa Ana, Costa Rica**

Length: 230–255 mm
Body mass: 145–170 g

Synonym:
- *Otus cooperi chiapensis* Moore 1947, Proc. Biol. Soc. Wash. 60: 13; Terra typica: Mazatán, Chiapas, southeast Mexico

Distribution: Southeast Mexico to northwest Costa Rica
Habitat: Dry woodland, semi-open country with scattered trees, giant cacti, palms and shrubs. Swampy woodland and mangroves. Lowland up to 330 m
Museum: AMNH, BMNH, M(oore)LZ, USNM (holotype)

Wing length: 163–183 mm
Tail length: 78–90 mm
Tarsus length: 33–37 mm
Length of bill: 22–25 mm
Body mass: 145–170 g

Illustration: D. Gardner in Stiles and Skutch 1991: Pl. 20; T. Boyer in Boyer and Hume 1991: 54; S. Webb in Howell and Webb 1995: Pl. 25; F. Weick in König et al. 1999: Pl. 18
Photograph: R. Behrstock in del Hoyo et al. 1999: 86; R. and N. Bowers in internet: Owl Pages, photo gallery 2003; M. Schropel in internet: Owl Pages, photo gallery 2004 (Gulf of Nicoya); W. Grummt in internet: Owl Pages, photo gallery 2003
Literature: Ridgway 1914: 710; Marshall 1967: 72 pp; Hekstra 1982a: 131 pp; Stiles and Skutch 1991: 190, 191; Boyer and Hume 1991: 54; Howell and Webb 1995: 357; del Hoyo et al. 1999: 173, 174; König et al. 1999: 255, 256; Johnsgard 2002: 98–103

Otus (Megascops) cooperi · Pacific Screech Owl · Mangroven-Kreischeule

Otus lambi **Moore and Marshall 1959**
Lamb's (or Oaxaca) Screech Owl · Oaxaca-Kreischeule · Petit-duc de Lamb · Tecolote de Oaxaca
Otus lambi **Moore and Marshall 1959, Condor 61: 224;**
Terra typica: **Rio Tehuantepec, 3 000 feet, 2 miles west of Nejapa, Oaxaca, Mexico**

Length: 200–220 mm
Body mass: 125–130 g

Distribution: Southern Mexico (Pacific Slope of Oaxaca)
Remarks: Probably subspecific treatment to *Otus cooperi*
Habitat: Thorny, dry woodland with candelabra cacti and palms, also coastal swamps with mangroves. From sea-level up to about 1 000 m
Museum: ChiMNH, LSUMZ (Marshall), M(oore)LZ, UADZ, UKMNH

Wing length: 148–166 mm
Tail length: 76–83 mm
Tarsus length: 28–32 mm
Length of bill: 19–22.5 mm
Body mass: ♂ 115–125 g, ♀ > 130 g

Illustration: F. Weick in König et al. 1999: Pl. 18
Photograph: B. Broder in Marshall 1967: Fig. 18, 19 (skin, b/w photograph)
Literature: Marshall 1967: 72 pp; Hekstra 1982a: 131 pp; del Hoyo et al. 1999: 173, 174; König et al. 1999: 254, 255; Johnsgard 2002: 98–102

Otus asio (Linné) 1758
Eastern Screech Owl · Ost-Kreischeule · Scops d'Amérique · Tecolote Oriental
(see Plate 3)

Length: 180–230 mm
Body mass: 125–250 g

Distribution: East of North America, from eastern Montana and south-central Canada (Great Lakes), south to Gulf of Mexico, northwest Mexico and from southern Ontario to Florida. Interbreeds with *Otus kennicotti* in areas of range overlap (near the US/Mexican border at the "Big Bend" of the Rio Grande); slightly sympatric in Colorado Springs
Habitat: Open deciduous and riparian woodland, suburban gardens and parks. Also in mixed forest, oak/juniper and pine forest and subtropical thorn-woodland. From sea-level up to about 1 500 m

Otus (Megascops) asio
Eastern Screech Owl
Ost-Kreischeule

- ***Otus asio asio*** (Linné) 1758
  Strix Asio von Linné 1758, Syst. Nat. ed. 10, 1: 92;
  Terra typica: America = South Carolina, excluding Catesby

Synonym:
- *Strix naevia* Gmelin 1788, Syst. Nat. 1(1): 289; Terra typica: New York

Remarks: Possibly a subspecies according to size and plumage!

Distribution: Northeastern United States, south to Oklahoma, east to South Carolina and Georgia
Museum: FNSF, MNBHU, NNML, SMNSt, SMTD, USNM, ZSBS

Wing length: 150–172 mm (162–177 mm *naev.*)
Tail length: 63–77 mm (74–89 mm)
Tarsus length: 28–31.5 mm (–35 mm)
Length of bill: 20–22 mm (23.5 mm)
Body mass: 99–235 g (140–252 g *naev.*)

Illustration: J. Audubon in Audubon 1827–1838: Pl. 235; R. Ridgway in Merriam and Fisher 1883: Pl. 23; L.A. Fuertes in Gilbert Pearson 1936: Pl. 56; L.A. Fuertes in Johnsgard 2002: Pl. 2; K. Karalus in Karalus and Eckert 1974: Pl. 15 and 17; L. Malick in Breeden et al. 1983: 242; T. Boyer in Boyer and Hume 1991: 50; F. Weick in König et al. 1999: Pl. 17; D. Sibley in Sibley 2000: 281
Photograph: Broder in Marshall 1967: Fig. 20, 21(skin, b/w); E. Hosking in Hosking and Flegg 1982: 152; McDonald in del Hoyo et al. 1999: 117; P. Johnsgard in Johnsgard 2002: Pl. 12; M. McCord in internet: Owl Pages, photo gallery 2002; J. Laurencelle in internet: Owl Pages, photo gallery 2003
Literature: Ridgway 1914: 691; Bent 1961: 243–263; Marshall 1967: 72 pp; Hekstra 1982a: 131 pp; Voous 1988: 64–68; Boyer and Hume 1991: 50; del Hoyo et al. 1999: 174; König et al. 1999: 253, 254

- ***Otus asio maxwelliae*** (Ridgway) 1877
  Scops asio var. *maxwelliae* Ridgway 1877, Field Forest 2: 213;
  Terra typica: Mountains of Colorado, Boulder County

Synonym:
- *Otus asio swenki* Oberholser 1937, Journ. Wash. Acad. Sci. 27: 354; Terra typica: Chadron, 3 450 feet, Dawner County, Nebraska

Distribution: Southern and central Canada and northern and central United States (west of the Great Lakes)

Museum: CMNH, SMTD, USNM

Wing length: 160–175 mm
Tail length: 76–100 mm
Tarsus length: 31–34 mm
Length of bill: 21.5–23 mm
Body mass: 219.7 g ($n = 12$)

Illustration: J. F. Landsdowne in Landsdowne and Livingston 1968: Pl. 31; K. Karalus in Karalus and Eckert 1974: Pl. 31; L. Malick in Breeden et al. 1983: 243; F. Weick in König et al. 1999: Pl. 17
Photograph: B. Broder in Marshall 1967: Fig. 20, 21 (skin, b/w photograph)
Literature: Ridgway 1914: 696; Bent 1961: 270–274; Marshall 1967: 72 pp; Landsdowne and Livingston 1968: 138; Hekstra 1982a: 131 pp; del Hoyo et al. 1999: 174; König et al. 1999: 253, 254; Johnsgard 2002: 83–91

- ***Otus asio hasbroucki*** Ridgway 1914
  Otus asio hasbroucki Ridgway 1914, US Nat. Mus. Bull. 50(6): 684, 694; Terra typica: Palo Pinto County, Texas

Distribution: Central Oklahoma to Texas
Museum: LSUMZ (Marshall), MNBHU, SMTD, USNM (holotype)

Wing length: 163–172 mm
Tail length: 74–97 mm
Tarsus length: 30–34.5 mm
Length of bill: 20.5–23 mm
Body mass: ∅ 199.3 g ($n = 7$)

Illustration: K. Karalus in Karalus and Eckert 1974: Pl. 23; D. Sibley in Sibley 2000: 281
Photograph: B. Broder in Marshall 1967: Fig. 22, 23 (skin, b/w photograph); A. Barzydlo in internet: Owl Pages, photo gallery 2003
Literature: Ridgway 1914: 684, 694; Bent 1961: 278–280; Marshall 1967: 72 pp; Hekstra 1982a: 131 pp; del Hoyo et al. 1999: 174; König et al. 1999: 253, 254; Johnsgard 2002: 83–91

- ***Otus asio floridanus*** (Ridgway) 1873
  Scops asio var. *Floridanus* Ridgway 1873, Bull. Essex Inst. 5: 200; Terra typica: Indian River, Florida

Distribution: Louisiana to Florida
Museum: AMNH, MNBHU, SMTD, USNM

Wing length: 145–156 mm
Tail length: 62–75 mm
Tarsus length: 27–30 mm
Length of bill: 18.5–22 mm
Body mass: ∅ 167.4 g ($n = 17$)

Illustration: P. Paillou 1745–1806 in *Catal. Sotheby's* 16.7.1992, no. 50 (probably another subspecies of *asio*); K. Karalus in Karalus and Eckert 1974: Pl. 21; F. Weick in König et al. 1999: Pl. 17
Photograph: S. Grossman in Grossman and Hamlet 1965: 128 and 414; B. Broder in Marshall 1967: Fig. 22, 23 (skin b/w photograph)
Literature: Ridgway 1914: 687; Bent 1961: 263–265; Marshall 1967: 72 pp; Hekstra 1982a: 131 pp; del Hoyo et al. 1999: 174; König et al. 1999: 253, 254; Johnsgard 2002: 83–91

- ***Otus asio mccallii*** (Cassin) 1854
  Scops McCallii Cassin 1854, Illustr. Bds. Cal. Texas etc. 6: 180; Terra typica: Texas and northern Mexico = Lower Rio Grande, Texas

Synonym:
- *Ephialtes ocreata* Lichtenstein 1854, Nomencl. Mus. Berol.: 7; Terra typica: "ex Mexico" (holotype NNML)
- *Otus asio semplei* Sutton and Burleigh 1939, Auk 56: 174; Terra typica: 6 miles south of Monterrey, Nuevo Leon, Mexico (holotype LSUMZ)

Distribution: Southern Texas and northeast Mexico
Museum: LSUMZ (Marshall), NNML, USNM (cotype)

Wing length: 150–166 mm
Tail length: 72–80 mm
Tarsus length: 30–34 mm
Length of bill: 20–22 mm
Body mass: ♂ 94–154 g, ♀ 115–162 g

Illustration: K. Karalus in Karalus and Eckert 1974: Pl. 28; F. Weick in König et al. 1999: Pl. 17; D. Sibley in Sibley 2000: 281
Photograph: B. Broder in Marshall 1967: Fig. 23, 24 (skin, b/w photograph); K. Horner in internet: Owl Pages, photo gallery 2003
Literature: Ridgway 1914: 693; Marshall 1967: 72 pp; Hekstra 1982a: 171 pp; Howell and Webb 1995: 355; del Hoyo et al. 1999: 174; König et al. 1999: 253, 254; Johnsgard 2002: 83–91

Otus trichopsis Wagler 1832
Whiskered Screech Owl · Flecken-Kreischeule · Scops tacheté (ou á moustaches) · Tecolote Bigotudo
(see Plate 3)

Length: 165–190 mm
Body mass: about 100 g

Distribution: Southeast Arizona through Mexico to northern and central Nicaragua
Habitat: Dense groves of oaks in pine/oak woodland, from 750 m up to 2 500 m above sea-level

- **_Otus trichopsis aspersus_ (Brewster) 1888**
  Megascops aspersus Brewster 1888, Auk 5: 87;
  Terra typica: El Carmen, Chihuahua

Distribution: Southeast Arizona to northern Mexico (Sonora and Chihuahua)
Museum: AMNH, MCZ (holotype), USNM

Wing length: 132–146 mm
Tail length: 62–79 mm
Tarsus length: 26–29 mm
Length of bill: 18–20.5 mm
Body mass: ♂ ∅ 83.6 g, ♀ ∅ 96.3 g

Illustration: G. M. Sutton in Phillips et al. 1964: Pl. 2; K. Karalus in Karalus and Eckert 1974: Pl. 34; L. Malick in Breeden et al. 1983: 243; A. Cameron in Voous 1988: 71; S. Webb in Howell and Webb 1995: Pl. 25; F. Weick in König et al. 1999: Pl. 18; D. Sibley in Sibley 2000: 280; M. E. Marcuson in Johnsgard 2002: Pl. 14
Photograph: B. Broder in Marshall 1967: Fig. 24, 25 (skin, b/w photograph); D. A. Rintoul in Johnsgard 2002: Pl. 17; R. and N. Bowers in internet: Owl Pages, photo gallery 2003
Literature: Ridgway 1914: 704–707; Bent 1961: 286–291; Marshall 1967: 72 pp; Hekstra 1982a: 131 pp; Breeden et al. 1983: 242; Voous 1988: 69–78; Boyer and Hume 1991: 52; del Hoyo et al. 1999: 174; König et al. 1999: 256, 257; Sibley 2000: 280; Johnsgard 2002: 104–107

- **_Otus trichopsis trichopsis_ (Wagler) 1832**
  Scops trichopsis Wagler 1832, Isis v. Oken, col. 276;
  Terra typica: Mexico

Synonym:
- _Megascops pinosus_ Nelson and Palmer 1894, Auk 11: 39; Terra typica: Northeastern base of the Cofre de Perote, 8 000 feet, near Las Vigas, Vera Cruz, Mexico (holotype USNM)
- _Megascops ridgwayi_ Nelson and Palmer 1894, Auk 11: 40; Terra typica: Michoacan, southwest Mexico (holotype USNM)
- _Otus trichopsis guerrensis_ van Rossem 1938, Condor 40: 258; Terra typica: Omilteme 8 000 feet, Guerrero, Mexico (holotype BMNH)

Distribution: Highlands of central Mexico (Durango, south to Veracruz, Oaxaca and Chiapas)
Museum: BMNH, LSUMZ (Marshall), MNBHU, USNM

Wing length: 145–160 mm
Tail length: 66–78 mm
Tarsus length: 25–28.5 mm
Length of bill: 18–20.5 mm
Body mass: ♂ 70–104 g, ♀ 79–121 g

Illustration: K. Karalus in Karalus and Eckert 1974: Pl. 33; T. Boyer in Boyer and Hume 1991: 52; S. Webb in Howell and Webb 1995: Pl. 25; F. Weick in König et al. 1999: Pl. 18
Photograph: ? in _Dover Nature Catalogue_ (no data): Frontcover; B. Broder in Marshall 1967: Fig. 24, 25 (skin); R. and N. Bowers in internet: Owl Pages, photo gallery 2003
Literature: Ridgway 1914: 704–707; Marshall 1967: 72 pp; Hekstra 1982a: 131 pp; Voous 1988: 69–78; Boyer and Hume 1991: 52; del Hoyo et al. 1999: 174; König et al. 1999: 256, 257; Johnsgard 2002: 104–107

- **_Otus trichopsis mesamericanus_ van Rossem 1932**
  Otus trichopsis mesamericanus van Rossem 1932, Trans. San Diego Soc. Nat. Hist. 7: 184; Terra typica: Los Esemiles, 8 000 feet, Chalatenango, El Salvador

Synonym:
- _Otus trichopsis pumilus_ Moore and Peters 1939, Auk 56: 47; Terra typica: Cerro Cantoral, Honduras (holotype MLZ)
- _Otus trichopsis inexpectus_ Hekstra 1982, Bull. Zool. Mus. Univ. Amsterdam 9(7): 58; Terra typica: Porto Jimenez, Costa Rica (holotype ChiMNH)

Distribution: Southeast Mexico (Chiapas) to northern and central Nicaragua
Museum: ChiMNH, LSUMZ (Marshall) (holotype coll. Dickey, Inst. Techn.)

Wing length: 139–154 mm
Tail length: 66–72 mm
Tarsus length: 26–28 mm
Length of bill: 18.5–20.5 mm
Body mass: ♂ ∅ 92.9 g, ♀ ∅ 102.7 g

Illustration: ? in Salvin and Godman 1897: Pl. 62; ? in Kelso 1934d: 22 (from Salvin and Godman); F. Weick in König et al. 1999: Pl. 18
Photograph: B. Broder in Marshall 1967: Fig. 26, 27 (skin, b/w photograph)
Literature: Kelso 1934d: 49; Marshall 1967: 72 pp; Land 1970: 135; Hekstra 1982a: 131 pp; Voous 1988: 69–78; del Hoyo et al. 1999: 174; König et al. 1999: 256, 257

Otus choliba (Vieillot) 1817
Tropical Screech Owl · Cholibaeule · Petit-duc (Scops) Choliba · Alicuco Común (Autillo Choliba)
(see Plate 3)

Length: 205–245 mm
Body mass: 100–160 g

Distribution: Costa Rica through Central America and large parts of South America east of the Andes. South to northern Argentina and Uruguay. Southernmost distrbution is the Province of Buenos Aires. Also in Trinidad. Absent from western Ecuador, western Peru and Chile
Habitat: Timbered savannas, galery forest, open forest and clearings, forest edges, secondary woodland with thickets. Plantations and parks. From sea-level up to 1 500 m (warm climates), locally up to 2 800 m
Remarks: Relationship unclear and probably no close affinities to any other *Otus* spp.

- ***Otus choliba luctisonus*** Bangs and Penard 1921
  Otus choloba luctisonus Bangs and Penard 1921, Proc. Biol. Soc. Wash. 34: 89; Terra typica: Escazú, Costa Rica

Distribution: Costa Rica to NW Colombia, also Pearl Island
Museum: MNBHU, USNM (holotype)

Wing length: 168–182 mm
Tail length: 86–96 mm
Tarsus length: 30–32 mm
Length of bill: 20–24.5 mm
Body mass: ?

Illustration: J. A. Gwynne in Ridgely and Gwynne 1989: Pl. 12; J. Fjeldsa in Fjeldsa and Krabbe 1990: Pl. 25; D. Gardner in Stiles and Skutch 1991: Pl. 20; F. Weick in König et al. 1999: Pl. 19
Photograph: E. Hosking in Hosking and Flegg 1982: 36
Literature: Ridgway 1914: 711–715; Hekstra 1982a: 131 pp; Ridgely and Gwynne 1989: 187; Fjeldsa and Krabbe 1990: 223, 224; del Hoyo et al. 1999: 175; König et al. 1999: 259–261

- ***Otus choliba margaritae*** Cory 1915
  Otus choliba margaritae Cory 1915, Field Mus. Nat. Hist. Publ. Orn. 1: 298; Terra typica: Margarita Island, Venezuela

Distribution: Margarita Island, off northern Venezuela
Museum: FMNH (holotype), SMTD

Wing length: 154–168 mm
Tail length: 78–90 mm
Tarsus length: 27–31 mm
Length of bill: 21–24 mm
Body mass: 135 g

Illustration: ?
Photograph: ?
Literature: del Hoyo et al. 1999: 175; König et al. 1999: 259–261; Hilty 2003: 359

- ***Otus choliba duidae*** Chapman 1929
  Otus choliba duidae Chapman 1929, Am. Mus. Novit. 380: 7; Terra typica: Mount Duida, 5 000 feet, Venezuela

Distribution: Duida Mountains in Venezuela
Museum: AMNH (holotype), BMNH, ChiMNH, SMNSt

Wing length: 165–175 mm
Tail length: 82–91 mm
Tarsus length: 27–31 mm
Length of bill: 21.5–22 mm
Body mass: ?

Illustration: F. Weick in König et al. 1999: Pl. 19
Photograph: ?
Literature: Hekstra 1982a: 131 pp; del Hoyo et al. 1999: 175; König et al. 1999: 259–261; Hilty 2003: 359

- **Otus choliba crucigerus** (Spix) 1824
  Strix crucigera Spix 1824, Av. Bras. 1: 22 and Pl. 9;
  Terra typica: Juxta flumen Amazonum

Synonym:
- *Otus choliba alticola* L. Kelso 1937, Biol. Leaflet 8: 1; Terra typica: Bogotá, Colombia
- *Otus choliba portoricensis* L. Kelso 1942, Biol. Leaflet 14: 2; Terra typica: Trinidad
- *Otus choliba montanus* Hekstra 1982, Bull. Zool. Mus. Univ. Amsterdam9(7): 61; Terra typica: Mérida, Montana Sierra Valle, Venezuela
- *Otus choliba kelsoi* Hekstra 1982, ibid.: 61 = new name for *O. c. portoricensis* Kelso (holotype AMNH)
- *Otus choliba caucae* Hekstra 1982, ibid.: 60; Terra typica: Rio Cauca, El Tambo, 5 100 feet, Colombia (holotype NNML)
- *Otus choliba guyanensis* Hekstra 1982, ibid.: 60; Terra typica: Mount Roraima, 3 500 feet, Guyana (holotype BMNH)

Distribution: Eastern Colombia and eastern Peru across to Venezuela, Trinidad, the Guianas and northeast Brazil
Museum: AMNH, ANSP, BMNH, ChiMNH, FNSF, LACM, MCZ, MHNG, MHNP, MHNUSM, NHRMSt, NNML, SMNSt, SMTD, USNM, ZFMK, ZSBS

Wing length:	162–181 mm
Tail length:	83–96 mm
Tarsus length:	31–34.5 mm
Length of bill:	21–24.5 mm
Body mass:	138–155 g

Illustration: ? in Spix 1824: Pl. 9; P. Barruel in Haverschmidt 1968: Pl. 13; F. Weick in König et al. 1999: Pl. 19; P. J. Greenfield in Ridgely and Greenfield 2001: II: Pl. 35(2); S. Webb in Hilty 2003: Pl. 24
Photograph: L. Schlawe in Eck and Busse 1973: Abb. 2; A. Ribeiro in internet: Owl Pages, photo gallery 2002
Literature: Haverschmidt 1968: 158; Hekstra 1982a: 131 pp; del Hoyo et al. 1999: 175; König et al. 1999: 259–261; Ridgely and Greenfield 2001: I:212, 213, II: 303, 304; Hilty 2003: 359

- **Otus choliba suturutus** L. Kelso 1941
  Otus choliba suturutus L. Kelso 1941, Biol. Leaflet 13: 1;
  Terra typica: Rio Suturutu, Buenavista, 400 m, Santa Cruz

Distribution: Bolivia
Museum: ZSBS

Wing length:	160–178 mm
Tail length:	87–93 mm
Tarsus length:	28–31 mm
Length of bill:	20 and 21 mm
Body mass:	?

Illustration: ?
Photograph: ?
Literature: Hekstra 1982a: 131 pp; König et al. 1999: 259–261

- **Otus choliba decussatus** (Lichtenstein) 1823
  Strix decussata Lichtenstein 1823, Verz. Doubl. Zool. Mus. Berlin: 59; Terra typica: Bahia

Synonym:
- *Otus choliba catingensis* Hekstra 1982, Bull. Zool. Mus. Univ. Amsterdam9(7): 59; Terra typica: Janáubu, North Minas Gerais (holotype ChiMNH)

Distribution: Central and eastern Brazil
Museum: AMNH, BMNH, ChiMNH, LACM, LSUMZ, MCZ, MNBHU, MZSBS, NHMWien, NNML, SMNSt, SMTD, USNM, UW(isc) ZM, ZIHeid, ZSBS

Wing length:	156–170 mm
Tail length:	83–91 mm
Tarsus length:	27–31 mm
Length of bill:	20.5–23 mm
Body mass:	?

Illustration: F. Weick in König et al. 1999: Pl. 19
Photograph: A. Ribeiro in internet: Owl Pages, photo gallery 2002
Literature: Hekstra 1982a: 131 pp; del Hoyo et al. 1999: 175; König et al. 1999: 259–261

- **Otus choliba choliba** (Vieillot) 1817
  Strix choliba Vieillot 1817, Nouv. Dict. Hist. Nat. 7: 39;
  Terra typica: Paraguay

Synonym:
- *Otus choliba chapadensis* Hekstra 1982, Bull. Zool. Mus. Univ. Amsterdam9(7): 59; Terra typica: Sierra de Chapada, Mato Grosso (holotype AMNH)

Distribution: Southern Mato Grosso, Sao Paulo to eastern Paraguay
Museum: AMNH, FNSF, NNML, SMNSt, USNM, ZFMK, ZSBS

Remarks: The additional subspecies described by Hekstra are probably only colour morphs or a result of individual variation. *Otus koepckeae* and *roboratus* are specifically distinct in morphology and vocalisation

Wing length: 158–172 mm
Tail length: 83–98 mm
Tarsus length: 28–32.5 mm
Length of bill: 20–23 mm
Body mass: 97–160 g

Illustration: T. Boyer in Boyer and Hume 1991: 55; F. Weick in König et al. 1999: Pl. 19
Photograph: E. Hosking in Hosking and Flegg 1982: 36; E. Hosking in Burton et al. 1992: 91; C. König 2001: 177
Literature: Kelso 1934c: 71–74; Eck and Busse 1973: 86, 87; Hekstra 1982a: 131 pp; Boyer and Hume 1991: 55; del Hoyo et al. 1999: 175; König et al. 1999: 259–261

- ***Otus choliba wetmorei*** Brodkorb 1937
Otus choliba wetmorei Brodkorb 1937, Proc. Biol. Soc. Wash. 50: 33; Terra typica: Puerto Casado, Paraguay Chaco

Synonym:
- *Otus choliba alilicuco* Hekstra 1982, Bull. Zool. Mus. Univ. Amsterdam9(7): 59; Terra typica: Argentina, Province Salta, Rosario (holotype BMNH)

Remarks: Sometimes regarded as a valid subspecies

Distribution: Western Paraguay and northern Argentina (south to Mendoza, west to Buenos Aires and north to Rio Negro)
Museum: AMNH, BMNH, ChiMNH, MCZ, NHRMSt, ZMKob, ZSBS

Wing length: 157–179 mm
Tail length: 82–94 mm
Tarsus length: 28–31.5 mm
Length of bill: 19–22.5 mm
Body mass: ?

Illustration: J. Fjeldsa in Fjeldsa and Krabbe 1990: Pl. 25 (*alilicuco*)
Photograph: C. König in collection Weick
Literature: Hekstra 1982a: 131 pp; Fjeldsa and Krabbe 1990: 223, 224; del Hoyo et al. 1999: 175; König et al. 1999: 259–261

- ***Otus choliba uruguaiensis*** Hekstra 1982
Otus choliba urugaii (sic) Hekstra 1982, Bull. Zool. Mus. Univ. Amsterdam9(7): 59; Terra typica: Misiones, Arroyo, River Uruguai

Distribution: Southeast Brazil, (Santa Catarina, Rio Grande do Sul), northeast Argentina (Misiones) and Uruguay
Museum: AMNH (holotype), BMNH, LACM, LSUMZ, MCZ, MHNP, NHRMSt, SMNSt, YPM, ZSBS

Wing length: 165–180 mm
Tail length: 86–98 mm
Tarsus length: 29–31 mm
Length of bill: ?
Body mass: ?

Illustration: ?
Photograph: ?
Literature: Hekstra 1982a: 131 pp; del Hoyo et al. 1999: 175; König et al. 1999: 259–263; Narosky and Yzurieta 2003: 141

Otus koepckeae Hekstra 1982
Koepcke's Screech Owl · Koepcke-Kreischeule · Scops de Koepcke · Urcututú de Koepcke
Otus choliba koepckeae **Hekstra 1982, Bull. Zool. Mus. Univ. Amsterdam9(7): 60; Terra typica: Dep. Ancash, Quebrada Yanganuco, Peru**
(see Figure 11)

Length: 240 mm
Body mass: 110–148 g

Distribution: High Andean slopes of northern and northwestern Peru, south to Lima. Perhaps farther south to western and central Bolivia
Habitat: Wooded areas or arid forest patches in Andean slopes above 2 500 m, up to 4 500 m above sea-level
Museum: AMNH (Koepcke coll.), AMSP, BMNH, MJPL, NHRMSt, ZFMK, ZMH

Wing length: 172–183 mm
Tail length: 86–104 mm
Tarsus length: 28–34 mm
Length of bill: ?
Body mass: 110–148 g

Illustration: J. Fjeldsa in Fjeldsa and Krabbe 1990: Pl. 25; F. Weick in König et al. 1999: Pl. 19; F. Weick unpublished (new edn. Owls: Pl. 67)
Photograph: ?
Literature: Hekstra 1982a: 131 pp; Fjeldsa and Krabbe 1990: 223, 224; Stotz et al. 1996; del Hoyo et al. 1999: 175; König et al. 1999: 261

❖

Otus roboratus Bangs and Noble 1918
Peruvian Screech Owl · Peru-Kreischeule · Scops du Pérou · Urcututú Peruano
Otus roboratus Bangs and Noble 1918, Auk 40: 448; Terra typica: Bellavista, Peru
(see Figure 11)

Length: 200–220 mm
Body mass: 144–162 g

Distribution: Southern Ecuador, northwest Peru
Habitat: Drainage of Chinchipe and Maranón Rivers, between the western and central Andes
Museum: FMNH, LSUMZ, MJPL, MVZ, SMTD
Remarks: The following species *Otus pacificus*, often regarded as a subspecies of *roboratus*, but is morphologically and vocally distinct

Wing length: 165–175 mm
Tail length: 88–96 mm
Tarsus length: 31–33 mm
Length of bill: 19.5–23 mm
Body mass: 144–162 g

Illustration: J. Fjeldsa in Fjeldsa and Krabbe 1990: Pl. 25; ? in Johnson and Jones 1990: Fig. 1 (head); T. Boyer in Boyer and Hume 1991: 54; F. Weick in König et al. 1999: Pl. 19; P. J. Greenfield in Ridgely and Greenfield 2001: II: Pl. 35(3); F. Weick unpublished (new edn. Owls: Pl. 67)
Photograph: R. S. Williams in Williams and Tobias 1996: 2 (colour photopraphs)
Literature: Kelso 1934d: 47; Eck and Busse 1973: 54; Hekstra 1982a: 131 pp; Johnson and Jones 1990: 199–212; Fjeldsa and Krabbe 1990: 223, 224; Williams and Tobias 1996: 76, 77; del Hoyo et al. 1999: 175; König et al. 1999: 262, 263; Ridgely and Greenfield 2001: I: 304, II: 213

Otus pacificus Hekstra 1982
Tumbes Screech Owl · Tumbes-Kreischeule · Petit-duc de Tumbes · Autillo de Tumbes
Otus guatemalae pacificus Hekstra 1982, Bull. Zool. Mus. Univ. Amsterdam 9(7): 58; Terra typica: Morropon, 140 m, Piura, northwest Peru
(see Figure 11)

Length: 200 mm
Body mass: ∅ 87.2 g ($n = 17$)

Synonym:
- *Otus guatemalae rufus* Hekstra 1982, Bull. Zool. Mus. Univ. Amsterdam 9(7): 58; Terra typica: Balzar Mountains, western Ecuador

Distribution: Southwestern Ecuador and extreme northwest of Peru (south to Dep. Lambayeque)
Habitat: Arid tropical coastal plains and foothills, below 500 m. Open dry scrub with bushes, cacti and scattered groups of trees
Museum: AMNH, ChiMNH, LSUMZ, MHNUSM, MVZ, NHMBas, SNM, ZFMK

Wing length: 139–150 mm
Tail length: 76–80 mm
Tarsus length: 27–30 mm
Length of bill: ?
Body mass: ∅ 87.2 g ($n = 17$)

Illustration: ? in Johnson and Jones 1990: Fig. 1 (head); F. Weick in König et al. 1999: 262 (b/w); F. Weick unpublished (new edn. Owls: Pl. 67)
Photograph: Johnson and Jones 1990: Fig. 2 (skins, b/w photograph)
Literature: Hekstra 1982a: 131 pp; Johnson and Jones 1990: 199–212; del Hoyo et al. 1999: 175; König et al. 1999: 262, 263

❖

Otus clarkii Kelso and Kelso 1935
Bare-shanked Screech Owl · Nacktbein-Kreischeule · Petit-duc de Clark · Tecolote de Clark
Otus clarkii Kelso and Kelso 1935, Biol. Leaflet 5: unpaged, new name for _Bubo nudipes_

Length: 230–245 mm
Body mass: 130–190 g

Synonym:
- *Bubo nudipes* Vieillot 1807, Ois. Am. Sept. 1: 53 Pl. 22; Terra typica: Greater Antilles (invalid name)

Distribution: Locally from Costa Rica to Panama and extreme northwest of Colombia
Habitat: Montane cloud forest, humid dense forest, also forest edges. From 900 m up to 2 350 m, ocassionally to 3 300 m
Museum: AMNH, BMNH, MVZ, USNM

Wing length:	173–190 mm
Tail length:	88–105 mm
Tarsus length:	31–38.5 mm
Length of bill:	21–25.5 mm
Body mass:	123–186 g

Illustration: D. Gardner in Stiles and Skutch 1989: Pl. 191; J. A. Gwynne in Ridgely and Gwynne 1989: Pl. 2; T. Boyer in Boyer and Hume 1991: 59; F. Weick in König et al. 1999: Pl. 18
Photograph: P. O'Neill in Burton et al. 1992: 87; R. and N. Bowers in internet: Owl Pages, photo gallery 2003
Literature: Sharpe 1875a: 121; Ridgway 1914: 727, 728; Moore and Peters 1939: 55; Marshall 1967: 72 pp; Eck and Busse 1973: 89; Weske and Terborgh 1981: 1–7; Hekstra 1982a: 131 pp; Hilty and Brown 1986: 226; Stiles and Skutch 1989: 191; Ridgely and Gwynne 1989: 12, 13; Boyer and Hume 1991: 59; del Hoyo et al. 1999: 175, 176; König et al. 1999: 258, 259

Otus barbarus (Sclater and Salvin) 1868
Santa Barbara (Bearded Screech) Owl · Tropfenkreischeule · Petit-duc bridé · Tecolote Barbudo
***Scops barbarus* Sclater and Salvin 1868, Proc. Zool. Soc. Lond.: 56; Terra typica: Santa Barbara, Vera Paz, Guatemala**

Length:	170–180 mm
Body mass:	69 g

Distribution: Highlands of southern Mexico (Chiapas) and northern/central Guatemala
Habitat: Humid pine/oak forest in the highlands. Cloud-forest, from 1 400 m up to 2 500 m, mostly about 1 800 m
Museum: AMNH, BMNH (holotype), SMNSt, USNM, ZMH
Remarks: Relationship uncertain. Monotypic

Otus (Megascops) barbarus
Bearded Screech Owl
Tropfenkreischeule

Wing length:	126–145 mm
Tail length:	62.5–76 mm
Tarsus length:	22.5–28 mm
Length of bill:	17–19.5 mm
Body mass:	69 g ($n = 1$)

Illustration: ? in Sclater and Salvin 1868b: 57 (b/w, foot); T. Boyer in Boyer and Hume 1991: 53; S. Webb in Howell and Webb 1995: Pl. 25; F. Weick in König et al. 1999: Pl. 18; F. Weick unpublished (new edn. Owls: Pl. 66 (juvenile))
Photograph: J. L. Rangel-Salazar in del Hoyo et al. 1999: 86; J. L. Rangel-Salazar in internet: Owl Pages, photo gallery 2002
Literature: Sharpe 1875a: 107; Ridgway 1914: 723, 724; Moore and Peters 1939: 55, 56; Eck and Busse 1973: 87; Weske and Terborgh 1981: 1–7; Hekstra 1982a: 131 pp; Boyer and Hume 1991: 53; Howell and Webb 1995: 358; del Hoyo et al. 1999: 176; König et al. 1999: 257

Otus ingens **(Salvin) 1897**
Rufescent Screech Owl · Salvin-Kreischeule · Petit-duc de Salvin · Tecolote de Salvin

Length: 250–280 mm
Body mass: 135–220 g

Distribution: Eastern slopes of Andes, locally from northern Venezuela to Peru and northern Bolivia
Habitat: Dense humid cloud forests rich in epiphytes and scrub of Andean slopes, from about 1 200 up to 2 500 m

- ***Otus ingens venezuelanus*** Phelps and Phelps 1954
  Otus ingens venezuelanus Phelps and Phelps 1954, Proc. Biol. Soc. Wash. 67: 103; Terra typica: Cerro Pechoaima, Rio Negro, Sierra de Perija, 1 700 m, Zulia

Distribution: Northern Colombia to northwest Venezuela
Museum: USNM (holotype, Phelps coll.)

Wing length: 203 mm
Tail length: 106 mm
Tarsus length: 28 mm
Length of bill: 23.5 mm
Body mass: 175 g

Illustration: S. Webb in Hilty 2003: Pl. 24
Photograph: ?
Literature: Hekstra 1982a: 119 (131 pp); Fitzpatrick and O'Neill 1986: 1–14; del Hoyo et al. 1999: 179; König et al. 1999: 265, 266; Hilty 2003: 359, 360

- ***Otus ingens ingens*** (Salvin) 1897
  Scops ingens Salvin 1897, Bull. Brit. Ornith. Cl. 6: 37; Terra typica: Jima, Ecuador

Synonym:
- *Ciccaba aequatorialis* Chapman 1922, Am. Mus. Novit. 31: 4–5; Terra typica: Ambato, east of Los Banos, Ecuador = *Otus albogularis*
- *Ciccaba minima* Carriker 1935, Proc. Ac. Nat. Sci. Phila. 87: 313; Terra typica: Santa Ana, 2 000 feet, Rio Coroico, Dep. La Paz Bolivia (holotype MHNUSM)

Distribution: Andes, from northeast Ecuador to western and central Bolivia
Museum: AMNH, BMNH, FMNH, LSUMZ, MHNUSM, USNM

Wing length: ♂ 183.5–200 mm, ♀ 188–212 mm
Tail length: 94–113 mm
Tarsus length: 31–36 mm
Length of bill: 22–24.5 mm
Body mass: ♂ 134–180 g, ♀ 140–223 g

Illustration: J. O'Neill in Fitzpatrick and O'Neill 1986: 11 (b/w, feet); J. Fjeldsa in Fjeldsa and Krabbe 1990: Pl. 25; T. Boyer in Boyer and Hume 1991: 56; F. Weick in König et al. 1999: Pl. 20; P. J. Greenfield in Ridgely and Greenfield 2001: II: Pl. 35(5)
Photograph: ?
Literature: Eck and Busse 1973: 90; Hekstra 1982a: 119 (131 pp); Fitzpatrick and O'Neill 1986: 1–14; Hilty and Brown 1986: 226; Fjeldsa and Krabbe 1990: 224; del Hoyo et al. 1999: 179; König et al. 1999: 265, 266; Ridgely and Greenfield 2001: I: 305, II: 213

Otus colombianus **Traylor 1952**
Colombian Screech Owl · Kolumbien-Kreischeule · Scops de Colombia · Autillo de Colombiano
Otus ingens colombianus **Traylor 1952, Nat. Hist. Misc. [Chicago] 99: 1–4; Terra typica: El Taumbo, Cauca, 5 000 feet, Colombia**

Length: 260–280 mm
Body mass: 150–210 g

Distribution: Western and central Colombia and northwestern Ecuador
Habitat: Cloud forest with dense understorey, from about 1 300 m up to 2 300 m above sea-level
Museum: AMNH, ChiMNH (holotype), FMNH, MHNP
Remarks: Relationship uncertain. Often included as race of *Otus ingens*. Sometimes considered as conspecific with *Otus petersoni* and *Otus marshalli*

Wing length: 175–189 mm (1 × 193 mm)
Tail length: 90–104 mm
Tarsus length: 34–37 mm
Length of bill: 22.5–25 mm
Body mass: ♂ 150 and 156 g, ♀ 210 g

Illustration: J. O'Neill in Fitzpatrick and O'Neill 1986: 11 (b/w, feet); J. Fjeldsa in Fjeldsa and Krabbe 1990: Pl. 25; T. Boyer in Boyer and Hume 1991: 57; F. Weick in König et al. 1999: Pl. 20; P. J. Greenfield in Ridgely and Greenfield 2001: II: Pl. 35(5)

Photograph: ? fide König from internet ? 4 colour photographs

Literature: Traylor 1952: 1–4; Hekstra 1982a: 120 (131 pp); Fitzpatrick and O'Neill 1986: 1–14; Hilty and Brown 1986: 226; Boyer and Hume 1991: 57; del Hoyo et al. 1999: 179; König et al. 1999: 267; Ridgely and Greenfield 2001: I: 305, II: 213

Otus petersoni Fitzpatrick and O'Neill 1986
Cinnamon Screech Owl · Zimt-Kreischeule · Petit-duc de Peterson · Urcututú acanelado (de Peterson)
Otus petersoni Fitzpatrick and O'Neill 1986, Wilson Bull. 98(1): 1–14; Terra typica: Cordillera de Cóndor, above San José de Lourdes, Dep. Cajamarca, Peru (see Figure 12)

Length: 210 mm
Body mass: 88–119 g

Distribution: Cordillera de Cutucú, southeast Ecuador, south to La Peca region, northwestern Peru

Habitat: Cloudforest with dense undergrowth, rich on epiphytes and mosses. 1700–2500 m above sea-level

Museum: AMNH (holotype), ANSP, FMNH, LSUMZ

Remarks: According to del Hoyo et al.: "formerly included under invalid name *O. huberi*", (p 179, no. 54) but this is an error: *O. huberi* is a synonym of *Otus vermiculatus*

Wing length: ♂ 153–165.5 mm, ♀ 155–161.5 mm
Tail length: 81–90.5 mm
Tarsus length: 25–29 mm
Length of bill: 21–22.5 mm
Body mass: ♂ 88–119 g, ♀ 92–105 g

Illustration: Peterson in Fitzpatrick and O'Neill 1986: Frontispiece; T. Boyer in Boyer and Hume 1991: 56; F. Weick in König et al. 1999: Pl. 20; P. J. Greenfield in Ridgely and Greenfield 2001: II: Pl. 35(6)

Photograph: M. Robbins in *Bull. ABC* 2000: 289

Literature: Fitzpatrick and O'Neill 1986: 1–14; Boyer and Hume 1991: 56; del Hoyo et al. 1999: 179; König et al. 1999: 268; Ridgely and Greenfield 2001: I: 305, 306, II: 213, 214

Otus marshalli Weske and Terborgh 1981
Cloudforest Screech Owl · Nebelwald-Kreischeule · Scops (Petit-duc) de Marshall · Urcututú de Marshall
Otus marshalli Weske and Terborgh 1981, Auk 98(1): 1–7; Terra typica: Prov. de la Convención, Dep. Cusco, Peru (see Figure 12)

Length: 200–230 mm
Body mass: 107–115 g

Distribution: Central and southern Peru, Departementos Pasco and Cusco

Habitat: Humid cloudforest, with epiphytes and mosses, and with dense understorey. 1900–2500 m above sea-level

Museum: AMNH (holotype), LSUMZ (1 ♂)

Remarks: According to del Hoyo et al.: "formerly included under invalid name *O. huberi*", (p 179, no. 55) but this is an error: *O. huberi* is a synonym of *Otus vermiculatus*

Wing length: 151.5–164 mm
Tail length: 84.5–91.5 mm
Tarsus length: 26–30 mm
Length of bill (cere): 10 and 11 mm
Body mass: ♂ 107 g, ♀ 115 g

Illustration: R. T. Peterson in Weske and Terborgh 1981: Frontispiece; J. Fieldsa in Fjeldsa and Krabbe 1990: Pl. 25; T. Boyer in Boyer and Hume 1991: 60; F. Weick in König et al. 1999: Pl. 20

Photograph: J. O'Neill in Burton et al. 1992: 86; S. Allen-Stotz in *Bull. ABC* 2000: 387; ? fide König from internet ? (Bolivia)

Literature: Weske and Terborgh 1981: 1–7; Fitzpatrick and O'Neill 1986: 1–14; Fjeldsa and Krabbe 1990: 224; Boyer and Hume 1991: 60; del Hoyo et al. 1999: 179; König et al. 1999: 266

Otus watsonii (Cassin) 1849
Tawny-bellied Screech Owl · Watson-Kreischeule · Scops de Watson · Urcututú del Amazonas

Length: 190–230 mm
Body mass: 114–155 g

Distribution: Eastern Colombia, south to northeast and eastern Peru, east to Surinam, Amazonian

Brazil, south to the lowland forests of northern Bolivia and North Mato Grosso

Habitat: Lowland rainforest, especially growth primary and mature secondary forest. From sea-level up to about 600 m

Remarks: According to Hilty 2003: 360, *O. watsonii* lives in elevations from 300 m (south of Orinoco) up to 2 000 m (north of Orinoco). Large vocal differences are noted. Probably a highland species, similar in plumage, possibly separated

- ***Otus watsonii watsonii*** (Cassin) 1849
Ephialtes Watsonii Cassin 1849, Proc. Acad. Nat. Sci. Phila. 1848(4): 123;
Terra typica: South America, Napo region of eastern Ecuador, error: i.e. Orinoco River, Venezuela, cf. Chapman (1928) Am. Mus. Novit. 332: 2

Synonym:
- *Scops lophotes* Lesson 1831, Traite d Orn. 1: 107 ("patrie in connue") = error: "provenant d'un échange avec Perrot de Cayenne, 1826"

Remarks: If this is a valid name, *Otus watsonii* is a synonym! See also Hekstra 1982a: 36 and 111

Distribution: Lowlands from northeast Colombia, extreme northwest Venezuela, south to northeast Peru and East Surinam (east of Amazon River) and Amazonian Brazil

Museum: AMNH, BMNH, FNSF, NNML, SMNSt, SMTD, USNM, ZFMK, ZMH

Wing length:	164–184 mm
Tail length:	85–99 mm
Tarsus length:	30–32 mm
Length of bill:	21.5–24.5 mm
Body mass:	114–155 g

Illustration: P. Barruel in Haverschmidt 1968: Pl. 13; T. Boyer in Boyer and Hume 1991: 58; F. Weick in König et al. 1999: Pl. 21; P. J. Greenfield in Ridgely and Greenfield 2001: II: Pl. 35(4); S. Webb in Hilty 2003: Pl. 24

Photograph: J. O'Neill in Burton et al. 1992: 89; R. Behrstock in del Hoyo et al. 1999: 86

Literature: Haverschmidt 1968: 139; Eck and Busse 1973: 89; Hekstra 1982a: 111; Boyer and Hume 1991: 58; del Hoyo et al. 1999: 179, 180; König et al. 1999: 268; Ridgely and Greenfield 2001: I: 306, II: 214; Hilty 2003: 360

- ***Otus watsonii usta*** (Sclater) 1858
Scops usta Sclater 1858, Trans. Zool. Soc. London 4: 265 and Pl. 61; Terra typica: Teffé on the Solimoes, Brazil

Synonym:
- *Otus atricapillus ater* Hekstra 1982, Bull. Zool. Mus. Univ. Amsterdam9(7): 61; Terra typica: Belem, Pará (holotype USNM)
- *Otus atricapillus morelius* Hekstra 1982, ibid.: 62; Terra typica: La Morelia, Rio Capueta, 600 feet, Colombia (holotype AMNH)
- *Otus atricapillus inambarii* Hekstra 1982, ibid.: 62; Terra typica: Huiyaumba, Inambari, Cuzco, 630 m, Peru (holotype ChiMNH)
- *Otus atricapillus fulvescens* Hekstra 1982, ibid.: 62; Terra typica: Chapada, Mato Grosso (holotype AMNH)

Distribution: Eastern Peru and southern Amazonian Brazil, south to the lowland forests of northern Bolivia and northern Mato Grosso

Remarks: Often regarded as a distinct species, but song differences to *O. w. watsonii* may represent a cline!

Museum: AMNH, ChiMNH, LSUMZ, MHNUSM, USNM

Wing length:	164–187 mm
Tail length:	82–96 mm
Tarsus length:	29–33 mm
Length of bill:	21–24.5 mm
Body mass:	115–141 g

Illustration: J. Wolf in Sclater 1858: Pl. 61; F. Weick in König et al. 1999. Pl. 21

Photograph: ? fide König from internet ? (eastern Peru)

Literature: Sclater 1858: 265; Sharpe 1875a: 111, 112; Eck and Busse 1973: 89, 90; Hekstra 1982a: 131 pp; Hilty and Brown 1986: 226; del Hoyo et al. 1999: 179, 180; König et al. 1999: 270; Hilty 2003: 360

Otus atricapillus Temminck 1822
Variable Screech Owl · Schwarzkappen-Kreischeule · Scops variable · Alicuco tropical
***Strix atricapilla* Temminck 1822, Pl. col. livr. 25: Pl. 145; Terra typica: Brazil**
(see Plate 3 and Figure 12)

Length:	220–230 mm
Body mass:	115–160 g

Synonym:
- *Ephialtes argentina* Lichtenstein 1854, Nomencl. Mus. Berol.: 7; Terra typica: Montevideo (holotype MNBHU)
- *Otus choliba pintoi* L. Kelso 1937, Biol. Leaflet 8: 1; Terra typica: Southern Brazil (holotype MHNUSM)

Distribution: Southeast Brazil (South Bahia and Goiás south to Santa Catarina), southeast Paraguay and extreme northeast of Argentina (North Misiones)
Habitat: Primary and secondary rainforest with dense undergrowth in warm climates. In south up to 600 m, in northern areas from sea-level up to 250 m
Museum: FNSF, MCNUS, MNBHU, MHNUSM, NNML (holotype), SMNSt, SMTD
Remarks: Specifically distinct from other neighbouring *Otus* species (vocalisations and DNA evidence)

Wing length: 170–184 mm
Tail length: 93–110 mm
Tarsus length: 30–33 mm
Length of bill: 21–22.5 mm
Body mass: ♂ 115–140 g, ♀ –160 g

Illustration: C. J. Temminck in Temminck 1822: Pl. 145; J. Marshall in Marshall 1991: Frontispiece; T. Boyer in Boyer and Hume 1991: 55; F. Weick in König et al. 1999: Pl. 21
Photograph: C. König in *Gefiederte Welt*: 2001(5): 178, 179; C. König in internet: Owl Pages, photo gallery 2003
Literature: Eck and Busse 1973: 89; Hekstra 1982a: 113, 114 (131 pp); Marshall 1991: 314, 315; Boyer and Hume 1991: 55; König et al. 1994: 1–35; del Hoyo et al. 1999: 181; König et al. 1999: 271, 272; König 2001: 177–181

Otus sanctae-catarinae (Salvin) 1897
Long-tufted Screech Owl · Santa Catarina-Kreischeule · Scops de Salvin · Alicuco de Santa Catarina
***Scops sanctae-catarinae* Salvin 1897, Bull. Brit. Ornith. Cl. 6: 37; Terra typica: Southern Brazil**

Length: 250–270 mm
Body mass: 155–211 g

Synonym:
- *Otus choliba maximus* Stolzmann 1926, Ann. Zool. Polon. Hist. Nat. 5: 124; Terra typica: Vermelho, Paraná, Brazil

Distribution: Southeast Brazil (Paraná, Santa Catarina, Rio Grande do Sul), northeast Argentina (Misiones) and Uruguay
Habitat: Semi-open woodland, open pastureland with wooded areas, upland moors. Above 300 m up to 1 000 m
Museum: BMNH, FNSF, MACN, MNBHU, NNML, SMNSt, ZSBS
Remarks: Formerly considered as conspecific with *O. atricapillus*, but vocally and morphologically distinct

Wing length: 182–210 (215) mm
Tail length: 97–121 (127) mm
Tarsus length: 33–40 mm
Length of bill: 22–25.5 mm
Body mass: ♂ 155–194 g, ♀ 174–211 g

Illustration: J. Marshall in Marshall 1991: Frontispiece; F. Weick in König et al. 1999: Pl. 21
Photograph: K. Zimmer in del Hoyo et al. 1999: 86
Literature: Hekstra 1982a: 115 (131 pp); König et al. 1994: 1–35; del Hoyo et al. 1999: 181; König et al. 1999: 272–274

Otus hoyi König and Straneck 1989
Montane Forest-Screech Owl · Bergwald-Kreischeule · Petit-duc de Hoy · Alicuco Fresco (yungueno)
***Otus hoyi* König and Straneck 1989, Stuttgarter Beitr. Naturkd. Ser. A 428: 4; Terra typica: "La Cornisa", between La Caldera and El Carmen, 40 km north of Salta, Argentina (see Plate 3 and Figure 12)**

Length: 230–240 mm
Body mass: 115–145 g

Distribution: Mountains of southern Bolivia (south Cochabamba), to Argentina (Salta, Jujuy, Tucumán and Catamarca)
Habitat: Montane forest and cloud forest on the Andean slopes and pre-Andean mountains, from 1 000 m up to 2 600 m
Museum: SMNSt (holotype), MACN

Wing length: 170–177 mm
Tail length: 88–98 mm
Tarsus length: 25 mm
Length of bill: 20 mm
Body mass: 115–145 g

Illustration: F. Weick in König et al. 1999: Pl. 20 and backcover; F. Weick in collection C. König, Ludwigsburg
Photograph: C. König 2001: 177–181; C. König in internet: Owl Pages, photo gallery 2003
Literature: König and Straneck 1989: 1–20; del Hoyo et al. 1999: 180, 181; König et al. 1999: 263, 264

Otus guatemalae **(Sharpe) 1875**
Guatemalan Screech Owl · Rotgesicht-Kreischeule · Petit-duc guatémaltèque · Tecolote Guatemalteco

Length: 200–230 mm
Body mass: 100–150 g

Distribution: From northwest Mexico south through Mexico, Yucatán and Cozumel Island to Honduras and northern Nicaragua
Habitat: Humid and semi-arid forest, semi-deciduous forest, thorn forest, scrub and plantations. From sea-level up to 1 500 m
Remarks: Sometimes considered as conspecific with *O. atricapillus*, but vocally distinct!

- ***Otus guatemalae hastatus*** (Ridgway) 1887
Megascops hastatus Ridgway 1887, Proc. US Nat. Mus. 10: 268; Terra typica: Mazatlan, Sinaloa, Mexico

Synonym:
- *Otus guatemalae tomlini* Moore 1937, Proc. Biol. Soc. Wash. 50: 65; Terra typica: La Guasimas, Sinaloa, Mexico

Distribution: Northwest Mexico to tropical western Mexico
Museum: USNM (holotype), M(oore) LZ (holotype *tomlini*), MCZ, UKMNH

Wing length: 153–165 mm
Tail length: 76–86 mm
Tarsus length: 28–30 mm
Length of bill: 20–22 mm
Body mass: ?

Illustration: J. Marshall in Marshall 1991: Frontispiece
Photograph: B. Broder in Marshall 1967: Fig. 26, 27 (skin, b/w)
Literature: Ridgway 1914: 718, 719; Marshall 1967: 72 pp; Hekstra 1982a: 131 pp; Marshall 1991: 314, 315; del Hoyo et al. 1999: 180; König et al. 1999: 276, 277; Johnsgard 2002: 98–103

- ***Otus guatemalae cassini*** (Ridgway) 1878
Scops brasilianus var. *cassini* Ridgway 1878, Proc. US Nat. Mus. 1: 102; Terra typica: Hacienda Mirador and Jalapa, Vera Cruz, Mexico

Synonym:
- *Otus guatemalae pettingilli* Hekstra 1982, Bull. Zool. Mus. Univ. Amsterdam9(7): 56; Terra typica: Tamaulipas, Mexico

Distribution: Eastern Mexico (southern Tamaulipas and northern Vera Cruz)
Museum: USNM (holotype), LSUMZ (holotype *pettingilli*), see photograph B. Broder in Marshall 1967: Fig. 26, 27

Wing length: 152–158.5 mm
Tail length: 68–81 mm
Tarsus length: 28.5–31 mm
Length of bill: 20–22.5 mm
Body mass: ?

Illustration: J. Marshall in Marshall 1991: Frontispiece; F. Weick in König et al. 1999: Pl. 22
Photograph: B. Broder in Marshall 1967: Fig. 26, 27 (skin, b/w)
Literature: Ridgway 1914: 320–322; Hekstra 1982a: 131 pp; Marshall 1991: 314, 315; del Hoyo et al. 1999: 180; König et al. 1999: 276, 277; Johnsgard 2002: 98–103

- ***Otus guatemalae guatemalae*** (Sharpe) 1875
Scops guatemalae Sharpe 1875, Cat. Birds Brit. Mus. 2: 112 and Pl. 9; Terra typica: Guatemala

Synonym:
- *Megascops marmoratus* Nelson 1897, Auk 14: 49; Terra typica: Catemaco, Veracruz
- *Otus choliba thompsoni* Cole 1906, Bull. Mus. Comp. Zool. 50: 125; Terra typica: Chichen Itza, Yucatan
- *Otus guatemalae fuscus* Moore and Peters 1939, Auk 56: 52; Terra typica: Motzorongo, Vera Cruz (holotype M(oore) LZ)

- *Otus guatemalae peteni* Hekstra 1982, Bull. Zool. Mus. Univ. Amsterdam9(7): 57; Terra typica: Laguna Perida, Peten, Guatemala (holotype UM(ich) MZ)

Distribution: Mexico from Veracruz, Yucatán and Cozumel Island to Guatemala and Honduras
Museum: BMNH (holotype), FNSF, MHNUSM, M(oore)LZ, MVZ, SMNSt, SMTD, UM(ich)MZ, ZFMK, ZMH

Wing length:	153–177 mm
Tail length:	81–94 mm
Tarsus length:	29–34 mm
Length of bill:	20–23.5 mm
Body mass:	91–123 g

Illustration: J. G. Keulemans in Sharpe 1875a: Pl. 9; D. Gardner in Stiles and Skutch 1989: Pl. 20; J. Marshall in Marshall 1991: Frontispiece; T. Boyer in Boyer and Hume 1991: 53; S. Webb in Howell and Webb 1995: Pl. 25; F. Weick in König et al. 1999: Pl. 22
Photograph: R. Behrstock in del Hoyo et al. 1999: 92; J. L. Rangel in Johnsgard 2002: Pl. 34; R. and N. Bowers in internet: Owl Pages, photo gallery 2003; R. A. Behrstock in Duncan 2003: 228
Literature: Sharpe 1875a: 112; Ridgway 1914: 715–718; Eck and Busse 1973: 88; Hekstra 1982a: 48–50 (131 pp); Ridgely and Gwynne 1989: 187; Stiles and Skutch 1989: 191; Marshall 1991: 314, 315; Boyer and Hume 1991: 53; Howell and Webb 1995: 358; del Hoyo et al. 1999: 180; König et al. 1999: 276, 277; Johnsgard 2002: 98–103

- ***Otus guatemalae dacrysistactus*** Moore and Peters 1939
Otus guatemalae dacrysistactus Moore and Peters 1939, Auk 56: 53; Terra typica: Jalapa, Nicaragua

Synonym:
- *Otus guatemalae centralis* Hekstra 1982, Bull. Zool. Mus. Univ. Amsterdam9(7): 57; Terra typica: Cerro Mali, 4 100 feet, Darien, Panama (holotype USNM)

Distribution: Northern Nicaragua to northern Panama
Museum: AMNH (holotype), USNM
Remarks: *Otus centralis*, by Ridgely and Greenfield, is a distinct species, very similar in plumage to *Otus roraimae*, but with a distinctive voice. (Tape-recorded by Robbins et al. 1987 and Coopmans 1992). Needs further research

Wing length:	164–175 mm
Tail length:	81–87 mm
Tarsus length:	29–31 mm
Length of bill:	22.5–23.5 mm
Body mass:	?

Illustration: ?
Photograph: B. Broder in Marshall 1967: Fig. 28, 29 (skin, b/w photograph)
Literature: Marshall 1967: 72 pp; Hekstra 1982a: 50 (131 pp); del Hoyo et al. 1999: 180; König et al. 1999: 276, 277; Ridgely and Greenfield 2001: 212, 213

Otus vermiculatus (Ridgway) 1887
Vermiculated Screech Owl · Marmor-(Kritzel-)Kreischeule · Scops vermiculé · Tecolote vermiculado
Megascops vermiculatus Ridgway 1887, Proc. US Nat. Mus. 10: 267; Terra typica: Costa Rica

Length:	200–230 mm
Body mass:	~107 g

Synonym:
- *Otus vermiculatus huberi* Kelso and Kelso 1943, Auk 60: 448; Terra typica: Bogotá, Colombia (holotype Acad. Nat. Sci. Bogotá (Rivoli coll.))
- *Otus guatemalae pallidus* Hekstra 1982, Bull. Zool. Mus. Univ. Amsterdam9(7): 57; Terra typica: Andes de Cumana, northern Venezuela (holotype AMNH)

Distribution: Costa Rica, Panama and extreme northwest Colombia and northern Venezuela
Habitat: Humid tropical forest in lowlands and foothills, up to 1 200 m
Museum: USNM (holotype), AMNH, ANSP, BMNH, ChiMNH, FNSF, MCZ, MNBHU, NNML
Remarks: del Hoyo et al. include *O. napensis* and *O. roraimae* with *vermiculatus*, but it is an isolated, endemic species

Wing length:	153–170 mm
Tail length:	75–83 mm
Tarsus length:	26–30 mm
Length of bill:	21–23 mm
Body mass:	~107 g

Illustration: A. Gwynne in Ridgely and Gwynne 1989: Pl. 12; J. Marshall in Marshall 1991: Frontispiece; F. Weick in König et al. 1999: Pl. 22

Photograph: B. Broder in Marshall 1967: Fig. 30, 31 (b/w); R. Behrstock in del Hoyo et al. 1999: 86; J. L. Salazar in del Hoyo et al. 1999: 92; R. and N. Bowers in internet: Owl Pages, photo gallery 2003
Literature: Ridgway 1914: 724, 726; Marshall 1967: 72 pp; Hekstra 1982a: 51 (131 pp); Hilty and Brown 1986: 225; Ridgely and Gwynne 1989: 187; Marshall 1991: 314, 315; del Hoyo et al. 1999: 180; König et al. 1999: 274, 275

❖

Otus napensis Chapman 1928
Rio Napo Screech Owl · Rio Napo-Kreischeule · Scops de Rio Napo · Tecolote del Rio Napo

Length: 200–230 mm
Body mass: ?

Distribution: Eastern Ecuador and eastern Colombia, along the eastern Andean slope to Peru and northern Bolivia
Habitat: Dense rainforest from about 250 up to 1 500 m above sea-level

▪ *Otus napensis napensis* Chapman 1928
Otus guatemalae napensis Chapman 1928, Am. Mus. Novit. 332: 3; Terra typica: Eastern Ecuador

Distribution: Eastern Ecuador to Peru and northern Bolivia
Museum: AMNH, MNBHU

Wing length: 156–174 mm
Tail length: 77–80 mm
Tarsus length: 26–28 mm
Length of bill: 20.5–23.5 mm
Body mass: ?

Illustration: J. Marshall in Marshall 1991: Frontispiece; F. Weick in König et al. 1999: Pl. 22
Photograph: ?
Literature: Chapman 1928: 3; Kelso 1934d: 48; Hekstra 1982a: 51, 51 (131 pp); Marshall 1991: 314, 315; del Hoyo et al. 1999: 180; König et al. 1999: 277, 278

▪ *Otus napensis helleri* L. Kelso 1940
Otus vermiculatus helleri L. Kelso 1940, Biol. Leaflet 12: 1; Terra typica: Rio Comberciato, 4 000 feet, Urubamba Valley, Dep. Cusco, Peru

Distribution: Peru
Museum: MHNUSM, ZMH
Remarks: In del Hoyo et al. (1999), *helleri* included with *napensis*

Wing length: 162–175 mm
Tail length: 80–82 mm
Tarsus length: 24–26.5 mm
Length of bill: 20.5 and 21 mm
Body mass: ?

Illustration: F. Weick in König et al. 1999: Pl. 22
Photograph: ?
Literature: Kelso 1940: 1; Hekstra 1982a: 52 (131 pp); König et al. 1999: 277, 278

▪ *Otus napensis bolivianus* Bond and Meyer de Schauensee 1941
Otus guatemalae bolivianus Bond and Meyer de Schauensee 1941, Not. Nat. 93: 2; Terra typica: Cochabamba, northern Bolivia

Remarks: Probably only a synonym; included in *O. n. helleri*?

Distribution: Northern Bolivia
Museum: ANSPh (holotype), USNM, ZFMK
Remarks: According to del Hoyo et al. (1999), *bolivianus* is merged with *napensis*

Wing length: 163–172 mm
Tail length: 79.5–92 mm
Tarsus length: 25–31.5 mm
Length of bill: 21–22.5 mm
Body mass: ?

Illustration: J. Marshall in Marshall 1991: Frontispiece
Photograph: ?
Literature: Bond and Meyer de Schauensee 1941: 2; Hekstra 1982a: 131 pp; Marshall 1991: 314, 315; König et al. 1999: 277, 278

❖

Otus roraimae (Salvin) 1897
Roraima (Foothill) Screech Owl · Roraima-Kreischeule · Scops de Roraima · Curucucú de Piedemonte
***Scops roraimae* Salvin 1897, Bull. Brit. Ornith. Cl. 6: 38; Terra typica: Roraima, British Guiana**

Length: 220 mm
Body mass: 105 g

Distribution: Endemic to southern Venezuela and northern Brazil (mountain regions of Roraima, Duida and Neblina tepuis.), eastern Ecuador, eastern Peru
Habitat: Rainforest in slopes of tepuis, about 1 000 m up to 1 800 m. Locally at lower elevation (300–1 000 m)
Museum: AMNH, BMNH, SMNSt
Remarks: According to Hilty (2003), birds from the tepuis may be distinct species!

Wing length: 150–168 mm
Tail length: 75–83 mm
Tarsus length: 24–27 mm
Length of bill: 20–22 mm
Body mass: 105 g

Illustration: J. Marshall in Marshall 1991: Frontispiece; F. Weick in König et al. 1999 Pl. 22; P. J. Greenfield in Ridgely and Greenfield 2001: II: Pl. 35(1); S. Webb in Hilty 2003: Pl. 24
Photograph: ?
Literature: Kelso 1934d: 48; Hekstra 1982a: 53 (131 pp); Marshall 1991: 314, 315; König et al. 1999: 275; Ridgely and Greenfield 2001: I: 302, II: 211, 212; Hilty 2003: 358, 359

Otus nudipes (Daudin) 1800

Puerto Rican Screech Owl · Nacktfußeule · Petit-duc de Puerto Rico · Tecolote Múcaro (de Puerto Rico)

Length: 200–230 mm
Body mass: 103–154 g

Remarks: Other generic name *Gymnasio* Bonaparte 1854: 543. Type *Strix nudipes* Daudin

Distribution: Puerto Rico, probably Isla de Culebra and Virgin Island, extinct on Isla de Vàsques
Habitat: Dense woodland, forest of all types, plantations. From sea-level up to 900 m

- ***Otus nudipes nudipes*** (Daudin) 1800
  Strix nudipes Daudin 1800, Traite d Orn. 2: 199; Terra typica: Puerto Rico

Synonym:
- *Gymnoglaux krugii* Gundlach 1874, J. Orn. 22: 310; Terra typica: Puerto Rico

Distribution: Puerto Rico
Museum: AMNH, BMNH, FNSF, MNBHU, ÜMB

Wing length: 153.5–171 mm
Tail length: 78–87 mm
Tarsus length: 36–41 mm
Length of bill: 21.5–24.5 mm
Body mass: 103–154 g

Illustration: 1868: 57 (feet *O. flammeolus, barbarus* and *O. nudipes*, b/w); ? in Sclater and Salvin 1868b: 328 (feet *O. nudipes* and *Gymnoglaux lawrencii)*; T. Boyer in Boyer and Hume 1991: 58; F. Weick in König et al. 1999: Pl. 22; K. Williams in Raffaele et al. 2003: Pl. 43
Photograph: M. Oberle in internet: Owl Pages, photo gallery 2002
Literature: Sclater and Salvin 1868b: 56, 57; Sclater and Salvin 1868c: 327–329; Sharpe 1875a: 149; Ridgway 1914: 677, 678; Hekstra 1982a: 57–59 (131 pp); Bond 1985: 120; Boyer and Hume 1991: 58; del Hoyo et al. 1999: 181; König et al. 1999: 84, 85

- ***Otus nudipes newtoni*** (Lawrence)
  Gymnoglaux Newtoni Lawrence 1860, Ann. Lyc. Nat. Hist. N.Y. 7: 259; Terra typica: St. Croix, Virgin Island (ex Ibis 1859, Pl. 1)

Distribution: Vieques Island (Isla de Vàsques, one record, probably extinct), Culebra and Virgin Island
Museum: BMNH, MNBHU, USNM

Wing length: 157–164 mm
Tail length: 79–85 mm
Tarsus length: 34–37 mm
Length of bill: 21 and 22 mm
Body mass: ?

Illustration: J. Wolf in *Ibis* 1859: Pl. 1; F. Weick in König et al. 1999: Pl. 22
Photograph: ?
Literature: Lawrence 1878a: 184–187; Ridgway 1914: 679; del Hoyo et al. 1999: 181; König et al. 1999: 278, 279

Subgenus
Macabra **Bonaparte 1854**
Macabra Bonaparte 1854, Rev. Mag. Zool. II, 6: 253.
Type by *Syrnium albo-gularis* Cassin 1849

Otus albogularis **Cassin 1849**
White-throated Screech Owl · Weißkehl-Kreischeule · Scops à gorge blanche · Curucucú Orejudo

Length:	200–270 mm
Body mass:	130–185 g

Distribution: Andean forests from Colombia and northwest Venezuela south through Ecuador and Peru to central Bolivia (Cochabamba)
Habitat: Humid montane evergreen and alpine forests (dense with epiphytes and bamboo), from 1 300 m up to 3 600 m

- ***Otus albogularis meridensis*** (Chapman) 1923
Ciccaba albogularis meridensis Chapman 1923, Am. Mus. Novit. 67: 1; Terra typica: Escorial, 2 300 m, near Mérida, Venezuela

Synonym:
- *Otus albogularis obscurus* Phelps and Phelps 1953, Proc. Biol. Soc. Wash. 66: 128; Terra typica: Sierra de Perijá, Northwest Venezuela

Distribution: Northwest and west Venezuela
Museum: AMNH, FNSF, MNBHU

Otus (Macabra) albogularis · White-throated Screech Owl · Weißkehleule

Wing length:	196–207 mm
Tail length:	114–125 mm
Tarsus length:	30–35 mm
Length of bill:	21–23 mm
Body mass:	♀ 185 g ($n = 1$)

Illustration: J. Fjeldsa in Fjeldsa and Krabbe 1990: Pl. 64; F. Weick in König et al. 1999: Pl. 23; S. Webb in Hilty 2003: Pl. 24
Photograph: ?
Literature: Chapman 1923: 67; Kelso 1934d: 58; Eck and Busse 1973: 90, 91; Hekstra 1982a: 123 (131 pp); Hilty and Brown 1986: 227; Fjeldsa and Krabbe 1990: 225, 226; del Hoyo et al. 1999: 181, 182; König et al. 1999: 279, 280; Hilty 2003: 360, 361

- ***Otus albogularis albogularis*** (Cassin) 1849
Syrnium albo-gularis Cassin 1849, Proc. Acad. Nat. Sci. Phila. 4, 1848: 124; Terra typica: "South America" = Coachi, 15 miles east of Bogota, Colombia, ex. Chapman (1917) Bull. Am. Mus. Nat. Hist. 36: 254

Synonym:
- *Ciccaba aequatorialis* Chapman 1922, Am. Mus. Novit. 31: 4; Terra typica: Ambato, East Ecuador (holotype AMNH)

Distribution: Eastern Andes of Colombia and northeast Ecuador
Museum: AMNH, BMNH, FNSF, MNBHU, MZUS, NNML, ZSBS

Wing length:	197–211 mm
Tail length:	114–130 mm
Tarsus length:	31–34 mm
Length of bill:	21–24.5 mm
Body mass:	130–185 g

Illustration: T. Boyer in Boyer and Hume 1991: 59; F. Weick in König et al. 1999: Pl. 23; P. J. Greenfield in Ridgely and Greenfield 2001: II: Pl. 35(7)
Photograph: ?
Literature: Kelso 1934d: 57, 58; Eck and Busse 1973: 90, 91; Hekstra 1982a: 123 (131 pp); Hilty and Brown 1986: 227; Fjeldsa and Krabbe 1990: 225, 226; Boyer and Hume 1991: 59; del Hoyo et al. 1999: 181, 182; König et al. 1999: 279, 280; Ridgely and Greenfield 2001: I: 306, II: 214, 215

- ***Otus albogularis macabrum*** (Bonaparte) 1850
  Syrnium macabrum Bonaparte 1850, Consp. Gen. Av. 1:53; Terra typica: Amerique méridionale "Columbie" by Verreaux

Distribution: Central and western Andes from Colombia and Ecuador to southern and northern Peru
Museum: AMNH, NNML (holotype)

Wing length:	190–213 mm
Tail length:	113–120 mm
Tarsus length:	32–34 mm
Length of bill:	?
Body mass:	?

Illustration: F. Weick in König et al. 1999: Pl. 23
Photograph: ?
Literature: Hekstra 1982a: 123, 124 (131 pp); Hilty and Brown 1986: 227; Fjeldsa and Krabbe 1990: 225, 226; del Hoyo et al. 1999: 181, 182; König et al. 1999: 279, 280

- ***Otus albogularis remotus*** Bond and Meyer de Schauensee 1941
  Otus albogularis remotus Bond and Meyer de Schauensee 1941, Not. Nat. 93:2; Terra typica: Incachaca, 10 000 feet, Dep. Cochabamba, Bolivia

Distribution: Eastern Andes from Peru south to central Bolivia (Cochabamba)
Museum: AMNH, MHNUSM, SMTD

Wing length:	196–209 mm
Tail length:	114–125 mm
Tarsus length:	30–35 mm
Length of bill:	21–23 mm
Body mass:	?

Illustration: J. Fjeldsa in Fjeldsa and Krabbe 1990: Pl. 25; F. Weick in König et al. 1999: Pl. 23
Photograph: J. O'Neill in Burton et al. 1992: 90; J. O'Neill in del Hoyo et al. 1999: 88
Literature: Hekstra 1982a: 124 (131 pp); Fjeldsa and Krabbe 1990: 225, 226; del Hoyo et al. 1999: 181, 182; König et al. 1999: 279, 280

Pyrroglaux

Genus
Pyrroglaux Yamashina 1938

Pyrroglaux Yamashina 1938, Tori 10: 1; by type *Noctua podargina* Hartlaub and Finsch 1872

Remarks: Often placed in genus *Otus*, but appears to be more distantly related

Pyrroglaux podarginus (Hartlaub and Finsch) 1872

Palau Owl · Palau-Zwergohreule · Petit-duc de Palau · Autillo de las Palau

Noctua podargina Hartlaub and Finsch 1872, Proc. Zool. Soc. Lond.: 90; Terra typica: Palau Islands

Length:	220 mm
Body mass:	?

Distribution: Palau Islands (Babelthuap, Koror, Peleliu and Angaur)

Habitat: Mangrove swamps, all types of forest, also near human settlements

Museum: AMNH, BMNH, MNBHU, NHRMSt, NNML, SMNSt, SMTD, USNM, ZMH

Wing length:	155–163 mm
Tail length:	78–87 mm
Tarsus length:	30–34 mm
Length of bill:	22–24.5 mm
Body mass:	?

Illustration: J. G. Keulemans in *Journ. Mus. Godeffroy* 8: Pl. 1 (adult and juvenile); B. Pratt in Pratt et al. 1987: Pl. 22; T. Boyer in Boyer and Hume 1991: 63; F. Weick in König et al. 1999: Pl. 23

Photograph: ?

Literature: Marshall 1949: 207, 208; Eck and Busse 1973: 80, 81; Pratt et al. 1987: 215; Boyer and Hume 1991: 63; del Hoyo et al. 1999; König et al. 1999: 281, 282

Pyrroglaux podarginus · Palau Owl · Palaueule

Genus

***Gymnoglaux* Cabanis 1855**

***Gymnoglaux* Cabanis 1855, J. Orn. 3: 466; by type *Noctua nudipes* Lembeye = *Gymnoglaux lawrencii* Sclater and Salvin**

Gymnoglaux lawrencii Sclater and Salvin 1868

Cuban Screech (Bare-legged) Owl · Kuba-Kreischeule · Petit-duc de Cuba · Cotunto

Length:	200–230 mm
Body mass:	80 g

Distribution: Cuba and Island of Pines

Habitat: Forest, thickets and semi-open limestone country with caves. Plantations

- ***Gymnoglaux lawrencii lawrencii*** Sclater and Salvin 1868
  Gymnoglaux lawrencii Sclater and Salvin 1868, Proc. Zool. Soc. Lond.: 327, Pl. 29; Terra typica: Cuba; ex Lawrence (1860)

Distribution: Central and eastern Cuba

Museum: FNSF, MHNP, USNM

Wing length:	137–154 mm
Tail length:	71–88 mm
Tarsus length:	31–36 mm
Length of bill:	19.5–22 mm
Body mass:	80 g ($n = 1$)

Illustration: J. Smit in Sclater and Salvin 1868c: Pl. 29 and 328 (foot); McManus in Grossman and Hamlet 1965: 444; D. Eckleberry in Bond 1985: 120, Fig. 93; T. Boyer in Boyer and Hume 1991: 123; F. Weick in König et al. 1999: Pl. 23; K. Williams in Raffaele et al. 2003: Pl. 43

Photograph: R. Piechocki in Eck and Busse 1973: Abb. 15 (b/w)

Literature: Sclater and Salvin 1868c: 327–329; Lawrence 1860: 257; Sharpe 1875a: 150, 151; Lawrence 1878a: 184–187; Ridgway 1914: 679, 680; Grossman ML and Hamlet J 1965: 444; Eck and Busse 1973: 146, 147; Hekstra 1982a: 42, 43 (131 pp); Bond 1985: 120; Boyer and Hume 1991: 123; del Hoyo et al. 1999: 182; König et al. 1999: 281, 282

- ***Gymnoglaux lawrencii exsul*** (Bangs) 1913
  Gymnasio lawrencei exsul Bangs 1913, Proc. New England Zool. Cl. 4: 91; Terra typica: Santa Sevilla, Island of Pines

Distribution: Western Cuba and Island of Pines

Museum: MCZ (Bangs coll.)

Remarks: Probably skin of a young bird; synonym to *lawrencii*

Wing length:	151 and 153 mm
Tail length:	82–85 mm
Tarsus length:	33 and 33.5 mm
Length of bill:	?
Body mass:	?

Illustration: F. Weick in König et al. 1999: Pl. 23

Photograph: ?

Literature: Lawrence 1878a: 184–187; Bangs 1913: 91; Ridgway 1914: 679, 680; Hekstra 1982a: 42, 43 (131 pp); del Hoyo et al. 1999: 182; König et al. 1999: 281, 282

Gymnoglaux lawrencii · Cuban Screech Owl · Kuba-Kreischeule

Ptilopsis

Genus
Ptilopsis Kaup 1848 (1851?)
Ptilopsis Kaup, Arch. Naturgesch. 17: 1. Type by *Strix leucotis* Temminck 1820

Ptilopsis leucotis (Temminck)
Northern White-faced Owl · Nordbüscheleule (Weißgesichteule) · Petit-duc à face blanche de Temminck · Autillo Cariblanco Norteno
Strix leucotis Temminck 1820, Pl. col. livr. 3: Pl. 16;
Terra typica: Senegal

Length: 240–250 mm
Body mass: ∅ 204 g ($n = 16$)

Synonym:
- *Asio leucotis nigrovertex* von Erlanger 1904, J. Orn. 52: 233 and Pl. 19; Terra typica: Gambo and Roba-Shalo, Ethiopia
- *Otus leucotis Margarethae* von Jordans and Neubaur 1932, Falco 28: 9; Terra typica: Zankab, Bahr el Abjad

Distribution: Africa, south of the Sahara, from Senegambia east to Sudan, Somalia, northern Uganda and northern and central Kenya
Habitat: Savanna and semi-desert with scattered trees and thorny shrubs, wooded desert water courses, forest edges and clearings. Absent in desert or dense tropical wood
Museum: BMNH, FNSF (holotype *nigrovertex*), MZUS, NHMWien, NMKN, NNML, SMNSt, SMTD, ÜMB, ZFMK (holotype *margarethae*), ZMA, ZMH, ZMUZ, ZSBS

Wing length: 170–205 mm (1 × 209 mm)
Tail length: 75–97 mm (1 × 102 mm)
Tarsus length: 30–34 mm
Length of bill: 25–26 mm
Body mass: ∅ 204 g ($n = 16$)

Illustration: J. J. Temminck in Temminck 1820b: Pl. 16; O. Kleinschmidt in von Erlanger 1904: Pl. 19; F. Neubaur in Koenig 1936: Pl. 50 (*margarethae*); M. Woodcock in Fry et al. 1988: Pl. 9; T. Boyer in Boyer and Hume 1991: 15; P. Hayman in Kemp and Kemp 1998: 305; J. Willis in del Hoyo et al. 1999: Pl. 9; F. Weick in König et al. 1999: Pl. 24; N. Borrow in Borrow and Demey 2001: Pl. 39; P. Hayman in Sinclair and Ryan 2003: 245; J. Gale in Stevenson and Fanshawe 2002: Pl. 100
Photograph: F. Sauer in Sauer 1985: 103; M. Goetz in Kemp and Kemp 1998: 304, 305; G. Lasley in internet: Owl Pages, photo gallery 2002
Literature: von Erlanger 1904: 233; Koenig 1936: 163–165; Vaurie 1965: 604; Fry et al. 1988: 119, 120; Boyer and Hume 1991: 60, 61; Zimmerman et al. 1996: 443; Kemp and Kemp 1998: 304, 305; del Hoyo et al. 1999: 182, 183; König et al. 1999: 282, 283; Borrow and Demey 2001: 493, 494

Ptilopsis granti (Kollibay) 1910
Southern White-faced Owl · Südbüscheleule (Weißgesichteule) · Petit-duc à face blanche de Grant · Autillo Cariblanco Soreno
Scops erlangeri Ogilvie-Grant 1906, Ibis: 660;
Terra typica: South Africa
Pisorhina leucotis granti Kollibay 1910, Orn. Monatsb. 18: 148;
Terra typica: Southwest Africa (new name for *Scops erlangeri* Ogilvie-Grant)
(see Plate 3)

Length: 220–240 mm
Body mass: ♂ 185–240 g, ♀ 215–275 g

Ptilopsis granti · Southern White-faced Owl · Südbüscheleule

Distribution: Africa from southeast Gabon, central Congo, southern Zaire, southern Uganda and southern Kenya, south to Namibia, North Cape Province and Natal

Habitat: Savanna with scattered trees, thorny shrubs and dry, open woods. Wooded areas along rivers, clearings and forest edges, but absent in desert

Museum: BMNH, FNSF, NHRMSt, NMKN, SMNSt, SMTD, ZFMK, ZMH, ZMK, ZSBS

Wing length:	191–206 mm
Tail length:	88–100 mm
Tarsus length:	30–36 mm
Length of bill:	23.5–28.5 mm
Body mass:	♂ 185–240 g, ♀ 225–275 g

Illustration: G. Arnott in Steyn 1982: Pl. 23; J. Willis in del Hoyo et al. 1999: Pl. 9; F. Weick in König et al. 1999: Pl. 24; N. Borrow in Borrow and Demey 2001: Pl. 39; P. Hayman in Sinclair and Ryan 2003: 245

Photograph: P. Ginn in König and Ertel 1979: 144; C. Laubscher in Ginn et al. 1989: 336, 337; C. Laubscher in Burton et al. 1992: 106; L. Hes and M. Goetz in Kemp and Kemp 1998: 304; J. Carlyon in Kemp and Kemp 1998: 305; R. J. Milne in internet: Owl Pages, photo gallery 2002

Literature: Priest 1939: 51–53; Steyn 1982: 259–261; Fry et al. 1988: 119, 120; Ginn et al. 1989: 336; Kemp and Kemp 1998: 304, 305; del Hoyo et al. 1999: 183; König et al. 1999: 283, 284; Borrow and Demey 2001: 493

Mimizuku

Genus
Mimizuku **Hachisuka 1934**
Mimizuku Hachisuka 1934, Bds. Philippine Is. 3: 50.
Type by *Pseutoptynx gurneyi* Tweeddale

Mimizuku gurneyi (Tweeddale) 1879
Giant Scops Owl · Rotohreule (Mindanao-Ohreule) · Hibou de Gurney · Búho de Mindanao
Pseudoptynx gurneyi of Tweeddale 1879, Proc. Zool. Soc. Lond. 1878: 940, Pl. 58; Terra typica: Zamboanga, Mindanao

Length: 300–350 mm
Body mass: ?

Distribution: Southern Philippines: Mindanao, Siargao, Dinagat, absent from Marinduque
Habitat: Lowland primary and secondary rainforest, from sea-level up to 1 200 m
Museum: BMNH (Everett = holotype and Gerrard), DelMNH, M(in)SUNHM, MNBHU
Remarks: DNA studies indicate that *Mimizuku* is closer to the genus *Otus* than to *Bubo*. Formerly placed in genus *Pseutoptynx* together with *Nesasio solomonensis* and *Bubo philippensis*. Placed in the expanded genus *Otus*, but is in fact morphologically distinct from this genus. No example in monotypic genus but further research is needed

Wing length: ♂ 217–242 mm, ♀ -274 mm
Tail length: ♂ 114–120 mm, ♀ -149 mm
Tarsus length: ♂ 39–45 mm, ♀ -54 mm
Length of bill: 29.5–38 mm
Body mass: ?

Illustration: J. Smit in of Tweeddale 1878: Pl. 58; G. Sandström in du Pont 1971: Pl. 38; F. Weick in König et al. 1999: Pl. 24; J. Willis in del Hoyo et al. 1991: Pl. 9; A. Sutherland in Kennedy et al. 2000: Pl. 35
Photograph: R. Steinberg in photo collection Weick (bird in captivity)
Literature: of Tweeddale 1878: 940; Ripley and Rabor 1971: Postilla 50: 4; Eck and Busse 1973: 92; Boyer and Hume 1991: 63; Dickinson et al. 1991: 227; del Hoyo et al. 1999: 183; König et al. 1999: 285; Kennedy et al. 2000: 179

Mimizuku gurneyi
Giant Scops Owl
Mindanao-Ohreule/Rotohreule

Tribus / Tribe
Bubonini · Eagle Owls and allies · Uhus und Verwandte

Genus
Nyctea **Stephens 1826**
Nyctea Stephens 1826, in Shaw's Gen. Zool., 1825, 13(2): 62; by type *Strix scandiaca* Linné 1758

Nyctea scandiaca **(Linné) 1758**
Snowy Owl · Schnee-Eule · Harfang des neiges · Búho Nival
Strix scandiaca von Linné 1758, Syst. Nat. ed. 10, 1: 92;
Terra typica: "Habitat in Alpibus Lapponica"
(see Plate 5)

Length: ♂ 525–640 mm, ♀ 590–660 mm
Body mass: ♂ 710–2500 g, ♀ 780–2950 g

Distribution: Circumpolar, from Greenland and Iceland across North Eurasia to Sakhalin, Alaska and northern Canada. Southern limit between 60° and 50° N, including Spitzbergen, western and northern Scandinavia, northern Russia, northern Siberia, Anadyr and Koryakland, Commandeur and Hall Islands, Aleutians, Alaska and Labrador. Winters south to United States, northern and central Europe, central Russia, northern China and Japan

Habitat: Open tundra, from near tree-line to edge of polar seas. Areas with lichens, mosses, some rocks and slight elevations as hummocks, etc. Locally coastal fields, open moorland and other open areas near seashores

Museum: AMNH, BMNH, FNSF, MHNP, SMNSt, SMTD, ZMH, ZSBS

Remarks: Closest to *Bubo*, but generically distinct on the basis of skull characters and much smaller eyes

Wing length: ♂ 384–429 mm, ♀ 428–462 mm
Tail length: ♂ 206–222 mm, ♀ 217–241 mm
Tarsus length: 48–62 mm
Length of bill: 40–48 mm
Body mass: ♂ 710–2500 g, ♀ 780–2950 g

Illustration: O. Rudbeck (1693–1710) in Krook 1988 (repr.): Pl. 27 and 30; J. J. Audubon (1785–1851) in Audubon 1981 (repr.): 237; J. F. Naumann in Naumann 1822: Pl. 41; E. Lear in Gould 1832–1837 (repr.): Pl. ?; E. Lear (1832–1837) in *Catal. Sotheby's* 1988: 54; E. Traviès in Buffon 1851: 15 and *Catal. Sotheby's* 1996: 97; J. Wolf and H. C. Richter in Gould 1862–1873: Pl. 3; J. G. Keulemans in Hennicke 1905: Pl. 7; P. Barruel in Sutter and

Nyctea scandiaca
Snowy Owl
Schnee-Eule

Barruel 1958: Pl. 22; L. A. Fuertes in Forbush and May 1955: Pl. 46; C. F. Tunnicliffe in Cusa 1984: 35; G. Petterson in Petterson 1984: 35–38; H. Delin in Cramp et al. 1985: Pl. 46; A. Cameron in Voous 1988: Frontispiece and 131; A. Thorburn in Thorburn 1990: Pl. 27; T. Boyer in Boyer and Hume 1991: 91; L. Jonsson in Jonsson 1992: 316; T. Worfolk in del Hoyo et al. 1999: Pl. 11; F. Weick in König et al. 1999: Pl. 36; Mullarney and Zetterström in Svensson et al. 1999: 209; D. Sibley in Sibley 2000: 275; L. A. Fuertes in Johnsgard 2002: Pl. 4; F. Weick in Weick 2004: Dezember

Photograph: E. Hosking in Hosking and Flegg 1982: 30, 71, etc.; E. Barth in Mikkola 1983: 25–32; T. W. Kitchin in MacKenzie 1986: 129; B. and P. Wood in MacKenzie 1986: 128; K. W. Fink in Burton et al. 1992: 81; E. and D. Hosking in Freethy 1992: 58; J. A. Barrie in *Living Bird* 1992(Winter): Frontcover and 17–21; T. and E. Bomford in del Hoyo et al. 1999: 137; T. Fitzharris in del Hoyo et al. 1999: 91; numerous photographs in Mebs and Scherzinger 2000 (by Clemens, Hafner, Löhr, Nill, Wothe etc.)

Literature: Naumann 1822: 417–422 and Nachtr. 168–173; Hartert 1912–1921: 958, 959; Ridgway 1914: 468–470; Dementiev and Gladkov 1951: 386–391; Vaurie 1965: 578, 579; Portenko 1972: 232 pp; Eck and Busse 1973: 113–115; Karalus and Eckert 1974: 185–194; Glutz von Blotzheim and Bauer 1980: 358–386; Mikkola 1983: 95–104; Bezzel 1985: 642–644; Cramp et al. 1985: 485–495; Voous 1988: 123–130; Boyer and Hume 1991: 90–92; Svensson et al. 1999: 208; del Hoyo et al. 1999: 194, 195; König et al. 1999: 317, 318; Sibley 2000: 275; Mebs and Scherzinger 2000: 167–183; Johnsgard 2002: 122–129

Genus
Bubo Duméril 1806
Bubo Duméril 1806, Zool. Anal.: 34; by type *Strix Bubo* Linné 1758

Bubo virginianus (Gmelin) 1788
Great Horned Owl · Virginiauhu · Grand-duc américain · Búho Americano (nacurutú)
(see Plate 4)

Length: 450–600 mm
Body mass: ~1 000 g

Distribution: America from Alaska to Central and South America, south to Brazil and central Argentina. Absent from the Pacific slopes and central parts of the Andes from Peru to Chile, also from Patagonia and Tierra del Fuego
Habitat: Habitat very variable: semi-open landscapes with trees, groves, open woodland and shrubs, deciduous, mixed and coniferous forest, secondary growth, swampy woodlands, open farmland or parks with patches of woodland. Semi-desert. From sea-level up to 4 000 m
Remarks: Formerly included *B. magellanicus* as a subspecies. Numerous geographical races are very poorly differentiated. For a fuller description of the specific – and subspecific – situation, see F. Weick (1999) with consideration of a mostly parallel variation within the polymorphic subspecies (with two colour plates and two distribution maps)

- ***Bubo virginianus saturatus*** Ridgway 1877
Bubo virginianus saturatus Ridgway 1877, U.S. Ged. Expl. 40, Parallel 4, Ornith.: 572; Terra typica: Sitka, Alaska
(see Figures 22 and 23)

Synonym:
- *Asio magellanicus lagophonus* Oberholser 1904, Proc. US Nat. Mus. 27: 178 (key); Terra typica: Fort Walla Walla, Washington (holotype USNM)
- *Asio magellanicus algistus* Oberholser 1904, Proc. US Nat. Mus. 27: 178 (key); Terra typica: St. Michael, Alaska
- *Bubo virginianus leucomelas* Bishop 1931, Proc. Biol. Soc. Wash. 44: 93–94; Terra typica: Southern parts of Vancouver Island, Victoria, British Columbia (holotype USNM (Bishop coll.))

Distribution: Coast of western Alaska to southeast Alaska, south to Washington, northeast and west Oregon, British Columbia, northern and central Alberta, northern California (Monterey)
Museum: CM(P), MNBHU, ROM, USNM (holotype)

Wing length: ♂ 345–370 mm, ♀ 350–400 mm
Tail length: ♂ 205–235 mm, ♀ 213–252 mm
Tarsus length: 62–70 mm
Length of bill: 35–43 mm
Body mass: ? (~ as *B. v. wapacuthu)*

Illustration: A. Brooks in Taverner 1943: Pl. 30; A. Brooks in Sprunt 1955: unpaged coloure plate; K. Karalus in Karalus and Eckert 1974: Pl. 50, 52, 56; T. Worfolk in del Hoyo et al. 1999: Pl. 10; F. Weick in König et al. 1999: Pl. 26; F. Weick in Weick 1999a: 373 c, d, f; D. Sibley in Sibley 2000: 274
Photograph: J. Templeton in MacKenzie 1986: 138; J. Henderson in *Living Bird* 1996(Summer): 8; R. Krahe in letter, photo collection Weick
Literature: Ridgway 1877b: 572; Sharpe 1875a: 19–23; Oberholser 1904: 177–192; Ridgway 1914: 739–754; Taverner 1942: 234–245; Bent 1961 (repr.): 330–345; Johnson and Earhardt 1970: 251–264; Eck and Busse 1973: 97–100; Karalus and Eckert 1974: 224–262; Voous 1988: 79–86; del Hoyo et al. 1999: 185; König et al. 1999: 289, 290; Weick 1999a: 363–387; Sibley 2000: 274; Johnsgard 2002: 113–121

- ***Bubo virginianus pacificus*** Cassin 1854
Bubo virginianus var. *pacificus* Cassin 1854, Illustr. Bds. California, Texas etc. 6: 178; Terra typica: Sacramento, California
(see Figure 23)

Synonym:
- *Asio magellanicus icelus* Oberholser 1904, Proc. US Nat. Mus. 27 (in key); Terra typica: San Luis Obisco, CA (a dark morph of *pacificus* or a *saturatus* × *pacificus* hybrid)

Distribution: Southwestern United States, California except the north, southern Oregon, western and central Nevada, south to northwest Baja, California
Museum: AMNH, ANSP, BYUU, ROM, USNM

Wing length: ♂ 313–353 mm, ♀ 335–375 mm
Tail length: ♂ 190–218 mm, ♀ 213–228 mm
Tarsus length: 57 mm
Length of bill: 34–41 mm
Body mass: ♂ 680–1 272 g, ♀ 825–1 668 g

Illustration: K. Karalus in Karalus and Eckert 1974: Pl. 54; F. Weick in Weick 1999a: 372, 373 h and j
Photograph: D. J. Boyle in *North Am. Wildlife* (Postcard)
Literature: Cassin 1854: 178; Oberholser 1904: 177–192; Ridgway 1914: 739–754; Bent 1961 (repr.): 333; Karalus and Eckert 1974: 224–262; Voous 1988: 79–86; del Hoyo et al. 1999: 185; König et al. 1999: 289, 290; Weick 1999a: 363–387

- ***Bubo virginianus wapacuthu*** (Gmelin) 1788
Strix wapacuthu Gmelin 1788, Syst. Nat. 1(1): 290;
Terra typica: Forests at Hudson Bay
(see Figure 22)

Synonym:
- *Bubo arcticus* Swainson 1832, Fauna Bor. Am. 2: 86 and Pl. 30; Terra typica: Bor. America (name preoccupied)
- *Bubo subarcticus* Hoy 1852, Proc. Acad. Nat. Sci. Phila. 6: 211; Terra typica: (name used for forms of western United States)
- *Bubo virginianus scalariventris* Snyder 1961, R. Ontario Mus. Publ. 54: 5; Terra typica: Algoma District, Ontario

Distribution: Canada, from James Bay to the Mackenzie Valley, north to the treeline, south to northern Alberta, Saskatchewan, central Manitoba and northern Ontario
Museum: BYUU, MNBHU, ROM, USNM, ZFMK

Wing length: ♂ 333–368 mm, ♀ 353–390 mm
Tail length: 215–235 mm
Tarsus length: 66 mm
Length of bill: 37–42 mm
Body mass: ♂ 1 035–1 389 g, ♀ 1 357–2 000 g

Illustration: A. Brooks in Taverner 1943: Pl. 30; J. Landsdowne in Landsdowne and Livingston 1966: Pl. 17; K. Karalus in Karalus and Eckert 1974: Pl. 57, 58, 59; A. Cameron in Voous 1988: 83; T. Worfolk in del Hoyo et al. 1999: Pl. 10; F. Weick in König et al. 1999: Pl. 25; F. Weick in Weick 1999a: 372, 373 a, b, d; D. Sibley in Sibley 2000: 274; R. Bateman in Dean 2004: 88
Photograph: E. Hosking in Hosking and Flegg 1982: 146, 147; K. Fink in del Hoyo et al. 1999: 83; B. Marcot in internet: Owl Pages, photo gallery 2002; H. Veith fide Steinberg in letter, photo collection Weick; L. Koerner in internet: Owl Pages, photo gallery 2002 (captured bird)
Literature: Ridgway 1914: 739–754; Bent 1961 (repr.): 330; Karalus and Eckert 1974: 224–262; Voous 1988: 79–86; del Hoyo et al. 1999: 185; König et al. 1999: 289, 290; Weick 1999a: 363–387; Dickerman 2004: 5, 6

- ***Bubo virginianus heterocnemis*** (Oberholser) 1904
Asio magellanicus heterocnemis Oberholser 1904, Proc. US Nat. Mus. 27: 178 (in key); Terra typica: Lance au Loupe, Labrador
(see Figures 22 and 23)

Synonym:
- *Bubo virginianus neochorus* Oberholser 1914, Proc. Biol. Soc. Wash. 27: 46; Terra typica: Fox Island River, Newfoundland (holotype USNM)
- *Bubo virginianus scalariventris* Snyder 1961, R. Ontario Mus. Publ. 54: 5; Terra typica: Algoma District, Ontario (holotype ROM)

Distribution: Eastern North America, Ungave Peninsula, Labrador, south to Nova Scotia, New Brunswick, Ontario and northern Maine
Museum: ROM, SMTD, USNM

Wing length: ♂ 350–365 mm, ♀ 360–390 mm
Tail length: ♂ 220–230 mm, ♀ 235–250 mm
Tarsus length: 62–70 mm
Length of bill: 38–44 mm
Body mass: ? (similar *wapacuthu*)

Illustration: K. Karalus in Karalus and Eckert 1974: Pl. 51, 54; A. Cameron in Voous 1988: 83; F. Weick in Weick 1999a: 372, 373 c–f
Photograph: D. Lockshaw in internet: Owl Pages, photo gallery 2002
Literature: Oberholser 1904: 177–192; Oberholser 1914: 178; Ridgway 1914: 739–754; Bent 1961 (repr.): 342; Snyder 1961: 5; Karalus and Eckert 1974: 224–262; Voous 1988: 79–86; del Hoyo et al. 1999: 186; König et al. 1999: 289, 290; Weick 1999a: 363–387; Johnsgard 2002: 213–221

- ***Bubo virginianus occidentalis*** Stone 1896
Bubo virginianus occidentalis Stone 1896, Auk 13: 155;
Terra typica: Mitchell, Iowa
(see Figure 22)

Distribution: Central Alberta, southern Saskatchewan and southern Manitoba, south to northeast California, Nevada, Colorado, Kansas and western Minnesota

Museum: ANSP (holotype), BYUU, SMTD, USNM, ZFMK
Remarks: Birds intermediate between *wapacuthu* and *virginianus* in appearance in north and south of distribution range

Wing length: ♂ 323–372 mm, ♀ 349–390 mm
Tail length: ♂ 200–225 mm, ♀ 220–240 mm
Tarsus length: ?
Length of bill: 35–43 mm
Body mass: ♂ 865–1 460 g, ♀ 1 112–2 046 g

Illustration: K. Karalus in Karalus and Eckert 1974: Pl. 53; F. Weick in Weick 1999a: 372, 373 e; D. Sibley in Sibley 2000: 274; J. Laurencelle in internet: Owl Pages, photo gallery 2003
Photograph: E. T. Jones in MacKenzie 1986: 114; Veith fide Steinberg in letter, photo collection Weick; D. Baccus in internet: Owl Pages, photo gallery 2003; P. Miller in internet: Owl Pages, photo gallery 2003
Literature: Oberholser 1904: 177–192; Ridgway 1914: 739–754; Bent 1961 (repr.): 348; Karalus and Eckert 1974: 224–262; Voous 1988: 79–86; del Hoyo et al. 1999: 185; König et al. 1999: 289, 290; Weick 1999a: 363–387

- ***Bubo virginianus virginianus*** (Gmelin) 1788
Strix virginianus Gmelin 1788, Syst. Nat. 1(1): 287;
Terra typica: Virginia
(see Figures 22 and 23)

Distribution: Southern Ontario, Quebec, New Brunswick, Nova Scotia, south through Oklahoma, eastern Kansas and eastern Texas to Florida
Remarks: Intergrating in north with *heterocnemis*, in west with *occidentalis* and *pallescens*
Museum: BYUU, MNBHU, ROM, USNM, ZFMK, ZSBS

Wing length: ♂ 319–355 mm, ♀ 343–382 mm
Tail length: ♂ 190–210 mm, ♀ 200–235 mm
Tarsus length: 56–58.5 mm
Length of bill: 38–43 mm
Body mass: ♂ 985–1 585 g, ♀ 1 417–2 503 g

Illustration: J. Audubon in Audubon 1827–1838 (1981 repr.): Pl. 236; R. Ridgway in Merriam and Fisher 1893: Pl. 24; L. A. Fuertes in Gilbert Pearson 1936: Pl. 57; L. A. Fuertes in MacCrecken-Peck 1952: 53; K. Karalus in Karalus and Eckert 1974: Pl. 49L; A. Wilson in Wilson 1975 (repr.): 50/1 (wood engraving); Malick in Breeden et al. 1983: 239; L. A. Fuertes in *Living Bird* 1990(Spring): 13T; Boyer in Boyer and Hume 1991: 67; T. Worfolk in del Hoyo et al. 1999: Pl. 10; F. Weick in König et al. 1999: Pl. 26; F. Weick in Weick 1999a: 372, 373 l, k; D. Sibley in Sibley 2000: 274; L. A. Fuertes in Johnsgard 2002: Pl. 3; R. Bateman in Dean 2004: 89
Photograph: S. Grossman in Grossman and Hamlet 1965: 32, 128; R. Ranford in MacKenzie 1986: 124; R. Robinson in MacKenzie 1986: 125; J. McDonald in *Living Bird* 1988(Spring): 23; P. Johnsgard in Johnsgard 2002: Pl. 18
Literature: Sharpe 1875a: 19–23; Oberholser 1904: 177–192; Ridgway 1914: 739–754; Bent 1961 (repr.): 295; Boyer and Hume 1991: 65–67; del Hoyo et al. 1999: 185; König et al. 1999: 289, 290; Weick 1999a: 363–387; Johnsgard 2002: 113–121

- ***Bubo virginianus pallens*** Stone 1897
Bubo virginianus pallens Stone 1897, Am. Nat. 31(363): 237;
Terra typica: Watson Ranch, 18 miles southwest of San Antonio, Texas
(see Figures 22 and 23)

Distribution: Southwestern United States, (southeast California, southern Arizona, southern New Mexico, South Texas) and Mexico (northeast Baja California, northern Mexico south to Guerrero, Morelos and west to Veracruz)
Remarks: Sympatric with *virginianus* in central and southeast Texas
Museum: BYUU, USNM (holotype)

Wing length: ♂ 312–368 mm, ♀ 332–381 mm
Tail length: ♂ 195–235 mm, ♀ 200–235 mm
Tarsus length: ?
Length of bill: 33–43 mm
Body mass: ♂ 724–1 257 g, ♀ 801–1 550 g

Illustration: K. Karalus in Karalus and Eckert 1974: Pl. 55; S. Webb in Howell and Webb 1995: Pl. 26; T. Worfolk in del Hoyo et al. 1999: Pl. 10; F. Weick in König et al. 1999: Pl. 26; F. Weick in Weick 1999a: 372, 373 h, k, m, n; D. Sibley in Sibley 2000: 274
Photograph: J. Natherton in del Hoyo et al. 1999: 101; J. Cancalori in del Hoyo et al. 1999: 124; B. Marcot in internet: Owl Pages, photo gallery 2003
Literature: Oberholser 1914: 177–192; Ridgway 1914: 739–754; Webster and Orr 1958: 134–142; Bent 1961 (repr.): 322; Karalus and Eckert 1974: 224–262; Howell and Webb 1995: 359; del Hoyo et al. 1999: 185; König et al. 1999: 289, 290; Weick 1999a: 363–387

- ***Bubo virginianus elachistus*** Brewster 1902
  Bubo virginianus elachistus Brewster 1902, Bull. Mus. Comp. Zool. 41(1): 96; Terra typica: Sierra de Laguna Baja California, Mexico
  (see Figure 23)

Distribution: Southern Baja, California, from 30° N south to Cabo San Lucas
Museum: USNM (coll. Brewster)

Wing length: ♂ 305–335 mm, ♀ 330–? mm
Tail length: 175–211 mm
Tarsus length: ?
Length of bill: 33–38 mm
Body mass: ?

Illustration: F. Weick in Weick 1999a: 372, 373 i, j
Photograph: ?
Literature: Brewster 1902: 96; Oberholser 1904: 177–192; Ridgway 1914: 739–754; Bent 1961 (repr.): 341; del Hoyo et al. 1999: 185; König et al. 1999: 289, 290; Weick 1999a: 363–387

- ***Bubo virginianus mayensis*** Nelson 1901
  Bubo virginianus mayensis Nelson 1901, Proc. Biol. Soc. Wash. 14: 170; Terra typica: Chichen Itza, Yucatan

Synonym:
- *Asio magellanicus mesembrinus* Oberholser 1904, Proc. US Nat. Mus. 27: 178 (in key); Terra typica: San Jose, Costa Rica (holotype USNM)
- *Asio magellanicus melancercus* Oberholser 1904, Proc. US Nat. Mus. 27: 178 (in key); Terra typica: Tehuantepec City, Oaxaca, Mexico (holotype USNM)

Distribution: Central America, from Mexico, Tehuantepec, through Yucatán Peninsula and Honduras south to western Panama
Museum: AMNH, BMNH, USNM (holotype *mayensis*)

Wing length: ♂ 297–340 mm, ♀ 303–357 mm
Tail length: ♂ 180–198 mm, ♀ 199–210 mm
Tarsus length: 54–65 mm
Length of bill: 39–41 mm
Body mass: ?

Illustration: D. Gardner in Stiles and Skutch 1991: Pl. 20; F. Weick in Weick 1999a: 372, 373 h, n–p
Photograph: ?
Literature: Nelson 1901b: 170; Oberholser 1904: 177–192; Ridgway 1914: 739–754; Griscom 1935: 546–547; Webster and Orr 1958: 134–142; Land 1970: 137; Stiles and Skutch 1991: 192; del Hoyo et al. 1999: 185; König et al. 1999: 289, 290; Weick 1999a: 363–387

- ***Bubo virginianus nacurutu*** (Vieillot) 1817
  Strix nacurutu Vieillot 1817, Nouv. Dict. Hist. Nat. 7: 44; Terra typica: ex Azara Paraguay
  (see Figure 23)

Synonym:
- *Bubo virginianus scotinus* Oberholser 1908, Sci. Bull. Brooklyn Inst. Arts Sci. 1: 371; Terra typica: Caicara, Rio Orinoco, Venezuela
- *Bubo virginianus elutus* Todd 1917, Proc. Biol. Soc. Wash. 30: 6; Terra typica: Lorica, Bolivar, Colombia (holotype USNM)

Distribution: Tropical lowlands east of the Andes, from Tucumán, Argentina through Paraguay, Bolivia, the Mato Grosso, British Guiana, Venezuela to Colombia. Probably also in eastern Argentina, Uruguay and eastern Brazil
Museum: FNSF, MNBHU, SMNSt, SMTD, USNM, ZFMK, ZMH, ZSBS

Wing length: ♂ 330–354 mm, ♀ 340–376 mm
Tail length: 184–217 mm
Tarsus length: ?
Length of bill: 43–52 mm
Body mass: ♂ 1011–1132 g, ♀ 1050 g

Illustration: P. Barruel in Haverschmidt 1968: Pl. 14; F. Weick in König et al. 1999: Pl. 26; F. Weick in Weick 1999a: 372, 373 h, o, p, q; S. Webb in Hilty 2003: Pl. 25
Photograph: E. Hosking in Hosking and Webb 1982: 145; E. Waldron in internet: Owl Pages, photo gallery 2002 (*magellanicus* = error); H. Wuchner in letter photo collection Weick; L. Koerner in internet: Owl Pages, photo gallery 2003
Literature: Oberholser 1904: 177–192; Traylor 1958: 143–149; Haverschmidt 1968: 159, 160; Fjeldsa and Krabbe 1990: 226; König et al. 1996: 1–9; del Hoyo et al. 1999: 185; König et al. 1999: 289, 290; Weick 1999a: 363–387; Hilty 2003: 361

- ***Bubo virginianus deserti*** Reiser 1905
  Bubo virginianus deserti Reiser 1905, Anz. Akad. Wiss. Wien 52(18): 324; Terra typica: Salitres near Joazeira, Bahia, Brazil (see Figure 23)

Distribution: Bahia and Santarem, southeast Brazil
Museum: AMNH, NHMWien (holotype), SMTD

Wing length: 340–380 mm
Tail length: 212 mm
Tarsus length: ?
Length of bill: 45 mm
Body mass: ?

Illustration: F. Weick in Weick 1999a: 372, 373 h, o
Photograph: only colour photograph skins
Literature: Traylor 1958: 143–149; König et al. 1999: 289, 290; Weick 1999a: 363–387

- ***Bubo virginianus nigrescens*** Berlepsch and Taczanowski 1884
  Bubo nigrescens von Berlepsch and Taczanowski 1884, Proc. Zool. Soc. Lond.: 309; Terra typica: Cechce, 1000 feet, western Ecuador
  (see Figure 23)

Synonym:
- *Bubo virginianus colombianus* Lehmann 1946, Auk 63: 218; Terra typica: Penablanca, east of Popayan, Cauca, Colombia (holotype MHNUC (coll. Lehmann))

Distribution: Andes from northwest Peru to Ecuador and Colombia
Museum: FNSF, MHNUC, MNBHU, SMTD, USNM

Wing length: ♂ 345–365 mm, ♀ 350–382 mm
Tail length: 185–217 mm
Tarsus length: 80 mm
Length of bill: 40–50 mm
Body mass: ?

Illustration: J. Fjeldsa in Fjeldsa and Krabbe 1990: Pl. 26; F. Weick in König et al. 1999: Pl. 26; F. Weick in Weick 1999a: 372, 373 r; P. J. Greenfield in Ridgely and Greenfield 2001: II: Pl. 36(3)
Photograph: ?
Literature: Traylor 1958: 143–149; Fjeldsa and Krabbe 1990: 226; del Hoyo et al. 1999: 185; König et al. 1999: 289, 290; Weick 1999a: 363–387; Ridgely and Greenfield 2001: I: 306, 307, II: 215

Bubo magellanicus Lesson 1828

Magellan Horned Owl · Magellan-Uhu · Grand-duc de Magellanie · Tucúquere / Búho Magellanico
Strix magellanicus **Lesson 1828, Man. Orn. 1: 116;**
Terra typica: **Terres Magellaniques, en Buffon Pl. 385 = Tierra del Fuego**
(see Plate 4 and Figures 22 and 23)

Length: ~450 mm
Body mass: ?

Synonym:
- *Bubo virginianus andicolus* L. Kelso 1941, Biol. Leaflet 13: 1; Terra typica: Ollantaytambo, Peru

Distribution: Central Peru, highlands of Bolivia, Chile, southern and western Argentina, south to Tierra del Fuego and Cape Horn
Habitat: Rocky upland pasture, semi-open *Nothofagus* forest and rocky semi-desert, from sea-level to montane areas, in Andes from 2 500 m up to 4 500 m
Museum: MNBHU, SMNSt
Remarks: Mostly regarded as conspecific with *Bubo virginianus*, but differs in vocalisations, morphology and also in DNA!

Wing length: ♂ 318–356 mm, ♀ 330–368 mm
Tail length: 180–209 mm
Tarsus length: ?
Length of bill: 37–42 mm
Body mass: ?

Illustration: J. Fjeldsa in Fjeldsa and Krabbe 1990: Pl. 26; T. Worfolk in del Hoyo et al. 1999: Pl. 10; F. Weick in König et al. 1999: Pl. 26; F. Weick in Weick 1999a: 372, 373 x, y; P. Burke in Jaramillo et al. 2003: Pl. 58
Photograph: T. Daskam in *Aves de Chile* (Calendar) 1977: Marzo; W. Mayr in *ZEITmagazin* 1992(46): Frontispiece; C. König in König et al. 1996: Abb. 1; G. Ziesler in del Hoyo et al. 1999: 115; H. Wuchner in photo collection Weick (numerous photographs); C. König in internet: Owl Pages, photo gallery 2004
Literature: Sharpe 1875a: 29, 30; Oberholser 1904: 177–192; Traylor 1958: 143–149; Fjeldsa and Krabbe 1990: 226; König et al. 1996: 1–9; del Hoyo et al. 1999: 185, 186; König et al. 1999: 290–292; Weick 1999a: 363–387; Jaramillo et al. 2003: 144

Bubo bubo (Linné) 1758
Eurasian Eagle Owl · Uhu · Hibou Grand-duc (d'Europe) · Búho Real
(see Plate 4)

Length: 580–710 mm
Body mass: ♂ 1 500–1 800 g, ♀ 1 750–4 200 g

Distribution: Widespread in continental Europe and North Africa, but locally very rare. Absent in Britain, Ireland and Iceland. From the Iberian Peninsula and Morocco to northern Scandinavia, Siberia, northern India, the Himalayas, east to Sakhalin. From eastern Siberia to southern China. Accidently and locally in northern Japan
Habitat: Rocky country with cliffs and ravines scattered with trees and bushes. Patches of woodland, open forest with clearings, cultivated open- and semi-open areas with rocky cliffs or quarries. River valleys with gorges and taiga forest. Locally near human settlements. From sea-level up to 2 000 m in Europe and 4 500 m in central Asia and the Himalayas

- ***Bubo bubo hispanus*** Rothschild and Hartert 1910
  Bubo bubo hispanus Rothschild and Hartert 1910, Novit. Zool. 17: 110; Terra typica: Aguilas, Spain

Distribution: Iberian Peninsula, formerly Atlas Mountains from Algeria and Morocco
Museum: AMNH (Rothsch. coll.), BMNH, SMTD, ZFMK

Wing length: ♂ 420–450 mm, ♀ 445–470 mm (1 × 495 mm)
Tail length: ?
Tarsus length: ?
Length of bill: ?
Body mass: ?

Illustration: H. Delin in Cramp et al. 1985: Pl. 45
Photograph: Marcord and Burgos in del Hoyo et al. 1999: 94; F. Márquez in del Hoyo et al. 1999: 119
Literature: Rothschild and Hartert 1910: 110; Vaurie 1965: 580, 581; Mikkola 1983: 69, 70; Cramp et al. 1985: 466–481; Voous 1988: 87–98; del Hoyo et al. 1999: 186; König et al. 1999: 292–294

- ***Bubo bubo bubo*** (Linné) 1758
  Strix Bubo von Linné 1758, Syst. Nat. ed. 10, 1: 92; Terra typica: Europe, restricted to Sweden

Synonym:
- *Bubo bubo norwegicus* Reichenow 1910, J. Orn. 58: 412; Terra typica: Norway (holotype MNBHU)
- *Bubo bubo hungaricus* Reichenow 1910, ibid.: 412; Terra typica: Hungary (holotype MNBHU)
- *Bubo bubo engadinensis* Burg 1921, Weidmann 9: 6; Terra typica: Engadin, Switzerland
- *Bubo bubo ognavi* Dementiev 1952, Bull. Mosc. Soc. Natur Biol. Ser. 57(2): 91; Terra typica: Scutari, Albania
- *Bubo bubo meridionalis* Orlando 1957, Riv. Ital. Orn. 27: 54; Terra typica: Southern Italy and Sicily

Distribution: Europe, from northern Spain and the Mediterranean, east to Bosporus and the Ukraine, north to Scandinavia and northern Russia
Museum: BMNH, FNSF, MNBHU, NNML, SMNSt, SMNKa, SMTD, ZFMK, ZMA, ZSBS

Wing length: ♂ 435–480 mm, ♀ 455–500 mm
Tail length: ♂ 231–252 mm, ♀ 248–288 mm
Tarsus length: 74–88 mm
Length of bill: 45–58 mm
Length of bill (cere): 30–40 mm
Body mass: ♂ 1 550–2 700 g (2 810), ♀ 2 280–4 200 g

Illustration: Rudbeck (1693–1710) in Krook 1988: portfolio, 8; J.F. Naumann in Naumann 1822: Pl. 44; E. Lear in Gould 1832–1837: Pl. 37; J. Wolf (and H. Richter) in Gould 1862–1873: *Bubo m maximus* (1866); P. Barruel in Sutter and Barruel 1958: Pl. 29; D. M. Henry in Fitter et al. 1973 (Niethammer): 162; P. A. Robert in Géroudet 1979: 341; C. Petterson in Petterson 1984: 33, 73, 78, 79; H. Delin in Cramp et al. 1985: Pl. 45; A. Thorburn in Thorburn 1990: Pl. 28; A. Thorburn in *Catal. Christie's* 1994: no. 125; T. Boyer in Boyer and Hume 1991: 69; L. Jonsson in Jonsson 1992: 162; Zetterström/Mullarney in Svensson et al. 1999: 207; R. Reboussin in Jeanson 1999: Pl. 51; T. Worfolk in del Hoyo et al. 1999: Pl. 10; F. Weick in König et al. 1999: Pl. 27; F. Weick in Weick 2004: Frontcover
Photograph: E. Hosking in Hosking and Flegg 1982: 137; H. Reinhard in Burton et al. 1992:

Bubo bubo bubo
Eurasian Eagle Owl
Uhu

63; numerous photographs in Mebs and Scherzinger 2000: 147 (Reinhard), 149 (Hecker), 157 and 161 (Nill), Frontcover (M. Danegger); J. Santana in internet: Owl Pages, photo gallery 2003

Literature: Sharpe 1875a: 14–17; Hartert 1912–1921: 960–962; Dementiev and Gladkov 1951: 391–396; Vaurie 1965: 580–582; März and Piechocki 1976: 119 pp; Glutz von Blotzheim and Bauer 1980: 303–357; Mikkola 1983: 69–90; Cramp et al. 1985: 466–481; Bezzel 1985: 638–642; Piechocki 1985: 128 pp; Voous 1988: 87–98; Boyer and Hume 1991: 68–70; Kemp and Kemp 1998: 254, 255; del Hoyo et al. 1999: 186; König et al. 1999: 292–294; Mebs and Scherzinger 2000: 147–166

- ***Bubo bubo ruthenus*** Zhitkov and Buturlin 1906
  Bubo bubo ruthenus Zhitkov and Buturlin 1906, Zapiski po Imp. Russh. Geogr. Obshcht. 41: 272; Terra typica: Prozino, Gouvern. of Ulyanovsk (Simbirsk), eastern Russia

Distribution: Central European Russia, east to foothills of the Ural Mountains, south to lower Volga Basin

Museum: SMTD, ZMA, ZMM, ZSBS

Wing length:	♂ 440–468 mm (1 × 430 mm), ♀ 471–490 mm (1 × 515 mm)
Tail length:	?
Tarsus length:	?
Length of bill:	?
Body mass:	?

Illustration: ?
Photograph: ?
Literature: Hartert 1912–1921: 962, 963; Dementiev and Gladkov 1951: 391, 396, 397; Vaurie 1965: 580, 582; Cramp et al. 1985: 466–481; del Hoyo et al. 1999: 186; König et al. 1999: 292–294

- ***Bubo bubo interpositus*** **Rothschild and Hartert 1910**
  Bubo bubo interpositus Rothschild and Hartert 1910, Novit. Zool. 17: 111; **Terra typica:** Eregli, southern Turkey

Synonym:
- *Bubo bubo aharonii* Rothschild and Hartert 1910, Novit. Zool. 17: 112; **Terra typica:** Wadi Suenut, Jordan Valley, Palestine. **Remarks:** Possibly *interpositus* × *ascalaphus* hybrid
- *Bubo bubo armeniacus* Nesterov 1912, Annuaire Mus. Zool. Sci. Pétersbourg 16(1911): 378; **Terra typica:** Armenia
- *Bubo bubo tauricus* Buturlin 1928, Opredelitel Ptits S.S.S.R: 114; **Terra typica:** Kara Aktachi, Crimea
- *Bubo bubo nativus* Gavrilenko 1928, Sbirnik Poltawsk Muz. 1: 279; **Terra typica:** Mirgorod, Poltava, Ukraine
- *Bubo bubo transcaucasicus* Tschchikwischwili 1930, Bull. Mus. Géorgie 5: 97; **Terra typica:** Transcaucasia

Distribution: From Romania and southern Ukraine, east to the Volga Delta. South to the Middle East (southern and central Israel, Jordan), also northwest Iran. Sympatric with *B. ascalaphus.* DNA evidence suggests specific distinctness from *B. bubo*!?
Museum: AMNH (Rothsch.), MNBHU, SMNSt, SMTD, USNM, ZFMK, ZMH *(aharonii)*

Wing length:	♂ 425–475 mm, ♀ 440–485 mm (1 × 502 mm)
Tail length:	240–290 mm
Tarsus length:	?
Length of bill:	?
Body mass:	?

Illustration: ?
Photograph: Eshbol in Shirihai 1996: Pl. 65
Literature: Hartert 1923: 382; Dementiev and Gladkov 1951: 391, 397, 398; Vaurie 1965: 580, 582, 583; Cramp et al. 1985: 466–481; del Hoyo et al. 1999: 186; König et al. 1999: 292–294

- ***Bubo bubo sibiricus*** **(Gloger) 1833**
  ? Strix sibirica "Licht." Gloger 1833, Abändern der Vögel durch Einfluss des Klimas: 142; **Terra typica:** Ural Mountains

Synonym:
- *Bubo bubo baschkirikus* Sushkin 1932, Alauda 4: 395; **Terra typica:** Environs d'Ufa, western Urals

Distribution: From the western foothills of the Ural Mountains, east to River Ob, south to western Altai, north to limits of the taiga. Sympatric with *ruthenus*
Museum: MNBHU, SMTD, ZFMK, ZMM

Wing length:	♂ 435–480 mm, ♀ 472–515 mm
Tail length:	?
Tarsus length:	?
Length of bill:	?
Body mass:	?

Illustration: J. Wolf in Susemihl 1846: Pl. 44; H. Delin in Cramp et al. 1985: Pl. 45; I. Willis in Mikkola 1983: Pl. 2; A. Cameron in Voous 1988: 95; T. Worfolk in del Hoyo et al. 1999: Pl. 10; F. Weick in König et al. 1999: Pl. 27
Photograph: J. Pukinski in Mikkola 1983: 22, 23 (b/w); L. Körner in internet: Owl Pages, photo gallery 2002; C. König in photo collection Weick
Literature: Sharpe 1875a: 17–19; Hartert 1912–1921: 963, 964; Dementiev and Gladkov 1951: 399, 400; Vaurie 1965: 580, 583; Mikkola 1983: 69–90; Cramp et al. 1985: 466–481; Voous 1988: 87–98; del Hoyo et al. 1999: 186; König et al. 1999: 292–294

- ***Bubo bubo yenisseensis*** **Buturlin 1911**
  Bubo bubo yenisseensis Buturlin 1911, Mess. Orn.: 26; **Terra typica:** Krasnoyarsk, Central Siberia

Synonym:
- *Bubo bubo zaissanensis* Khakhlov 1915, Mess. Orn. 6: 224; **Terra typica:** Saur Range, on the border of Dzungaria and Kazakhstan, southeast of Zaisan Nor
- *Bubo bubo auspicabilis* Dementiev 1931, Alauda 3: 364; **Terra typica:** Alexandrowski Mountains

Distribution: Central Siberia, from River Ob to Lake Baikal, south to the Altai and northern Mongolia. Status as subspecies uncertain, probably a synonym of *jakutensis*
Museum: SMTD, ZFMK, ZMH

Wing length: ♂ 435–470 mm, ♀ 473–518 mm
Tail length: ?
Tarsus length: ?
Length of bill: ?
Body mass: ?

Illustration: ?
Photograph: ?
Literature: Hartert 1912–1921: 964; Dementiev and Gladkov 1951: 391, 400, 401; Vaurie 1965: 580, 583, 584; Cramp et al. 1985: 466–481; del Hoyo et al. 1999: 186; König et al. 1999: 292–294

- ***Bubo bubo jakutensis*** **Buturlin 1908**
Bubo bubo jakutensis Buturlin 1908, J. Orn. 56: 287; Terra typica: Yakutsk

Distribution: Northeast Siberia, east of *yenisseensis* and north of *ussuriensis*
Museum: ZMM

Wing length: ♂ 455–490 mm, ♀ 480–503 mm
Tail length: ?
Tarsus length: ?
Length of bill: ?
Body mass: ?

Illustration: F. Weick in König et al. 1999: Pl. 27 (from a bird in captivity)
Photograph: F. Weick in photo collection Weick (bird from Leipold aviary)
Literature: Hartert 1912–1921: 965; Hartert 1923: 382; Dementiev and Gladkov 1951: 391; Vaurie 1965: 580, 584; Cramp et al. 1985: 466–481; del Hoyo et al. 1999: 186; König et al. 1999: 292–294

- ***Bubo bubo ussuriensis*** **Poliakov 1915**
Bubo bubo ussuriensis Poliakov 1915, Mess. Orn. 6: 44; Terra typica: Nikolsk-Ussuriysk, South Ussuriland

Synonym:
- *Bubo bubo borissowi* Hesse 1915, J. Orn. 63: 366; Terra typica: Tymi River, East Sakhalin
- *Bubo bubo dauricus* Stegmann 1929, Annuaire Mus. Zool. Acad. Sci. U.R.S.S. 29(1928): 178; Terra typica: Sokotui and Vicinity of Aga, southeast Transbaikalia (based on Sushkin manuscript name)
- *Bubo bubo yamashinai* Momiyama 1930, Dobutsu. Zasshi 42: 329; Terra typica: Obihiro, Hokkaido

Distribution: Southeast Siberia to northeast China, Sakhalin, Hokkaido and South Kuril Islands
Museum: MNBHU, SMTD, ZFMK

Wing length: ♂ 430–475 mm, ♀ 460–502 mm
Tail length: ?
Tarsus length: ?
Length of bill: ?
Body mass: ?

Illustration: ?
Photograph: ?
Literature: Hartert 1912–1921: 965; Hartert 1923: 383; Dementiev and Gladkov 1951: 391, 401, 402; Vaurie 1965: 584, 585; Cramp et al. 1985: 466–481; del Hoyo et al. 1999: 186; König et al. 1999: 292–294

- ***Bubo bubo kiautschensis*** **Reichenow 1903**
Bubo bubo kiautschensis Reichenow 1903, Orn. Monatsb. 11: 85; Terra typica: Kiaochow, Shantung, China

Synonym:
- *Bubo bubo setschuanus* Reichenow 1903, Orn. Monatsb. 11: 86; Terra typica: Szechuan (holotype MNBHU)
- *Bubo bubo tenuipes* Clark 1907, Proc. US Nat. Mus. 32: 470; Terra typica: Fusan, South Korea (holotype USNM)
- *Bubo bubo swinhoei* Hartert 1913, Die Vögel der paläarktischen Fauna 1912–1921: 966; Terra typica: Kiukiang, North Kiangsi (holotype BMNH)
- *Bubo bubo jarlandi* La Touche 1921, Bull. Brit. Ornith. Cl. 42: 14; Terra typica: Mengtsz, South Yunnan
- *Bubo bubo inexpectatus* Bangs 1932, in La Touche's Handb. Bds. E. China 2(2): 113; Terra typica: Chiu Lung Shan, Chihli
- *Bubo bubo inexpectatus* Dementiev 1933 (ex Bangs), Alauda 2(4): 394; Terra typica: Manchuria

Distribution: Western, central and southeast China, eastern Korea. Migratory to Japan
Museum: AMNH, BMNH, MNBHU (holotype), SMTD, USNM

Wing length: ♂ 410–448 mm, ♀ 440–485 mm
Tail length: ?
Tarsus length: ?
Length of bill: ?
Body mass: ?

Illustration: T. Miyamoto in Kobayashi 1965: Pl. 164; S. Takano in Sonobe et al. 1983: 189; T. Worfolk in del Hoyo et al. 1999: Pl. 10
Photograph: Ehlers, Leipzig in photo collection Weick (Zoo Peking)
Literature: Hartert 1912–1921: 966, 967; Vaurie 1965: 580, 586; Sonobe et al. 1983; 188; del Hoyo et al. 1999: 186; König et al. 1999: 292–294

- ***Bubo bubo turcomanus*** Eversmann 1835
Strix turcomana Eversmann 1835, Addenda Pallas Zoogr., fasc. 1: 3;
Terra typica: inter mare Caspium et lacum Aral ansem = Ust Urt Plateau

Synonym:
- *Bubo bubo tarimensis* Buturlin 1928, Opredelitel Ptits S.S.S.R.: 114; Terra typica: Lop Nor, Tarim Basin, Sinkiang
- *Bubo bubo eversmanni* Dementiev 1931, Alauda 3: 361; Terra typica: Tourangly, Aral-Sea and Kazakhstan
- *Bubo bubo gladkovi* Zaletaev 1962, Ornitologiia 4: 190; Terra typica: Capes Melovoy and Skalistoy, East Caspian Sea

Distribution: Steppe between lower Volga River and Ural River, east to Transbaikalia
Museum: ZFNK, ZNM, ZSBS

Wing length: ♂ 440–470 mm, ♀ 445–512 mm
Tail length: 260–310 mm
Tarsus length: 77–81 mm
Length of bill: 45–47 mm
Body mass: ?

Illustration: ?
Photograph: M. Mehner in *Jagd und Hund* 1998(November), Neumann
Literature: Sharpe 1875a: 17–19; Baker 1927: 413; Dementiev and Gladkov 1951: 391, 402, 403; Vaurie 1965: 581, 588; Ali and Ripley 1969: 271, 272; Cramp et al. 1985: 466–481; del Hoyo et al. 1999: 186; König et al. 1999: 292–294

- ***Bubo bubo omissus*** Dementiev 1933
Bubo bubo omissus Dementiev 1933, Alauda 4(1932): 392;
Terra typica: Ashkabad, South Transcaspia

Synonym:
- *Bubo bubo paradoxus* Domaniewski 1933, Acta Orn. Zool. Mus. Polonici 1: 79; Terra typica: Harirud, Transcaspia

Distribution: Transcaspia, Turkmenia and adjacent Iran and Chinese Turkestan
Museum: AMNH (Rothsch.)

Wing length: ♂ 420–450 mm (1 × 404 mm), ♀ 445–460 mm (1 × 425 mm)
Tail length: ?
Tarsus length: ?
Length of bill: ?
Body mass: ?

Illustration: F. Weick in König et al. 1999: Pl. 28
Photograph: W. Wozniak in Eck and Busse 1973: Abb. 3 (b/w); E. Hosking in Hosking and Flegg 1982: 59, 139; K. Rudloff in Piechocki 1985: Abb. 13
Literature: Dementiev 1933a: 392; Dementiev and Gladkov 1951: 391, 403, 404; Vaurie 1965: 581, 589; H. Busse 1968, *Der Falke* 1968(15): 178, 179; del Hoyo et al. 1999: 186; König et al. 1999: 292–294; Rasmussen and Anderton 2005: 239

- ***Bubo bubo nikolskii*** Zarudny 1905
Bubo bubo nikolskii Zarudny 1905, Orn. Jahrb. 16: 142;
Terra typica: Djebel Trüe, Arabistan = Zagros, Southwest Iran

Distribution: Eastern Iraq and Iran, Afghanistan, North and West Pakistan
Museum: ZFMK

Wing length: ♂ 405–430 mm (1 × 378 mm), ♀ 410–465 mm (1 × 394 mm)
Tail length: ?
Tarsus length: ?
Length of bill: ?
Body mass: ?

Illustration: F. Weick in König et al. 1999: Pl. 28; McQueen in Rasmussen and Anderton 2005: Pl. 76
Photograph: Fink in Burton et al. 1992: 62
Literature: Hartert 1912–1921: 963; Vaurie 1965: 581, 589; Cramp et al. 1985: 466–481; del Hoyo et al. 1999: 186; König et al. 1999: 292–294; Rasmussen and Anderton 2005: 239

- ***Bubo bubo hemachalana*** Hume 1873
  B(ubo) Hemachalana A.O.H. = Hume 1873, Str. Feath. 1: 315; Terra typica: Kulu, 12 000 feet, India (northern Punjab)

Synonym:
- *Bubo bubo tibetanus* Bianchi 1906, Bull. Brit. Ornith. Cl. 16: 69; Terra typica: Upper Yangtse River, Southeast Tibet

Distribution: From Pamir and North Tienshan south to the Himalayas
Museum: BMNH, SMTD

Wing length: ♂ 450–485 mm (420 mm *B. b. tibetanus*), ♀ 470–505 mm
Tail length: ?
Tarsus length: ?
Length of bill: 42–45 mm
Body mass: ?

Illustration: D. Cole in Grimmett et al. 1998: Pl. 29; F. Weick in König et al. 1999: Pl. 27 (Zoo Heidelberg)
Photograph: Ehlers, Leipzig in letter, Zoo Peking: photo collection Weick
Literature: Hartert 1912–1921: 965; Baker 1927: 414; Vaurie 1965: 581, 586, 587; Ali and Ripley 1969: 272, 273; Etchécopar and Hüe 1978: 468, 469; del Hoyo et al. 1999: 186; König et al. 1999: 292–294

Bubo bengalensis **(Franklin) 1831**
Rock Eagle Owl · Bengalenuhu · Grand-duc des Indes · Búho Bengali
Otus bengalensis **Franklin 1831, Proc. Comm. Zool. Soc. London 1830–1831: 15; Terra typica: Ganges, between Calcutta, Benares and the Vindhyan Hills**

Length: 500–560 mm
Body mass: ♂ 1 100 g ($n = 1$)

Distribution: Indian Subcontinent except Sri Lanka, north to the foothills of the Himalayas and western Myanmar
Habitat: Rocky hills and wooded country, semi-desert areas with rocks and scrub. In plantations and orchards, also near human settlements. From lowland up to 2 400 m
Museum: MNBHU, SMNSt, SMTD, ZSBS
Remarks: Often considered as conspecific with *B. bubo* due to distribution-range overlap, but differs in DNA and voice

Wing length: ♂ 358–391 mm, ♀ 375–433 mm
Tail length: ♂ 185–195 mm, ♀ 205–227 mm
Tarsus length: 69–76 mm
Length of bill: 42–47 mm
Body mass: ♂ 1 100 g ($n = 1$)

Illustration: ? in Gould 1831–1832: Pl. 3; D. Cole in Grimmett et al. 1998: Pl. 29; T. Worfolk in del Hoyo et al. 1999: Pl. 10; F. Weick in König et al. 1999: Pl. 28; H. Burn in Robson 2000: Pl. 22; McQueen in Rasmussen and Anderton 2005: Pl. 76
Photograph: K. Simmons in Simmons 1976: 136 (b/w); V. Cavale in internet: Owl Pages, photo gallery 2002; P. Steyn in Duncan 2003: 235; J. Bird in internet: Owl Pages, photo gallery 2003
Literature: Sharpe 1875a: 25–27; Baker 1927: 414, 415; Vaurie 1965: 581; Ali and Ripley 1969: 273–275; Simmons 1976: 135–138; Grimmett et al. 1998: Pl. 432; del Hoyo et al. 1999: 186, 187; König et al. 1999: 295, 296; Robson 2000: 257; Rasmussen and Anderton 2005: 239
Remarks: See also Glutz von Blotzheim and Bauer 1980: 303, 304 (Rassengliederung)

Bubo ascalaphus **Savigny 1809**
Pharaoh Eagle Owl · Wüstenuhu · Grand-duc du désert · Búho del Sáhara (Búho desértico)
Bubo Ascalaphus **Savigny 1809, Descr. Égypte 1(1): 110, Oiseaux, Pl. 3; Terra typica: Upper Egypt**
(see Plate 4)

Length: 450–500 mm
Body mass: ♂ ∅ 1 900 g, ♀ ∅ 2 300 g

Synonym:
- *Bubo ascalaphus barbarus* von Erlanger 1897, Orn. Monatsb. 5: 192; Terra typica: Qued Kasserine, central Tunisia (holotype MNBHU)
- *Bubo ascalaphus desertorum* von Erlanger 1897, ibid.: 192; Terra typica: Sidi Ali bin Aouin, southern Tunisia (holotype MNBHU)

Bubo ascalaphus · Pharaoh Eagle Owl (light morph) · Wüstenuhu (helle Morphe)

Distribution: North and northwest Africa, from Rif and the Atlas Mountains, through most of the Sahara (south to Chad), Mauretania, Mali, Niger, northern Egypt, Sudan and western Ethiopia, Arabia, Syria, Israel and Palestine to eastern and western Iraq

Habitat: Rocky deserts and semi-deserts, hills, wadi with cliffs, extending south into dry savanna. Outcrops of oases

Museum: BMNH, MNBHU, SMNSt, SMTD, ÜMB, ZFMK, ZMH

Remarks: Locally sympatric with *B. b. interpositus*

Wing length:	♂ 324–368 mm (1 × 411 mm), ♀ 340–416 mm (430 mm Shirihai!)
Tail length:	♂ 160–224 mm, ♀ 188–233 mm
Tarsus length:	65–83 mm
Length of bill:	42–46 mm
Body mass:	♂ ~1 900 g, ♀ ~2 300 g

Illustration: Savigny 1809: Pl. 3; E. de Maes in *J. Orn.* 1898: Pl. 12 and 13; F. Neubaur in Koenig 1936: Pl. 51; P. Barruel in Etchécopar and Hüe 1967: Pl. 11; H. Delin in Cramp et al. 1985: Pl. 45; M. Woodcock in Fry et al. 1988: Pl. 8; A. Cameron in Voous 1988: 99; P. Hayman in Kemp and Kemp 1998: 253; T. Worfolk in del Hoyo et al. 1999: Pl. 10; F. Weick in König et al. 1999: Pl. 28; Zetterström and Mullarney in Svensson et al. 1999: 207; N. Borrow in Borrow and Demey 2001: Pl. 60

Photograph: Mulsow 1964: 247 (b/w); E. Hosking in Hosking and Flegg 1982: 136; L. Boom in Shirihai 1996: Pl. 68; Anonymus in *Natural Emirats* 1997(Jan.): Wildl. and Environments; M. Gunther and X. Eichaker in Kemp and Kemp 1998: 252/3; A. Dragesco in del Hoyo et al. 1999: 99

Literature: Sharpe 1875a: 24, 25; von Erlanger 1897: 192; Koenig 1936: 166–172; Mulsow 1964: 246–248; Vaurie 1965: 581, 589, 590; Etchécopar and Hüe 1967: 340, 341; Glutz von Blotzheim and Bauer 1980: 304; Mikkola 1983: 68–90; Cramp et al. 1985: 466–481; Fry et al. 1988: 122, 123; Voous 1988: 87–98; Shirihai 1996: 313, 314; Kemp and Kemp 1998: 252, 253; del Hoyo et al. 1999 186, 187; König et al. 1999: 294, 295; Borrow and Demey 2001: 494

Bubo capensis **A. Smith 1834**
Cape Eagle Owl · Kapuhu · Grand-duc du Cap (de montagne) · Búho de El Cabo
(see Plate 4)

Length:	460–580 mm
Body mass:	900–1 800 g

Distribution: Eastern and southern Africa, from Eritrea and Ethiopia south to Kenya, Tanzania, Zimbabwe, Mozambique and Cape Province to southern Namibia. Distribution patchy and very local

Habitat: Woody valleys, cliffs and ravines in mountainous regions. Hilly, rocky areas, from 2 000 m up to 4 200 m, but in south also in flat and dry open country

Remarks: Sometimes regarded as conspecific with *Bubo bubo*, but seems more distantly related

- ***Bubo capensis dilloni*** des Murs and Prévost 1846
Bubo Dilloni des Murs and Prévost 1846, Rev. Zool.: 242;
Terra typica: Abyssinia, Quodgerate = Northeast Ethiopia

Distribution: Ethiopian highlands and southern Eritrea

Museum: BMNH, MNBHU, NMZB, SMTD, ZFMK

Wing length:	♂ 341–391 mm, ♀ 380–417 mm
Tail length:	♂ 175–208 mm, ♀ 181–241 mm
Tarsus length:	?
Length of bill:	?
Body mass:	?

Illustration: M. Woodcock in Fry et al. 1988: Pl. 8; F. Weick in König et al. 1999: Pl. 29
Photograph: ?
Literature: Fry et al. 1988: 124–126; del Hoyo et al. 1999: 187; König et al. 1999: 297, 298

- ***Bubo capensis mackinderi*** Sharpe 1899
  Bubo mackinderi Sharpe 1899, Bull. Brit. Ornith. Cl. 10: 28;
  Terra typica: Mount Kenya, 3 000 feet

Distribution: From west-central Kenya south to Zimbabwe and western Mozambique (but locally rare or absent)
Museum: BMNH (holotype), NMKN, NMP, NMZB, SAMC, TMBD, USNM
Remarks: Sometimes considered as distinct species

Wing length:	♂ 375–402 mm, ♀ 406–428 mm
Tail length:	♂ 184–205 mm, ♀ 200–238 mm (1 × 256 mm)
Tarsus length:	75 mm
Length of bill:	48 mm
Body mass:	♂ 1 221 and 1 387 g, ♀ 1 400–1 800 g

Illustration: G. E. Lodge in *Proc. Zool. Soc. Lond.*: Pl. 43; M. Woodcock in Fry et al. 1988: Pl. 8; P. Hayman in Kemp and Kemp 1998: 251; T. Worfolk in del Hoyo et al. 1999: Pl. 10; F. Weick in König et al. 1999: Pl. 29; J. Gale in Stevenson and Fanshawe 2002: Pl. 101
Photograph: Benson and Stuart-Irwin in Benson and Stuart-Irwin 1967: 10, 11 (b/w); E. Hosking in Hosking and Flegg 1982: 141; Dean/Coleman in Burton et al. 1984: 71 (top) and 1992: 69; G. Budich in *Der Falke* 1988: 315
Literature: Benson and Stuart-Irwin 1967: 1–19; H. Busse 1988, *Der Falke* 1988: 314; Fry et al. 1988: 124–126; Kemp and Kemp 1998: 250, 251; del Hoyo et al. 1999: 187; König et al. 1999: 297, 298; Stevenson and Fanshawe 2002: 202

- ***Bubo capensis capensis*** A. Smith 1834
  Bubo Capensis A. Smith 1834, S. Afr. Q. J. 2: 317;
  Terra typica: Near Cape Town, South Africa

Distribution: South Africa and extreme south of Namibia
Museum: BMNH, NMZB, SAMC, TMBD, ZMH

Wing length:	♂ 330–357 mm, ♀ 363–392 mm
Tail length:	♂ 155–215 mm, ♀ 169–240 mm
Tarsus length:	71–75 mm
Length of bill:	42–50 mm
Body mass:	♂ 905–960 g, ♀ 1 240–1 400 g

Illustration: A. Smith in Smith 1839: Pl. 70; G. Arnott in Steyn 1982: Pl. 24; M. Woodcock in Fry et al. 1988: Pl. 8; T. Boyer in Boyer and Hume 1991: 71; D. Zimmerman in Zimmerman et al. 1996: Pl. 55; P. Hay-man in Kemp and Kemp 1998: 251; T. Worfolk in del Hoyo et al. 1999: Pl. 10; F. Weick in König et al. 1999: Pl. 29
Photograph: K. Kussmann in Eck and Busse 1973: Abb. 4 (b/w); P. Steyn in Steyn 1982: 271 (b/w photograph juvenile); E. Hosking in Hosking and Flegg 1982: 143; R. A. C. Jensen in Ginn et al. 1989: 340; Taylor in Burton et al. 1984: 69 and 1992: 71 (left); J. J. Brooks in Kemp and Kemp 1998: 250; N. Dennis in Kemp and Kemp 1998: 251
Literature: Sharpe 1875a: 27–29; Benson and Stuart-Irwin 1967: 1–19; Steyn 1982: 266–272; Fry et al. 1988: 124–126; Ginn et al. 1989: 340; Boyer and Hume 1991: 70–72; Kemp and Kemp 1998: 250, 251; del Hoyo et al. 1999: 187; König et al. 1999: 297, 298

Bubo africanus (Temminck) 1821

Spotted Eagle Owl · Fleckenuhu · Grand-duc africain · Búho Africano
(see Plate 4)

Length:	400–450 mm
Body mass:	550–850 g

Distribution: Sub-Saharan Africa from Gabon east to Zaire (south of the rainforest), southern Uganda, southeast and central Kenya, south to the Cape
Habitat: Habitats very variable, open- or semi-open woodland, savanna with scrub and scattered trees, also semi-desert (Kalahari). From sea-level up to 2 100 m. Avoids dense forest

- ***Bubo africanus africanus*** (Temminck) 1821
  Strix africana Temminck 1821, Pl. col. livr. 9: Pl. 50;
  Terra typica: Cape of Good Hope

Synonym:
- *Asio maculosus amerimnus* Oberholser 1905, Proc. US Nat. Mus. 28: 856; **Terra typica:** Durban, Natal, S Africa
- *Bubo ascalaphus trotha* Reichenow 1906, Orn. Monatsb. 14: 10; **Terra typica:** Keetmanskoop, SW Afr. Protect.

Distribution: Gabon, east to Zaire, southern Uganda, central Kenya, south to the Cape

Museum: BMNH, FNSF, MHNP, NMKN, NMZB, SMNKa, USNM, ZFMK, ZMH, ZSBS

Wing length: ♂ 323–348 mm, ♀ 314–360 mm
Tail length: 184–222 mm
Tarsus length: 62–82 mm
Length of bill: 37–43 mm
Body mass: ♂ 487–620 g, ♀ 640–850 g

Illustration: C.J. Temminck in Temminck 1821c: Pl. 50; G. Arnott in Steyn 1982: Pl. 24; M. Woodcock in Fry et al. 1988, Pl. 8; T. Boyer in Boyer and Hume 1991: 72; D. Zimmerman in Zimmerman et al. 1996: Pl. 55; P. Hayman in Kemp and Kemp 1998: 257; T. Worfolk in del Hoyo et al. 1999: Pl. 10; F. Weick in König et al. 1999: Pl. 30; N. Borrow in Borrow and Demey 2001: Pl. 60
Photograph: P. Steyn in Steyn 1982: 277 (b/w photograph juvenile); E. Hosking in Hosking and Flegg 1982: 67 and 140; J.D. Llewellyn in Hosking and Flegg 1982 (photograph from 1854!); R.E. Viljoen in Ginn et al. 1989: 341; M. Craig-Cooper in Ginn et al. 1989: 341; J. v. Jaarsveld in Ginn et al. 1989: 341; Richards in Burton et al. 1992: 70; Haagner in Burton et al. 1992: 71; P. Pickford in Kemp and Kemp 1998: 256; P. Chadwick in Kemp and Kemp 1998: 257; N. Myburgh in Kemp and Kemp 1998: 256; P. Steyn in del Hoyo et al. 1999: 105
Literature: Sharpe 1875a: 30–32; Eck and Busse 1973: 100, 101; Steyn 1982: 273–278; Fry et al. 1988: 126–128; Voous 1988: 100–104; Ginn et al. 1989: 341; Zimmerman et al. 1996: 443; Kemp and Kemp 1998: 256, 257; del Hoyo et al. 1999: 187, 188; König et al. 1999: 299–300; Borrow and Demey 2001: 494; Stevenson and Fanshawe 2002: 202

- ***Bubo africanus tanae*** **Keith and Twomey 1968**
  Bubo africanus tanae Keith and Twomey 1968, Ibis 110: 538, 539; Terra typica: Central and lower Tanae River and Lali Hills

Distribution: Central and lower Tanae River and Lali Hills, southeast Kenya
Museum: BMNH?

Wing length: 290–315 mm ($n = 5$)
Tail length: ?
Tarsus length: ?
Length of bill: ?
Body mass: ?

Illustration: ?
Photograph: ?
Literature: Keith and Twomey 1968: 538, 539; Voous 1988: 100–104; del Hoyo et al. 1999: 187, 188; König et al. 1999: 300

❖

Bubo (africanus) milesi Sharpe 1886

Miles' Spotted Eagle Owl · Milesuhu (Arabien-Fleckenuhu)
Bubo milesi **Sharpe 1886, Ibis: 163;** Terra typica: **Muscat (Masqat), Arabia**

Length: 420 mm
Body mass: ?

Distribution: Southwest Arabia, Yemen, Oman
Habitat: Rocky outcrops in deserts, semi-desert with scrubs and thornbushes, rocky hills
Museum: BMNH
Remarks: Possibly a distinct species (allopatric), but taxonomic status uncertain

Wing length: 302–330 mm
Tail length: ?
Tarsus length: ?
Length of bill: ?
Body mass: ?

Illustration: J.G. Keulemans in Sharpe 1886: Pl. 6; A. Cameron in Voous 1988: 102; F. Weick in König et al. 1999: Pl. 30
Photograph: ?
Literature: Sharpe 1886: 163; Meinertzhagen 1954: 314; Eck and Busse 1973: 100, 101; Voous 1988: 100–104; del Hoyo et al. 1999: 187, 188; König et al. 1999: 300

❖

Bubo cinerascens Guérin Méneville 1843

Vermiculated Eagle Owl · Sprenkeluhu (Grauuhu) · Grand-duc vermiculé · Búho Ceniciento
Bubo cinerascens **Guérin Mèneville 1843, Rev. Zool.: 321;** Terra typica: **Abyssinia (Adowa, Ethiopia)**

Length: 430 mm
Body mass: ~500 g

Synonym:
- *Bubo africanus kollmanspergeri* Niethammer 1957, Bonn. Zool. Beitr. 8: 278; Terra typica: Ennedi Mountains, Sahara, Chad

Distribution: Sub-Saharan Africa, from Senegambia and Cameroon east to Ethiopia (north of the rainforest)
Habitat: Dry rocky desert and semi-desert, hillsides with scrub or scattered trees. Open and semi-open savanna, with thorn bushes and trees. Avoids dense forests, but occurs in lowland woodland in Somalia
Museum: SMTD, ÜMB, ZFMK, ZMH
Remarks: Often regarded as conspecific with *B. africanus*, but morphologically distinct and not known to interbreed in the overlapping area (sympatric)

Wing length:	♂ 284–333 mm, ♀ 298–338 mm
Tail length:	170–200 mm
Tarsus length:	45 mm
Length of bill:	38 mm
Body mass:	~500 g

Illustration: F. Neubaur in Koenig 1936: Pl. 53; M. Woodcock in Fry et al. 1988: Pl. 8; D. Zimmerman in Zimmerman et al. 1996: Pl. 95; T. Disley in Barlow et al. 1997: Pl. 25; P. Hayman in Kemp and Kemp 1998: 257; T. Worfolk in del Hoyo et al. 1999: Pl. 10; F. Weick in König et al. 1999: Pl. 30; N. Borrow in Borrow and Demey 2001: Pl. 60; J. Gale in Stevenson and Fanshawe 2002: Pl. 101
Photograph: E. Hosking in Hosking and Flegg 1982: 138; M. Fodgen in Burton et al. 1992: 72; "Mick" in internet: Owl Pages, photo gallery 2002; J. Bird in internet: Owl Pages, photo gallery 2002; L. Koerner in internet: Owl Pages, photo gallery 2003
Literature: Sharpe 1875a: 32, 33; Koenig 1936: 174–178; Vaurie 1965: 590, 591; Eck and Busse 1973: 100, 101; Fry et al. 1988: 126–128; Zimmerman et al. 1996: 443, 444; Kemp and Kemp 1998: 556, 557; del Hoyo et al. 1999: 188; König et al. 1999: 301, 302; Borrow and Demey 2001: 494, 495; Stevenson and Fanshawe 2002: 202; Sinclair and Ryan 2003: 240, 241

Bubo poensis Fraser 1854
Fraser's Eagle Owl · Guinea-Uhu · Grand-duc à aigrettes · Búho de Guinea
***Bubo Poensis* Fraser 1854, Proc. Zool. Soc. Lond. 1853: 13;**
Terra typica: Fernando Póo
(see Plate 4)

Length:	390–440 mm
Body mass:	585–815 g

Synonym:
- *Bubo fasciolatus* Hartlaub 1855, J. Orn.: 354, 360; Terra typica: West Africa

Distribution: From tropical to forested West Africa (Liberia), east to western Uganda, south through the Congo Basin to central Zaire and northwest Angola. Also Bioko Island (Fernando Póo)
Habitat: Lowland with evergreen primary and secondary forest, clearings, forest edges and plantations. From sea-level up to 1 600 m
Museum: AMNH (Correia coll.), BMNH, MHNP, MNBHU, SMNSt, ZFMK

Wing length:	♂ 276–318 mm, ♀ 296–333 mm
Tail length:	♂ 133–155 mm, ♀ 153–185 mm
Tarsus length:	51–61 mm
Length of bill:	41–46 mm
Body mass:	♂ 575 g, ♀ 685–815 g

Bubo poensis · Fraser's Eagle Owl · Guinea-Uhu (immature and adult)

Illustration: J. G. Keulemans in *Ibis* 1869: Pl. 4 (adult and juvenile); J. Wolf in Palmer 1895, facing p. 76; M. Woodcock in Fry et al. 1988: Pl. 8; P. Hayman in Kemp and Kemp 1998: 261; T. Worfolk in del Hoyo et al. 1999: Pl. 10; F. Weick in König et al. 1999: Pl. 30; N. Borrow in Borrow and Demey 2001: Pl. 60; J. Gale in Stevenson and Fanshawe 2002: Pl. 101; P. Hayman in Sinclair and Ryan 2003: 241

Photograph: H. Busse in Eck and Busse 1973: Abb. 6 (b/w photograph); H. Busse in *Der Falke* 1976(11): 395 (b/w photograph); Zool. Soc. London in Olney 1984: 131; Soc. Roy. de Zool. d'Anvers in Burton et al. 1992: 74; D. Robertson in Kemp and Kemp 1998: 261; A. Schoonbee in internet: Owl Pages, photo gallery 2004 (adult and juvenile)

Literature: Sharpe 1875a: 42; Eck and Busse 1973: 101, 102; Busse 1976, *Der Falke* 1976(11): 394; Fry et al. 1988: 129, 130; Boyer and Hume 1991: 74; Kemp and Kemp 1998: 260; del Hoyo et al. 1999: 188; König et al. 1999: 302, 303; Borrow and Demey 2001: 495; Stevenson and Fanshawe 2002: 202; Sinclair and Ryan 2003: 240

Bubo vosseleri Reichenow 1908
Usambara or Nduk Eagle Owl · Usambara-Uhu · Grand-duc des Usambara · Búho de Usambara
***Bubo vosseleri* Reichenow 1908, J. Orn. 56: 139;**
Terra typica: **Amani, Tanganyika Terr. = Tanzania**
(see Plate 4)

Length: 450–480 mm
Body mass: 770–1 053 g

Distribution: Usambara Mountains of northeast Tanzania, also Ulguru Mountains and, possibly, Nguru Mts.

Habitat: Evergreen montane forest, forest edges and plantations from 900 to 1 500 m, movement in cold weather down to 200 m from sea-level

Museum: BMNH, MNBHU (holotype)

Remarks: Sometimes regarded as a subspecies of *B. poensis*, but separated by distinct vocalisations and isolated distribution

Wing length: 331–365 mm
Tail length: 176–189 mm
Tarsus length: 51–60 mm
Length of bill: 45–48 mm
Body mass: ♂ 770 g, ♀ 850, 875 and 1 052 g

Illustration: T. Boyer in Boyer and Hume 1991: 74; M. Woodcock in Fry et al. 1988: Pl. 8; P. Hayman in Kemp and Kemp 1998: 261 (in flight); T. Worfolk in del Hoyo et al. 1999: Pl. 10; F. Weick in König et al. 1999: Pl. 30; J. Gale in Stevenson and Fanshawe 2002: Pl. 101; P. Hayman in Sinclair and Ryan 2003: 241

Photograph: K. Kussmann in *Der Falke* 1975(10): 350 (b/w); Zool. Soc. London in Everett 1977: 148 (adult and juvenile); Zool. Soc. London in Olney 1984: 131 (adult and juvenile); E. Hosking in Hosking and Flegg 1982: 142; E. and D. Hosking in del Hoyo et al. 1999: 144; R. Williams in Duncan 2003: 237

Literature: Reichenow 1908: 139; Eck and Busse 1973: 101, 102; Olney 1984: 129–134; Fry et al. 1988: 129, 130; Boyer and Hume 1991: 74; Evans et al. 1994: 42–47; del Hoyo et al. 1999: 188; König et al. 1999: 302, 303; Borrow and Demey 2001: 495; Stevenson and Fanshawe 2002: 202; Sinclair and Ryan 2003: 240

Bubo nipalensis Hodgson 1836
Forest Eagle Owl · Nepaluhu · Grand-duc de Nepal · Búho Nepali

Length: 510–630 mm
Body mass: ?

Distribution: Himalayas, from northern Uttar Pradesh east to southwest China, south to Cambodia and Vietnam, southern India, western Ghats, Tamil Nadu and Sri Lanka

Habitat: Dense evergreen and humid deciduous forest, usually near water, montane wet temperate forest and dense riparian gallery forest. From 300 m up to 2 100 m, locally up to about 3 000 m

- ***Bubo nipalensis nipalensis*** Hodgson 1836
  Bubo nipalensis Hodgson 1836, As. Res. 19: 172;
  Terra typica: Nepal

Distribution: Himalayas, from northern Uttar Pradesh east to southwest China, south to Cambodia and Vietnam, southern India, western Ghats and Tamil Nadu

Museum: BMNH, BNHS, FNSF, ZMH

Wing length:	425–470 mm
Tail length:	229–250 mm (1 × 260 mm)
Tarsus length:	60–62 mm
Length of bill:	52–54 mm
Body mass:	?

Illustration: G. M. Henry in Ali 1949: Pl. 52; P. Barruel in Ali and Ripley 1969: Pl. 17; A. Cameron in Voous 1988: 197; T. Boyer in Boyer and Hume 1991: 74; W. Monkol in Lekagul and Round 1991: Pl. 61; D. Cole in Grimmett et al. 1998: Pl. 29; T. Worfolk in del Hoyo et al. 1999: Pl. 11; F. Weick in König et al. 1999: Pl. 32; H. Burn in Robson 2000: Pl. 22; McQueen in Rasmussen and Anderton 2005:Pl. 76

Photograph: D. Messner in photo collection Weick (Vogelpark Walsrode); H. Ehlers in photo collection Weick, Singapore Zoo

Literature: Sharpe 1875a: 37–39; Baker 1927: 418, 419; Delacour and Jabouille 1931: 124, 125; Ali 1949: 154–156; Whistler 1963 (repr.): 342, 343; Ali and Ripley 1969: 276, 277; Smythies and Hughes 1984 (repr.): 317; Voous 1988: 105–108; Boyer and Hume 1991: 74; Lekagul and Round 1991: 180; Grimmett et al. 1998: 432, 433; del Hoyo et al. 1999: 191; König et al. 1999: 306, 307; Robson 2000: 291; Rasmussen and Anderton 2005: 240

- ***Bubo nipalensis blighi*** Legge 1878
  Bubo nipalensis blighi Legge 1878, A history of the birds of Ceylon 1: 133; **Terra typica:** Ceylon

Distribution: Sri Lanka
Museum: BMNH, NMC

Wing length:	370–412 mm
Tail length:	184–220 mm
Tarsus length:	?
Length of bill:	50.5–55.5 mm
Body mass:	?

Illustration: G. M. Henry in Henry 1955: Pl. 16; F. Weick in König et al. 1999: Pl. 32; Smit in *Proc. Zool. Soc. Lond.* 1884: Pl. 52; McQueen in Rasmussen and Anderton 2005: Pl. 76 (head and chest)

Photograph: Peters in *Gefiederte Welt* 1988(8): 235 (Colombo Zoo)

Literature: Legge 1878a: 133; Henry 1955: 196, 197; Ali and Ripley 1969: 277, 278; del Hoyo et al. 1999: 191; König et al. 1999: 306, 307; Rasmussen and Anderton 2005: 240

Bubo nipalensis blighi · Forest Eagle Owl · Nepaluhu

Bubo sumatranus **(Raffles) 1822**
Barred (Malay) Eagle Owl · Malaienuhu · Grand-duc bruyant (de Malaise) · Búho Malayo

Length:	400–460 mm
Body mass:	620 g

Distribution: Extreme southern Myanmar, peninsular Thailand, south to Sumatra, Borneo, Java, Bali and Banka

Habitat: Lowland evergreen forest, semi-evergreen forest, forest edges and clearings, secondary wood, plantations, Botanical Gardens, groves in cultivated country. From sea-level to 600 m, locally to 1 000 m and in western Java recorded at about 1 500 m

- ***Bubo sumatranus sumatranus*** (Raffles) 1822
  Strix sumatrana Raffles 1822, Trans. Linn. Soc. London 13(2): 279; **Terra typica:** Sumatra

Synonym:
- *Bubo orientalis minor* Schlegel 1863, Mus. Pays-Bas 2, Oti: 13; **Terra typica:** Banka (holotype NNML)

Distribution: Extreme southern Myanmar and peninsular Thailand, south to Sumatra and Banka Island
Museum: BMNH, FNSF, NNML, SMNSt, SMTD, ZMH

Wing length:	323–358 mm
Tail length:	183–190 mm
Tarsus length:	51–58 mm
Length of bill:	41–45 mm
Body mass:	620 g (Berlin Zoo)

Illustration: T. Boyer in Boyer and Hume 1991: 75; W. Mongkol in Lekagul and Round 1991: Pl. 61; T. Worfolk in del Hoyo et al. 1999: Pl. 11; F. Weick in König et al. 1999: Pl. 32; H. Burn in Robson 2000: Pl. 22
Photograph: Balgooy in Pengemar Alam 1957: 37/122: 23; G. F. Mees in Mees 1964b: 118 (b/w photograph skins); Frederic in Burton et al. 1984: 82; Lenton in Burton et al. 1992: 80; ? in *ZEITmagazin* 1992(46): 17; Budich in *Gefiederte Welt* 1993(7): 242; F. Lambert in del Hoyo et al. 1999: 98; D. G. Joyce in internet: Owl Pages, photo gallery 2002; L. Poh in internet: Owl Pages, photo gallery 2003 (2 photographs)
Literature: Sharpe 1875a: 39–41; Baker 1927: 419, 420; Mees 1964b: 116–119; Eck and Busse 1973: 106, 107; Boyer and Hume 1991: 75; Lekagul and Round 1991: 180; Blaskiewitz 1993, *Gefiederte Welt* 1993(7): 241; del Hoyo et al. 1999; König et al. 1999: 305, 306; Robson 2000: 291

▪ ***Bubo sumatranus strepitans*** (Temminck) 1823
Strix strepitans Temminck 1823, Pl. col. livr. 30, Pl. 18;
Terra typica: Java

Synonym:
- *Strix orientalis* Horsfield 1821, Trans. Linn. Soc. London 13(1): 140; Terra typica: Java (invalid name)

Distribution: Java and Bali
Museum: BMNH, MNBHU, NNML (holotype), SMNSt, SMTD, ZFMK

Wing length:	370–417 mm
Tail length:	186–200 mm
Tarsus length:	58–65.5 mm
Length of bill:	42–53 mm
Body mass:	?

Illustration: Syamsudin et al. in McKinnon 1990: Pl. 18; F. Weick in König et al. 1999: Pl. 32
Photograph: G. M. Mees in Mees 1964b: 118 (b/w photo of skins); K. Kussmann in Eck and Busse 1973: Abb. 8 (b/w photo, Berlin Zoo)
Literature: Mees 1964b: 116–119; Eck and Busse 1973: 106, 107; McKinnon 1990: 223; del Hoyo et al. 1999: 191; König et al. 1999: 305, 306

▪ ***Bubo sumatranus tenuifasciatus*** Mees 1964
Bubo sumatranus tenuifasciatus Mees 1964, Zool. Meded. Rijksmus. Nat. Hist. Leiden 40(13): 116–119;
Terra typica: Rantan, southeastern Borneo

Distribution: Borneo
Museum: FNSF, NNML (holotype), SMTD

Wing length:	323–345 mm
Tail length:	?
Tarsus length:	48–51 mm
Length of bill:	?
Body mass:	?

Illustration: Hughes in Smythies 1968: Pl. 15; K. Phillipps in McKinnon and Phillipps 1993: Pl. 39; F. Weick in König et al. 1999: Pl. 32
Photograph: G. M. Mees in Mees 1964b: 118 (b/w photo, skins)
Literature: Mees 1964b: 116–119; Smythies 1968: 278, 279; McKinnon and Phillipps 1993: 193; del Hoyo et al. 1999: 191; König et al. 1999: 305, 306

Bubo shelleyi (Sharpe and Ussher) 1872
Shelley's Eagle Owl · Bindenuhu (Shelley-Uhu) · Grand-duc de Shelley · Búho Barrado
***Huhua shelleyi* Sharpe and Ussher 1872, Ibis: 182;**
Terra typica: Fanti, Gold Coast
(see Plate 5)

Length:	530–610 mm
Body mass:	1257 g (1 ♂?)

Distribution: Western and central Africa; Upper and Lower Guinea, Liberia, Ghana, Gabon to eastern and northern Zaire. A very rare bird, fewer than 20 specimens known
Habitat: Primary rainforest in lowland, also forest edges and clearings
Museum: BMNH (holotype), SMNSt (adult and juvenile)

Wing length: 420–470 mm (1 × 492 mm)
Tail length: 240–266 mm (1 × 233 mm)
Tarsus length: 76–84 mm
Length of bill: 56–62 mm
Body mass: 1 257 g (1 ♂?)

Illustration: J. G. Keulemans in Sharpe 1875a: Pl. 2; H. Grönvold in Bannerman 1953: 546 (b/w); M. Woodcock in Fry et al. 1988: Pl. 8; T. Boyer in Boyer and Hume 1991: 76; P. Hayman in Kemp and Kemp 1998: 327; T. Worfolk in del Hoyo et al. 1999: Pl. 11; F. Weick in König et al. 1999: Pl. 31; N. Borrow in Borrow and Demey 2001: Pl. 60; P. Hayman in Sinclair and Ryan 2003: 239
Photograph: S. Grossman in Grossman and Hamlet 1965: 424 (b/w); L. Schlawe in Eck and Busse 1973: 36 (b/w); S. Schlawe in Eck 1975, *Der Falke* 1975(10): 351 (b/w, Zoo Antwerpe
Literature: Sharpe and Ussher 1872: 182; Sharpe 1875a: 37; Bannerman 1953: 545, 546; Eck and Busse 1973: 101; Fry et al. 1988: 130; Boyer and Hume 1991: 76; Kemp and Kemp 1998: 326, 327; del Hoyo et al. 1999: 191; König et al. 1999: 304, 305; Borrow and Demey 2001: 495; Sinclair and Ryan 2003: 238

❖

Bubo lacteus **Temminck 1820**
Verreaux's Eagle Owl · Blaßuhu (Milchuhu) · Grand-duc de Verreaux · Búho Lechoso
Strix lactea **Temminck 1820, Pl. col. livr. 1: Pl. 4;**
Terra typica: Senegal
(see Plate 5)

Length: 600–650 mm
Body mass: 1 615–3 115 g

Distribution: West Africa, patchily from Senegal and central Mali east to Cameroon, from central Sudan, northern Ethiopia and Somalia south to the Cape. Absent from Namibia (desert) and the rainforests of west and central Africa
Habitat: Dry savanna and semi-desert with thornbushes and scattered trees. Riparian areas with groups of trees and small semi-open woods with adjacent savanna. Absent from dense forest and desert. From sea-level up to 3 000 m
Museum: BMNH, NNML, SMNSt, SMTD, ÜMB, ZFMK, ZMH, ZSBS

Wing length: ♂ 420–480 mm (∅ 448 mm, $n = 18$), ♀ 447–490 mm (∅ 465, $n = 22$)
Tail length: ♂ 220–275 mm, ♀ 230–273 mm
Tarsus length: 73–86 mm
Length of bill: 51–54 mm
Body mass: ♂ 1 615–1 960 g (1 × 2 200 g, zoo), ♀ 2 475–3 115 g

Illustration: C. J. Temminck 1820a: Pl. 4; Neubaur in Koenig 1936: 172–174; G. Arnott in Steyn 1982: Pl. 24; M. Woodcock in Fry et al. 1988: Pl. 8; T. Boyer in Boyer and Hume 1991: 77; D. Zimmerman in Zimmerman et al. 1996: Pl. 55; T. Disley in Barlow et al. 1997: Pl. 25; P. Hayman in Kemp and Kemp 1998: 249; T. Worfolk in del Hoyo et al. 1999: Pl. 11; F. Weick in König et al. 1999: Pl. 31; J. Gale in Stevenson and Fanshawe 2002: Pl. 101; P. Hayman in Sinclair and Ryan 2003: 239; R. Bateman in Dean 2004: 151
Photograph: L. Schlawe in Eck and Busse 1973: Abb. 7 (b/w); P. Johnson in Everett 1977: 25 and 103; E. Hosking in Hosking and Flegg 1982: 144; P. Steyn in Steyn 1982: 282, 283 (b/w); K. Rudloff in *Der Falke* 1985(11): 395; C. H. Haagner in Ginn et al. 1989: 342; P. Steyn in Burton et al. 1992: 73; P. Funston in Kemp and Kemp 1998: 248; D. Balfour in Kemp and Kemp 1998: 248; L. Hes in Kemp and Kemp 1998: 249; R O'Connor in del Hoyo et al. 1999: 127; L. Koerner in internet: Owl Pages, photo gallery 2002; B. Marcot in internet: Owl Pages, photo gallery 2002; G. Schultz in Duncan 2003: 196
Literature: Sharpe 1875a: 33–35; Koenig 1936: 172–174; Eck and Busse 1973: 102, 103; Steyn 1982: 278–283; Dathe 1985, *Der Falke* 1985(11): 394; Fry et al. 1988: 131–133; Ginn et al. 1989: 342; Boyer and Hume 1991: 77; Zimmerman et al. 1996: 444; Kemp and Kemp 1998: 248, 249; del Hoyo et al. 1999: 191, 192; König et al. 1999: 303, 304; Stevenson and Fanshawe 2002: 495, 496; Sinclair and Ryan 2003: 238

❖

Bubo coromandus **(Latham) 1790**
Dusky Eagle Owl · Koromandeluhu · Grand-duc sombre (Coromandel) · Búho de Coromandel

Length: 480–530 mm
Body mass: ?

Distribution: Eastern Pakistan, northern and central India, southern Nepal to Assam, Bangladesh, western and southern Myanmar, western Thailand and eastern China(?)
Habitat: Open wooded country near water. Riparian forest, plantations, dense foliaged groves. From sea-level to 250 m
Remarks: Possibly not closely related to other members of genus *Bubo*

- ***Bubo coromandus coromandus*** (Latham) 1790
Strix coromanda Latham 1790, Index Orn. 1: 53;
Terra typica: Coromandel Coast

Distribution: Eastern Pakistan, northern and central India and southern Nepal to Assam and Bangladesh. Eastern China?
Museum: BMNH, FNSF, MNBHU, SMNSt, ZMH

Wing length:	380–435 mm
Tail length:	196–224 mm
Tarsus length:	(55) 60–71 mm
Length of bill:	41–49 mm
Body mass:	?

Illustration: D. Watson in Ali and Ripley 1969: Pl. 43; T. Boyer in Boyer and Hume 1991: 78; D. Cole in Grimmett et al. 1998: Pl. 29; T. Worfolk in del Hoyo et al. 1999: Pl. 11; F. Weick in König et al. 1999: Pl. 33; H. Burn in Robson 2000: Pl. 22; L. McQueen in Rasmussen and Anderton 2005: Pl. 76
Photograph: E. H. N. Lowther 1949: A Bird Photographer in India: Pl. 76 (adult and juvenile); L. Poh in internet: Owl Pages, photo gallery 2001; R. A. Behrstock in Duncan 2003: 240
Literature: Sharpe 1875a: 35–37; Baker 1927: 416, 417; Whistler 1963 (repr.): 344, 345; Ali and Ripley 1969: 278, 279; Eck and Busse 1973: 103, 104; Boyer and Hume 1991: 78; Grimmett et al. 1998: 433; del Hoyo et al. 1999: 192; König et al. 1999: 307, 308; Robson 2000: 291, 292; Rasmussen and Anderton 2005: 240

- ***Bubo coromandus klossi*** Robinson 1911
Bubo coromandus klossi Robinson 1911, J. Fed. Malay St. Mus. 4: 246; Terra typica: Gunong Semanggol, Perak, Malay States

Distribution: Western and southern Myanmar, western Thailand
Museum: BMNH, ZMH?
Remarks: Known only from three museum skins!

Wing length:	381–398 mm
Tail length:	187–220 mm
Tarsus length:	64–67 mm
Length of bill:	44 and 45 mm
Body mass:	?

Illustration: W. Mongkol in Lekagul and Round 1991: Pl. 61
Photograph: ?
Literature: Robinson 1911: 246; Baker 1927: 417; Smythies and Hughes 1984: 316; Lekagul and Round 1991: 180; del Hoyo et al. 1999: 192; König et al. 1999: 307, 308

Bubo leucostictus Hartlaub 1855

Akun Eagle Owl · Gelbfuß- oder Käferuhu · Grand-duc tacheté · Búho de Ákun
Bubo leucostictus "Temm." Hartlaub 1855, J. Orn. 3: 354;
Terra typica: Dabocrom, Gold Coast
(see Plate 4)

Length:	400–460 mm
Body mass:	500–600 g

Distribution: West Africa, patchily from Guinea, Sierra Leone, Liberia, Ivory Coast, Ghana and Nigeria to Cameroon. South to mouth of Congo River. East to Zaire, Cabinda and probably northwest Congo. South to western Angola
Habitat: Lowland primary and old secondary rainforest. Also forest edges, clearings. Forest along rivers and wooded river-islands. Sometimes in swampy country and groups of trees near farmland. Only in lowlands
Museum: BMNH, FNSF, MHNP, MNBHU, SMNSt
Remarks: No obvious relatives in genus *Bubo*. Monotypic

Wing length:	♂ 292–338 mm, ♀ 310–332 mm
Tail length:	190–219 mm
Tarsus length:	41–48 mm
Length of bill:	30–35 mm
Body mass:	♂ 486 and 536 g, ♀ 524 and 607 g

Illustration: F. W. Frohawk in Bannerman 1953: 547 (b/w); M. Woodcock in Fry et al. 1988: Pl. 8; T. Boyer in Boyer and Hume 1991: 79; P. Hayman in Kemp and Kemp 1998: 259 (adult and juvenile); T. Worfolk in del Hoyo et al. 1999: Pl. 11; F. Weick in König et al. 1999: Pl. 29; N. Borrow in Borrow and Demey 2001: Pl. 60

Photograph: ?

Literature: Sharpe 1875a: 41; Bannerman 1953: 546, 547; Jellicoe 1954: 154–167; Eck and Busse 1973: 102; Fry et al. 1988: 133, 134; Boyer and Hume 1991: 79; Kemp and Kemp 1998: 258, 259; del Hoyo et al. 1999: 192; König et al. 1999: 298, 299; Borrow and Demey 2001: 496

Bubo philippensis (Kaup) 1851
Philippine Eagle Owl · Streifen- oder Philippinen-Uhu · Grand-duc des Philippines · Búho Filipino

Length: 400–430 mm
Body mass: ?

Distribution: Luzon, Catanduanes, Samar, Leyte, Mindanao, Bohol

Habitat: Forest near rivers and lakes, at lower elevations

Remarks: Possibly not closely related to other species of genus *Bubo*

- ***Bubo philippensis philippensis*** (Kaup) 1851
  Pseudoptynx philippensis Kaup 1851, Arch. Naturgesch. 17/1: 110; Terra typica: Philippine Islands

Distribution: Luzon and Catanduanes
Museum: BMNH (holotype), FNSF, MHNP, SMTD

Wing length: 341 and 343 mm
Tail length: 162 and 170 mm
Tarsus length: 70 mm
Length of bill: 47 mm
Body mass: ?

Illustration: J. Smit in *Trans. Zool. Soc. London* 1877(9): Pl. 25; G. Sandström in du Pont 1971: Pl. 39; T. Boyer in Boyer and Hume 1991: 79; T. Worfolk in del Hoyo et al. 1999: Pl. 11; F. Weick in König et al. 1999: Pl. 33; A. Sutherland in Kennedy et al. 2000: Pl. 35

Bubo philippensis mindanensis · Philippine Eagle Owl · Philippinen-Uhu

Photograph: Wirth in *SCRO Mag.* 1997(1): 32

Literature: Sharpe 1875a: 43; du Pont 1971: 174; Eck and Busse 1973: 107; Dickinson et al. 1991: 227, 228; Boyer and Hume 1991: 79; Steinberg 1997b: 31, 32; del Hoyo et al. 1999: 192, 193; König et al. 1999: 308, 309; Kennedy et al. 2000: 179, 180

- ***Bubo philippensis mindanensis* (Ogilvie-Grant) 1906**
  ***Pseudoptynx mindanensis* Ogilvie-Grant 1906, Bull. Brit. Ornith. Cl. 16: 99; Terra typica: Davao, Mindanao**

Distribution: Samar, Leyte, Mindanao, Bohol
Museum: BMNH (holotype), SMTD?

Wing length: 341–360 mm
Tail length: 170–178 mm
Tarsus length: 69 and 70 mm
Length of bill: 52 mm
Body mass: ?

Illustration: F. Weick in König et al. 1999: Pl. 33

Photograph: C. and D. Frith in del Hoyo et al. 1999: 142; D. Allen in internet: Owl Pages, photo gallery 2002; G. Ehlers in Weick, photo collection (Luzon Zoo)
Literature: McGregor 1909: 250–252; du Pont 1971: 174; Eck and Busse 1973: 107; Dickinson et al. 1991: 227, 228; del Hoyo et al. 1999: 192, 193; König et al. 1999: 308, 309; Kennedy et al. 2000: 179, 180

❖

Bubo blakistoni Seebohm 1884

Blakiston's Fish or Eagle Owl · Riesenfischuhu · Grand-duc de Blakiston · Búho Manchú
(see Plate 5)

Length: 600–710 mm
Body mass: ?

Distribution: West Manchuria, southeast Siberia and extreme northeast China to border of Korea. Sakhalin Island, Hokkaido and Kuril Island
Habitat: Dense broadleaf or mixed broadleaf/coniferous forest along clear and slow-flowing rivers and streams. Steep-sided, wooded valleys near rivers, islands in fast-flowing rivers and streams, partly ice-free in winter. In Kuril Island, dense fir/spruce forest with some deciduous trees, bordering lakes, rivermouth and sea coast. Sometimes also on rocky coasts. Lowland
Remarks: Often placed in genus *Ketupa*, but skeletal details differ, being closer to those of *Bubo*

▪ *Bubo blakistoni blakistoni* Seebohm 1884

Bubo Blakistoni Seebohm 1884, Proc. Zool. Soc. Lond. 1883: 466; Terra typica: Hokkaido, type from Hakodate

Synonym:
- *Bubo blakistoni karafutonis* Kuroda 1931, Tori 7: 4; Terra typica: Sakhalin

Distribution: Sakhalin, Hokkaido and Kuril Islands
Museum: BMNH, NSMT

Wing length: 498–534 mm
Tail length: 243–286 mm
Tarsus length: 81–102 mm
Length of bill (cere): 34–38 mm
Body mass: ?

Illustration: J. G. Keulemans in *Ibis* 1884: Pl. 6; T. Miyamoto in Kobayashi 1965: 63; S. Takano in Massey et al. 1983: 189; A. Cameron in Voous 1988: 111; T. Boyer in Boyer and Hume 1991: 81; Yabuuchi in Brazil 1991: (b/w); T. Worfolk in del Hoyo et al. 1999: Pl. 11; F. Weick in König et al. 1999: Pl. 34; ? on Stamp "Nippon, *Ketupa blakistoni*", 60
Photograph: Takano et al. 1991: 338, 339 (adult and juvenile); S. Kaufman in del Hoyo et al. 1999: 147
Literature: Hartert 1912–1921: 970, 971; Hartert 1923: 85; Kuroda 1931: 41, 42 (Japanese and English); Kobayashi 1965: 62; Massey et al. 1983: 188; Voous 1988: 109–113; Boyer and Hume 1991: 80–82; del Hoyo et al. 1999: 193; König et al. 1999: 311, 312

▪ *Bubo blakistoni doerriesi* Seebohm 1895

Bubo blakistoni doerriesi Seebohm 1895, Bull. Brit. Ornith. Cl. 5: 4; Terra typica: Sidemi on the lower Ussuri

Synonym:
- *Bubo blakistoni piscivorus* Meise 1933, Orn. Monatsb. 41: 169; Terra typica: "Jakschi" west of Great Khinghan = Yakoshih, northwest Manchuria

Remarks: Subspecies *piscivorus* known only from the holotype, which was destroyed in World War II. But description similar to subspecies *doerriesi*
Distribution: West Manchuria, southeast Siberia and northeast China to the border with Korea
Museum: AMNH, BMNH (holotype), MNBHU, SMTD

Wing length: ♂ 510–550 mm, ♀ 538–560 mm
Tail length: 285–305 mm
Tarsus length: 85 mm
Length of bill (cere): 34 mm
Length of bill: 55–65 mm
Body mass: ?

Illustration: ? in Dementiev and Gladkov 1951: 408 (b/w); R. D. Digby in Sayers 1976b: 60; F. Weick in König et al. 1999: Pl. 34
Photograph: J. Pukinski in Pukinski 1983: Abb. 64–69 and 78; J. Sibnev in Knystaustas and Sibnev 1987: 45; J. Sibnev in *Der Falke* 2000: 69–72
Literature: Hartert 1912–1921: 971; Hartert 1923: 385; Meise 1933: 169–173; Dementiev 1933b: 383–388; Dementiev and Gladkov 1951: 406–408; Sibnev 1963: 486; Vaurie 1965: 591, 592; Pukinski 1973: 40–47; Pukinski 1975: 128 pp; Sayers 1976b: 60–63; Etchécopar and Hüe 1978: 471; Meyer de Schauensee 1984: 268; Knystaustas and Sibnev 1985: 188 pp; del Hoyo et al. 1999: 193; König et al. 1999: 311, 312; Mikhailov 2000: 68–73; Sibnev 2000: 69–72

Genus

Ketupa **Lesson 1830**

Ketupa Lesson 1830, Traite d Orn. livr. 2: 114. Type by *Ketupa javanensis* Lesson = *Strix Ketupu* Horsfield

Synonym:
- *Strigonax* W. Miller 1915, Bull. Am. Mus. Nat. Hist. 34:515

Remarks: Sometimes merged in genus *Bubo*, but skeletal features (skull osteology) suggest it is distinct

Ketupa zeylonensis **(Gmelin) 1788**

Brown Fish Owl · (Brauner) Fischuhu · Hibou pêcheur brun (Kétoupa brun) · Pescador de Ceilán

Length: 480–580 mm
Body mass: ♂ 1 105 g, ♀ 1 308 g

Distribution: Locally distributed from southwest Asia Minor, Iraq and parts of Iran and northwest Pakistan, India (south of the Himalayas), Sri Lanka, Assam, Myanmar, Thailand, Vietnam and southeast China

Habitat: Deciduous, semi-deciduous and open woodland near water. Old mango groves or plantations, densely foliaged, evergreen trees along canals, rivers, streams and reservoirs. Ravines, steep river banks. Often near human habitations. From lowlands up to 1400 m, sometimes up to 1900 m, above sea-level

- ***Ketupa zeylonensis semenowi*** Zarudny 1905
  Ketupa semenowi Zarudny 1905, Orn. Jahrb. 16: 141;
  Terra typica: Eastern slopes of Zagros Mountains, Pers. Arabistan

Synonym:
- *Strix hardwickii* Gray 1834, in Hardwick Ill. Zool. 2, Pl. 31; Terra typica: Futteghur, northwest India

Distribution: Southern Turkey, Israel (extinct), northern Syria to northwest India

Museum: LivCM (Tristram coll.), MNBHU, NNML

Remarks: Barely distinguishable from *K. z. leschenault* – possibly should be united with this subspecies

Wing length: ♂ 396–429 mm (1 × 434 mm), ♀ 399–404 mm
Tail length: 197–214 mm
Tarsus length: 74–80 mm
Length of bill: 49–54 mm
Body mass: ?

Illustration: Gray in Hardwick Ill. Zool. 2: Pl. 31; J. Willis in Mikkola 1983: Pl. unpaged; H. Delin in Cramp et al. 1985: Pl. 45; A. Harris in Shirihai 1996: 315 (b/w); T. Worfolk in del Hoyo et al. 1999: Pl. 11; Zetterström andMullarney in Svensson et al. 1999: 207

Photograph: A. Boldo in Mikkola 1983: Pl. 24; L. Koerner in internet: Owl Pages, photo gallery 2002

Literature: Hartert 1912–1921: 972, 973; Baker 1927: 408, 409; Dementiev and Gladkov 1951: 406–408; Vaurie 1965: 592; Eck and Busse 1973: 104, 105; Mikkola 1983: 91–94; Cramp et al. 1985: 481–484; Voous 1988: 114–118; Shirihai 1996: 315; del Hoyo et al. 1999: 193; König et al. 1999: 312, 313; Rasmussen and Anderton 2005: 240, 241

- ***Ketupa zeylonensis leschenault*** (Temminck) 1820
  Strix leschenault Temminck 1820, Pl. col. livr. 4: Pl. 20;
  Terra typica: Eastern provinces of India = Chandranagore

Synonym:
- *Cultrungius Nigripes* Hodgson 1836, J. As. Soc. Beng. 5: 364; Terra typica: Nepal

Distribution: India south of the Himalayas, Nepal, Sikkim and Bhutan, east to Myanmar (except the northeast) and Thailand

Museum: BMNH, MNBHU, ZMH

Remarks: According to Dementiev, a synonym of the nominate

Wing length: 370–430 mm
Tail length: 186–210 mm
Tarsus length: 71–90 mm
Length of bill: 49–54 mm
Body mass: ♂ 1 105 g, ♀ 1 308 g

Illustration: Hughes in Ali and Ripley 1969: Pl. 46; H. Grönvold in Whistler 1963 (repr.): Pl. 18; Hughes in Smythies and Hughes 1984: Pl. 18; W. Mongkol in Lekagul and Round 1991: Pl. 61; T. Boyer in Boyer and Hume 1991: 82; T. Worfolk in del Hoyo et al. 1999: Pl. 11; F. Weick in König et al. 1999: Pl. 34; McQueen in Rasmussen and Anderton 2005: Pl. 76

Photograph: C. Everett in Everett 1977: 28; A. and M. Shah in del Hoyo et al. 1999: 102; B. Marcot in internet: Owl Pages, photo gallery 2002

Literature: Sharpe 1875a: 4, 5; Baker 1927: 409; Dementiev and Gladkov 1951: 406–408; Whistler 1963 (repr.): 340, 341; Ali and Ripley 1969: 280–282; Smythies and Hughes 1984: 314, 315; Voous 1988: 114–118; Lekagul and Round 1991: 180; Grimmett et al. 1998: 433; del Hoyo et al. 1999: 193; König et al. 1999: 312, 313; Robson 2000: 292; Rasmussen and Anderton 2005: 240, 241

- ***Ketupa zeylonensis zeylonensis*** (Gmelin) 1788
Strix zeylonensis Gmelin 1788, Syst. Nat. 1(1): 287;
Terra typica: Ceylon

Distribution: Sri Lanka

Museum: BMNH, MNBHU, MZUS, SMTD, ÜMB, ZFMK, ZSBS

Wing length:	355–403 mm
Tail length:	175–206 mm
Tarsus length:	85–90 mm
Length of bill:	42–48 mm
Body mass:	♂ 1 105 g ($n = 1$)

Illustration: G.M. Henry in Henry 1955: Pl. 16; A. Cameron in Voous 1988: 115; T. Boyer in Boyer and Hume 1991: 82; T. Worfolk in del Hoyo et al. 1999: Pl. 11; F. Weick in König et al. 1999: Pl. 34

Photograph: K. W. Fink in Campbell 1974: Pl. 439 (*ketupu* = error); Auscape Internat. in Duncan 2003: 292

Literature: Sharpe 1875a: 4, 5; G. M. Henry 1955: 195, 196; Ali 1964: 62; Ali and Ripley 1969: 282; Eck and Busse 1973: 104; Boyer and Hume 1991: 82, 83; del Hoyo et al. 1999: 193; König et al. 1999: 312, 313

- ***Ketupa zeylonensis orientalis*** Delacour 1926
Ketupa ceylonensis (sic.) *orientalis* Delacour 1926, Bull. Brit. Ornith. Cl. 47: 11; Terra typica: Dakto, Annam

Distribution: Northeast Myanmar to southeast China (Guangsi, Guangdong), south to the Malay Peninsula, Indochina and Hainan

Museum: BMNH, USNM

Wing length:	♂ 365–400 mm, ♀ -457 mm
Tail length:	195–210 mm
Tarsus length:	67–70 mm
Length of bill:	?
Body mass:	?

Illustration: F. Berille in Etchécopar and Hüe 1978: Pl. 18

Photograph: ?

Literature: Baker 1927: 406–409; Dementiev and Gladkov 1951: 406; Eck and Busse 1973: 104; Etchécopar and Hüe 1978: 469, 470; Meyer de Schauensee 1984: 268; del Hoyo et al. 1999: 193; König et al. 1999: 312, 313

Ketupa flavipes (Hodgson) 1836

Tawny Fish Owl · Himalaja-Fischuhu · Kétoupa roux · Búho Pescador Leonado

Cultrungius Flavipes **Hodgson 1836, J. As. Soc. Beng. 5: 364, Pl. 26;** Terra typica: **Nepal**

Length:	480–580 mm
Body mass:	?

Distribution: The Himalayas, from northwest India, Nepal and Bhutan to northeast India, east to central China and Taiwan, south to northern Bangladesh, northeast Myanmar and southern Indochina

Habitat: Dense forests bordering streams and pools. Prefers running water. From 250 m up to 1 500 m, in India sometimes up to 2 450 m

Museum: BMNH, BNHS, MNBHU

Ketupa flavipes · Tawny Fish Owl · Himalaja-Fischuhu

Wing length: 410–477 mm
Tail length: 215–227 mm (1 × 240 mm)
Tarsus length: 60–67 mm
Length of bill: 48–50 mm
Body mass: ?

Illustration: Hodgson 1836b: Pl. 26; A. Cameron in Voous 1988: 122; T. Boyer in Boyer and Hume 1991: 83; D. Cole in Grimmett et al. 1998: Pl. 29; T. Worfolk in del Hoyo et al. 1999: Pl. 11; F. Weick in König et al. 1999: Pl. 33; H. Burn in Robson 2000: Pl. 22; McQueen in Rasmussen and Anderton 2005: Pl. 76
Photograph: ? in *ZEITmagazin* 1992(46)
Literature: Sharpe 1875a: 5–8; Baker 1927: 411, 412; Delacour and Jabouille 1931: 122, 123; Vaurie 1965: 592; Ali and Ripley 1969: 283, 284; Eck and Busse 1973: 107, 108; Smythies and Hughes 1984: 315; Voous 1988: 119–121; Boyer and Hume 1991. 83, 84; Grimmett et al. 1998: 433; del Hoyo et al. 1999: 193, 194; König et al. 1999: 310, 311; Robson 2000: 292; Rasmussen and Anderton 2005: 241

Ketupa ketupu (Horsfield) 1821
Buffy (Malay) Fish Owl · Sunda-Fischuhu · Kétoupa (Hibou pêcheus) malais · Búho Pescador Malayo (see Plate 5)

Length: 400–480 mm
Body mass: 1 000–2 100 g

Distribution: Southern Myanmar, southern Assam, south and east to peninsular Thailand and Annam, Malay Peninsula, Riau Archipelago, Sumatra, Banka, Belitung and neighboring islands on the western side (Nias, etc.), Java, Bali, Borneo
Habitat: Wooded areas bordering streams, rivers, lakes, fishponds, rice paddies. Old plantations and parks near wetlands and mangroves. Often near human habitation. From lowlands at sea-level up to 1 100 m, locally 1 600 m

- ***Ketupa ketupu aagaardi*** Neumann 1935
  Bubo ketupu aagaardi Neumann 1935, Bull. Brit. Ornith. Cl. 55: 138; Terra typica: Bang Nara, Peninsular Siam = Thailand

Distribution: Southern Assam to southern Thailand and Vietnam
Museum: BMNH (holotype)
Remarks: Intergrades with the nominate

Wing length: 315–345 mm (354 mm?)
Tail length: 160 mm
Tarsus length: 70 mm
Length of bill: 40 mm
Body mass: ?

Illustration: W. Mongkol in Lekagul and Round 1991: Pl. 61; H. Burn in Robson 2000: Pl. 22; L. McQueen in Rasmussen and Anderton 2005: 76
Photograph: R. Suryadi in internet: Owl Pages, photo gallery 2002
Literature: Neumann 1935: 138; Baker 1927: 410, 411; Smythies and Hughes 1984: 315; Lekagul and Round 1991: 182; del Hoyo et al. 1999: 194; König et al. 1999: 309, 310; Robson 2000: 292; Rasmussen and Anderton 2005: 242

- ***Ketupa ketupu ketupu*** (Horsfield) 1821
  Strix ketupu Horsfield 1821, Trans. Linn. Soc. London 13(1): 141; Terra typica: Java

Distribution: Malay Peninsula, Riau Archipelago, Sumatra, Banka, Belitung, Java, Bali, Borneo (except range *pageli*)
Museum: BMNH, MHNP, MZUS, SMNSt, SMTD, ZFMK, ZMH, ZSBS

Wing length: 335–390 mm
Tail length: 160–181 mm
Tarsus length: 70–80 mm
Length of bill: 40–42 mm
Body mass: 1 028–2 100 g

Illustration: A. Hughes in Smythies 1960 (repr.): Pl. 15; Syamsudin et al. in McKinnon 1990: Pl. 18; T. Boyer in Boyer and Hume 1991: 84; K. Phillipps in McKinnon and Phillipps. 1993. Pl. 39; D. Cole in Grimmett et al. 1998: Pl. 29; T. Worfolk in del Hoyo et al. 1999: Pl. 11; F. Weick in König et al. 1999: Pl. 33; H. Burn in Robson 2000: Pl. 22
Photograph: G.E. Kirkpatrick in Everett 1977: 31 (b/w); H. Busse 1984: 143; J. Steinbacher in *Gefiederte Welt* 1989(8): 229 (b/w); ? in Gosler 1991: 255; Stagsden Bd. Garden in Burton et al. 1992: 55/56 (not *zeyl*); Plage in Burton et al. 1992: 55; G. Chuen Hang in *SCRO Mag.* 1999(2): 54; G. Ziesler in del Hoyo et al. 1999: 81; A. B. van den Berg in del Hoyo et al. 1999: 120; Auscape Internat. in Duncan 2003: 293; L. Koerner in internet: Owl Pages, photo gallery 2003; M. Wood in internet: Owl Pages, photo

gallery 2003; R. Suryadi in internet: Owl Pages, photo gallery 2003

Literature: Baker 1927: 410, 411; Delacour and Jabouille 1931: 121, 122; Smythies 1960 (repr.): 279, 280; Eck and Busse 1973: 107, 108; Busse 1984: 142; McKinnon 1990: 181; Boyer and Hume 1991: 84; Gosler 1991: 254; McKinnon and Phillipps 1993: 193, 194; Grimmett et al. 1998: 434; del Hoyo et al. 1999: 194; König et al. 1999: 309, 310

- ***Ketupa ketupu minor*** Büttikofer 1896
  Ketupa minor Büttikofer 1896, Notes Leyden Mus. 18: 165; Terra typica: Nias Island
  Bubo ketupu büttikoferi Chasen 1935, Bull. Raffles Mus. 11: 84 (new name for *K. minor*)

Distribution: Nias Island off western Sumatra
Museum: NNML

Wing length: 295–300 mm
Tail length: ?
Tarsus length: ?
Length of bill: ?
Body mass: ?

Illustration: ?
Photograph: ?
Literature: Büttikofer 1896: 165; Eck and Busse 1973: 107, 108; del Hoyo et al. 1999: 194; König et al. 1999: 309, 310

- ***Ketupa ketupu pageli*** Neumann 1935
  Bubo ketupu pageli Neumann 1935, Bull. Brit. Ornith. Cl. 55: 138; Terra typica: Marudo Bay, Bengkoke River, eastern coast of northern Borneo

Distribution: Northwest Borneo, coast of Sarawak
Museum: BMNH

Wing length: 310–330 mm
Tail length: ?
Tarsus length: ?
Length of bill: ?
Body mass: ?

Illustration: ?
Photograph: ?
Literature: Neumann 1935: 138; del Hoyo et al. 1999: 194; König et al. 1999: 309, 310

Genus
Scotopelia **Bonaparte 1850**
Scotopelia Bonaparte 1850, Consp. Gen. Av. 1: 44.
Type by *Strix peli* Temminck

Scotopelia peli **(Bonaparte) 1850**
Pel's Fishing Owl · Bindenfischeule (Pelfischeule) · Chouette pêcheuse de Pel · Cárabo Pescador Común
Strix peli "Temm." Bonaparte 1850, Consp. Gen. Av. 1: 44;
Terra typica: Ashanti
(see Plate 5)

Length: 510–610 mm
Body mass: 2 000–2 300 g

Synonym:
- *Scotopelia oustaleti* de Rochebrune 1883, Bull. Sci. Soc. Philom (7), 7: 165; Terra typica: Senegambia
- *Scotopelia peli fischeri* Zedlitz 1908, Orn. Monatsb. 16: 172/173; Terra typica: Kau, Kenya (holotype MNBHU)
- *Scotopelia peli Salvagoraggii* Zedlitz 1908, Orn. Monatsb. 16: 172/173; Terra typica: Tacazze River, northern Ethiopia (holotype MNBHU)

Distribution: Locally from Senegambia east to Benin and from Nigeria south across the Congo Basin and central Africa to Botswana, Mozambique and north-eastern South Africa. Also southeast Sudan, Ethiopia, southern Somalia, Kenya and Tanzania. Probably also in southern Mali, Bukina Faso and southern Niger

Habitat: Forest along rivers and lakes, swamps and estuaries. Favours large riparian trees on river islands. From sea-level up to about 1 700 m

Museum: BMNH, MHNP, MNBHU, ÜMB

Wing length: ♂ 423–447 mm, ♀ 407–445 mm
Tail length: 207–243 mm (1 × 254 mm)
Tarsus length: 68–77 mm
Length of bill: 53–60 mm
Body mass: ♀ 2 055–2 325 g ($n = 4$)

Illustration: J. Wolf in *Ibis* 1859: Pl. 15; K. Lilly in Everett 1977: 27; G. Arnott in Steyn 1982: Pl. 24; M. Woodcock in Fry et al. 1988: Pl. 7; T. Boyer in Boyer and Hume 1991: 85; D. Zimmerman in Zimmerman et al. 1996: Pl. 55; T. Disley in Barlow and Wacher 1997: Pl. 25; P. Hayman in Kemp and Kemp 1998: 263; T. Worfolk in del Hoyo et al. 1999: Pl. 11; F. Weick in König et al. 1999: Pl. 35; N. Borrow in Borrow and Demey 2001: Pl. 58; J. Gale in Stevenson and Fanshawe 2002: Pl. 102; P. Hayman in Sinclair and Ryan 2003: 239

Photograph: S. Grossman in Grossman and Hamlet 1965: 428 (juvenile); L. Schlawe in Eck and Busse 1973: Abb. 9 (b/w); E. Bomford and D. Reucassel in Ginn et al. 1989: 343; Survival Anglia in Bellamy 1989: 57; A. Bannister in Kemp and Kemp 1998: 263; D. Balfour in Kemp and Kemp 1998: 262; E. Bomford in del Hoyo et al. 1999: 121; B. Marcot in internet: Owl Pages, photo gallery 2003

Literature: Sharpe 1875a: 10, 11; von Zedlitz 1908: 172–174; Bannerman 1953: 348, 349; Eck and Busse 1973: 108, 109; Steyn 1982: 284–286; Fry et al. 1988: 134–136; Ginn et al. 1989: 343; Boyer and Hume 1991: 85, 86; Zimmerman et al. 1996: 444; Claffey 1996: 135, 136; Barlow and Wacher 1997: 238; Kemp and Kemp 1998: 262, 263; del Hoyo et al. 1999: 195; König et al. 1999: 314, 315; Borrow and Demey 2001: 496; Stevenson and Fanshawe 2002: 204

Scotopelia peli · Pel's Fishing Owl · Bindenfischeule

Scotopelia ussheri Sharpe 1871
Rufous Fishing Owl · Rotrücken- (Rote) Fischeule · Chouette pêcheuse rousse · Cárabo Pescador Rojizo
Scotopelia ussheri Sharpe 1871, Ibis: 101, 417;
Terra typica: Fanti, Gold Coast
(see Plate 5)

Length: 460–510 mm
Body mass: ♂ 743 g, ♀ 834 g

Distribution: Sierra Leone, Liberia, Ivory Coast, Ghana, possibly Guinea?
Habitat: Lowland forest, mainly primary, bordering rivers and lagoons, sometimes secondary and degraded forest, plantations and coastal mangroves
Museum: BMNH (holotype)

Wing length: 330–345 mm
Tail length: 166–205 mm
Tarsus length: 60–63 mm
Length of bill: 44 and 45 mm
Body mass: ♂ 743 g, ♀ 834 g

Illustration: J. G. Keulemans in *Ibis* 1871: Pl. 12; H. Grönvold in Bannerman 1953: 550; M. Woodcock in Fry et al. 1988: Pl. 7; T. Boyer in Boyer and Hume 1991: 87; P. Hayman in Kemp and Kemp 1998: 267; T. Worfolk in del Hoyo et al. 1999: Pl. 11; F. Weick in König et al. 1999: Pl. 35; N. Borrow in Borrow and Demey 2001: Pl. 58
Photograph: E. Hosking in Hosking and Flegg 1982: 155; E. Hosking in Freethy 1992: 62; M. Gore in Burton et al. 1992: 59; Birdlife Int. in Kemp and Kemp 1998: 267; A. Compost in *Bull. ABC* ? (b/w with prey); G. Rondeau in internet: Owl Pages, photo gallery 2003
Fry et al. 1988: 136, 137
Literature: Sharpe 1871: 101, 417; Sharpe 1875a: 11; Bannerman 1953: 549, 550; Eck and Busse 1973: 109, 110; Boyer and Hume 1991: 87; Kemp and Kemp 1998: 266, 267; del Hoyo et al. 1999: 195; König et al. 1999: 315, 316; Borrow and Demey 2001: 496

Scotopelia bouvieri Sharpe 1875
Vermiculated Fishing Owl · Marmorfischeule · Chouette pêcheuse de Bouvier · Cárabo Pescador Marmorata
Scotopelia bouvieri Sharpe 1875, Ibis: 261;
Terra typica: Lopé, Ogowe River, Gaboon
(see Plate 5)

Length: 460–510 mm
Body mass: ?

Distribution: Central African forest from near the Atlantic coast (Congo Basin in South Cameroon, Gabon), to north-eastern Zaire and northwest Angola, possibly to southeast Nigeria
Habitat: Gallery forest along rivers or bordering lakes, also on small rivers, flooded areas of primary forest at some distance from water. Presence not dependent on water and fish
Museum: BMNH (holotype), MNBHU, ZMH
Remarks: Relationship of *S. ussheri* and *bouvieri* to *S. peli* uncertain!

Wing length: 302–330 mm
Tail length: 167–203 mm
Tarsus length: 56–65 mm
Length of bill: 35–40 mm
Body mass: ♀ 637 g ($n = 1$)

Illustration: J. G. Keulemans in Sharpe 1875a: Pl. 1; M. Woodcock in Fry et al. 1988: Pl. 7; T. Boyer in Boyer and Hume 1991: 87; P. Hayman in Kemp and Kemp 1998: 265; T. Worfolk in del Hoyo et al. 1999: Pl. 11; F. Weick in König et al. 1999: Pl. 35; N. Borrow in Borrow and Demey 2001: Pl. 58
Photograph: L. Schlawe in Eck and Busse 1973: Abb. 10 (b/w); Zool. Soc. London in Everett 1977: 35 (b/w); E. Hosking in Hosking and Flegg 1982: 154; ? in Gosler 1991: 213; E. Hosking in Freethy 1992: 63
Literature: Sharpe 1875b: 261; Sharpe 1875a: 11, 12; Bannerman 1953: 550, 551; Eck and Busse 1973: 110; Fry et al. 1988: 137, 138; Boyer and Hume 1991: 88; Kemp and Kemp 1998: 264, 265; del Hoyo et al. 1999: 195; König et al. 1999: 316; Borrow and Demey 2001: 497

Tribus / Tribe
Strigini · Wood Owls and allies · Waldkäuze und Verwandte

Genus
Strix Linné 1758
Strix von Linné 1758, Syst. Nat. ed. 10, 1: 92.
Type by *Strix stridula* = *Strix aluco* Linné

Synonym:
- *Ciccaba* Wagler 1832, Isis v. Oken, col. 1222. Type by *Ciccaba huhula*
- *Scotiaptex* Swainson 1836, Class. Birds I: 327. Type by *Strix cinerea* = *Strix nebulosa*

Strix seloputo Horsfield 1821
Spotted Wood Owl · Pagodenkauz · Chouette obscure (ou des pagodes) · Cárabo de las Pagodes

Length: 440–480 mm
Body mass: ♂ 1 011 g ($n = 1$)

Distribution: Southern Myanmar and southern Thailand, through Cambodia, South Vietnam and the Malay Peninsula south to Sumatra (Jambi) and Java. Also Bawean Island off Java and the western Philippines (Calamian Island and Palawan)
Habitat: Open and partially cleared lowland forest, forest edges, plantations, orchards and parks. Sometimes near human habitations. Paddy fields, mangrove forest, open deforested areas and swampy forest. From sea-level near coasts up to about 1 000 m

- ***Strix seloputo seloputo*** Horsfield 1821
Strix Selo-puto Horsfield 1821, Trans. Linn. Soc. London 13(1): 140; Terra typica: Java

Synonym:
- ? *Strix Orientalis* Shaw 1809, Gen. Zool. 7(1): 257 – "China Owl" Latham (not identifiable)
- *Strix pagodorum* Temminck 1823, Pl. col. livr. 39: Pl. 230; Terra typica: India and Java

Distribution: Southern Myanmar and southern Thailand, through Cambodia, South Vietnam and the Malay Peninsula, to Sumatra (Jambi) and Java
Museum: BMNH, FNSF, MHNP, SMTD, ZFMK

Wing length: 338–369 mm (1 × 376 mm)
Tail length: 188–198 mm
Tarsus length: 58–60 mm
Length of bill: 39–45 mm
Body mass: ♂ 1 011 g ($n = 1$)

Illustration: C. J. Temminck in Temminck 1823b: Pl. 230; W. Mongkol in Lekagul and Round 1991: Pl. 61; T. Boyer in Boyer and Hume 1991: 137; K. Phillipps in McKinnon and Phillipps 1993: Pl. 39; T. Worfolk in del Hoyo et al. 1999: Pl. 12; F. Weick in König et al. 1999: Pl. 38; H. Burn in Robson 2000: Pl. 10
Photograph: L. Schlawe in Eck and Busse 1973: Abb. 18 (b/w); E. Hosking in Everett 1977: 44 (b/w); E. Hosking in Hosking and Flegg 1982: 67 and 158; K. de la Motte in *SCRO Mag.* 1999(2): 55; L. Poh in internet: Owl Pages, photo gallery 2003; K. W. Fink in Duncan 2003: 245
Literature: Sharpe 1875a: 261, 262; Stresemann 1924: 111; Delacour and Jabouille 1931: 130, 131; Eck and Busse 1973: 159, 160; Smythies and Hughes 1984: 314; Boyer and Hume 1991: 137; Lekagul and Round 1991: 182; McKinnon and Phillipps 1993: 195; del Hoyo et al. 1999: 197; König et al. 1999: 323, 324; Robson 2000: 292; Rasmussen and Anderton 2005: 243

- ***Strix seloputo baweana*** Oberholser 1917
Strix baweana Oberholser 1917, Proc. US Nat. Mus. 52: 184; Terra typica: Bawean Island

Distribution: Bawean Island, off northern Java
Museum: USNM (holotype)

Wing length: 297 mm
Tail length: ?
Tarsus length: ?
Length of bill: ?
Body mass: ?

Illustration: ?
Photograph: ?
Literature: Oberholser 1917: 184; del Hoyo et al. 1999: 197; König et al. 1999: 323, 324

- ***Strix seloputo wiebkeni*** (Blasius) 1888
  Syrnium Wiebkeni Blasius 1888, Braunschweig. Anz. 52(1) March: 467; Terra typica: Near Puerto Princesa, Palawan

Synonym:
- *Syrnium Whiteheadi* Sharpe 1888, Ibis: 196; Terra typica: Near Puerto Princesa, Palawan

Distribution: Calamian Islands and Palawan Island, The Philippines
Museum: AMNH (lectotype), SNMB (holotype = Platen coll.)

Wing length: ♂ 320 mm, ♀ 330 mm
Tail length: ♂ 175 mm, ♀ 190 mm
Tarsus length: 51–58 mm
Length of bill: 34 and 35 mm
Body mass: ?

Illustration: J. G. Keulemans in *Ibis* 1888: Pl. 3; G. Sandström in du Pont 1971: Pl. 39; T. Worfolk in del Hoyo et al. 1999: Pl. 12; F. Weick in König et al. 1999: Pl. 38; A. Sutherland in Kennedy et al. 2000: Pl. 35
Photograph: ?
Literature: Blasius 1888b: 467; Sharpe 1888b: 196; du Pont 1971: 176, 177; Dickinson et al. 1991: 231; König et al. 1999: 323, 324

Strix ocellata **(Lesson) 1839**
Mottled Wood Owl · Mangokauz · Chouette ocellée (indienne) · Cárabo Ocelado

Length: 405–480 mm
Body mass: ?

Distribution: From the Pakistan Himalayas and peninsular India to lower Bengal, south to Nligiris and Pondicherry. Also in western Myanmar, but subspecies undetermined
Habitat: Patchily wooded plains, open woodland, groves of old and densely foliaged trees, also near human habitats

- ***Strix ocellata grisescens*** Koelz 1950
  Strix ocellata grisescens Koelz 1950, Am. Mus. Novit. 1452: 4; Terra typica: Nichlaul, United Provinces, northern India

Distribution: From the Himalayas in Pakistan, south to about Radjasthan and east to Bibar
Museum: AMNH (holotype)
Remarks: Intergrades in northern India with the nominate *ocellata*

Wing length: 338–346 mm
Tail length: 184–192 mm
Tarsus length: ?
Length of bill: ?
Body mass: ?

Illustration: H. Burn in Robson 2000: Pl. 22
Photograph: ?
Literature: Koelz 1950: 1–10; Ali and Ripley 1969: 305; Eck and Busse 1973: 160; del Hoyo et al. 1999: 197; König et al. 1999: 324, 324; Robson 2000: 292

- ***Strix ocellata grandis*** Koelz 1950
  Strix ocellata grandis Koelz 1950, Am. Mus. Novit. 1452: 4; Terra typica: Sasan, Junagadh, Kathiawar, western India

Distribution: Southern Gujarat (Saurashtra = Kathiawar Peninsula, western India)
Museum: AMNH (holotype)

Wing length: ♂ 360–372 mm
Tail length: ♂ 197–203 mm, ♀ 215 mm
Tarsus length: ?
Length of bill: ?
Body mass: ?

Illustration: F. Weick in König et al. 1999: Pl. 38
Photograph: ?
Literature: Koelz 1950: 1–10; Ali and Ripley 1969: 304, 305; Eck and Busse 1973: 160; del Hoyo et al. 1999: 197; König et al. 1999: 324, 325

- ***Strix ocellata ocellata*** (Lesson) 1839
  Syrnium ocellatum Lesson 1839, Rev. Zool.: 289; Terra typica: Pondicherry

Distribution: Peninsular India (southern Kerala and Tamil Nadu, north through Mysore, Andhra Pradesh, Maharashtra, Gujarat, Madhya Pradesh and Orissa to Bangladesh)
Museum: AMNH, BMNH, FNSF, MHNP, MZUSSMNSt, SMTD, ZMH

Wing length: 320–345 mm (333–357 mm)
Tail length: 174–210 mm
Tarsus length: 54–56 mm (61–65 mm)
Length of bill: 38–42 mm
Body mass: ?
Remarks: () = birds from Gujarat and southern India

Illustration: D. Watson in Ali and Ripley 1969: Pl. 43; T. Boyer in Boyer and Hume 1991: 137; D. Cole in Grimmett et al. 1998: Pl. 29; T. Worfolk in del Hoyo et al. 1999: Pl. 12; F. Weick in König et al. 1999: Pl. 38; L. McQueen in Rasmussen and Anderton 2005: Pl. 77
Photograph: ?
Literature: Sharpe 1875a: 263; Ali and Ripley 1969: 305, 306; Eck and Busse 1973: 160; Boyer and Hume 1991: 138; Grimmett et al. 1998: 434; del Hoyo et al. 1999: 197; König et al. 1999: 324, 325; Rasmussen and Anderton 2005: 242

Strix leptogrammica Temminck 1831
Brown Wood Owl · Malaienkauz · Chouette leptogramme · Cárabo Oriental

Length: 340–450 mm
Body mass: 500–700 g

Distribution: Peninsular India, Sri Lanka, southern Myanmar, southern Thailand, Malay Peninsula, Sumatra, Mentawai Island, Nias and Banyak Islands, Borneo
Habitat: In heavy tropical forest, along coast, in lowlands and lower hills. In Sunda region, lowland primary forest

- ***Strix leptogrammica indranee*** **Sykes 1832**
  Strix Indranee **Sykes 1832, Proc. Comm. Zool. Soc. London: 82; Terra typica: The Ghauts = Lonauli, western Ghats (see Plate 6)**

Synonym:
- *Syrnium ochrogenys* Hume 1873, Str. Feath. 1: 431; Terra typica: Ceylon
- *Strix leptogrammica connectens* Koelz 1950, Am. Mus. Novit. 1452: 3, 4; Terra typica: Central Provinces

Distribution: Peninsular India and Sri Lanka
Museum: AMNH, BMNH, FNSF, MCZ, ÜMB, USNM

Wing length: ♂ (291) mm 320–372 mm, ♀ 330–400 mm
Tail length: 170–229 mm
Tarsus length: 50–60 mm
Length of bill: 41–50 mm
Body mass: ?

Illustration: J. G. Keulemans in Legge 1878a: Pl. 6 (*indranee*); G. M. Henry in Henry 1955: Pl. 16; D. Watson in Ali and Ripley 1969: Pl. 43; D. Cole in Grimmett et al. 1998: Pl. 29; T. Worfolk in del Hoyo et al. 1999: Pl. 12; F. Weick in König et al. 1999: Pl. 42; L. McQueen in Rasmussen and Anderton 2005: Pl. 77
Photograph: De Zylva in Burton et al. 1992: 132; B. Marcot in internet: Owl Pages, photo gallery 2004; L. Koerner in internet: Owl Pages, photo gallery 2003
Literature: Sharpe 1875a: 282, 283; Henry 1955: 194, 195; Ali and Ripley 1969: 308–310; Eck and Busse 1973: 157–160; Ripley 1977: 993–1001; Grimmett et al. 1998: 434, 435; del Hoyo et al. 1999: 197, 198;

Strix leptogrammica · Brown Wood Owl · Malaienkauz

König et al. 1999: 236, 237; Rasmussen and Anderton 2005: 242, 243

- ***Strix leptogrammica maingayi*** (Hume) 1878
  Syrnium maingayi Hume 1878, Str. Feath. 6: 27, 28;
  Terra typica: Malacca

Synonym:
- *Strix indranee rileyi* Kelso 1937, Auk 54: 305; Terra typica: Khaw Nok Ram, Trong, Lower Siam (holotype AMNH)

Distribution: Southern Myanmar, southern Thailand and the Malay Peninsula
Museum: AMNH, BMNH (holotype), USNM, ZMH

Wing length: ♂ 328–332 mm, ♀ 364–373 mm
Tail length: 186–210 mm
Tarsus length: 50–60 mm
Length of bill: 36–39 mm
Body mass: ?

Illustration: W. Mongkol? in Lekagul and Round 1991: Pl. 61
Photograph: D. Messner in photo collection Weick; F. Weick in photo collection from Lakus aviary; L. Poh in internet: Owl Pages, photo gallery 2003 (2 photographs)
Literature: Baker 1927: 401, 402; Eck and Busse 1973: 157–160; Ripley 1977: 993–1001; Smythies and Hughes 1984: 313, 314; Lekagul and Round 1991: 182; del Hoyo et al. 1999: 197, 198; König et al. 1999: 336, 337; Robson 2000: 292, 293

- ***Strix leptogrammica myrtha*** (Bonaparte) 1850
  Ciccaba myrtha Bonaparte 1850, Consp. Gen. Av. 1: 44;
  Terra typica: Sumatra

Synonym:
- *Strix leptogrammica nyctiphasma* Oberholser 1924, Journ. Wash. Acad. Sci. 14: 302; Terra typica: Banyak Island
- *Strix leptogrammica chaseni* Hoogerwerf and de Boer 1947, Chronica Nat. 103(7): 140; Terra typica: Belitung Island

Distribution: Sumatra, Mentawai Islands, Banyak Islands (off western Sumatra) and Belitung Island (off south-eastern Sumatra)
Museum: ANSP, MNBHU, NNML, SMTD, USNM, YPM

Wing length: 297–310 mm
Tail length: 158–164 mm
Tarsus length: ?
Length of bill: ?
Body mass: ?

Illustration: H. Quintscher in Eck and Busse 1973: Abb. 39 (b/w)
Photograph: ?
Literature: Oberholser 1924: 302; Hoogerwerf and de Boer 1947: 140; Ripley 1977: 993–1001; Eck and Busse 1973: 157–160; del Hoyo et al. 1999: 197, 198; König et al. 1999: 336, 337

- ***Strix leptogrammica niasensis*** (Salvadori) 1887
  Syrnium niasense Salvadori 1887, Ann. Mus. Civ. Genova 24: 526; Terra typica: Nias Island, off western Sumatra

Distribution: Nias Island, off western Sumatra
Museum: AMNH, FNSF, MGD (holotype), MNBHU

Wing length: ♂ 273 and 286 mm, ♀ 279 and 280.5 mm
Tail length: 151–156.5 mm
Tarsus length: ?
Length of bill: ?
Body mass: ?

Illustration: H. Quintscher in Eck and Busse 1973: Abb. 39 (b/w); F. Weick in König et al. 1999: Pl. 39
Literature: Ripley 1977: 993–1001; Eck and Busse 1973: 157–159; del Hoyo et al. 1999: 197, 198; König et al. 1999: 336, 337

- ***Strix leptogrammica leptogrammica*** Temminck 1831
  Strix leptogrammica Temminck 1831, Pl. col. livr. 88: Pl. 525;
  Terra typica: Borneo

Distribution: Central and southern Borneo
Museum: AMNH, BMNH, FNSF, MNBHU, SMTD, YPM

Wing length: ♂ 286–314 mm, ♀ -334 mm
Tail length: ♂ 151–164 mm, ♀ 155–190 mm
Tarsus length: 47 mm?
Length of bill: 40 mm
Body mass: ?

Illustration: A. Thorburn in *Catal. Christie's* 15.12.1981, no. 236; ? T. Boyer in Boyer and Hume 1991: 138; K. Phillipps in McKinnon and Phillipps 1993: Pl. 39; T. Worfolk in del Hoyo et al. 1999: Pl. 12; F. Weick in König et al. 1999: Pl. 42 (adult and juvenile)
Photograph: ? L. Poh in internet: Owl Pages, photo gallery 2003; ? in *ZEITmagazin* 1992(46): 16
Literature: Sharpe 1875a: 264, 265; Eck and Busse 1973: 157–159; Ripley 1977: 993–1001; Voous 1988: 205–208; Boyer and Hume 1991: 138; McKinnon and Phillipps 1993: 195, 196; del Hoyo et al. 1999: 197, 198; König et al. 1999: 336, 337

- ***Strix leptogrammica vaga*** Mayr 1938
  Strix leptogrammica vaga Mayr 1938, Bull. Raffles Mus. 14: 15;
  Terra typica: Benkoka, northern Borneo

Distribution: Northern Borneo in Sabah (Benkoker and Sandakan areas)
Museum: AMNH (holotype), SMTD

Wing length: ♂ 312.5–329 mm, ♀ 339 mm ($n = 1$)
Tail length: ♂ 160–185 mm, ♀ 195 mm ($n = 1$)
Tarsus length: ?
Length of bill: ?
Body mass: ?

Illustration: F. Weick in König et al. 1999: Pl. 42
Photograph: ?
Literature: Mayr 1938: 15; Eck and Busse 1973: 157–159; Ripley 1977: 993–1001; del Hoyo et al. 1999: 197, 198; König et al. 1999: 336, 337

❖

Strix bartelsi Finsch 1906

Bartel's Wood Owl · Bartelskauz · Chouette de Bartels · Cárabo de Bartels
Syrnium bartelsi **Finsch 1906, Bull. Brit. Ornith. Cl. 16: 63;**
Terra typica: **Pasis Datar, Mount Pangerango, 2 600 feet, Java**

Length: 390–430 mm
Body mass: 500–700 g

Distribution: Western and central Java (Pangerango, Gede, Salak and Ciremai Mountains)
Habitat: Undisturbed mountain forest and forest edges, from 700 up to about 2 000 m
Museum: BMNH, MNBHU, USNM
Remarks: Status uncertain, but differing in vocalisations, plumage and range (higher elevations)

Wing length: 360–376 mm (1 × 335 mm)
Tail length: 200–230 mm (1 × 191 mm)
Tarsus length: ?
Length of bill: ?
Body mass: 500–700 g

Illustration: J. G. Keulemans in Finsch 1906a: Pl. 17; F. Weick in König et al. 1999: Pl. 42
Photograph: ?
Literature: Finsch 1906a: 1, 2; Eck and Busse 1973: 157–159; Ripley 1977: 993–1001; del Hoyo et al. 1999: 197, 198; König et al. 1999: 337, 338

❖

Strix newarensis (Hodgson) 1836

Himalayan Wood Owl · Himalaja-Braunkauz · Chouette des Himalayas

Length: 460–550 mm
Body mass: ?

Distribution: The Himalayas from Pakistan (Punjab) to Nepal, Sikkim, north and south Myanmar, northern Thailand, Laos, North Vietnam, southern China, and Hainan and Taiwan Islands
Habitat: Dense evergreen forest, prefers undisturbed mountain forest, in Nepal broadleaf forest. Avoids human areas. From 750 m up to about 2 750 m a.s.l.
Remarks: This is a northern form, living in montane habitats, with larger size and distinct vocalisations

- ***Strix newarensis newarensis*** (Hodgson) 1836
  Ulula newarensis Hodgson 1836, As. Res. 19: 168;
  Terra typica: Nepal

Distribution: The Himalayas from Pakistan (Punjab) to Nepal, Sikkim and Bhutan
Museum: AMNH, BMNH, MNBHU, SMTD, USNM, ZMH, ZSBS

Wing length: 395–442 mm
Tail length: 243–262 mm
Tarsus length: 53–56 mm
Length of bill: 40–45 mm
Body mass: ?

Illustration: T. Worfolk in del Hoyo et al. 1999: Pl. 12; F. Weick in König et al. 1999: Pl. 42; L. McQueen in Rasmussen and Anderton 2005:Pl. 77
Photograph: B. Marcot in internet: Owl Pages, photo gallery 2002
Literature: Baker 1927: 400, 401; Ali and Ripley 1969: 307, 308; Eck and Busse 1973: 157–159; Voous 1988: 205–208; del Hoyo et al. 1999: 197, 198; König et al. 1999: 338, 339; Rasmussen and Anderton 2005: 242, 243

- ***Strix newarensis ticehursti*** Delacour 1930
  Strix leptogrammica ticehursti Delacour 1930, Ois. 11: 654
  new name for invalid name *S. l. orientalis*

Synonym:
- *Strix leptogrammica orientalis* Delacour and Jabouille 1930, Ois. 11: 406; Terra typica: Pakha, Tonkin
- ? *Strix indranee shanensis* Baker 1935, Bull. Brit. Ornith. Cl. 56: 36; Terra typica: Sintaung, 6 000 feet, Shan States

Distribution: Northern and central Myanmar, east to southeast China (northeast to Anhui), south to northwest Thailand, northern Laos and North Vietnam
Museum: AMNH, MCZ
Remarks: According to Ripley (1977), a synonym of the nominte *newarensis*!

Wing length: (355) 377–395 mm
Tail length: 234–249 mm
Tarsus length: 60 mm
Length of bill: 40 mm
Body mass: ?

Illustration: T. Worfolk in del Hoyo et al. 1999: Pl. 12
Photograph: ?
Literature: Delacour and Jabouille 1930: 406; Delacour 1930: 654; Ripley 1977: 993–1001; Eck and Busse 1973: 157–159; del Hoyo et al. 1999: 197, 198; König et al. 1999: 338, 339

- ***Strix newarensis laotiana*** Delacour 1926
  Strix newarensis laotianus (sic) Delacour 1926, Bull. Brit. Ornith. Cl. 47: 11; Terra typica: Xieng-Khouang, Laos

Distribution: Southern Laos and central Vietnam (Annam)
Museum: AMNH, MNBHU
Remarks: According to Ripley (1977), a synonym; Ripley suggests uniting with *newarensis* (also *ticehursti* and *caligata*)!

Wing length: 377–406 mm
Tail length: 217–223 mm (1 × 254 mm)
Tarsus length: 56 mm
Length of bill: 39 mm
Body mass: ?

Illustration: H. Burn in Robson 2000: Pl. 22 (adult and juvenile)
Photograph: ?
Literature: Delacour 1926: 11; Delacour and Jabouille 1931: 134; Ripley 1977: 993–1001; Voous 1988: 205–208; del Hoyo et al. 1999: 197, 198; König et al. 1999: 338, 339

- ***Strix newarensis caligata*** (Swinhoe) 1863
  Bubo caligatus Swinhoe 1863, Ibis: 218;
  Terra typica: Formosa

Distribution: Hainan and Taiwan
Museum: BMNH, USNM
Remarks: Hartert (1910) wrote in Novit. Zool. 17: 205, 206: "... hardly [to] separable from Himalayan birds!". Ripley (1977) unites *ticehursti*, *laotiana* and *caligata* with *newarensis*. Ripley was unable to separate birds from Tonkin (*ticehursti*) and Taiwan (*caligata*) from *newarensis*! However, it is hard to understand why only one taxon would emerge from such an enormous geographical area! For example, a specimen in MNBHU, labelled *laotiana*, has a much darker face and darker chestband than all examined *newarensis* skins. Four subspecies are recognised. The whole species complex *leptogrammica*, *bartelsi* and *newarensis* requires further studies

Wing length: 367–390 mm (1 × 401 mm)
Tail length: 223–234 mm
Tarsus length: ?
Length of bill: ?
Body mass: ?

Illustration: A. Cameron in Voous 1988: 207
Photograph: ?
Literature: Ripley 1977: 993–1001; Voous 1988: 205–208; del Hoyo et al. 1999: 197, 198; König et al. 1999: 338, 339

Strix aluco Linné 1758
Tawny Owl · Waldkauz · Chouette hulotte · Cárabo Común
(see Plate 6)

Length: 360–400 mm
Body mass: ♂ 440 g, ♀ 553 g

Distribution: Locally in North Africa (Morocco to Tunisia), Great Britain (except Northern Ireland) and continental Europe, from the Iberian Peninsula in the south and Scandinavia to western Siberia in the north. From Greece, Asia Minor and the Middle East to the Caspian Sea and Turkestan to eastern China and Taiwan. Present in Sardinia and Sicily, rare in Mallorca, absent from Ireland, Corsica, Crete, Rhodos and Cyprus
Habitat: Open and semi-open forest, woodland and open landscapes with wooded patches. Riverine forest, parks, gardens with old trees, etc. Rocky areas with trees and bushes. Locally near human settlements. From lowlands up to 4200 m in the Himalayas

- ***Strix aluco mauritanica*** (Witherby) 1905
Syrnium aluco mauritanicum Witherby 1905, Bull. Brit. Ornith. Cl. 15: 36; Terra typica: Algeria, type from Les Glacières, near Blida

Distribution: Northern northwest Africa (Morocco to Tunisia)
Museum: BMNH, NNML, SMTD, ZFMK, ZMA

Wing length: 272–305 mm
Tail length: 173–189 mm
Tarsus length: 54–61 mm
Length of bill: 28–31 mm
Body mass: ♂ 325–470 g, ♀ 390–575 g

Illustration: H. Delin in Cramp et al. 1985: Pl. 49; M. Woodcock in Fry et al. 1988: Pl. 7; P. Hayman in Kemp and Kemp 1998: 293; T. Worfolk in del Hoyo et al. 1999: Pl. 12; F. Weick in König et al. 1999: Pl. 41
Photograph: P. Steyn and R. Tidman in Kemp and Kemp 1998: 292
Literature: Hartert 1912–1921: 1025; Dementiev and Gladkov 1951: 466; Vaurie 1965: 620; Etchécopar and Hüe 1967: 342, 343; Cramp et al. 1985: 526–546; Voous 1988: 209–219; Kemp and Kemp 1998: 292, 293; del Hoyo et al. 1999: 198; König et al. 1999: 333–335

- ***Strix aluco sylvatica*** Shaw 1809
Strix sylvatica Shaw 1809, Gen. Zool. 7(1): 253;
Terra typica England

Synonym:
- *Strix aluco clanceyi* von Jordans 1950, Syllegomena biologica: 176; Terra typica: Linares, Salamanca, Spain

Distribution: Britain, France and Iberia. Probably southern Italy, Greece, western and central Turkey and the Middle East
Museum: BMNH, MHNP, NNML, ZFMK
Remarks: Intergrades with *sanctinicolai*

Wing length: ♂ 248–268 mm, ♀ 255–278 mm
Tail length: 155 mm
Tarsus length: ?
Length of bill: 30 mm
Body mass: ♂ 352–465 g, ♀ 435–716 g

Illustration: T. Bewick in Bewick 1809: 91 (wood engraving); B. Fawcett in Jackson 1978: 107 (wood engraving, 1808–1893) 134, 135; P. Paillou in *Catal. Sotheby's* 22.11.1979; C. Tunnicliffe in *Catal. Sotheby's* 21.9.1983: 237; C. Atkinson 1785 in *Catal. Christie's* 1983: no. 69; C. Tunnicliffe in Cusa 1984: 37; H. Delin in Cramp et al. 1985: Pl. 49; A. Cameron in Voous 1988: 211; T. Boyer in Boyer and Hume 1991: 139; R. Bateman in Dean 2004: 176
Photograph: E. Hosking in Hosking and Flegg 1982: 98; N. Blake in internet: Owl Pages, photo gallery 2003
Literature: Hartert 1912–1921: 1022–1025; Dementiev and Gladkov 1951: 466; Vaurie 1965: 620, 621; Glutz von Blotzheim and Bauer 1980: 580–610; Bezzel 1985: 651–654; Cramp et al. 1985: 526–546; Voous 1988: 209–219; del Hoyo et al. 1999: 198; König et al. 1999: 333–335

- ***Strix aluco aluco*** Linné 1758
Strix Aluco von Linné 1758, Syst. Nat. ed. 10, 1: 93;
Terra typica: Europe, type locality = Sweden

Synonym:
- *Strix aluco volhyniae* Dunajewski 1948, Bull. Brit. Ornith. Cl. 68: 130; Terra typica: Dolsk, Volhynia

Distribution: Northern and eastern Europe, east to western Russia (Ural Mountains), south to the Alps, the Balkans and the Black Sea

Museum: BMNH, MNBHU, NNML, SMNKa, SMNSt, SMTD, ZFMK, ZMH, ZAM, ZSBS
Remarks: Intergrades with *siberiae*

Wing length: ♂ 259–275 mm, ♀ 269–287 mm
Tail length: ♂ 148–166 mm, ♀ 154–171 mm
Tarsus length: 45–53 mm
Length of bill: 28.5–34.5 mm
Body mass: ♂ 342–540 g, ♀ 301–620 g (1 × 685 g)

Illustration: O. Rudbeck 1693–1710 in Krook 1988 (repr.): Pl. 94, 96; J. F. Naumann in Naumann 1822: Pl. 46, 47; R. C. Richter in Gould 1862–1873: *Syrn. aluco* (1864); J. G. Keulemans in Hennicke 1905: unpaged; P. Barruel in Sutter and Barruel 1958: Pl. 20; P. A. Robert in Géroudet 1979: 381; H. Delin in Cramp et al. 1985: Pl. 49; T. Worfolk in del Hoyo et al. 1999: Pl. 12; F. Weick in König et al. 1999: Pl. 41; F. Weick in Weick 2004: Januar; F. Weick in Kröher and Weick 2004: 156, 157
Photograph: M. C. Noailles in Everett 1977: 37; E. Hosking in Hosking and Flegg 1982: 61; W. Layer in Mebs 1987: 29; H. Reinhard in Burton et al. 1992: 125; R. Groß in *Der Falke* 1999(10): 299; Archiv Vogelschutzwarte Karlsruhe in *Eulen verstehen* 1999: 52; photographs in Mebs and Scherzinger 2000: 13 (Pott), 59, 238 (Löhr), 81 (Zeininger), 227 (Reinhard), 228 (Schendel), 228, 233 (Hopf), 229 (Giel), 235 (Limbrunner), 244 (Scherzinger)
Literature: Naumann 1822: 473–483, Nachtr. 178, 179; Hartert 1912–1921: 1022–1025; Hartert 1923: 393, 394; Dementiev and Gladkov 1951: 465–470; Vaurie 1965: 621; Glutz von Blotzheim and Bauer 1980: 580–610; Bezzel 1985: 651–654; Cramp et al. 1985: 526–546; Voous 1988: 209–219; Boyer and Hume 1991: 139–141; del Hoyo et al. 1999: 198; König et al. 1999: 333–335; Mebs and Scherzinger 2000: 226–246

- ***Strix aluco siberiae*** Dementiev 1933
  Strix aluco siberiae Dementiev 1933, Alauda 5: 339;
  Terra typica: Sokolowa, near Tobolsk, Siberia

Distribution: From the Ural Mountains to western Siberia, south to the Blaya River, Miass River and region of Chelyabinsk, to the region of Ishim on the Irtysh River
Museum: BMNH, SMTD, ZFMK
Remarks: Intergrades with the nominate *aluco*

Wing length: ♂ 280–300 mm, ♀ 301–311 mm
Tail length: –175 mm
Tarsus length: ?
Length of bill: 33 mm
Body mass: ♂ 450–490 g, ♀ 590–680 g

Illustration: T. Worfolk in del Hoyo et al. 1999: Pl. 12; F. Weick in König et al. 1999: Pl. 41
Photograph: ?
Literature: Hartert 1923: 393, 394; Dementiev 1933b: 339; Dementiev and Gladkov 1951: 466, 470, 471; Vaurie 1965: 621; Cramp et al. 1985: 526–546; Voous 1988: 209–219; del Hoyo et al. 1999: 198; König et al. 1999: 333–335

- ***Strix aluco willkonskii*** (Menzbier) 1896
  Syrnium willkonskii (sic) Menzbier 1896, Bull. Brit. Ornith. Cl. 6: 6; Terra typica: Transcaspia; *willkouskii* = typographical error accord.: ibid.: 24. Types from western Georgia and Azerbaijan Republic according to Dementiev (1933b) Alauda 5: 340

Synonym:
- *Strix aluco obscurata* Stegmann 1926, Bull. Brit. Ornith. Cl. 47: 39; Terra typica: Lenkoran, Talych

Distribution: Norhteast Turkey, Caucasus, northwest Iran, east to Turkmenia
Museum: BMNH, ZFMK (Kleinschmidt coll.), ZMH (*willkonskii*)

Wing length: ♂ 266–296 mm, ♀ 282–305 mm
Tail length: ?
Tarsus length: ?
Length of bill: ?
Body mass: ♂ 510 g, ♀ 582 g

Illustration: ?
Photograph: ?
Literature: Menzbier 1896: 6 and 24; Hartert 1923: 394; Stegmann 1926: 39; Dementiev 1933b, *Alauda 5*: 340; Dementiev and Gladkov 1951: 466, 471, 472; Vaurie 1965: 622; Cramp et al. 1985: 526–546; Voous 1988: 209–219; del Hoyo et al. 1999: 198; König et al. 1999: 333–335

- ***Strix aluco sanctinicolai*** (Zarudny) 1905
  Syrnium sancti-nicolai Zarudny 1905, Orn. Monatsb. 13: 49; Terra typica: Schalil River and Gamdalkal, Bachtiari region, Iran

Distribution: Northeast Iraq and western Iran
Museum: AMNH, BMNH, MNBHU

Wing length: ♂ 255–273 mm, ♀ 270–285 mm
Tail length: ?
Tarsus length: ?
Length of bill: ?
Body mass: ?

Illustration: A. Cameron in Voous 1988: 227; F. Weick in König et al. 1999: Pl. 41
Photograph: ?
Literature: Hartert 1912–1921: 1025; Hartert 1923: 394; Vaurie 1965: 622; Eck and Busse 1973: 161–163; Cramp et al. 1985: 526–546; Voous 1988:209–219; del Hoyo et al. 1999: 198; König et al. 1999: 333–335

- ***Strix aluco härmsi*** (Zarudny) 1911
  Syrnium härmsi Zarudny 1911, Orn. Monatsb. 19: 14; Terra typica: Chirchik Basin, Russian Turkestan

Distribution: Turkestan (altitude of 2000–2275 m in western Tien Shan)
Museum: MNBHU?

Wing length: ♂ (296) 303–316 mm, ♀ (315) 318–332 (335) mm
Tail length: ?
Tarsus length: ?
Length of bill: ?
Body mass: ?

Illustration: ?
Photograph: ?
Literature: Hartert 1912–1921: 1026; Hartert 1923: 394; Dementiev and Gladkov 1951: 472, 473; Vaurie 1965: 622, 623; Eck and Busse 1973: 161–163; del Hoyo et al. 1999: 198; König et al. 1999: 333–335

- ***Strix aluco biddulphi*** Scully 1881
  Syrnium biddulphi Scully 1881, Ibis: 423 and Pl. 14; Terra typica: Gilgit

Distribution: Pakistan and northwest India (1500 to 3800 m above sea-level)
Museum: BMNH (Hume coll.), BNHS

Wing length: ♂ (285) 294–323 mm, ♀ 320–343 mm
Tail length: 191–210 mm
Tarsus length: 50 and 51 mm
Length of bill: 33–35 mm
Body mass: ?

Illustration: J. G. Keulemans in Scully 1881: Pl. 14; D. Cole in Grimmett et al. 1998: Pl. 29; F. Weick in König et al. 1999: Pl. 41; L. McQueen in Rasmussen and Anderton 2005: Pl. 77
Photograph: ?
Literature: Scully 1881: 423; Hartert 1912–1921: 1025, 1026; Baker 1927: 397, 398; Vaurie 1965: 623; Ali and Ripley 1969: 310, 311; Grimmett et al. 1998: 435; del Hoyo et al. 1999: 198; König et al. 1999: 333–335; Rasmussen and Anderton 2005: 243

- ***Strix (aluco) nivicola*** (Blyth) 1845
  Syrnium nivicolum "Hodgson" Blyth 1845, J. As. Soc. Beng. 141(1): 185; Terra typica: Himalaya = Nepal

Synonym:
- *Syrnium blanfordi* Zarudny 1911, Orn. Monatsb. 19: 34; Terra typica: Northwest Himalayas, ex Blanford (1895) Fauna Brit. India, Birds 3: 274
- *Strix aluco harterti* La Touche 1919, Bull. Brit. Ornith. Cl. 40: 50; Terra typica: Changlo hsien, Hupeh
- *Strix aluco nivipetens* Riley 1925, Proc. Biol. Soc. Wash. 38: 10; Terra typica: Lichiang Mountains, Yunnan
- *Strix aluco obrieni* Koelz 1954, Contrib. Inst. Reg. Expl. 1: 27; Terra typica: Blue Mountains, Lushai Hills, Assam

Distribution: Nepal, east to southeast China, south to northern Myanmar and northern India
Museum: BMNH, ZFMK, ZMH, ZSBS

Wing length: ♂ (282) 290–305 mm, ♀ 304–320 mm
Tail length: 168–190 mm
Tarsus length: 45–48 mm
Length of bill: 33 mm
Body mass: ♂ 375 g (392), ♀ 375–392 g

Illustration: D. Watson in Ali and Ripley 1969: Pl. 43; D. Cole in Grimmett et al. 1998: Pl. 29; T. Worfolk in del Hoyo et al. 1999: Pl. 12; H. Burn in Robson 2000: Pl. 22 (adult and juvenile); L. McQueen in Rasmussen and Anderton 2005: Pl. 77
Photograph: ?
Literature: Sharpe 1875a: 250, 251; Hartert 1912–1921: 1026, 1027 and 1923: 394; Baker 1927a: 298, 299;

Vaurie 1965: 623; Ali and Ripley 1969: 311, 312; Etchécopar and Hüe 1978: 487, 488; Grimmett et al. 1998: 435; del Hoyo et al. 1999: 198; König et al. 1999: 333–335; Robson 2000: 293; Rasmussen and Anderton 2005: 244

- ***Strix (aluco) ma*** (Clark) 1907
  Syrnium ma Clark 1907, Proc. US Nat. Mus. 32: 471;
  Terra typica: Fusan, Korea

Distribution: Northern China (Jilin) and Korea
Museum: USNM (holotype), SMTD

Wing length: ♂ 272, 278 mm, ♀ 280–297 mm
Tail length: ?
Tarsus length: ?
Length of bill: ?
Body mass: ?

Illustration: ?
Photograph: ?
Literature: Hartert 1923: 394, 395; Vaurie 1965: 623; del Hoyo et al. 1999: 198; König et al. 1999: 333–335

- ***Strix (aluco) yamadae*** Yamashina 1936
  Strix nivicola yamadae Yamashina 1936, Tori 9: 220;
  Terra typica: Takata, Tainan district, Taiwan

Distribution: Taiwan
Museum: NSMT, BMNH?
Remarks: A geographical variation is well marked. The European group (*aluco* including *mauritanica*) and the Asiatic (*nivicola* including *ma* and *yamadae*) are geographical isolated and may be specifically distinct. Also different in vocalisations and plumage from the *aluco* group

Wing length: ♂ (280) 290–305 mm, ♀ 304–320 mm
Tail length: ?
Tarsus length: ?
Length of bill: ?
Body mass: ?

Illustration: ?
Photograph: ?
Literature: Yamashina 1936: 220; Mees 1971: 229; Voous 1988: 209–219; del Hoyo et al. 1999: 198; König et al. 1999: 333–335

Strix butleri (Hume) 1878
Hume's (Tawny) Owl · Fahlkauz, Wüstenkauz · Chouette de Butler · Cárabo Árabe
Asio butleri Anonymus = Hume 1878, Str. Feath. 7: 316;
Terra typica: Omara (Ormara), Mekran Coast, southern Baluchistan
(see Plate 6)

Length: 300–340 mm
Body mass: about 220 g

Distribution: Eastern and southern Israel, Jordan, Sinai Peninsula and eastern Egypt (Red Sea Mountains). Patchily in Arabian Peninsula (Saudi Arabia, Yemen, Oman). Possibly southern Pakistan (Makran Coast), perhaps also Iran?
Habitat: Rocky gorges, canyons and ravines in deserts and semideserts. Arid, rocky mountains, near acacia or palm groves, ruined buildings, etc., with springs and rain pools. Sometimes near human settlements
Museum: BMNH, DML, MNBHU, ZFMK

Wing length: 237–256 mm
Tail length: 134–140 mm (1 × 150 mm)
Tarsus length: 50–56 mm
Length of bill: 26.5–29 mm
Body mass: 162, 214, 220 and 225 g

Illustration: E. N. Fischer in *Auk* 1915, 32: Pl. 17; Meinertzhagen 1930: Pl. 16 (false colour irides); P. Barruel in Etchécopar and Hüe 1967: Pl. 11 (false irides); H. Delin in Cramp et al. 1985: Pl. 49, 53; A. Cameron in Voous 1988: 223; M. Woodcock in Fry et al. 1988: Pl. 7; T. Boyer in Boyer and Hume 1991: 136; A. Harris in Shirihai 1996: 319 (b/w); D. Cole in Grimmett et al. 1998: Pl. 29; P. Hayman in Kemp and Kemp 1998: 291; T. Worfolk in del Hoyo et al. 1999: Pl. 12; F. Weick in König et al. 1999: Pl. 41; Zetterström/Mullarney in Svensson et al. 1999: 215; L. McQueen in Rasmussen and Anderton 2005: Pl. 77
Photograph: Y. Leshem in Leshem 1979: 376, 377; M. C. Jennings in Mikkola 1983: Pl. 45 (b/w); Y. Eshbol in Bds. in land of the Bible, Palphot (head); Y. Eshbol in Shirihai 1996: Pl. 66; Y. Eshbol in prospectus (? year) Middle Eastern Specialities, Hadoram Experiences; M. Gunther in Kemp and Kemp 1998: 291 (subadult); E. and D. Hosking in del Hoyo et al. 1999: 100

Literature: Hume 1878: 316; Hartert 1912–1921: 1027, 1028; Meinertzhagen 1930: 361, 362; Aharoni 1931: 171–173; Vaurie 1965: 619; Etchécopar and Hüe 1967: 343, 344; Leshem 1974: 66, 67; Mendelsson 1975: 110, 111; Leshem 1979: 375–377; Aronson 1980: 18, 19; Leshem 1981b: 100–107; Mikkola 1983: 157–161; Goodman and Sabry 1984: 79–84; Cramp et al. 1985: 547–550; Voous 1988: 220–224; Fry et al. 1988: 146, 147; Boyer and Hume 1991: 135, 136; Shirihai 1996: 319, 320; Grimmett et al. 1998: 435; Kemp and Kemp 1998: 290, 291; del Hoyo et al. 1999: 198, 199; König et al. 1999: 335, 336; Svensson et al. 1999: 214; Rasmussen and Anderton 2005: 244

❖

Strix woodfordi (A. Smith) 1834
African Wood Owl · Afrika-Waldkauz (Woodfordkauz) · Chouette africaine · Cárabo Africano
(see Plate 6)

Length: 305–350 mm
Body mass: 240–350 g

Distribution: Africa, south of Sahel Zone, from Senegambia to Ethiopia, south to Angola, Botswana, Zimbabwe, Mozambique and South Africa
Habitat: Forest and woodland, from dense montane forest to bush and riparian areas. Forest edges and plantations. From sea-level up to about 3 700 m

- ***Strix woodfordi umbrina*** (Heuglin)
  Surnium ? (sic) *umbrinum* von Heuglin 1863, J. Orn. 11: 12; Terra typica: Prov. of Begemeder, 9 000 feet, Ethiopia

Distribution: Ethiopia and southeast Sudan
Museum: SMNSt

Wing length: 246 mm
Tail length: ?
Tarsus length: ?
Length of bill: ?
Body mass: ?

Illustration: P. Hayman in Kemp and Kemp 1998: 279
Photograph: ?
Literature: von Heuglin 1863: 12; Fry et al. 1988: 145–150; del Hoyo et al. 1999: 204; König et al. 1999: 339–341

- ***Strix woodfordi nigricantior*** (Sharpe) 1897
  Syrnium nigricantius Sharpe 1897, Bull. Brit. Ornith. Cl. 6: 247; Terra typica: Mpwapwa, Tanganyika

Synonym:
- *Syrnium woodfordi* var. *suahelicum* Reichenow 1898, in Werther (1998): Die mittleren Hochländer des nördlichen Deutsch-Ost-Afrika: 272; Terra typica: German East Africa, Ukami, Tanganyika (holotype MNBHU)
- *Syrnium woodfordi* var. *sansibaricum* Reichenow 1898, ibid.: 272; Terra typica: Zanzibar (holotype MNBHU)

Distribution: Southern Somalia, Kenya, Tanzania, Zanzibar and eastern Zaire
Museum: BMNH, MNBHU, SMNKa, SMTD, ZFMK, ZSBS

Wing length: 234–258 mm
Tail length: 172 mm
Tarsus length: ?
Length of bill: ?
Body mass: 220 and 240 g

Illustration: M. Woodcock in Fry et al. 1988: Pl. 7; D. Zimmerman in Zimmerman et al. 1996: Pl. 55; P. Hayman in Kemp and Kemp 1998: 279; T. Worfolk in del Hoyo et al. 1999: Pl. 13; F. Weick in König et al. 1999: Pl. 43 (light morph); J. Gale in Stevenson and Fanshawe 2002: Pl. 99
Photograph: E. Hosking in Everett 1977: 44; E. Hosking in Hosking and Flegg 1982: 149
Literature: Sharpe 1897: 247; Fry et al. 1988: 148–150; Zimmerman et al. 1996: 443; Kemp and Kemp 1998: 278, 279; del Hoyo et al. 1999: 204; König et al. 1999: 339–341; Stevenson and Fanshawe 2002: 446

- ***Strix woodfordi nuchalis*** (Sharpe) 1870
  Syrnium nuchale Sharpe 1870, Ibis: 487; Terra typica: Fanti

Synonym:
- *Syrnium Bohndorffi* Sharpe 1884, J. Linn. Soc. Lond. Zool. 17: 439; Terra typica: Semmio, Niam-Niam Country, Ubangi-Shari (holotype BMNH, cotype MNBHU)

Distribution: Senegambia east to southern Sudan and Uganda, south to northern Angola and Zaire (except in the south and east), also Bioko Island

Museum: BMNH, MNBHU, SMNSt, SMTD, ZFMK, ZMH, ZSBS

Wing length: 231–273 mm
Tail length: 123–145 mm
Tarsus length: 42–55 mm
Length of bill: 28–33 mm
Body mass: 252–305 g

Illustration: H. Grönvold in Bannerman 1953: 532 (b/w); P. Hayman in Kemp and Kemp 1998: 279; T. Worfolk in del Hoyo et al. 1999: Pl. 13; F. Weick in König et al. 1999: Pl. 43; N. Borrow in Borrow and Demey 2001: Pl. 60
Photograph: L. Koerner in internet: Owl Pages, photo gallery 2002; P. Morris in Duncan 2003: 66
Literature: Sharpe 1870: 487; Sharpe 1875a: 265–267; Bannerman 1953: 532, 533; Fry et al. 1988: 148–150; Kemp and Kemp 1998: 278, 279; del Hoyo et al. 1999: 204; König et al. 1999: 339–341; Borrow and Demey 2001: 499

- ***Strix woodfordi woodfordi*** (A. Smith) 1834
  Noctua Woodfordi A. Smith 1834, S. Afr. Quat. Journ. 2: 312; Terra typica: South Africa

Distribution: Southern Angola and southern Zaire, east to southwest Tanzania, south to northern Botswana and the Cape (Transvaal and along the East Coast)
Museum: BMNH, SMNKa, TMBD, ZFMK, ZSBS

Wing length: ♂ 222–269 mm, ♀ 235–264 mm
Tail length: 135–176 mm
Tarsus length: 43–52 mm
Length of bill: 28–33 mm
Body mass: ♂ 242–269 g, ♀ 285–350 g

Illustration: A. Smith in Smith 1839–1849: Pl. 71; G. Arnott in Steyn 1982: Pl. 23; M. Woodcock in Fry et al. 1988: Pl. 7; T. Boyer in Boyer and Hume 1991: 134; P. Hayman in Kemp and Kemp 1998: 279; T. Worfolk in del Hoyo et al. 1999: Pl. 12; F. Weick in König et al. 1999: Pl. 43
Photograph: ? in Campbell 1974: Pl. 442; P. Steyn in Steyn 1982: 253 (b/w); R. E. Viljoen in Ginn et al. 1989: 333 (adult and juvenile); Kemp in Burton et al. 1992: 130; N. Dennis, J. Carlyon in Kemp and Kemp 1998: 278; P. Pickford in Kemp and Kemp 1998: 279; J. J. Brooks in del Hoyo et al. 1999: 134
Literature: Sharpe 1875a: 267–269; Eck and Busse 1973: 162, 163; Steyn 1982: 251–254; Fry et al. 1988: 148–150; Ginn et al. 1989: 332; Boyer and Hume 1991: 133, 134; del Hoyo et al. 1999: 204; König et al. 1999: 339–341

Strix virgata (Cassin) 1849
Mottled Owl · Sprenkelkauz · Chouette striée · Cárabo Café (Lechuza Café)
(see Plate 6)

Length: 300–380 mm
Body mass: 175–320 g

Distribution: From Mexico south to Middle America and forested South America, from Venezuela and Ecuador to northeast Argentina and southeast Brazil. Absent from the Pacific slope of the Andes south of Ecuador
Habitat: Different varieties of forest: primary and secondary humid forest, drier woodland, thorn forest and gallery forest, plantations. Often near human habitation. From sea-level up to 2 500 m

- ***Strix virgata squamulata*** (Bonaparte) 1850
  Syrnium squamulatum Bonaparte 1850, Consp. Gen. Av. 1: 53; Terra typica: Mexico, Tehuantepec City, Oaxaca (by Kelso 1933)

Synonym:
- *Ciccaba virgata amplonotata* L. Kelso 1933, Proc. Biol. Soc. Wash. 46: 151; Terra typica: Mazatlan, Sinaloa

Distribution: Western Mexico (Sonora to Guerrero and Guanajuato to Morelos)
Museum: AMNH, FMNH, NNML (holotype)

Wing length: ♂ 239–254 mm, ♀ 240–265 mm
Tail length: ♂ 139–151 mm, ♀ 147–163 mm
Tarsus length: ?
Length of bill (cere): 18.5–21 mm
Body mass: ♂ 177–142 g, ♀ 251–345 g

Illustration: T. E. Tibbitts in Blake 1963: 219 (b/w); S. Webb in Howell and Webb 1995: Pl. 26; T. Worfolk in del Hoyo et al. 1999: Pl. 13; F. Weick in König et al. 1999: Pl. 43

Photograph: K. W. Fink in Burton et al. 1992: 112; R. Gerhardt in Johnsgard 2002: Pl. 38; R. and N. Bowers in internet: Owl Pages, photo gallery 2003 (2 photographs)
Literature: Ridgway 1914: 763–767; Kelso and Kelso 1934: 25, 26; Kelso 1934d: 56; Eck and Busse 1973: 170, 171; Howell and Webb 1995: 365; del Hoyo et al. 1999: 204, 205; König et al. 1999: 341, 342; Johnsgard 2002: 170–172

- ***Strix virgata tamaulipensis*** Phillips 1911
  Strix virgata tamaulipensis Phillips 1911, Auk 28: 76; Terra typica: Rio Martinez, Tamaulipas

Remarks: Probably a synonym of *S. v. centralis*

Distribution: Northeast Mexico (Nuevo León and Tamaulipas)
Museum: MCZ (holotype)

Wing length: 235–238 mm
Tail length: 137–148.5 mm
Tarsus length: ?
Length of bill (cere): 18–19.5 mm
Body mass: ?

Illustration: ?
Photograph: ?
Literature: Phillips 1911: 76; Ridgway 1914: 763, 764, 767; Eck and Busse 1973: 170, 171; del Hoyo et al. 1999: 204, 205

- ***Strix virgata centralis*** (Griscom) 1929
  Ciccaba virgata centralis Griscom 1929, Bull. Mus. Comp. Zool. 69: 159; Terra typica: Chivela, Oaxaca, Mexico

Synonym:
- *Ciccaba virgata eatoni* L. and E.H. Kelso 1936, Auk 53: 215; Terra typica: Apazote, Campeche, Mexico

Distribution: Eastern and southern Mexico to western Panama
Museum: FMNH, FNSF, MCZ, USNM, ZMH

Wing length: 221–233 (243) mm
Tail length: 128–140 mm
Tarsus length: ?
Length of bill (cere): 18.5–20.5 mm
Body mass: ♀ 356 g

Strix virgata squamulata · Mottled Owl · Sprenkelkauz

Illustration: D. Gardner in Stiles and Skutch 1989: Pl. 20; S. Webb in Howell and Webb 1995: Pl. 26; F. Weick in König et al. 1999: Pl. 43
Photograph: ? R. Gerhardt in Johnsgard 2002: Pl. 38
Literature: Ridgway 1914: 763, 764; Kelso 1934d: 57; Blake 1963: 218, 219; Eck and Busse 1973: 170, 171; Stiles and Skutch 1989: 190–191; Ridgely and Gwynne 1989: 190, 191; Howell and Webb 1995: 365; del Hoyo et al. 1999: 204, 205; König et al. 1999: 341, 342; Johnsgard 2002: 170–172

- ***Strix virgata virgata*** (Cassin) 1849
  Syrnium virgatum Cassin 1849, Proc. Acad. Nat. Sci. Phila. 1848(4): 124; Terra typica: South America, restricted to Bogotá by von Berlepsch (1908) Novit. Zool. 15: 288

Synonym:
- *Ciccaba virgata minuscula* L. Kelso 1940, Biol. Leaflet 12: 1; Terra typica: Western Colombia
- *Ciccaba virgata occidentalis* Sneidern 1955, Noved. Colomb. 2: 35; Terra typica: Narino, Colombia
- *Strix virgata sneiderni* Eck 1972, Zool. Abh. Staatl. Mus. Tierk. Dresden 1971, 30(15): 211; new name for *C. v. occidentalis* Sneidern

Distribution: Eastern Panama, Colombia, Ecuador, Venezuela and Trinidad
Museum: SMNSt, SMTD, ÜMB, ZFMK, ZSBS

Illustration: D. Eckleberry in *Wilson Bull.* 78: 1966: Pl. 1 (dark morph); T. Boyer in Boyer and Hume 1991: 130; T. Worfolk in del Hoyo et al. 1999: Pl. 13 (two morphs); F. Weick in König et al. 1999: Pl. 43 (dark morph); P. J. Greenfield in Ridgely and Greenfield 2001: II: Pl. 36(8); S. Webb in Hilty 2003: Pl. 92 (adult and pullus)
Photograph: R. Hoppe 1967: 310, 311 (b/w); A. Hillman in *Avic. Mag.* 1977: 65 (b/w); R. and N. Bowers in internet: Owl Pages, photo gallery 2003
Literature: Sharpe 1875a: 273–275; Hoppe 1967: 310, 311; Buchanan 1971: 103–106; Wylie 1976: 64, 65 (subspecies unknown); Voous 1988: 200–204; Boyer and Hume 1991: 130; del Hoyo et al. 1999: 204, 205; König et al. 1999: 341, 342; Ridgely and Greenfield 2001: I: 312, 313, II: 220; Hilty 2003: 365

- ***Strix virgata macconnelli*** (Chubb) 1916

Ciccaba superciliaris macconnelli Chubb 1916, Bds. Brit. Guiana 1: 290; Terra typica: Sturibisi River, British Guiana

Distribution: The Guianas
Museum: MCZ, MZUS, NNML

Wing length:	230 mm
Tail length:	?
Tarsus length:	?
Length of bill:	?
Body mass:	♀ 307 g

Illustration: P. Barruel in Haverschmidt 1968: Pl. 13
Photograph: ?
Literature: Haverschmidt 1968: 161; Eck and Busse 1973: 170, 171; del Hoyo et al. 1999: 204, 205; König et al. 1999: 341, 342

- ***Strix virgata superciliaris*** (Pelzeln) 1863

Syrnium superciliare "Natterer" von Pelzeln 1863, Verh. Zool.-Bot. Ges. Wien 15: 1125; Terra typica: Brazil, i.e. Villa Bella de Mato Grosso, River Gueporé, c.f. Hellmayr (1912) Abh. k. Bayr. Akad. Wiss. Wien 26: 78

Distribution: Northern, central and north-eastern Brazil
Museum: BMNH, MCZ, SMNSt, ÜMB, ZSBS

Wing length:	274 mm
Tail length:	170–178 mm
Tarsus length:	47 mm
Length of bill:	40 mm
Body mass:	?

Illustration: ?
Photograph: ?
Literature: Sharpe 1875a: 271–273; Eck and Busse 1973: 170, 171; del Hoyo et al. 1999: 204, 205; König et al. 1999: 341, 342

- ***Strix virgata borelliana*** (W. Bertoni) 1901

Syrnium Borellianum W. Bertoni 1901, Aves Nuevas Parag.: 177, 178

Synonym:
- *Strix suinda* Vieillot 1817 = *Asio flammeus suindus* (Vieillot); see Kelso 1934d: 39, 56

Distribution: Southeast Brazil, eastern Paraguay and northeast Argentina (Misiones)
Museum: FNSF, MZUS, NNML, SMNSt, SMTD,

Wing length:	256 mm
Tail length:	?
Tarsus length:	?
Length of bill:	?
Body mass:	?

Illustration: A. Cameron in Voous 1988: 200 (possibly *squamulata*?); T. Worfolk in del Hoyo et al. 1999: Pl. 13 (head); F. Weick in König et al. 1999: Pl. 43
Photograph: ?
Literature: Kelso 1934a: 6–28; Kelso 1934d: 32–68; Eck and Busse 1973: 170, 171; Voous 1988: 200–204; del Hoyo et al. 1999: 204, 205; König et al. 1999: 342, 343

Strix albitarsis (Bonaparte) 1850
Rufous-banded Owl · Rötelkauz · Chouette fasciée (striée rouge) · Cárabo Patiblanco
Syrnium albitarse "Gr" Bonaparte 1850, Consp. Gen. Av. 1: 52; Terra typica: South America, i.e. Bogotá. (The identity of the name is based on the type in BMNH and Sclater's description and plate in Trans. Zool. Soc. London 1859: 263 and Pl. 9)

Length: 300–350 mm
Body mass: ?

Synonym:
- *Ciccaba albitarse goodfellowi* Chubb 1916, Bull. Brit. Ornith. Cl. 36: 46; Terra typica: North of Quito, 11 000 feet, Ecuador
- *Ciccaba albitarse opaca* Peters 1943, Bull. Mus. Comp. Zool. 92: 297; Terra typica: Maraynioc, Dept. Junin, 12 000 feet, Peru
- *Ciccaba albitarse tertia* Todd 1947, Proc. Biol. Soc. Wash. 60: 95; Terra typica: Incachaca, Bolivia (without measurements) (holotype CMP)

Distribution: Patchily distributed on the Andean slope from Venezuela, Colombia and Ecuador to northernmost northwest Peru and from northern Peru south to Bolivia
Habitat: Dense, humid, montane evergreen and cloud forest, but also in open areas with scattered trees between wooded parts. From 1 700 m up to 3 700 m a.s.l.
Museum: BMNH, FNSF, SMTD, ZSBS

Wing length: 274 mm
Tail length: ?
Tarsus length: ?
Length of bill: ?
Body mass: ?

Illustration: J. Wolf in Sclater 1859: Pl. 9; J. Fjeldsa in Fjeldsa and Krabbe 1990: Pl. 25; T. Boyer in Boyer and Hume 1991: 132; T. Worfolk in del Hoyo et al. 1999: Pl. 13; F. Weick in König et al. 1999: Pl. 44; P. J. Greenfield in Ridgely and Greenfield 2001: II: Pl. 36(9); S. Webb in Hilty 2003: Pl. 25
Photograph: ?
Literature: Peters 1938: 179–186; Voous 1964: 471–478; Fjeldsa and Krabbe 1990: 228; Boyer and Hume 1991: 132, 133; del Hoyo et al. 1999: 205; König et al. 1999: 345, 346; Ridgely and Greenfield 2001: I: 313, II: 221; Hilty 2003: 365, 366

Strix chacoensis Cherrie and Reichenberger 1921
Chaco Owl · Chaco-Kauz · Chouette du Chaco · Cárabo Chaqueno
Strix chacoensis Cherrie and Reichenberger 1921, Am. Mus. Novit. 27: 1; Terra typica: Fort Wheeler, Paraguayan Chaco

Length: 350–380 mm
Body mass: 420–500 g

Distribution: Chaco of southern Bolivia, northern Paraguay and northern Argentina, south to Cordoba and Buenos Aires province
Habitat: Semi-open, rather dry landscape with cacti and thornscrub or with small groups of trees. Sometimes in areas near water. In Argentina up to 1 200 m above sea-level
Museum: AMNH (holotype), BMNH, SMNKa, ZFMK, ZSBZ
Remarks: Vocalisation very different from other *Strix*, but not a reason for a separate genus. Some other South American members of *Strix* (e.g. *hylophila*), also have very different vocalisations

Wing length: ♂ 251–163 mm, ♀ 281–291 mm
Tail length: ♂ 146–150 mm, ♀ 162–182 mm
Tarsus length: 50–53 mm
Length of bill: 32 and 33 mm
Body mass: ♂ 360–425 g, ♀ 420–500 g

Illustration: T. Worfolk in del Hoyo et al. 1999: Pl. 12; F. Weick in König et al. 1999: Pl. 43
Photograph: R. Steinberg 1999: 50 (adult, juvenile and skins); W. Pitterman 2005: 170–173 (adult and juvenile)
Literature: Eck and Busse 1973: 172; Fjeldsa and Krabbe 1990: 228, 229; Steinberg 1999: 45–52; del Hoyo et al. 1999: 201; König et al. 1999: 343, 344; W. Pitterman 2005: 170–173

Strix rufipes King 1828
Rufous-legged Owl · Rotfußkauz · Chouette masquée (à pieds rouge) · Cárabo Bataraz
Strix rufipes King 1828, Zool. J. 3: 426;
Terra typica: Port Famine, Straits of Magellan
(see Plate 6)

Length: 330–380 mm
Body mass: similar *chacoensis*

Synonym:
- *Strix rufipes sanborni* Wheeler 1938, Field Mus. Nat. Hist. Publ. Zool. Ser. 20: 479; Terra typica: Quellon, Chiloe Island off Chile (holotype FMNH)

Remarks: Plumage similar to dark male from nominate. Sanborni was referred by Hellmayr as typical rufipes, without comment! (See also paintings in König et al. 1999 and Jaramillo et al. 2003!)

Distribution: From central Chile and extreme western and central Argentina south to Tierra del Fuego, also Chiloe Island, off south-central Chile
Habitat: Dense, moist forest with mosses and lichens, also old secondary forest. Sometimes semi-open woodland. From lowland to mountain slopes, up to around 2 000 m
Museum: AMNH, FMNH, FNSF, MNBHU, SMNSt, ZMH, ZSBS
Remarks: Possibly most related to *chacoensis* and *hylophila*, but not generically distinct from other *Strix*!

Wing length:	♂ 250–264 mm, ♀ 261–275 mm (241 mm *sanborni*, a young bird)
Tail length:	(141) 151–161 mm
Tarsus length:	(44) 47–51 mm
Length of bill (cere):	(25) 28–30 mm
Body mass:	?

Illustration: des Murs 1846: Icon. Ornith. Pl. 37, but yellow irides; N. Quintscher in Eck and Busse 1973: Abb. 41; J. Fjeldsa in Fjeldsa and Krabbe 1990: Pl. 25; T. Boyer in Boyer and Hume 1991: 145; T. Worfolk in del Hoyo et al. 1999: Pl. 12; F. Weick in König et al. 1999: Pl. 43; P. Burke in Jaramillo et al. 2003: Pl. 59
Photograph: R. Steinberg 1999: 51; F. Gohier in del Hoyo et al. 1999: 88
Literature: Sharpe 1875a: 261; Wheeler 1938: 479–482; Philippi 1940: 147–152; Eck and Busse 1973: 172; Fjeldsa and Krabbe 1990: 228, 229; Boyer and Hume 1991: 145; del Hoyo et al. 1999: 200, 201; König et al. 1999: 242, 243; Jaramillo et al. 2003: 58

Strix hylophila Temminck 1825
Rusty-barred Owl · Brasilkauz, Rostkauz · Chouette de Brésil (dryade) · Cárabo Brasileno
Strix hylophila Temminck 1825, Pl. col. livr. 63: Pl. 373; Terra typica: Brazil, i.e. Ypanema, Sao Paulo, fide Hartert (1908) Novit. Zool. 15: 288

Length:	350–360 mm
Body mass:	285–395 g

Synonym:
- *Nyctale Bergiana* W. Bertoni 1901, An. Cient. Paraguayos 1(1): 173; Terra typica: Rio Mondaih, Paraguay
- ? *Nyctale fasciata* M. and W. Bertoni 1901, ibid.: 174; Terra typica: Djaguarasapá, Alto Paraná, Paraguay

Distribution: Eastern and southern Paraguay, southeast Brazil (Minas Geraës to Rio Grande do Sul). Extreme north of Argentina (Misiones)
Habitat: Lowland and montane, tropical, evergreen forest, often with dense undergrowth, bamboo and creepers. Sometimes secondary woodland near human settlements. From lowlands up to about 1 000 m
Museum: AMNH, BMNH, MNBHU, NNML

Wing length:	280 mm
Tail length:	173 mm
Tarsus length:	53 mm
Length of bill:	38 mm
Body mass:	♂ 285–340 g, ♀ 345–395 g

Illustration: C. J. Temminck 1825: Pl. 373; T. Boyer in Boyer and Hume 1991: 145; T. Worfolk in del Hoyo et al. 1999: Pl. 12; F. Weick in König et al. 1999: Pl. 44 (adult and juvenile)
Photograph: H. Busse in Eck and Busse 1973: Abb. 19 (juvenile); E. Hosking in Hosking and Flegg 1982: 157; K. Rudloff in *Der Falke* 1989(11): 325 (Berlin Zoo); L. Koerner in internet: Owl Pages, photo gallery 2002
Literature: Sharpe 1875a: 269, 270; Eck and Busse 1973: 172, 173; Busse 1989: 324; Boyer and Hume 1991: 145; del Hoyo et al. 1999: 200; König et al. 1999: 344, 345

Strix nigrolineata (Sclater) 1859
Black and White Owl · Bindenhalskauz · Chouette noire et blanche · Cárabo Blanquinegro
Ciccaba nigrolineata Sclater 1859, Proc. Zool. Soc. Lond.: 131; Terra typica: South Mexico, Oaxaca (by Kelso 1932: 11)

Length:	350–400 mm
Body mass:	♂ ~435 g, ♀ ~535 g

Synonym:
- (*Syrnium nigrolineatum*) Subspecies a *Syrnium spilinotum* Sharpe 1875, Cat. Birds Brit. Mus. 2: 277; Terra typica: Bogotá, Colombia (holotype BMNH)

Distribution: Central Mexico to northwest Venezuela, western Colombia and western Ecuador, also extreme northwest of Peru
Habitat: Rain forest, semi-deciduous forest, clearings, forest edges and semi-open, swampy and flooded woodland. Gallery forest, mangroves, plantations, sometimes near human habitation. From sea level up to about 1 200 m in Mexico, 2 100 m in Panama and 2 400 m in Colombia
Museum: AMNH, BMNH, FNSF, MHNP, MNBHU, SMNSt, SMTD

Wing length: ♂ 272–285 mm, ♀ 255–293 mm
Tail length: ♂ 161–171.5 mm, ♀ 154–179.5 mm
Tarsus length: 48–49.5 mm
Length of bill (cere): 18.5–22 mm
Length of bill: 35.5–45.5 mm
Body mass: ♂ 405–436 g, ♀ 468–535 g

Illustration: J. Wolf in *Trans. Zool. Soc. London* 1862: Pl. 63; D. E. Tibbitts in Blake 1963: 280 (b/w); J. A. Gwynne in Ridgely and Gwynne 1989: Pl. 12; D. Gardner in Stiles and Skutch 1989: Pl. 20; T. Boyer in Boyer and Hume 1991: 131; S. Webb in Howell and Webb 1995: Pl. 26 and Fig. 6; T. Worfolk in del Hoyo et al. 1999: Pl. 13; F. Weick in König et al. 1999: Pl. 44 (adult and juvenile); P. J. Greenfield in Ridgely and Greenfield 2001: II: Pl. 36(6); S. Webb in Hilty 2003: Pl. 25
Photograph: M. and P. Fodgen in del Hoyo et al. 1999: 80; M. Kasprzyk in Johnsgard 2002: Pl. 39; J. L. Rangel in Johnsgard 2002: Pl. 40; R. and N. Bower in internet: Owl Pages, photo gallery 2002
Literature: Sharpe 1875a: 276, 277; Ridgway 1914: 760–762; Blake 1963: 219; Land 1970: 140, 141; Eck and Busse 1973: 171, 172; Ridgely and Gwynne 1989: 191; Stiles and Skutch 1989: 195; Boyer and Hume 1991: 131; Howell and Webb 1995: 365, 366; del Hoyo et al. 1999: 205; König et al. 1999: 346, 347; Johnsgard 2002: 173–175; Ridgely and Greenfield 2001: I: 311, 312, II: 219, 220; Hilty 2003: 364

Strix nigrolineata · Black and White Owl · Bindenhalskauz

Strix huhula **Daudin 1800**
Black-banded Owl · Zebrakauz · Chouette huhul (obscure) · Lechuza Negra, Cárabo Negro

Length: 305–350 mm
Body mass: ~370 g

Distribution: South America, east of the Andes. From Colombia, Venezuela, Ecuador and the Guianas, south to Bolivia, Paraguay, northern Argentina and the southeast and northeast of Brazil
Habitat: Tropical and subtropical rainforest and clearings, plantations. In south, in the Araucaria range forest. Lowlands up to 500 m above sea-level. Higher in Andean slopes, but rarely at 1 100–1 400 m

- ***Strix huhula huhula*** Daudin 1800
  Strix huhula Daudin 1800, Traite d Orn. 2: 190;
  Terra typica: Cayenne

Distribution: Eastern Colombia, southern Venezuela and the Guianas to northeast Brazil, southern and eastern Peru, northwest Argentina, northern Paraguay and Bolivia
Museum: BMNH, FNSF, MHNP, MNBHU, NHMWien, ÜMB, SMNSt, SMTD, ZSBS

Wing length:	243–270 mm
Tail length:	133–168 mm
Tarsus length:	48 mm
Length of bill:	38 mm
Body mass:	~370 g (Dunning), 397 g (Haverschmidt)

Illustration: P. Barruel in Haverschmidt 1968: Pl. 13; T. Boyer in Boyer and Hume 1991: 132; T. Worfolk in del Hoyo et al. 1999: Pl. 13; F. Weick in König et al. 1999: Pl. 44; P. J. Greenfield in Ridgely and Greenfield 2001: II: Pl. 36(7); Ridgely and Greenfield 2001: I: 312, II: 220; S. Webb in Hilty 2003: Pl. 25
Photograph: R. Behrstock in del Hoyo et al. 1999: 112 (Ecuador)
Literature: Sharpe 1875a: 275, 276; Haverschmidt 1968: 161; Eck and Busse 1973: 171, 172; Hilty and Brown 1986: 251; Boyer and Hume 1991: 132; del Hoyo et al. 1999: 205; König et al. 1999: 347, 348; Hilty 2003: 365

- ***Strix huhula albomarginata*** Spix 1824
  Strix albomarginata Spix 1824, Av. Bras. 1: 23, lám. 10a; Terra typica: "in sylvis prov. Rio de Janeiro"

Distribution: Southeast Brazil, eastern Paraguay and northeast Argentina (Misiones)
Museum: BMNH, ZSBS (holotype)

Wing length:	265–280 mm
Tail length:	165–171 mm
Tarsus length:	?
Length of bill:	?
Body mass:	?

Illustration: F. Weick in König et al. 1999: Pl. 44
Photograph: W. H. Patridge in Patridge 1956: 146 (b/w)
Literature: W. H. Patridge 1956: 143–148; Eck and Busse 1973: 171, 172; del Hoyo et al. 1999: 205; König et al. 1999: 347, 348

❖

Strix fulvescens **(Sclater and Salvin) 1868**
Fulvous Owl · Gilbkauz · Chouette fauve · Cárabo Guatemalteco (fulvo)
Syrnium fulvescens **Sclater and Salvin 1868, Proc. Zool. Soc. Lond.: 58; Terra typica: Guatemala**

Length:	410–440 mm
Body mass:	?

Distribution: Southern Mexico (south of the Isthmus), to Guatemala, Honduras and El Salvador
Habitat: Humid montane pine-oak and upper tropical forest; also cloud forest, from 1 200 m up to 3 100 m
Museum: AMNH, BMNH, FNSF, MNBHU, MZUS, SMNSt, ZMH
Remarks: Often considered to be conspecific with *S. varia*, but differs in morphology and vocalisations. Possibly forms superspecies with *S. occidentalis* and *S. varia*, but this, without DNA studies, remains speculation

Wing length:	300–333 mm
Tail length:	185–203 mm
Tarsus length:	56 mm
Length of bill (cere):	22.5–24.5 mm
Body mass:	?

Illustration: ? in Sharpe 1875a: 259 (foot, b/w); ? in Salvin and Godman 1897: Pl. 61 and in Kelso and Kelso 1934: 6; H. Quintscher in Eck and Busse 1973: Pl. 2; T. Boyer in Boyer and Hume 1991: 144; S. Webb in Howell and Webb 1996: Pl. 26; T. Worfolk in del Hoyo et al. 1999: Pl. 12; F. Weick in König et al. 1999: Pl. 39
Photograph: G. Lasley in internet: Owl Pages, photo gallery 2003
Literature: Sclater and Salvin 1868a: 58, 59; Sharpe 1875a: 258–260; Ridgway 1914: 647, 648; Land 1970: 141; Eck and Busse 1973: 163–168; Boyer and Hume 1991: 144; Howell and Webb 1995: 367; del Hoyo et al. 1999. 200; König et al. 1999: 326, 327

Strix occidentalis **(Xantus) 1860**
Spotted Owl · Fleckenkauz · Chouette tachetée · Cárabo Californiano (Búho Manchado)
(see Plate 6)

Length:	405–480 mm
Body mass:	520–760 g

Distribution: West of North America, from southwest Canada (British Columbia) south to Baja California, New Mexico, south-western Texas and northern and central Mexico
Habitat: Dark coniferous and mixed forest, in montane regions. Prefers forest slopes and deep gorges with old wooded areas and cool, moist climate. Vicinity of water seems to be important

Strix occidentalis · Spotted Owl · Fleckenkauz

- ***Strix occidentalis caurina*** (Merriam) 1898
  Syrnium occidentale caurinum Merriam 1898, Auk 15: 39, 40; Terra typica: Vernon, Skagit Valley, Washington

Distribution: Southwest Canada (British Columbia), south to west United States coast ranges (western Washington to western Oregon, central California)
Museum: MVZ, UCal, USNM

Wing length:	304–323 mm
Tail length:	198–220 mm
Tarsus length:	54 mm
Length of bill (cere):	21.5 mm
Body mass:	♂ ~579 g, ♀ ~662 g

Illustration: K. E. Karalus in Karalus and Eckert 1974: Pl. 6; D. Sibley in Sibley 2000: 276
Photograph: E. Bull in *Gefiederte Welt* 1993(6): 206; G. Vaughn in *Living Bird* 1994(Winter): 11
Literature: Merriam 1898: 39, 40; Ridgway 1914: 650, 651; Oberholser 1915: 251–257; Bent 1961 (repr.): 208, 209; Karalus and Eckert 1974: 37–43; Eck and Busse 1973: 168; Voous 1988: 231–236; del Hoyo et al. 1999: 199; König et al. 1999: 325, 326; Sibley 2000: 276; Johnsgard 2002: 176–185

- ***Strix occidentalis occidentalis*** (Xantus) 1860
  Syrnium occidentale Xantus 1860, Proc. Acad. Nat. Sci. Phila. 1859: 193; Terra typica: Fort Tejon, California

Distribution: Central and southern California, coast ranges and western slope of Sierra Nevada, south to northern Baja California
Museum: BMNH, MHSciArt, UCalMVZ, USNM (holotype)

Wing length:	301–328 mm
Tail length:	193–220 mm
Tarsus length:	51–54.5 mm
Length of bill:	29.5–33 mm
Body mass:	♂ 518–694 g, ♀ 548–760 g

Illustration: ? in Baird et al. 1860: Pl. 66 (yellow irides); A. Brooks in Grosvenor and Wetmore 1937: 15; G. Coheleach in *Living Bird* 1972 frontcover; K. E. Karalus in Karalus and Eckert 1974: Pl. 5; L. Malick in Breeden et al. 1983: 241; A. Cameron in Voous 1988: 234; T. Boyer in Boyer and Hume 1991: 142; T. Worfolk in del Hoyo et al. 1999: Pl. 12; F. Weick in König et al. 1999: Pl. 38 (adult and juvenile); D. Sibley in Sibley 2000: 276
Photograph: G. Ladd in MacKenzie 1986: 132; C. A. Morgan in *Living Bird* 1986 (Winter): 15; W. Lankinen in *Living Bird* 1986(Winter): 16; D. Plummer in *Living Bird* 1988(Autumn): 33; W. Lankinen in Burton et al. 1992: 118; G. Vaughn in *Living Bird* 1993(Summer): 35; G. Vaughn in *Living Bird* 1994(Winter): 11; K. W. Fink in Johnsgard 2002: Pl. 24; J. Hobbs in internet: Owl Pages, photo gallery 2004
Literature: Sharpe 1875a: 260; Ridgway 1914: 648–650; Oberholser 1915: 251–257; Grosvenor and Wetmore 1937: 14; Marshall 1942: 66, 67; Bent 1961 (repr.): 202–207; Gould 1977: 131–146; Breeden et al. 1983: 240; Voous 1988: 231–236; Boyer and Hume 1991: 142; del Hoyo et al. 1999: 199; König et al. 1999: 325, 326; Sibley 2000: 276; Johnsgard 2002: 176–185

- ***Strix occidentalis lucida*** (Nelson) 1903
  Syrnium occidentale lucidum Nelson 1903, Proc. Biol. Soc. Wash. 16: 152; Terra typica: Mount Tacitaro, Michoacan, Mexico (holotype USNM)

Synonym:
- *Strix occidentale huachucae* Swarth 1910, Univ. Cal. Publ. Zool. 7: 3; Terra typica: Huachuca Mountains, Arizona (holotype UCalMVZ)
- ? *Strix occidentalis juanaphillipsae* Dickerman 1997, Bull. Brit. Ornith. Cl. 117; Terra typica: Central Mexico, west to the Puebla boundary

Distribution: From northern Arizona, southeast Utah and southern Colorado, south to northern and central Mexico (Michoacan and Guanajuato)
Museum: USNM, UCalMVZ, MSWBUNM (holotype *juanaphillipsae*)
Remarks: Spotted and Barred Owls hybridise where range overlaps

Wing length: ♂ 302–309 mm, ♀ 302–328 mm
Tail length: ♂ 191–206 mm, ♀ 196–226 mm
Tarsus length: 51.5–55.5 mm
Length of bill: 27–33 mm
Body mass: ?

Photograph: J. Cancalosi in del Hoyo et al. 1999: 143 (adult and juvenile)
Illustration: K. E. Karalus in Karalus and Eckert 1974: Pl. 7; T. Worfolk in del Hoyo et al. 1999: Pl. 12; F. Weick in König et al. 1999: Pl. 38; D. Sibley in Sibley 2000: 276; D. Lockshaw in internet: Owl Pages, photo gallery 2002
Literature: Ridgway 1914: 652; Oberholser 1915: 251–257; Bent 1961 (repr.): 209–213; Blake 1963: 220; Karalus and Eckert 1974: 45–47; Breeden et al. 1983: 239; Howell and Webb 1995: 366; Dickerman 1997: Mus. SW Biol., Dep. Biol.; del Hoyo et al. 1999: 199; Univ. of New Mexico: 117; König et al. 1999: 325, 326; Johnsgard 2002: 176–185

Strix varia **Barton 1799**
Barred Owl · Streifenkauz · Chouette rayée (barrée) · Cárabo Barrado (Búho Barrado)
(see Plate 6)

Length: 480–550 mm
Body mass: ♂ ∅ ~630 g, ♀ ∅ ~800 g

Distribution: North America, from southeast Alaska south to northern California. Also extends from about the Great Lakes in a narrow belt across the Rocky Mountains to southwest Canada (east to Nova Scotia) and central and eastern United States south to Texas, Florida, southern and central Mexico
Habitat: Coniferous and mixed forest, usually heavy, aged wood with nearby open country, riverine and swampy areas, woodland bordering lakes and rivers. Sometimes in parks. From sea-level up to around 1 500–2 500 m

- ***Strix varia varia*** Barton 1799
  Strix varius Bartram MS. Barton 1799, Fragm. Nat. Hist. Pennsylvania: 11; Terra typica: Pennsylvania

Synonym:
- *Strix varia brunnescens* Bishop 1931, Proc. Biol. Soc. Wash. 44: 94; Terra typica: Lake of Woods County, Minnesota
- *Strix varia albescens* Bishop 1931, ibid.: 95; Terra typica: Atelante, Quebec

Distribution: From southeast Alaska and southwest Canada (east to Nova Scotia), south to northern California and central and eastern United States (south to northern Texas and North Carolina)
Museum: AMNH, MNBHU, SMTD, USNM, ZMH

Wing length: ♂ 312–340 mm, ♀ 330–352 mm
Tail length: ♂ 215–230 mm, ♀ 224–257 mm
Tarsus length: 63.5 mm
Length of bill (cere): 23.5–30 mm
Body mass: ♂ 468–774 g (∅ 630 g), ♀ 610–1 051 g (∅ 800 g)

Illustration: J. J. Audubon in Audubon 1827–1838 (repr.): Pl. 239; R. Ridgway in Merriam and Fisher 1893: Pl. 22; A. Brooks in Grosvenor and Wetmore 1937: 15; A. Brooks in Sprunt 1955: unpaged colour plate; J. F. Landsdowne in Landsdowne and Livingston 1968: Pl. 33; H. Quintscher in Eck and Busse 1973: Pl. 2; K. E. Karalus in Karalus and Eckert 1974: Pl. 2; L. Malick in Breeden et al. 1983: 241; M. J. Rauzon in *Living Bird* 1986(Winter): 31; A. Cameron in Voous 1988: 217; T. Boyer in Boyer and Hume 1991: 143; T. Worfolk in del Hoyo et al. 1999: Pl. 12; F. Weick in König et al. 1999: Pl. 39; D. Sibley in Sibley 2000: 276; L. A. Fuertes in Johnsgard 2002: Pl. 6

Photograph: S. Grossman in Grossman and Hamlet 1965: unpaged; E. Hosking in Hosking and Flegg 1982: 159; B. R. Ranford in MacKenzie 1986: 134; B. and P. Wood in MacKenzie 1986: 135; S. D. Faccio in *Living Bird* 1996(Autumn): Backcover; P. Johnsgard in Johnsgard 2002: Pl. 25; Numerous colour photographs in internet: Owl Pages, photo gallery (Axia Wildlife)(C. Tanner, F. Bednar, etc.) 2003, 2004

Literature: Sharpe 1875a: 257, 258; Ridgway 1914: 641–644; Grosvenor and Wetmore 1937: 14; Sprunt 1955: 203, 204; Eck 1968: 283–289; Landsdowne and Livingston 1968: 142; Eck and Busse 1973: 163–168; Karalus and Eckert 1974: 20–31; Breeden et al. 1983: 240; Voous 1988: 225–230; Boyer and Hume 1991: 142–144; Howell and Webb 1995: 366, 367; del Hoyo et al. 1999: 199, 200; König et al. 1999: 327, 328; Sibley 2000: 276; Johnsgard 2002: 186–193

- ***Strix varia helveola*** (Bangs) 1899
  Syrnium nebulosum helveolum Bangs 1899, Proc. New England Zool. Cl. 1: 31; **Terra typica:** Corpus Christi, Texas

Synonym:
- *Strix varia albogilva* Bangs 1908, Auk 25: 316; new name for *helveola*

Distribution: Texas and adjacent lowlands of Mexico
Museum: AMNH, BMNH?, USNM

Wing length:	330–355 mm
Tail length:	210–254 mm
Tarsus length:	?
Length of bill (cere):	25–28 mm
Body mass:	?

Illustration: K. E. Karalus in Karalus and Eckert 1974: Pl. 4; F. Weick in König et al. 1999: Pl. 39; D. Sibley in Sibley 2000: 276

Photograph: R. Austing in Everett 1977: 36; R. Austing in Burton et al. 1992: 117; numerous colour photographs in internet: Owl Pages, photo gallery 2003

Literature: Ridgway 1914: 646; Bent 1961 (repr.): 182–197; Karalus and Eckert 1974: 35, 36; Voous 1988: 225–230; del Hoyo et al. 1999: 199, 200; König et al. 1999: 327, 328; Sibley 2000: 276

- ***Strix varia georgica*** Latham 1801
  Strix Georgica Latham 1801, Index Orn. Suppl. XV; **Terra typica:** Georgia, Americana = Southern Georgia

Synonym:
- *Strix nebulosa alleni* Ridgway 1880, Proc. US Nat. Mus. 3(8): 191; **Terra typica:** Clearwater, southwest Florida (holotype USNM)
- *Syrnium nebulosum sablei* Nicholson 1838, Fla. Nat. 17: 99; **Terra typica:** Munroe County, Florida

Distribution: South-eastern United States, south of North Carolina to Georgia and Florida
Museum: SMTD, USNM, ZFMK

Wing length:	(315–340 mm) 341–357 mm
Tail length:	205–231 mm
Tarsus length:	?
Length of bill (cere):	23–28 mm
Body mass:	?

Illustration: M. Sutton in Burleigh 1958: Pl. p 258; K. E. Karalus in Karalus and Eckert. 1974: Pl. 3; F. Weick in König et al. 1999: Pl. 39

Photograph: Numerous colour photographs in internet: Owl Pages, photo gallery 2004 (Axia Wildlife, etc.)

Literature: Ridgway 1914: 644, 645 (*alleni*); Bent 1961 (repr.): 197–201 (*alleni*); Karalus and Eckert 1974: 32–35; del Hoyo et al. 1999: 199, 200; König et al. 1999: 327, 328; Johnsgard 2002: 186–193

- ***Strix varia sartorii*** (Ridgway) 1873
  Syrnium nebulosum var. *Sartorii* Ridgway 1873, Bull. Essex Inst. 5: 200; **Terra typica:** Mirador, Vera Cruz, Mexico

Distribution: Mountains of central Mexico (Durango to Oaxaca), from 1 500 m up to 2 500 m
Museum: USNM, MNBHU
Remarks: *Strix fulvescens* is sometimes included as a subspecies of *Strix varia*, but possibly forms a superspecies also including *Strix occidentalis.* Kleinschmidt and Eck included *Strix fulvescens, varia, uralensis* and *davidi* in one "Formenkreis" (see Eck and Busse 1973: 163–168)

Wing length:	342–380 mm
Tail length:	220–252 mm
Tarsus length:	56 mm
Length of bill (cere):	24.5–28.5 mm
Body mass:	?

Illustration: F. Weick in König et al. 1999: Pl. 39
Photograph: ?

Literature: Ridgway 1914: 646, 647; Kelso 1934d: 54; Blake 1963. 220; Howell and Webb 1995: 366, 367; del Hoyo et al. 1999: 199, 200; König et al. 1999: 327, 328

Strix uralensis Pallas 1771
Ural Owl · Habichtskauz · Chouette de l'Oural · Cárabo Uralense

Length: 510–610 mm
Body mass: 500–1 300 g

Distribution: Northern Europe, from Norway, Sweden, Finland and the Baltic Republics, through northern Russia and Siberia to Korea, coast of the Okhotsk Sea, Sakhalin and Japan. Locally in central and south-eastern Europe
Habitat: Boreal forest and mixed woodland, not too dense and with clearings. From 450 m up to 1 600 m. In Japan, from sea-level up to around 1 600 m

- ***Strix uralensis liturata*** Lindroth 1788
Strix Liturata Lindroth 1788, Mus. Natural. Grillianum Söderforss.: 5; Terra typica: "Elfkarlely" = Älvkarleby, Uppland, Svecia
(see Plate 6)

Synonym:
- *Strix liturata* Tengmalm 1793, Kongl. Vet. Acad. Nya. Handl. 14: 267; Terra typica: Sweden

Distribution: Northern Europe and northwest Russia, east to Archangelsk region. South to northern Poland, Belarus and middle Volga River
Museum: BMNH, NNML, SMTD, ZFMK, ZMA, ZMH

Wing length: ♂ 342–368 mm, ♀ 349–382 mm
Tail length: 253–282 mm
Tarsus length: 50–56 mm
Length of bill: 38–45 mm
Body mass: ♂ 451–825 g, ♀ 520–1 020 g

Illustration: O. Kleinschmidt in Kleinschmidt 1958: Pl. 58; P. Barruel in Sutter and Barruel 1958: Pl. 21; L. Binder in Wüst 1970: 240; G. Pettersson in Pettersson 1984: 55, 56; H. Delin in Cramp et al. 1985: Pl. 50; A. Cameron in Voous 1988: 239; T. Boyer in Boyer and Hume 1991: 147; F. Weick in König et al. 1999: Pl. 40; F. Weick in Weick 2004: April
Photograph: O. Hedvall in Everett 1977: 118; P. Helo in Mikkola 1983: 47 (b/w); C. E. Ekman in Mikkola 1983: 48 (b/w); H. Reinhard in Mebs 1987: 95; ? in Delin and Svensson 1988: 163 (adult and juvenile); O. Hedvall in Burton et al. 1992: 123; B. Volmer in *Der Falke* 1999(10): 300; D. Forsman in del Hoyo et al. 1999: 95; Hautala in Mebs and Scherzinger 2000: 217
Literature: Sharpe 1875a: 255, 256; Hartert 1912–1921: 1017–19; Hartert 1923: 392, 393; Dementiev and Gladkov 1951: 459–462; Kleinschmidt 1958: 38; Vaurie 1965: 624, 625; Eck and Busse 1973: 163–168; Glutz von Blotzheim and Bauer 1980. 611; Mikkola 1983: 162–177; Cramp et al. 1985: 550–560; Mebs 1987: 94–99; Boyer and Hume 1991: 146–148; del Hoyo et al. 1999: 203; König et al. 1999. 329–331

- ***Strix uralensis macroura*** Wolf 1810
Strix uralensis macroura Wolf 1810, Meyer and Wolf's Taschenbuch der deutschen Vogelkunde 1: 84; Terra typica: Gebirgswälder Österreichs

Synonym:
- (*Strix uralensis*) *carpathica* Dunajewski 1940, Ann. Mus. Natl. Hungar. 33, zool.: 99; Terra typica: Northern slopes of the Carpathian Mountains

Distribution: Central and southeast Europe, from Carpathian Mountains south to Bulgaria, western Balkans, locally Bohemian Forest
Museum: NHMW (holotype), NNML, ZFMK, LRUR, NMP, ZMA
Remarks: Vaurie (1965) ignored subspecies *macroura* and included birds from central and southeast Europe in subspecies *liturata*

Wing length: 358–400 mm
Tail length: 282–315 mm
Tarsus length: ?
Length of bill: ?
Body mass: ♂ 503–950 g, ♀ 568–1 307 g (1 × 1 454 g)

Illustration: J. F. Naumann 1822: Pl. 42; A. Frisch in J. Orn. 7 1859: Pl. 2 (dark morph); H. Quintscher in Eck and Busse 1973: Pl. 2; F. Weick in König et al. 1999: Pl. 40 (adult, juvenile and moor); Zetterström/Mullarney in Svensson et al. 1999: 209
Photograph: ? in *Der Falke* 1969: 315; ? in Kohl 1977: 309–334 (b/w skins); L. Simak and J. Svelik in Mikkola 1983: 49–51; Scherzinger in Mebs and

Scherzinger 2000: 17; numerous photographs in Mebs and Scherzinger 2000: 205–223 (Brosette, Danko, Hecker, Partsch, Siegel, etc.)
Literature: Naumann 1822: 422–427; Hartert 1912–1921: 1017–1019; Dementiev and Gladkov 1951: 459–462; Eck and Busse 1973: 163–168; Kohl 1977: 309–334; Glutz von Blotzheim and Bauer 1980: 611–629; Mikkola 1983: 162–177; Bezzel 1985: 654–656; Cramp et al. 1985: 550–560; Voous 1988: 237–243; del Hoyo et al. 1999: 203; König et al. 1999: 329–331; Svensson et al. 1999: 208; Mebs and Scherzinger 2000: 205–225

- ***Strix uralensis uralensis*** Pallas 1771
Stryx (sic) *uralensis* Pallas 1771, Reise durch versch. Prov. Russ. Reichs 1: 455; Terra typica: circa Alpes Uralensis

Synonym:
- *Syrnium uralense sibiricum* Tschusi zu Schmidhoffen 1903, Orn. Jahrb. 14: 166; Terra typica: Tomsk, Siberia
- *Strix uralensis buturlini* Dementiev 1951, Ptitsy Sovietskogo Soiuza 1: 417; Terra typica: Lake Enerdekh, Markha River, western central Yakutia

Distribution: From eastern European Russia, east to the Okhotsk Coast
Museum: BMNH, SMTD, ZFMK, ZMH, ZSBS
Remarks: Intergrades with *liturata*

Wing length: ♂ 334–375 mm, ♀ 348–380 mm
Tail length: ♀ 317 mm
Tarsus length: 58.5 mm
Length of bill: ?
Body mass: ♂ 500–712 g, ♀ 950 g

Illustration: J. G. Keulemans in Hennicke 1905: unpaged; T. Worfolk in del Hoyo et al. 1999: Pl. 13; F. Weick in König et al. 1999: Pl. 40
Photograph: ?
Literature: Dresser 1902: 477–478; Hartert 1912–1921: 1017–1020; Dementiev and Gladkov 1951: 459, 462–464; Vaurie 1965: 625, 626; Eck and Busse 1973: 163–168; Cramp et al. 1985: 550–560; Voous 1988: 237–243; del Hoyo et al. 1999: 203; König et al. 1999: 329–331

- ***Strix uralensis yenisseensis*** Buturlin 1915
Strix uralensis yenisseensis Buturlin 1915, Mess. Orn. 6: 133; Terra typica: Krasnoyarsk, Siberia

Distribution: Central Siberian Plateau. In winter recorded from Transbaikalia and north-eastern Mongolia
Museum: ZFMK

Remarks: Intergrades with *uralensis*

Wing length: ♂ 328–350 mm, ♀ 348–370 mm
Tail length: 292 mm
Tarsus length: 56 mm
Length of bill: ?
Body mass: ?

Illustration: ?
Photograph: ?
Literature: Buturlin 1915: 133; Hartert 1923: 392; Dementiev and Gladkov 1951: 459, 464; Vaurie 1965: 625, 626; Eck and Busse 1973: 163–168; del Hoyo et al. 1999: 203; König et al. 1999: 329–331

- ***Strix uralensis nikolskii*** (Buturlin) 1907
Syrnium uralense nikolskii Buturlin 1907, J. Orn. 55: 235; Terra typica: South-eastern Siberia

Synonym:
- *Syrnium uralense coreensis* Momiyama 1927, J. Chosen Nat. Hist. Soc. 4: 1; Terra typica: Taianzan, northeast Korea
- *Strix uralensis morii* Momiyama 1927, Bull. Brit. Ornith. Cl. 48: 21; Terra typica: Korea
- *Strix uralensis tatibanai* Momiyama 1927, ibid.: 21; Terra typica: Sakhalin
- *Strix uralensis jinkou* Momiyama 1928, Auk 45: 182; Terra typica: Southeast Manchuria
- *Strix uralensis dauricus* Stegmann 1929, Annuaire Mus. Zool. Acad. Sci. URSS [Leningrad] 29(1828): 181, ex Sushkin in MS., vicinity of Chita, Transbaikalia

Distribution: Transbaikalia east to Sakhalin, south to northeast China and Korea
Museum: MNBHU, NSMT, SMTD, ZMM
Remarks: Specimens from west to east and north to south vary slightly in plumage; winglength decreases in a cline from west to east and from north to south

Wing length: ♂ 293–335 mm, ♀ 317–350 mm
Tail length: 209–254 mm
Tarsus length: ?
Length of bill: ?
Body mass: ♂ 630 g, ♀ 608–842 g

Illustration: H. Quintscher in Eck and Busse 1973: Pl. 2 (*daurica*)

Photograph: Knystaustas and Sibnev 1987: 76
Literature: Nikolski 1889: Sapiski Imper. Akad. Nauk. St. Petersburg; Hartert 1912–1921: 1020; Hartert 1923: 392; Murata 1914: 45; Momiyama 1927a: 1; Momiyama 1927b: 21; Momiyama 1928: 177–185; Stegmann 1929: 181; Dementiev and Gladkov 1951: 459, 465; Vaurie 1965: 625–627; Voous 1988: 237–243; del Hoyo et al. 1999: 203; König et al. 1999: 329–331

- ***Strix uralensis japonica*** **(A.H. Clark) 1907**
  Syrnium uralense japonicum A.H. Clark 1907, Proc. US Nat. Mus. 32: 471; Terra typica: Hokkaido

Distribution: Hokkaido
Museum: AMNH, BMNH, USNM, ZFMK (Kleinschm. coll.)

Wing length:	♂ 267 mm, ♀ 295–326 mm
Tail length:	201–235 mm
Tarsus length:	?
Length of bill:	?
Body mass:	?

Illustration: S. Takano in Massey et al. 1983: 191
Photograph: ?
Literature: Whitely 1867: 194; Blakiston and Pryer 1878: 246; Clark 1907: 471; Hartert 1912–1921: 1020, 1021; Hartert 1923: 393; Momiyama 1928: 177–185; Vaurie 1965: 625, 627; del Hoyo et al. 1999: 203; König et al. 1999: 329–331

- ***Strix uralensis hondoensis*** **(A.H. Clark) 1907**
  Syrnium uralense hondoensis A.H. Clark 1907, Proc. US Nat. Mus. 32: 472; Terra typica: Northern Hondo

Synonym:
- *Strix uralensis media* Momiyama 1928, Auk 45: 183; Terra typica: Hondo
- *Strix uralensis momiyamae* Taka-Tsukasa 1931, Tori 7(31): 14; Terra typica: Shinano, Hondo

Distribution: Northern and central Hondo
Museum: AMNH, USNM

Wing length:	♂ 302–322 mm, ♀ 319–347 mm
Tail length:	♂ 220–229 mm, ♀ 224–244 mm
Tarsus length:	44–50 mm
Length of bill (cere):	23–25 mm
Body mass:	?

Illustration: T. Miyamoto in Kobayashi 1956: Pl. 24
Photograph: ?
Literature: Clark 1907: 472; Hartert 1923: 393; Momiyama 1928: 177–185; Kobayashi 1956: 65; Vaurie 1965: 625, 627; del Hoyo et al. 1999: 203; König et al. 1999: 329–331

- ***Strix uralensis fuscescens*** **Temminck and Schlegel 1847**
  Strix rufescens Temminck and Schlegel 1847, in Siebold's Fauna Japonica, Aves: 30; Terra typica: Kyushu
  Strix fuscescens Temminck and Schlegel 1847, ibid.: Pl. 10; restricted by Hartert (1912–1921): 1021

Synonym:
- *Strix uralensis pacifica* Kuroda 1924, "On an apparently new form of Ural Owl ..."; Terra typica: Central Hondo, Prov. Idzu
- *Strix uralensis nigra* Momiyama 1927, Bull. Brit. Ornith. Cl. 48: 21; Terra typica: Prov. Ohsumi, Kyushu

Distribution: Southern Honshu, south to Kyushu
Museum: BMNH, FNSF, ZMH

Wing length:	♂ 301–311 mm, ♀ 315–330 mm
Tail length:	223–232 mm
Tarsus length:	51 mm
Length of bill:	?
Body mass:	?

Illustration: J. Wolf in Temminck and Schlegel 1847: Pl. 10; T. Worfolk in del Hoyo et al. 1999: Pl. 13; F. Weick in König et al. 1999: Pl. 40
Photograph: ?
Literature: Sharpe 1875a: 256, 257; Hartert 1912–1921: 1021; Uchida 1922a: 101; Uchida 1922b: 406, 407; Hachisuka 1925: 39; Uchida 1926: 362–364; Momiyama 1928: 184, 185; Vaurie 1965: 625, 627, 628; del Hoyo et al. 1999: 203; König et al. 1999: 329–331

Strix (uralensis) davidi **(Sharpe) 1875**
Sichuan (Wood) Owl · Davidkauz · Chouette de (Sitchouan) David · Cárabo de Sichuán
Syrnium davidi **Sharpe 1875, Ibis: 256;** Terra typica: **Mupin, Szechwan**

Length:	580 mm
Body mass:	?

Synonym:
- *Ptynx fulvescens* "David" Sharpe 1875, Ibis: 256; Terra typica: Mupin, Szechwan

Distribution: Central China: southeast Quinghai, western and central Sichuan (Lianhuashan Nature Reserve, Gansu Province)
Habitat: Coniferous and mixed forest with adjacent open areas (alpine meadows with low vegetation). From about 2 700 m up to 4 200 m (sometimes 5 000 m)
Museum: MHNP (holotype), SMTD, Museum Lianhuashan Nat. Reserve
Remarks: According to Scherzinger possibly conspecific to *Strix uralensis* (vocalisations)

Wing length: 371 and 372 mm
Tail length: 266 mm (290 mm?)
Tarsus length: 53 mm
Length of bill: ?
Body mass: ?

Illustration: David and Oustalet 1877: Pl. 3; H. Quintscher in Eck and Busse 1973: Pl. 2; J. H. Dick in Meyer de Schauensee 1984: Pl. 5; T. Worfolk in del Hoyo et al. 1999: Pl. 13; F. Weick in König et al. 1999: Pl. 39
Photograph: Jia Chen-xi in Scherzinger 2005: Fig. 3, 4
Literature: Eck and Busse 1973: 163–165; Etchécopar and Hüe 1978: 489, 490; Meyer de Schauensee 1984: 271; Tsohsin 1987; Sung 1998; del Hoyo et al. 1999: 203; König et al. 1999: 328, 329; Scherzinger 2005: 1–10

Strix nebulosa J.R. Forster 1772

Great Grey Owl · Bartkauz · Chouette lapone · Cárabo Lapón

Length: 575–670 mm
Body mass: 800–1 700 g

Distribution: North America, from central Alaska east to southwest Quebec, south to the Rocky Mountains of northern California, Idaho, Wyoming and Minnesota. Eurasia from Fenno-Scandia east to western Koryakland, south to Lithuania, northern Mongolia, northeast China and northern Sakhalin
Habitat: Dense boreal or coniferous, lichen-covered forest. Mixed woodland with birch, larch and poplars. Often near clearings, forest edges or open fields. Rarely near human settlements. From sea-level up to 1 000 m, in California up to 2 400 m and in Utah to around 3 200 m

▪ ***Strix nebulosa nebulosa*** J.R. Forster 1772
Strix nebulosa J.R. Forster 1772, Philos. Trans. 62: 424;
Terra typica: Hudson Bay

Synonym:
- *Strix cinerea* Gmelin 1788, Syst. Nat. 1(1): 291; Terra typica: Hudson Strait

Distribution: North America, from central Alaska east to southwest Quebec, south to Sierra Nevada of northern California, northern Idaho, western Montana, Wyoming and northeast Minnesota. Winters irregularly through southern Canada and the northern United States
Museum: AMNH, BMNH, MHNP, MNBHU, SMTD, ÜMB, USNM, ZMH

Wing length: ♂ 387–447 mm, ♀ 408–465 mm
Tail length: ♂ 300–323 mm, ♀ 310–347 mm
Tarsus length: 55–60 mm
Length of bill (cere): 23–28.5 mm
Body mass: ♂ 790–1 030 g (1 × 1 050 g),
♀ 1 144–1 454 g (1 × 1 700 g)

Illustration: J. J. Audubon in Audubon 1827–1838: Pl. 240; A. Brooks in Grosvenor and Wetmore 1937: 13; L. A. Fuertes in Forbush and May 1955: Pl. 46; L. A. Fuertes in Sprunt 1955: unpaged colour Pl.; K. E. Karalus in Karalus and Eckert 1974: Pl. 8; Landsdowne in Landsdowne and Livingston 1967: Pl. 19; L. Malick in Breeden et al. 1983: 241; T. Worfolk in del Hoyo et al. 1999: Pl. 13; F. Weick in König et al. 1999: Pl. 40; D. Sibley in Sibley 2000: 275
Photograph: B. and P. Wood in MacKenzie 1986: 116; J. D. Taylor in MacKenzie 1986: 117; Survival Anglia in Bellamy 1989: 55; T. and P. Leeson in *Living Bird* 1990(Spring): 15; W. Scherzinger in *Gefiederte Welt* 1997(10): 353; A. Morrin in *Living Bird* 1998(Winter): Frontcover; Bull in Mebs and Scherzinger 2000: 189; P. Johnsgard in Johnsgard 2002: Pl. 26, 29; G. Lasley in internet: Owl Pages, photo gallery 2003; M. Moon in internet: Owl Pages, photo gallery 2003; D. Roberson in internet: Owl Pages, photo gallery 2003
Literature: Sharpe 1875a: 252–254; Hartert 1912–1921: 1016, 1017; Ridgway 1914: 635–639; Grosvenor and Wetmore 1937: 12; Sprunt 1955: 206, 207; Bent 1961 (repr.): 213–220; Karalus and Eckert 1974: 48–55; Breeden et al. 1983: 240; Voous 1988: 244–251; del Hoyo et al. 1999: 203, 204; König et al. 1999: 331, 332; Johnsgard 2002: 194–201

- ***Strix nebulosa lapponica*** Thunberg 1798
  Strix lapponica Thunberg 1798, Kongl. Vetensk. Akad. Nya Handlingar 19: 184; Terra typica: Lapland
  (see Plate 6)

Synonym:
- *Strix barbata* Latham 1790!, Index Orn. 1: 62; Terra typica: Mountains of eastern Siberia
- *Syrnium cinereum sakhalinense* Buturlin 1907, Psovaya I Ruzheinaya Ochota: 87 (in Russian); Terra typica: Upper Sugnur River, north of Ulan Bator, northeast Mongolia

Distribution: Eurasia, from Fenno-Scandia east to western Koryakland, south to Lithuania, northern Mongolia, northeast China and northern Sakhalin
Museum: BMNH, NNML, SMTD, ZFMK, ZMH, ZSBS

Wing length: ♂ 430–477 mm, ♀ 438–483 mm
Tail length: ♂ 285–303 mm, ♀ 287–323 mm
Tarsus length: 52–58 mm
Length of bill: 38–45 mm
Body mass: ♂ 568–1 100 g, ♀ (680) 977–1 900 g

Illustration: J. F. Naumann in Naumann 1822: Nachtrag Pl. 349; J. Wolf in Dresser 1871–1896: Abb. 308; J. G. Keulemans in Hennicke 1905: Pl. 8; O. Kleinschmidt in Kleinschmidt 1958: Pl. 37; P. Barruel in Sutter and Barruel 1958: Pl. 21; H. Delin in Cramp et al. 1985: Pl. 50; G. Pettersson in Pettersson 1984: 23, 59; A. Cameron in Voous 1988: 247; T. Boyer in Boyer and Hume 1991: 149; T. Worfolk in del Hoyo et al. 1999: Pl. 1; F. Weick in König et al. 1999: Pl. 40; Mullarney and Zetterström in Svensson et al. 1999: 209
Photograph: O. Hedvall in Everett 1977: 72; E. Kemilä in Mikkola 1981: 17, 19, 57, 61, 88; P. Helo in Mikkola 1981: 24; E. Kemilä in Mikkola 1983: 54, 55; Survival Anglia in Bellamy 1989: 56; P. Helo in Delin and Svensson 1988: 163; W. Scherzinger in *Gefiederte Welt* 1990(1): 22; O. Hedvall in Burton et al. 1992: 121; G. Schleussner in *Gefiederte Welt* 1994(6): 198; B. Lundberg in del Hoyo et al. 1999: 129; numerous colour photographs in Mebs and Scherzinger 2000: 69, 85, 184–202 (by Hafner, Hautala, Hecker, Reinhard, Stefanson, etc.)
Literature: Sharpe 1875a: 254, 255; Naumann 1822, Nachtrag: 180–188; Hartert 1912–1921: 1014–1016; Hartert 1923: 392; Dementiev and Gladkov 1951: 455–459; Kleinschmidt 1958: 37; Vaurie 1965: 628, 629; Eck and Busse 1973: 169, 170; Etchécopar and Hüe 1978, 490, 491; Mikkola 1981: 124 pp; Mikkola 1983: 180–212; Glutz von Blotzheim and Bauer 1980: 629–639; Bezzel 1985: 656, 657; Cramp et al. 1985: 561–571; Voous 1988: 244–251; Boyer and Hume 1991: 148–150; del Hoyo et al. 1999: 203, 204; König et al. 1999: 331, 332; Svensson et al. 1999: 208; Mebs and Scherzinger 2000: 184–204

Strix nebulosa lapponica
Great Grey Owl
Bartkauz (adult and juvenile)

Genus
***Jubula* Bates 1929**
***Jubula* Bates 1929, Bull. Brit. Ornith. Cl. 49: 90.**
Type by *Bubo lettii* Büttikofer, 1889

Jubula lettii (Büttikofer) 1889

Maned Owl · Mähneneule · Hibou à crinière (Duc à crinière) · Búho de Crin
***Bubo lettii* Büttikofer 1889, Notes Leyden Mus. 11: 34;**
Terra typica: Liberia
(see Plate 9)

Length: 340–400 mm
Body mass: ♂ 183 g ($n = 1$)

Distribution: Africa, Liberia, Ivory Coast, Ghana and patchily from southern Cameroon and northern Gabon to eastern Zaire
Habitat: Primary lowland and gallery forest with creepers in the vicinity of rivers and lakes
Museum: BMNH, MHNG, MHNP, MNBHU, NNML (holotype), YPM (Cooper coll.)

Wing length: ♂ 263–277 mm, ♀ 241–285 mm
Tail length: 147–179 mm
Tarsus length: 34–41 mm (1 × 31 mm)
Length of bill: 27–30 mm
Body mass: ♂ 183 g ($n = 1$)

Illustration: ? in *Notes Leyden Mus.* 1889: 11, Pl. 6; H. Grönvold in Bannerman 1953: 537 (b/w); H. Quintscher in Eck and Busse 1973: Abb. 33 (b/w); M. Woodcock in Fry et al. 1988: Pl. 7; M. Andrews in *Bull. ABC* 1996(3)2: 135 (b/w); T. Boyer in Boyer and Hume 1991: 64; J. Rignall in Burton et al. 1992: 107; P. Hayman in Kemp and Kemp 1998: 287; T. Worfolk in del Hoyo et al. 1999: Pl. 13; F. Weick in König et al. 1999: Pl. 25 (adult and juvenile); N. Borrow in Borrow and Demey 2001: Pl. 60
Photograph: ?
Literature: Büttikofer 1889: 34; Bannerman 1953: 537, 538; Eck and Busse 1973: 92, 93; Fry et al. 1988: 121; Boyer and Hume 1991: 64; Dowsett-Lemaire 1995: 134, 135; Kemp and Kemp 1998: 286, 287; del Hoyo et al. 1999: 206; König et al. 1999: 285, 286; Borrow and Demey 2001: 494

Jubula lettii · Maned Owl · Mähneneule

Lophostrix

Genus
Lophostrix Lesson 1836

Lophostrix Lesson 1836, Comp. Œuvres Buffon 7: 261. Type by *Lophostrix griseata* Lesson = *Strix cristata* Daudin 1800

Lophostrix cristata (Daudin) 1800

Crested Owl · Haubenkauz · Hibou à casque (Duc à aigrettes) · Búho Corniblanco
(see Plate 9)

Length: 380–430 mm
Body mass: 400–600 g

Distribution: Locally from southern Mexico through Middle America to Venezuela, Surinam, the Guianas, Amazonian Colombia, Ecuador, Peru, Brazil and Bolivia
Habitat: Lowland rainforest with undergrowth, tall secondary forest, gallery woodland. Mostly near water. From sea-level up to 1 200 m in Guatemala and to 1 950 m in the cloudforest of Honduras

Lophostrix cirstata · Crested Owl · Haubenkauz

- ***Lophostrix (cristata) stricklandi*** Sclater and Salvin 1859
Lophostrix stricklandi Sclater and Salvin 1859, Ibis: 221; Terra typica: Vera Paz, Guatemala

Synonym:
- *Scops cristata* var. H. E. Strickland 1852, Contrib. Orn.: 60 and Pl. 10

Distribution: Southern Mexico through Guatemala and Honduras to western Panama and western Colombia
Museum: BMNH, MNBHU, NHMWien, SMNSt, ZFMK
Remarks: Possibly specifically distinct from *Lophostrix cristata*

Wing length: (287) 298–325 mm
Tail length: 171–203 mm
Tarsus length: 45–50 mm
Length of bill (cere): 18–20 mm
Body mass: ♂ 425 and 510 g, ♀ 620 g

Illustration: CDMS? in *Contrib. Orn.* 1852: Pl. 10 (*Scops cristata* var. Strickland); D. E. Tibbitts in Blake 1963: 214 (b/w); H. Quintscher in Eck and Busse 1973: Abb. 94; D. Gardner in Stiles and Skutch 1991: Pl. 20; T. Boyer in Boyer and Hume 1991: 64; J. Rignall in Burton et al. 1992: 107; S. Webb in Howell and Webb 1995: Pl. 26 and Fig. 26; T. Worfolk in del Hoyo et al. 1999: Pl. 13 (head); F. Weick in König et al. 1999: Pl. 25; P. Johnsgard in Johnsgard 2002: 108 (b/w)
Photograph: J. L. Rangel in del Hoyo et al. 1999: 79; J. L. Rangel in Johnsgard 2002: Pl. 36; R. and N. Bowers in internet: Owl Pages, photo gallery 2003 (orange irides!)
Literature: Sharpe 1875a: 122–125; Ridgway 1914: 732–736; Slud 1960: 96 pp; Blake 1963: 214, 215; Slud 1964: 430 pp; Land 1970: 137; Storer 1972: 452–455; Eck and Busse 1973: 94, 95; Stiles and Skutch 1991: 192; Boyer and Hume 1991: 64; Howell and Webb 1995: 358, 359; del Hoyo et al. 1999: 206; König et al. 1999: 286, 287; Johnsgard 2002: 108, 109

- ***Lophostrix cristata wedeli*** Griscom 1932
Lophostrix cristata wedeli Griscom 1932, Bull. Mus. Comp. Zool. 72: 326; Terra typica: Permé, Darien, Panama

Distribution: Eastern Panama to northeast Colombia and northwest Venezuela (from northern Venezuela: Aragua, 1 specimen)
Museum: MCZ (holotype)

Wing length: 280–312 mm
Tail length: ?
Tarsus length: ?
Length of bill: ?
Body mass: 545 g (Hilty 2003)

Illustration: G. Tudor in Hilty and Brown 1986: Pl. 9; J. A. Gwynne in Ridgely and Gwynne 1989: Pl. 12 (with yellow irides!)
Photograph: ?
Literature: Hilty and Brown 1986: 227; Ridgely and Gwynne 1989: 188; del Hoyo et al. 1999: 206; König et al. 1999: 286, 287; Ridgely and Greenfield 2001: I: 310, II: 218; Hilty 2003: 363

- ***Lophostrix cristata cristata*** (Daudin) 1800
  Strix cristata Daudin 1800, Traite d Orn. 2: 207;
  Terra typica: Guiana

Synonym:
- *Lophostrix cristata amazonica* L. Kelso 1940, Biol. Leaflet 12: 1; Terra typica: Yquitos, upper Amazonian River

Distribution: Southern Venezuela and the Guianas to northern Brazil (west Pará). South through Amazonia to northern Bolivia and northern Mato Grosso. West to southwest Colombia, Ecuador and eastern Peru
Museum: BMNH, MCZ, MHNP, MNBHU, MZUS, SMTD, ZFMK
Remarks: Possibly in all taxon of *L. cristata*, light morphs have yellow (orange) irides, dark morphs brown (buff) irides, but further studies are needed

Wing length: 295–319 mm
Tail length: 170–190 mm (1 × 216 mm)
Tarsus length: 40–46 mm
Length of bill: ?
Body mass: ? 545 g (*wedeli*?)

Illustration: P. Barruel in Haverschmidt 1968: Pl. 13; H. Quintscher in Eck and Busse 1973: Abb. 34; T. Worfolk in del Hoyo et al. 1999: Pl. 13; F. Weick in König et al. 1999: Pl. 25; P. J. Greenfield in Ridgely and Greenfield 2001: II: Pl. 36(2); S. Webb in Hilty 2003: Pl. 92 (with yellow irides)
Photograph: R. A. Behrstock in Duncan 2003: 28
Literature: Haverschmidt 1968: 210; Eck and Busse 1973: 94, 95; Hilty and Brown 1986: 227; Boyer and Hume 1991: 64; del Hoyo et al. 1999: 206; König et al. 1999: 286, 287

Genus
***Pulsatrix* Kaup 1848**
Pulsatrix Kaup 1848, Isis v. Oken, col. 771.
Type by *Strix torquata* Daudin = *Strix perspicillata* Latham

Synonym:
- *Novipulsatrix* L. Kelso 1933, Biol. Leaflet 1. Type by *Pulsatrix sharpei* von Berlepsch = *P. koeniswaldiana*

***Pulsatrix perspicillata* (Latham) 1790**
Spectacled Owl · Brillenkauz · Chouette à lunettes · Lechuzón de Anteojos (Búho de Anteojos)
(see Plate 6)

Length:	430–520 mm
Body mass:	600–980 g

Distribution: From southern Mexico through Middle America, Venezuela, the Guianas, Colombia, Ecuador, eastern Peru, Amazonian Brazil, south to Bolivia and northern Argentina. Also Trinidad Island
Habitat: Rather dense tropical and subtropical forest and forest edges, dense rainforest with clearings. Savanna woodland and also drier forest. Plantations and groves. Subtropical montane forest. Gallery forest along rivers. From sea-level up to about 1 500 m

Pulsatrix perspicillata · Spectacled Owl · Brillenkauz

- ***Pulsatrix perspicillata saturata*** Ridgway 1914
Pulsatrix perspicillata saturata Ridgway 1914, US Nat. Mus. Bull. 50(6): 758; Terra typica: Santo Domingo, Oaxaca, Mexico

Synonym:
- *Pulsatrix perspicillata austini* L. Kelso 1938, Biol. Leaflet 10: 3; Terra typica: Costa Rica

Distribution: Southern Mexico (Veracruz and Oaxaca) to western Panama (Pacific Slope to Chiriqui)
Museum: BMNH, MZUS, USNM (holotype), ZMH

Wing length:	♂ 314–347 mm, ♀ 317–360 mm
Tail length:	164–215 mm
Tarsus length:	?
Length of bill (cere):	27–32.5 mm
Body mass:	♂ 591–761 g, ♀ 765–982 g

Illustration: S. Webb in Howell and Webb 1995: Pl. 26 (lacks fine barring on belly!); T. Worfolk in del Hoyo et al. 1999: Pl. 13; F. Weick in König et al. 1999: Pl. 37
Photograph: K. W. Fink in Johnsgard 2002: Pl. 36 (adult), 37 (juvenile)
Literature: Ridgway 1914: 758, 759; Blake 1963: 215; Land 1970: 138; Howell and Webb 1995: 359; del Hoyo et al. 1999: 206, 207; König et al. 1999: 318–320; Johnsgard 2002: 110–112

- ***Pulsatrix perspicillata chapmani*** Griscom 1932
Pulsatrix perspicillata chapmani Griscom 1932, Bull. Mus. Comp. Zool. 72: 325; Terra typica: Permé, Caribbean slope of Darien, eastern Panama

Distribution: Caribbean slope of Costa Rica. Through eastern Panama and to Colombia, western Ecuador and northwest Peru
Museum: MCZ (holotype), ZMH
Remarks: Probably should be united with the nominate *perspicillata*

Wing length:	326–346 mm
Tail length:	183–193 mm
Tarsus length:	59.5 mm
Length of bill (cere):	29–30 mm
Body mass:	750 g (Stiles and Skutch 1989)

Illustration: D. Gardner in Stiles and Skutch 1989: Pl. 8 (adult with traces of barring on belly); J. A. Gwynne in Ridgely and Gwynne 1989: Pl. 12

Photograph: ?

Literature: Ridgway 1914: 356–358; Kelso 1934d: 43, 44; Hilty and Brown 1986: 228; Stiles and Skutch 1989: 192; Ridgely and Gwynne 1989: 188; del Hoyo et al. 1999: 206, 207; König et al. 1999: 318–320; Ridgely and Greenfield 2001: I: 310, II: 218

- ***Pulsatrix perspicillata perspicillata*** (Latham) 1790
Strix perspicillata Latham 1790, Index Orn. 1: 58;
Terra typica: Cayenne

Synonym:
- *Pulsatrix perspicillata trinitatis* Bangs and Penard 1918, Bull. Mus. Comp. Zool. 62: 51; Terra typica: Trinidad

Distribution: Venezuela to the Guianas, Brazil, and from eastern Colombia south to northern Bolivia. Trinidad (rare or extinct)

Museum: AMNH, BMNH, FNSF, MCZ (holotype = *trinitatis*), MHH, SMNSt, ÜMB, ZFMK, ZMH, ZSBS

Wing length:	♂ 305–335 mm, ♀ 318–350 mm
Tail length:	173.5–196 mm
Tarsus length:	?
Length of bill (cere):	26.5–31.5 mm
Body mass:	571–761 g (850 and 980 g)

Illustration: P. Barruel in Haverschmidt 1968: Pl. 13; E. Lear in Hayman 1980: 83; G. Tudor in Hilty and Brown 1986: Pl. 9 (b/w); T. Boyer in Boyer and Hume 1991: 88; T. Worfolk in del Hoyo et al. 1999: Pl. 13; F. Weick in *SCRO Mag.* 1999(2) (backcover, adult and juvenile); F. Weick in König et al. 1999: Pl. 37 (adult and juvenile); P. J. Greenfield in Ridgely and Greenfield 2001: II: Pl. 36(4); S. Webb in Hilty 2003: Pl. 25 (adult and juvenile)

Photograph: S. Grossman in Grossman and Hamlet 1965: 145 and unpaged; H. Busse 1969: 257 (b/w); Zool. Soc. London in Everett 1977: 34; M. Fodgen in Burton et al. 1992: 109 (adult and juvenile); R. and N. Bowers in internet: Owl Pages, photo gallery 2004; D. Messner (from Walsrode Zoo), photo collection Weick

Literature: Ridgway 1914: 756–758; Kelso 1934a: 234–236; Kelso 1934d: 43–45 and 89–96; Kelso 1946: 1–13; Haverschmidt 1968: 160, 161; H. Busse 1968: 256; Herklots 1969: 129; Eck and Busse 1973: 111, 112; Hilty and Brown 1986: 228; Boyer and Hume 1991: 89; del Hoyo et al. 1999: 206, 207; König et al. 1999: 318–320; Ridgely and Greenfield 2001: I: 310, 311, II: 218, 219; Hilty 2003: 364

- ***Pulsatrix perspicillata boliviana*** L. Kelso 1933
Pulsatrix perspicillata boliviana L. Kelso 1933, Biol. Leaflet 2: 1; Terra typica: Carapari, 1 000 m, Bolivia

Distribution: Southern Bolivia and northern Argentina

Museum: ZFMK

Remarks: In plumage nearest to *P. p. chapmani*, but with much longer flank feathers (150 mm)

Wing length:	>335 mm
Tail length:	?
Tarsus length:	?
Length of bill (cere):	28 mm
Body mass:	?

Illustration: ?

Photograph: ?

Literature: Kelso 1933: 1; Kelso 1934a: 234–236; Kelso 1934d: 44; del Hoyo et al. 1999: 206, 207; König et al. 1999: 318–320

❖

Pulsatrix pulsatrix (Wied) 1820
Brown (or Short-browed) Spectacled Owl · Brauner (oder Kurzbrauen-) Brillenkauz
***Strix pulsatrix* zu Wied-Neuwied 1820, Reise Bras. 1: 366;**
Terra typica: **Rio Grande do Belmonte, Bahia**

Length:	510–525 mm
Body mass:	1 050–1 250 g

Distribution: Eastern Brazil (Bahia south to Rio Grande do Sul), Paraguay, northern Argentina (Misiones)

Habitat: Semi-open primary and secondary forest, especially Araucaria forest, clearings, edges. Sometimes near human settlements. This is Wied's Knocking Owl!

Museum: AMNH, FNSF, MNH, SMNSt, SMTD

Remarks: Distinct from *P. perspicillata* in size, plumage and vocalisations

Wing length:	363–384 mm
Tail length:	211–226 mm
Tarsus length:	59.5 mm
Length of bill:	?
Body mass:	1 050–1 250 g

Illustration: ? in A. du Bois, Synopsis Generum Avium, reprint in Kelso 1934d: 18; F. Weick in König et al. 1999: Pl. 37

Photograph: P. Moore in internet: Owl Pages, photo gallery 2002 (2 photographs of capt. bird)

Literature: Kelso 1934a: 234–236; Kelso 1934d: 18, 19 and 43–45; Eck and Busse 1973: 111, 112; Belton 1984: 178 (4); del Hoyo et al. 1999: 206, 207; König et al. 1999: 320, 321

❖

Pulsatrix koeniswaldiana **Bertoni and Bertoni 1901**

Tawny-browed Owl · Gelbbrauenkauz · Chouette de sourcils jaune · Urucurcá Chico

Syrnium Keniswaldianum **M. and W. Bertoni 1901, An. Cient. Paraguayos 1(1): 175; Terra typica: Probably from Puerto Bertoni, on the Alto Paraná, Paraguay**

(see Plate 6)

Length: 440 mm
Body mass: 480 g

Synonym:

- *Pulsatrix sharpei* von Berlepsch 1901, Bull. Brit. Ornith. Cl. 12: 6 (10 months later!); Terra typica: Stato of Espirito Santo, Brazil

Distribution: Eastern Paraguay, extreme northern Argentina (Misiones) and southern Brazil (Espirito Santo south to Santa Catarina)

Habitat: Strictly nocturnal. Humid tropical and subtropical woodland, often mixed with Araucaria in montane regions. Also in degraded and marginal forest. From sea-level up to 1 500 m

Museum: AMNH, BMNH, SMNSt, SMTD, ZSBS

Wing length: 300–320 mm
Tail length: 172 mm
Tarsus length: ?
Length of bill: ?
Body mass: 480 g

Illustration: D. Eckleberry in *Living Bird* 1965: 149 (from alive); H. Quintscher in Eck and Busse 1973: Pl. 3 (irides too light); S. Frisch in Frisch 1981: 121; T. Boyer in Boyer and Hume 1991: 89 (with yellow irides); T. Worfolk in del Hoyo et al. 1999: Pl. 13; F. Weick in König et al. 1999: Pl. 37

Photograph: J. Tobias in del Hoyo et al. 1999: 150 (Brazil)

Literature: Kelso 1934a: 233–236; Kelso 1934d: 45; Eck and Busse 1973: 112, 113; Boyer and Hume 1991: 89; del Hoyo et al. 1999: 207; König et al. 1999: 321, 322

Pulsatrix melanota **(Tschudi) 1844**

Band-bellied (or Rusty-barred) Owl · Bindenkauz · Chouette à collier · Lechuzón Barrado

Noctua melanota **von Tschudi 1844, Arch. Naturgesch. 10(1): 266; Terra typica: Peru**

Length: 440–480 mm
Body mass: ?

Synonym:

- *Pulsatrix fasciaventris* Salvadori and Festa 1900, Boll. Mus. Zool. Anat. Comp. Torino 15(368): 32; Terra typica: Valle del Zamora, Ecuador
- *Pulsatrix melanota philoscia* Todd 1947, Proc. Biol. Soc. Wash. 60: 95; Terra typica: San José, Yungas de Cochabamba, Bolivia (holotype CMP)

Distribution: Patchy distribution: southeast Colombia, eastern Ecuador, north and southeast Peru (east of Andes) and west and central Bolivia

Habitat: Humid tropical rainforest, locally more open woodland. From lowlands up to about 1 600 m, mainly upwards of 700 m above sea-level

Museum: BMNH, CMP, FNSF, MHNP, MNBHU

Wing length: 275–307 mm (1 × 325 mm)
Tail length: 163–185 (192) mm
Tarsus length: (47) 50–55 mm
Length of bill (cere): 22–26.5 mm
Body mass: ?

Illustration: H. Quintscher in Eck and Busse 1973: Pl. 3 (light irides); T. Boyer in Boyer and Hume 1991: 89 (yellow irides); T. Worfolk in del Hoyo et al. 1999: Pl. 13; F. Weick in König et al. 1999: Pl. 37; P. J. Greenfield in Ridgely and Greenfield 2001: II: Pl. 36 (5)

Photograph: ?

Literature: Sharpe 1975: 280; Salvadori and Festa 1900: 32; Kelso 1934a: 233–236; Kelso 1934d: 45; Todd 1947: 95; Eck and Busse 1973: 112, 113; Hilty and Brown 1986: 228; Boyer and Hume 1991: 89; del Hoyo et al. 1999: 207; König et al. 1999: 322; Ridgely and Greenfield 2001: I: 311, II: 219

Subfamilia / Subfamily
Surniinae · Hawk-Owls · Falkeneulen

Tribus / Tribe
Surniini

Genus
Surnia **Duméril 1806**
Surnia Duméril 1806, Zool. Anal.: 34. Type by *Strix funerea* Gmelin = *Strix Ulula* Linné 1758 (Gray, List Gen. Bds. 1840: 5)

Surnia ulula **(Linné) 1758**
Northern Hawk Owl · Sperbereule · Chouette épervière · Cárabo Gavilán
(see Plate 8)

Length: 360–410 mm
Body mass: 220–390 g

Distribution: Boreal zones of Eurasia. From Norway, Sweden and Finland, east through Siberia to Kamchatka, Sakhalin and northern China. In central Asia to Tien Shan. Boreal North America, from Alaska east to Labrador, south to extreme northern United States

Habitat: Rather open boreal coniferous forest with clearings and moors in lowlands and mountains. Winters in open heathlands and prairies in the northern United States. From sea-level up to 1 500 m, in Tian Shan to about 3 000 m

- ***Surnia ulula ulula*** (Linné) 1758
  Strix Ulula von Linné 1758, Syst. Nat. ed. 10, 1: 93;
  Terra typica: Europe, restricted to Sweden

Synonym:
- *Surnia ulula pallasi* Buturlin 1907, Orn. Monatsb. 15: 100; Terra typica: Siberia
- *Surnia ulula orokensis* Stachanov 1931, Koscag 4: 21; Terra typica: Sakhalin Island

Distribution: Northern Eurasia, from Norway, Sweden and Finland, east to Sakhalin, central Siberia and south to Tarbagatay

Museum: BMNH, NNML, SMTD, ZMA, ZMH, ZFMK

Wing length: 220–249 mm
Tail length: 164–191 mm
Tarsus length: 24–27 mm
Length of bill: 23–26.5 mm
Body mass: ♂ (215) 247–375 g, ♀ 323–380 g

Illustration: J. F. Naumann in Naumann 1822: Pl. 42; H. C. Richter in Gould 1832–1837: (repr) unpaged; J. G. Keulemans in Dresser 1871–1896: Pl. 311; J. G. Keulemans in Hennicke 1905: (repr) unpaged; P. Barruel in Sutter and Barruel 1958: Pl. 21; O. Kleinschmidt in Kleinschmidt 1958: Pl. 36; G. Pettersson in Pettersson 1984: 40, 42; H. Delin in Cramp et al. 1985: Pl. 47 (adult and juvenile); A. Cameron in Voous 1988: 134; A. Thorburn in Thorburn 1990: Pl. 27; T. Boyer in Boyer and Hume 1991: 94; C. Byers in del Hoyo et al. 1999: Pl. 14; F. Weick in König et al. 1999: Pl. 63 (adult and juvenile); Mullarney and Zetterström in Svensson et al. 1999: 211; F. Weick in Weick 2004: März

Photograph: H. Busse in *Der Falke* 1966(13): 251; P. Helo in Mikkola 1983: Fig. 33, 34; F. Sauer in Mebs 1987: 104; P. Helo in Delin and Svensson 1988: 165; numerous photographs in Mebs and Scherzinger 2000: 71, 357–371 (by Hautala, Scherzinger, Siegel, Wothe, Zeininger, etc.); A. V. Krechmar in Krechmar 2005

Literature: Naumann 1822: 427–433, Nachtr. 173, 174; Sharpe 1875a: 129, 130; Buturlin 1907c: 100; Hartert 1912–1921: 1010–1012; Stachanow 1931: 21; Dementiev and Gladkov 1951: 451–454; Sutter and Barruel 1958: 48; Kleinschmidt 1958: 36; Vaurie 1965: 616; Busse 1966, *Der Falke* 1966(13): 250, 251; Eck and Busse 1973: 115, 116; Etchécopar and Hüe 1978: 485; Glutz von Blotzheim and Bauer 1980: 453–463; Bezzel 1985: 644, 645; Mikkola 1983: 105–112; Meyer de Schauensee 1984: 268, 269; Cramp et al. 1985: 496–505; Mebs 1987: 100–104; Boyer and Hume 1991: 93; del Hoyo et al. 1999: 209; König et al. 1999: 421, 422; Svensson et al. 1999: 210; Mebs and Scherzinger 2000: 357–373; Krechmar 2005: 330–337

- ***Surnia ulula tianschanica*** Smallbones 1906
  Surnia ulula tianschanica Smallbones 1906, Orn. Monatsb. 14: 27; Terra typica: Tian Shan

Synonym:
- *Surnia ulula korejewi* Zarudny and Loudon 1907, Orn. Monatsb. 15: 2; Terra typica: Tian Shan

Distribution: Central Asia and northwest and northeast China, possibly northern Mongolia
Museum: AMNH, BMNH, ZFMK
Remarks: Doubtful if distinct from nominate *S. u. ulula*

Wing length:	238–258 mm
Tail length:	185–204 mm
Tarsus length:	?
Length of bill:	?
Body mass:	?

Illustration: F. Berille in Etchécopar and Hüe 1978: Pl. 18
Photograph: ?
Literature: Hartert 1912–1921: 1012; Dementiev and Gladkov 1951: 451–455; Vaurie 1965: 617; Etchécopar and Hüe 1978: 483; Meyer de Schauensee 1984: 268, 269; Cramp et al. 1985: 496–505; del Hoyo et al. 1999: 209; König et al. 1999: 421, 422

- ***Surnia ulula caparoch*** (P.L.S. Müller) 1776
  Strix caparoch P.L.S. Müller 1776, Natursyst. Suppl.: 69; Terra typica: Europe, error = Hudson Bay, ex Edwards

Distribution: Alaska through Canada to Newfoundland and Labrador, south to extreme northern United States
Museum: BMNH, FNSF, SMNKa, SMTD, USNM

Wing length:	♂ 218–235 mm, ♀ 223–251 mm
Tail length:	160–191 mm
Tarsus length:	?
Length of bill (cere):	17–20.5 mm
Body mass:	♂ 273–326 g, ♀ 306–392 g

Illustration: Edwards in Müller 1776: Suppl. Pl. 62; J. J. Audubon in Audubon 1827–1838: Pl. 238; J. G. Keulemans in Dresser 1871–1896: Pl. 312; Hartert 1912–1921: 1012 (flank feathers); R. Ridgway in Ridgway 1914: Pl. 36; L. A. Fuertes in Gilbert Pearson 1936: Pl. 55; A. Brooks in Grosvenor and Wetmore 1937: 20; L. A. Fuertes in Forbush and May 1955: Pl. 47; F. Landsdowne in Landsdowne and Livingston 1967: 18; K. E. Karalus in Karalus and Eckert 1974: Pl. 41; A. Wilson in Wilson 1975 (repr.): 50/2 (wood engraving); L. Malick in Breeden et al. 1983: 247; C. Byers in del Hoyo et al. 1999: Pl. 14; F. Weick in König et al. 1999: Pl. 62; D. Sibley in Sibley 2000: 283 (adult and juvenile)
Photograph: R. H. Rauch in Bent 1961 (repr.): Pl. 85; K. Christoffersen in Bent 1961 (repr.): Pl. 85; K. Kussmann in Eck and Busse 1973: Abb. 13; H. H. Valega in Johnsgard 2002: Pl. 19; J. Santana in internet: Owl Pages, photo gallery 2003; J. Cook in internet: Owl Pages, photo gallery 2003
Literature: Sharpe 1875a: 131; Hartert 1912–1921: 1013; Ridgway 1914: 772–779; Grosvenor and Wetmore 1937: 21; Sprunt 1955: 193–195; Bent 1961 (repr.): 375–384; Landsdowne and Livingston 1967: 18; Eck and Busse 1973: 115, 116; Karalus and Eckert 1974: 195–203; Breeden et al. 1983: 246; Cramp et al. 1985: 496–505; del Hoyo et al. 1999: 209; König et al. 1999: 421, 422; Sibley 2000: 283; Johnsgard 2002: 130–136

Surnia ulula caparoch · Northern Hawk Owl · Sperbereule

Genus
Glaucidium **Boie 1826**
Glaucidium **Boie 1826, Isis v. Oken 2, col. 970.**
Type by *Strix passerina* Linné (Gray, List Gen. Bds. 1840: 6)

Synonym:
- *Phalaenopsis* Bonaparte 1854, Rev. Mag. Zool. 2: 544. Type by *Glaucidium nanum*

Glaucidium passerinum **(Linné) 1758**
Eurasian Pygmy Owl · Sperlingskauz · Chouette chevêchette · Mochuelo Chico (Alpino)
(see Plate 7)

Length: 160–190 mm
Body mass: 50–80 g

Distribution: Central and northern Europe, east to eastern Siberia, Sakhalin and northern China
Habitat: Coniferous forest of the boreal zone, also mixed forest (taiga and montane). Generally above 200 m, up to 700 m in the Black Forest and to 1 000 m in the Alps. Needs clearings, moors, meadows or avalanche pathways

- ***Glaucidium passerinum passerinum*** (Linné) 1758
  Strix passerina von Linné 1758, Syst. Nat. ed. 10, 1: 93;
  Terra typica: Europe, restricted to Sweden

Synonym:
- *Strix torquata* Fischer 1812, Mem. Soc. Imp. Nat. Moscou 3: 276; Terra typica: Vicinity of Moscow
- *Glaucidium setipes* von Madarász 1900, Magyar Madarai: 203; Terra typica: Hungary

Distribution: From Scandinavia and mountains of southern central and eastern Europe, east across northwest and central Russia and Siberia to Sakhalin and northeast China
Museum: BMNH, FNSF, NNML, SMTD, ÜMB, ZMA, ZMH, ZMM

Wing length: ♂ 93–100 mm, ♀ 101–109 mm
Tail length: ♂ 53–60 mm, ♀ 58–65 mm
Tarsus length: 16–18.5 mm
Length of bill: 13.5–16 mm
Body mass: ♂ 47–72 g, ♀ 67–83 g

Illustration: Krook 1988 (repr.) O. Rudbeck (1693–1710): 90; J. F. Naumann 1822, vol I: Pl. 43; J. G. Keulemans in Hennicke 1905: (repr.), unpaged; P. A. Robert in Géroudet 1979: 359; O. Kleinschmidt in Kleinschmidt 1958: Pl. 35; P. Barruel in Sutter nad Barruel 1958: 53; L. Binder in Wüst 1970: Abb. 127; D. M. Reid-Henry in Fitter et al. 1973: 163; G. Pettersson in Pettersson 1984: 45; H. Delin in Cramp et al. 1985: Pl. 47 (adult and juvenile); A. Cameron in Voous 1988: 142; T. Boyer in Boyer and Hume 1991: 97; L. Jonsson in Jonsson 1992: 322, 323; R. Reboussin in Jeanson 1999: Pl. 48; C. Byers in del Hoyo et al. 1999: Pl. 14; F. Weick in König et al. 1999: Pl. 45 (adult and juvenile); Mullarney andZetterström in Svensson et al. 1999: 211; F. Weick in Weick 2004: September; F. Weick in Kröher and Weick 2004: 186
Photograph: Feuerstein in *Der Falke* 1982(10): Frontcover; H. Hautala in Mikkola 1983: 36, 37; E. Jussila in Mikkola 1983: 37; H. Reinhard in Mebs 1987: 69; P. Zeininger in Mebs 1987: 71; H. Limbrunner and F. Sauer in Mebs 1987: 73; K. Menning in *Gefiederte Welt* 1989(5): 145, 146; B. Volmer in *Der Falke* 1999(10): 298; numerous fine colour photographs in Mebs and Scherzinger

Glaucidium passerinum · Eurasian Pygmy Owl · Sperlingskauz

2000: 19, 34, 40–44, 65, 334–354 (by Brosette, Essler, Giel, Hortig, Partreh, Schalter, Scherzinger, Schmidt, Stengel and Wiesner)

Literature: Naumann 1822: 434–439, Nachtr. 174; Sharpe 1875a: 191, 193; Hartert 1912–1921: 1007–1009; Dementiev and Gladkov 1951: 447–450; Kleinschmidt 1958: 35; Vaurie 1965: 614; Eck and Busse 1973: 119, 120; Schönn 1978: 123 pp; Glutz von Blotzheim and Bauer 1980: 464–501; Mikkola 1983: 113–123; Cramp et al. 1985: 505–513; Bezzel 1985: 646–648; Mebs 1987: 68–73; Voous 1988: 138–145; Boyer and Hume 1991: 96–98; Jonsson 1992: 322; del Hoyo et al. 1999: 209, 210; König et al. 1999: 348-351; Svensson et al. 1999: 210; Mebs and Scherzinger 2000: 334–356

- ***Glaucidium passerinum orientale*** Taczanowski 1891
  Glaucidium passerinum orientale Taczanowski 1891, Mem. Acad. Imp. Sci. St. Petersb. 7, Sci. Math. Phys. Nat. 39: 128; **Terra typica:** Eastern Siberia

Distribution: Central and eastern Siberia: north to about 60° N latitude, region of Olekminsk, Okhotsk (coast), south to Sayans, Transbaikalia to Mongolia, central Manchuria, Ussuriland and Sakhalin

Museum: MNBHU, ZMM

Wing length:	♂ 94–106 mm, ♀ 103–108 (110) mm
Tail length:	?
Tarsus length:	?
Length of bill:	?
Body mass:	♂ 69 g ($n = 1$)

Illustration: F. Weick in König 1999a: Pl. 45

Photograph: ?

Literature: Hartert 1912–1921: 1009; Dementiev and Gladkov 1951: 448, 450; Vaurie 1965: 614, 615; Etchécopar and Hüe 1978: 482, 483; Meyer de Schauensee 1984: 269; König et al. 1999: 349–351

Glaucidium perlatum (Vieillot) 1818

Pearl Spotted Owl · Perlkauz · Chevêchette perlée · Mochuelo Perlado
(see Plate 7)

Length:	170–200 mm
Body mass:	60–140 g

Distribution: Africa, south of Sahara: from Senegambia to Ethiopia, western Somalia and western Sudan. South to northern and southern South Africa. In the west through Namibia to northwest Angola. Absent from deserts and dense rainforest in western and central Africa

Habitat: Open savanna (bushveld) and open woodland. Dry, semi-open woodland and open or semi-open riverine forest with adjacent savanna

- ***Glaucidium perlatum perlatum*** (Vieillot) 1818
  Strix perlata Vieillot 1818, Nouv. Dict. Hist. Nat. 7: 26; **Terra typica:** Senegal

Synonym:
- *Glaucidium albiventer* Alexander 1901, Bull. Brit. Ornith. Cl. 12: 10; **Terra typica:** Kwobia, Gold Coast (leucozistic bird)

Distribution: Senegambia to western Sudan, possibly Liberia

Museum: BMNH, SMNKa, SMNSt, ÜMB, ZFMK, ZMH, ZSBS

Wing length:	♂ 105.5–113 mm, ♀ 107–118 mm
Tail length:	64–81 mm
Tarsus length:	22–26 mm
Length of bill:	15–16 mm
Body mass:	♂ 61–86 g, ♀ 77–147 g

Illustration: H. Grönvold in *Ibis* 1902: Pl. 9 (*G. albiventer*); J. W. Frohawk in Bannerman 1953: 539 (b/w); T. Boyer in Boyer and Hume 1991: 102; T. Disley in Barlow et al. 1997: Pl. 25; P. Hayman in Kemp and Kemp 1998: 299; C. Byers in del Hoyo et al. 1999: Pl. 14; F. Weick in König et al. 1999: Pl. 45 (adult and juvenile); N. Borrow in Borrow and Demey 2003: Pl. 59

Photograph: E. Hosking in Hosking and Flegg 1982: 76; W. Scherzinger 1986b: 305, 306; M. Goetz in Kemp and Kemp 1998: 298; C. Paterson-Jones in Kemp and Kemp 1998: 299

Literature: Sharpe 1875a: 209–211; ? 1902, *Ibis:* Birds of Gold Coast Colony: 371; Bannerman 1953: 538, 539; Scherzinger 1986b: 305, 306; Fry et al. 1988: 139, 140; Boyer and Hume 1991: 102; Kemp and Kemp 1998: 298, 299; del Hoyo et al. 1999. 210, 211; König et al. 1999: 351–353; Borrow and Demey 2003: 497; Sinclair and Ryan 2003: 244

- ***Glaucidium perlatum licua*** (Lichtenstein) 1842
Strix licua Lichtenstein 1842, Verz. Säuget. and Vögel d. Kaffernlandes: 12; Terra typica: Liqua River, Northeast Capeland = Cape Province

Synonym:
- *Glaucidium kilimense* Reichenow 1893, Orn. Monatsb. 1: 178; Terra typica: Kilimanjaro, East Africa
- *Glaucidium perlatum diurnum* Clancey 1968, Durban Mus. Novit. 8(11): 119; Terra typica: Devuli, Birchenough Bridge, Sabi Valley, Southeast Tanzania

Distribution: Eastern Sudan, Ethiopia, Uganda. South to northern and eastern South Africa and Angola
Museum: SMNKa, SMNSt, ZFMK

Wing length: ♂ 100–109 mm, ♀ 103–111 mm
Tail length: 66–82 mm
Tarsus length: 21–26 mm
Length of bill: 14–18 mm
Body mass: ♂ 36–86 g, ♀ 61–99 g

Illustration: G. Arnott in Steyn 1982: Pl. 23; M. Woodcock in Fry et al. 1988: Pl. 9; D. Zimmerman in Zimmerman et al. 1996: Pl. 56; C. Byers in del Hoyo et al. 1999: Pl. 14; F. Weick in König et al. 1999: Pl. 45
Photograph: C. König in König and Ertel 1979: 145; E. Hosking in Hosking and Flegg 1982: 76; P. Steyn in Steyn 1982: 261, 263 (b/w); Mc Illeron in Ginn et al. 1989: 328; E. Hosking in Burton et al. 1992: 176; N. Dennis in Kemp and Kemp 1998: 298; S. Porter in internet: Owl Pages, photo gallery 2003
Literature: Sharpe 1875a: 209–211; Clancey 1968: 119; Steyn 1982: 261–264; Fry et al. 1988: 139, 140; Ginn et al. 1989: 338; Zimmerman et al. 1996: 445; Kemp and Kemp 1998: 298, 299; del Hoyo et al. 1999: 210, 211; König et al. 1999: 351–353

Glaucidium californicum P.L. Sclater 1857
Northern Pygmy Owl · Rocky Mountains-Sperlingskauz · Chevêchette de Rocheuses · Mochuelo Californiano (Norteamericano)
(see Plate 7)

Length: 170–190 mm
Body mass: 62–73 g

Distribution: Western North America, from southeast Alaska and British Columbia south to the southwestern United States, northwest Mexico. Vancouver Island
Habitat: Coniferous and mixed forest in mountainous regions, from 1 200 m up to around 2 200 m

- ***Glaucidium californicum grinnelli*** Ridgway 1914
Glaucidium gnoma grinnelli Ridgway 1914, US Nat. Mus. Bull. 50(6): XVI and 781; Terra typica: Humboldt Bay, California

Distribution: Southeast Alaska, through coastal British Columbia, south to coastal western United States (Washington, Oregon, California)
Museum: USNM (holotype), SMTD

Wing length: 88.5–100.5 mm
Tail length: 61–72.5 mm
Tarsus length: 21 mm
Length of bill (cere): 10–11.5 mm
Body mass: ?

Illustration: K. E. Karalus in Karalus and Eckert 1974: Pl. 45; A. Cameron in Voous 1988: 147; D. Sibley in Sibley 2000: 282
Photograph: S. Roberts in del Hoyo et al. 1999: 116; J. Hobbs in internet: Owl Pages, photo gallery 2004; D. Metz in internet: Owl Pages, photo gallery 2004
Literature: Ridgway 1914: 791–793; Kelso 1934d: 63; Bent 1961 (repr.): 430; Karalus and Eckert 1974: 213, 214; Voous 1988: 146–151; del Hoyo et al. 1999: 211; König et al. 1999: 354, 355; Sibley 2000: 282; Johnsgard 2002: 137–144

- ***Glaucidium californicum swarthi*** Grinnell 1913
Glaucidium gnoma swarthi Grinnell 1913, Auk 30: 224; Terra typica: Errington, Vancouver Island

Distribution: Vancouver Island, British Columbia
Museum: MVZ (holotype)

Wing length: ♂ 86.5–95.5 mm, ♀ 92–96 mm
Tail length: ♂ 60–66 mm, ♀ 65–67.5 mm
Tarsus length: 18–20 mm
Length of bill (cere): 10.5–11.5 mm
Body mass: ?

Illustration: K. E. Karalus in Karalus and Eckert 1974: Pl. 46; F. Weick in König et al. 1999: Pl. 46
Photograph: A. G. Nelson in Johnsgard 2002: Pl. 23?

Literature: Grinnell 1913: 224; Ridgway 1914: 781, 793; Bent 1961 (repr.): 428; Karalus and Eckert 1974: 214, 215; del Hoyo et al. 1999: 211; König et al. 1999: 354, 355

Museum: BMNH, USNM (holotype)
Remarks: Possibly not a valid subspecies; could unite with nominate *californicum*

Wing length: ♂ 94–96.5 mm, ♀ 98–105 mm
Tail length: ♂ 62.5–68 mm, ♀ 66–78.5 mm
Tarsus length: 23 mm
Length of bill (cere): 10.5–12 mm
Body mass: ?

Illustration: K. E. Karalus in Karalus and Eckert 1974: Pl. 42; F. Weick in König et al. 1999: Pl. 46
Photograph: ?
Literature: Ridgway 1914: 781, 789, 790; Bent 1961 (repr.): 401; Karalus and Eckert 1974: 204–210; Voous 1988: 146–151; König et al. 1999: 354, 355

- ***Glaucidium californicum californicum*** Sclater 1857
  Glaucidium californianum Sclater 1857, Proc. Zool. Soc. Lond.: 4; Terra typica: Oregon and California

Synonym:
- *Glaucidium gnoma vigilante* Grinnell 1913, Auk 30: 224; Terra typica: Foothills at 2 250 feet, 4 miles north of Pasadena (holotype ANSP)

Distribution: British Columbia and Alberta to western United States (south to Nevada and California) and northwest Mexico: north to Sonora and northwest to Chihuahua
Museum: BMNH, SMNSt, SMTD, ZFMK

Wing length: ♂ 89.5–97 mm, ♀ 92.5–102 mm
Tail length: ♂ 61–68.5 mm, ♀ 63.5–72.5 mm
Tarsus length: 20.5 and 21 mm
Length of bill (cere): 10–12 cm
Body mass: ♂ 54–80 g, ♀ 64–87 g

Illustration: J. J. Audubon in Audubon 1827–1838: Pl. 44; J. G. Keulemans in *Ibis* 1875: Pl. 1; A. Brooks in Grosvenor and Wetmore 1937: 32; K. E. Karalus in Karalus and Eckert 1974: Pl. 44; L. Malick in Breeden et al. 1983: 245; T. Boyer in Boyer and Hume 1991: 98; C. Byers in del Hoyo et al. 1999: Pl. 14; F. Weick in König et al. 1999: Pl. 46; D. Sibley in Sibley 2000: 282
Photograph: W. C. Shuster in Johnsgard 2002: Pl. 20; G. Lasley in internet: Owl Pages, photo gallery 2002
Literature: Sharpe 1875a: 194–196; Sharpe 1875b: 35–59; Grinnell 1913: 222–224; Ridgway 1914: 781, 790, 791; Grosvenor and Wetmore 1937: 23; Sprunt 1955: 195, 196; Bent 1961 (repr.): 410; Karalus and Eckert 1974: 210, 211; Breeden et al. 1983: 244; Voous 1988: 146–151; Boyer and Hume 1991: 98, 99; del Hoyo et al. 1999: 211; König et al. 1999: 354, 355; Sibley 2000: 282; Johnsgard 2002: 137–144

- ***Glaucidium californicum pinicola*** Nelson 1910
  Glaucidium gnoma pinicola Nelson 1910, Proc. Biol. Soc. Wash. 23: 103; Terra typica: Alma, New Mexico

Distribution: Western Unites States: Idaho, Montana, south to Arizona and New Mexico, east to Colorado

Glaucidium hoskinsii Brewster 1888
Baja or Cape Pygmy Owl · Hoskins-Sperlingskauz · Chevêchette de Hoskins · Tecolotito de Hoskins
***Glaucidium gnoma hoskinsii* Brewster 1888, Auk 5: 136;**
Terra typica: Sierra de la Laguna, Baja California, Mexico

Length: 150–170 mm
Body mass: 50–65 g

Distribution: Baja California, Mexico, from Sierra de la Gigante down to the Cape
Habitat: Pine and pine/oak forest, from 1 500–2 100 m. In winter at lower altitudes in deciduous woodland
Museum: BMNH, MCZ (holotype, Brewster collection)
Remarks: Not closely related to *G. californicum* (DNA evidence)

Wing length: 86–89 mm
Tail length: 61–65.5 mm
Tarsus length: ?
Length of bill (cere): 9.5–11 mm
Body mass: 50–65 g

Illustration: S. Webb in Howell and Webb 1995: Pl. 25; C. Byers in del Hoyo et al. 1999: Pl. 14; F. Weick in König et al. 1999: Pl. 46
Photograph: ?
Literature: Brewster 1888: 136; Ridgway 1914: 781, 788, 789; Bent 1961 (repr.): 434; Blake 1963: 215, 216; Howell and Webb 1995: 261; del Hoyo et al. 1999: 212; König et al. 1999: 355, 356; Johnsgard 2002: 137–146

Glaucidium gnoma Wagler 1832
Mountain Pygmy Owl · Gnomen-(Sperlings)kauz · Chevêchette cabouré (ou des montagnes) · Tecolotito serrano (Mochuelo Gnomo)

Length: 150–170 mm
Body mass: 54–73 g

Distribution: Southernmost United States (southeast Arizona), through interior highlands of Mexico (Chihuahua and Coahuila, Nuevo León and Tamaulipas) south to Guatemala, Costa Rica, Honduras and Panama
Habitat: Pine/oak, pine and humid pine/evergreen forest. From ca. 900 m (*costaricanum*), 1 500 m up to 3 000 m (*gnoma*) and about 400 m to 2 600 m (*cobanense*)
Remarks: Taxa *cobanense* and *costaricanum* are probably separate species, but vocalisations and DNA evidence not available

- ***Glaucidium gnoma gnoma*** Wagler 1832
Glaucidium Gnoma Wagler 1832, Isis v. Oken, col.: 275;
Terra typica: Mexico

Synonym:
- *Glaucidium fisheri* Nelson and Palmer 1894, Auk 11: 41; Terra typica: Nueva León and Tamaulipas, south to Guerrero, Mexico and Puebla

Distribution: Southeast Arizona, south through interior highlands of Mexico (Chihuahua, Coahuila south to Oaxaca) and central Honduras
Museum: BMNH, FMNH, LSUMZ, MVZ, SMNSt, USNM (holotype)

Wing length: ♂ 82–89 mm, ♀ 87–98 mm
Tail length: 57–63.5 mm
Tarsus length: ?
Length of bill (cere): 9.5–10.5 mm
Body mass: ♂ 48–54 g, ♀ 59.5–73 g

Illustration: K. E. Karalus in Karalus and Eckert 1974: Pl. 44; S. Webb in Howell and Webb 1995: Pl. 25(2a); C. Byers in del Hoyo et al. 1999: Pl. 14; F. Weick in König et al. 1999: Pl. 46 (adult and juvenile)
Photograph: ?
Literature: Sharpe 1875b: 35–59; Ridgway 1914: 781, 785–788; Blake 1963: 315, 316; Karalus and Eckert 1974: 211, 212; Voous 1988: 146–151; Howell and Webb 1995: 360; del Hoyo et al. 1999: 211; König et al. 1999: 256, 257; Johnsgard 2002: 137–146

- ***Glaucidium (gnoma) cobanense*** Sharpe 1875
Glaucidium cobanense Sharpe 1875, Ibis: 259, 260;
Terra typica: Coban, Alta Vera Paz, Guatemala

Distribution: Southernmost Mexico to Guatemala and Honduras
Museum: BMNH, MHNP (holotype), ZMH

Wing length: 82–98 mm
Tail length: 65 mm
Tarsus length: 19 mm
Length of bill (cere): ?
Body mass: ?

Illustration: J. G. Keulemans in Sharpe 1875b: Pl. 13; S. Webb in Howell and Webb 1995: Pl. 25(2b); C. Byers in del Hoyo et al. 1999: Pl. 14; F. Weick in König et al. 1999: Pl. 46
Photograph: ?
Literature: Sharpe 1875a: 199, 200; Ridgway 1914: 787, 788; Kelso 1934d: 64; Land 1970: 138, 139; Howell and Webb 1995: 360; del Hoyo et al. 1999: 211, 212; König et al. 1999: 356, 357

- ***Glaucidium (gnoma) costaricanum*** L. Kelso 1937
Glaucidium jardinii costaricanum L. Kelso 1937, Auk 54: 304;
Terra typica: Costa Rica
(see Figure 14)

Distribution: Central Costa Rica to western (and eastern?) Panama
Museum: AMNH, BMNH, LSUMZ, UCR, USNM

Wing length: ♂ 89.5–93.5 mm, ♀ 96–99 mm
Tail length: 51.5–57.5 mm
Tarsus length: ?
Length of bill (cere): ?
Body mass: ♂ 53–70 g, ♀ 99 g

Illustration: D. Gardner in Stiles and Skutch 1991: Pl. 20 (*jardinii*); C. Byers in del Hoyo et al. 1999: Pl. 15; F. Weick in König et al. 1999: Pl. 46; F. Weick in *Gefiederte Welt* 2005(5): 20
Photograph: ?
Literature: Kelso 1937: 304; Voous 1988: 146–151; Ridgely and Gwynne 1989: 189; Stiles and Skutch 1991: 193; Robbins and Stiles 1999: 305–315; del Hoyo et al. 1999: 215; König et al. 1999: 356, 357; Weick 2003–2005, *Gefiederte Welt* 2005(1): 20

Glaucidium nubicola Robbins and Stiles 1999
Cloudforest Pygmy Owl · Nebelwald-Sperlingskauz · Chevêchette des nuages · Tecolotito Ecuatoriano
Glaucidium nubicola Robbins and Stiles 1999, Auk 116(2): 305–315; Terra typica: Provincia Carchi, Ecuador
(see Figure 14)

Length: 160 mm
Body mass: 56–65 g

Distribution: Western slope of Andes in Colombia (Cordillera Central) and Ecuador (probably northernmost Peru?)
Habitat: Wet primary cloudforest, on steep slopes, at elevations from 1 400 m to 2 000 m above sea-level
Museum: ANSP (holotype), AMNH, CMC, ICN, LSUMZ, MVZ, SMNSt, UCR

Wing length: ♂ 90–95 mm, ♀ 96 mm
Tail length: 44.5–50 mm
Tarsus length: ?
Length of bill (cere): 11.3 mm
Body mass: 72.5–80 g

Illustration: Pedersen in Robbins and Stiles 1999: Frontcover (Auk); C. Byers in del Hoyo et al. 1999: Pl. 15; F. Weick in *Gefiederte Welt* 2005(1): 20; F. Weick in unpublished (new edn. Owls): Pl. 65; P. J. Greenfield in Ridgely and Greenfield 2001: II: Pl. 35(8)
Photograph: G. Stiles in Robbins and Stiles 1999: 306; G. Stiles in del Hoyo et al. 1999: 85; S. Blain in internet: Owl Pages, photo gallery 2005
Literature: Robbins and Stiles 1999: 305–315; del Hoyo et al. 1999: 215; König et al. 1999: 357; Weick 2003–2005, *Gefiederte Welt* 2005(1): 20, 21; Ridgely and Greenfield 2001: I: 307, II: 216

❖

Glaucidium jardinii (Bonaparte) 1855
Andean Pygmy Owl · Anden-Sperlingskauz · Chevêchette des Andes · Tecolotito (Mochuelo) Andino
Phalaenopsis jardinii Bonaparte 1855, Compt. Rend. Acad. Sci. Paris 41: 654; Terra typica: Andes of Quito, Ecuador
(see Figure 14)

Length: 145–160 mm
Body mass: 56–74.8 g

Remarks: Formerly considered to include *G. costaricanum*, but the latter very distinct monotypic or subspecies of *G. gnoma*
Distribution: From northern Colombia and western Venezuela south through Ecuador to central Peru. Entirely absent from the western slopes of the Andes
Habitat: Semi-open montane and cloud forest, patches of *Polylepsis* woodland and elfin forest near the Páramos. Wooded ravines with transitions of swampy or grassy habitat. From elevations about 2 000 m up to the treeline (3 500 m), sometimes to 4 000 m
Museum: AMNH, ANSP, BMNH, FMNH, LSUMZ, MHNP, SMNSt, SMTD, ÜMB, ZFMK, ZSBS

Wing length: 94.5–101 mm
Tail length: ♂ 55–62 mm, ♀ 59.5–64.5 mm
Tarsus length: 18–21.5 mm
Length of bill (cere): ?
Body mass: ♂ 56–63 g, ♀ 64.5–74.8 g

Illustration: J. G. Keulemans in *Ibis* 1875: Pl. 11; J. Fjeldsa in Fjeldsa and Krabbe 1990: Pl. 25; T. Boyer in Boyer and Hume 1991: 100; C. Byers in del Hoyo et al. 1999: pl 15; F. Weick in König et al. 1999: Pl. 48; S. Webb in Hilty 2003: Pl. 24; F. Weick in *Gefiederte Welt*: 2005(1): 20; P. J. Greenfield in Ridgely and Greenfield 2001: II: Pl. 35(8)
Photograph: ?
Literature: Sharpe 1875a: 207–209; Sharpe 1875b: 43–45 and 57; Carriker 1910: 479 and 657; Kelso 1937: 304; Eck and Busse 1973: 123, 124; Hilty and Brown 1986: 229; Ridgely and Gwynne 1989: 189; Fjeldsa and Krabbe 1990: 226, 227; König 1991b: 23–35; Boyer and Hume 1991: 100; del Hoyo et al. 1999: 215; König et al. 1999: 364, 365; Hilty 2003: 361, 362; Weick 2003–2005, *Gefiederte Welt* 2005(1): 20; Ridgely and Greenfield 2001: I: 307, II: 215, 216

Glaucidium bolivianum König 1991
Yungas Pygmy Owl · Yungas-Sperlingskauz · Chevêchette des yungas · Mochuelo (Caburé) yungueno
Glaucidium bolivianum König 1991, Okol. Vogel 13(1): 36; Terra typica: Salta, Argentina
(see Plate 7 and Figure 13)

Length: 160 mm
Body mass: 53–58 g

Distribution: Slopes of the eastern Andes, from Peru to Bolivia and northern Argentina (Juguy, Salta, Tucumán)
Habitat: Montane and cloud forest, with epiphytes and creepers and dense undergrowth. From about 1 000 m up to 3 000 m
Museum: FMNH, FNSF, IML, LSUMZ, SMNSt (holotype), ZSBS

Wing length: 94–103 mm
Tail length: 67–72.5 mm
Tarsus length: ?
Length of bill (cere): ?
Body mass: ♂ 53–58 g

Illustration: C. Byers in del Hoyo et al. 1999: Pl. 15; F. Weick in König et al. 1999: Pl. 48 (grey and red)
Photograph: ?
Literature: König 1991b: 36; del Hoyo et al. 1999: 215; König et al. 1999: 365–367

❖

Glaucidium peruanum **König 1991**
Peruvian Pygmy Owl · Peru-Sperlingskauz · Chevêchette du Pérou · Mochuelo Peruano (de Pacifico)
Glaucidium peruanum **König 1991, Okol. Vogel 13(1): 56;**
Terra typica: **Apurimae, Peru**
(see Figure 13)

Length: 160–170 mm
Body mass: 58–65 g

Distribution: From western Ecuador: Manabi, south through western Peru to northern Chile. Also east of Andes in extreme southeast Ecuador: Zamora, Chinchipa and Maranón drainage of Peru
Habitat: Riparian woodland, thickets, semi-arid bushland, agricultural country with trees, plantations and parks. From sea-level up to 3 000 m
Museum: BMNH (holotype), LSUMZ, MJPL, SMNSt, ZFMK

Wing length: 98–104 mm
Tail length: 67.5–75 mm
Tarsus length: ?
Length of bill (cere): ?
Body mass: 58–65 g

Illustration: C. Byers in del Hoyo et al. 1999: Pl. 15 (grey and red); F. Weick in König et al. 1999: Pl. 48 (grey and red); P. J. Greenfield in Ridgely and Greenfield 2001: II: Pl. 35(12); P. Burke in Jaramillo et al. 2003: Pl. 59
Photograph: L. Sheldon in *Bull. ABC* 1999: 291; R. Behrstock in del Hoyo et al. 1999: 8 (Ecuador); C. Quested in internet: Owl Pages, photo gallery 2003
Literature: König 1991b: 15–76; del Hoyo et al. 1999: 218; König et al. 1999: 367, 368; Ridgely and Greenfield 2001: I: 309, II: 217, 218; Jaramillo et al. 2003: 145

❖

Glaucidium nanum **(King) 1828**
Austral Pygmy Owl · Patagonien-(Araukaner-)Sperlingskauz · Chevêchette australe · Caburé Patagón
Strix nana **King 1828, Zool. J. 3(1827): 427;**
Terra typica: **Port Famine, Straits de Magellan**

Length: 170–210 mm
Body mass: 55–100 g

Synonym:
- *Glaucidium nanum vafrum* Wetmore 1922, Journ. Wash. Acad. Sci. 12: 323; Terra typica: Concon, Intendencia de Valparaiso, Chile (holotype USNM)

Glaucidium nanum · Austral Pygmy Owl · Patagonien-Sperlingskauz

Distribution: Argentine Patagonia, from Rio Negro to Tierra del Fuego. Chile from south of the Atacama Desert to Tierra del Fuego. Some birds wintering farther north in Chile and Argentina
Habitat: Open landscape with scrub and groups of trees. Warm desert areas with oases and puna. Also in rather open temperate forest and cold beech forest. Sometimes in city parks and gardens. From sea-level up to 2 000 m
Museum: AMNH, BMNH, FNSF, NHMWien, SMNSt, SMTD, ZFMK, ZMH, ZSBS

Wing length: ♂ 95–104 mm, ♀ 97–108 mm
Tail length: ♂ 65–73 mm, ♀ 68–76 mm
Tarsus length: 23 and 24 mm
Length of bill (cere): ?
Body mass: ♂ 55–75 g, ♀ 70–100 g

Illustration: J. Fjeldsa in Fjeldsa and Krabbe 1990: Pl. 25 (adult and juvenile); C. Byers in del Hoyo et al. 1999: Pl. 15; F. Weick in König et al. 1999: Pl. 48 (grey and red); P. Burke in Jaramillo et al. 2003: Pl. 59 (3 morphs)
Photograph: T. Daskam in *Aves de Chile* (Calendar) 1977: Noviembre; C. König in König 1991b: Abb. 25; C. König in Burton et al. 1992: 173; G. Ziesler in del Hoyo et al. 1999: 95, 109 and 136; C. König in internet: Owl Pages, photo gallery 2003; J. Gorsfield in internet: Owl Pages, photo gallery 2004; R. and N. Bowers in internet: Owl Pages, photo gallery 2004
Literature: Sharpe 1875a: 190, 191; Sharpe 1875b: 41–43 and 57; Fjeldsa and Krabbe 1990: 227; König 1991b: 62–70; del Hoyo et al. 1999: 218; König et al. 1999: 368–370; Jaramillo et al. 2003: 146

Glaucidium siju (d'Orbigny) 1839

Cuban Pygmy Owl · Kuba-Sperlingskauz · Chevêchette de Cuba · Mochuelo Siju

Length: 170 mm
Body mass: 55–90 g

Distribution: Endemic to Cuba and the Isle of Pines (Isla de la Juventud)
Habitat: Coastal deciduous and montane forest, forest edges, clearings and second growth. Also open country with bushes and trees or plantations and large parks. From sea-level up to 1 500 m

- ***Glaucidium siju siju*** (d'Orbigny) 1839
Noctua siju d'Orbigny 1839, in de la Sagra's Hist. fis., pol. y nat. Isla de Cuba 3, Aves: 33 and 41, Atlas Aves Pl. 3; Terra typica: Island of Cuba

Distribution: Endemic to Cuba
Museum: BMNH, FNSF, MNBHU, MNHP, SMNSt, SMTD, ÜMB

Wing length: ♂ 87–92.5 mm, ♀ 97–104 mm
Tail length: ♂ 54–60 mm, ♀ 59.5–67 mm
Tarsus length: 20.5 mm
Length of bill (cere): 10–13 mm
Body mass: ♂ 55–57 g, ♀ 66–90 g

Illustration: ? in de la Sagra's Atlas "*Aves*" 1839: Pl. 3; L. Poole in Bond 1986: Fig. 95 (line drawing); T. Boyer in Boyer and Hume 1991: 99; C. Byers in del Hoyo et al. 1999: Pl. 15; F. Weick in König et al. 1999: Pl. 46; K. Williams in Raffaele et al. 2003: Pl. 43
Photograph: G. Budich in *Der Falke* 1978(1): 35; G. H. Harrison in *Living Bird* 1982 (Autumn): 22; D. Wechsler in del Hoyo et al. 1999: 126
Literature: Sharpe 1875a: 193, 194; Ridgway 1914: 782, 804, 805; Eck and Busse 1973: 121, 122; H. Busse 1978: 34; Bond 1986 (repr.): 121, 122; Boyer and Hume 1991: 99; del Hoyo et al. 1999: 218; König et al. 1999: 358; Raffaele et al. 2003: 100

- ***Glaucidium siju vittatum*** Ridgway 1914
Glaucidium siju vittatum Ridgway 1914, US Nat. Mus. Bull. 50: 782; Terra typica: Isle of Pines

Distribution: Isle of Pines = Isla de la Juventud, off Cuba
Museum: USNM (holotype)

Wing length: ♂ 94.5 mm, ♀ 101.5–109.5 mm
Tail length: ♂ 61.5 mm, ♀ 68.5–73 mm
Tarsus length: ?
Length of bill (cere): 11.5–12.5 mm
Body mass: ♂ 65–68 g, ♀ 84–92 g

Illustration: F. Weick in König et al. 1999: Pl. 46
Photograph: ?
Literature: Ridgway 1914: 782, 805, 806; Eck and Busse 1973: 121, 122; Dunning 1993: 94–100; del Hoyo et al. 1999: 218; König et al. 1999: 358

Glaucidium ridgwayi Sharpe 1875
Ridgway's Pygmy Owl · Ridgway-Sperlingskauz · Chevêchette de Ridgway · Caburé (Tecolotito) de Ridgway (see Plate 7)

Length: 165–185 mm
Body mass: 46–102 g

Distribution: Southern United States, Mexico and south to central America, Panama (possibly also extreme northwest Colombia)
Habitat: Semi-open country with thorny scrub and giant cacti, also patches of woodland or plantations in open areas. Dry woodland and evergreen secondary growth. From sea-level up to about 1500 m

- *Glaucidium ridgwayi cactorum* van Rossem 1937
  Glaucidium brasilianum cactorum van Rossem 1937, Proc. Biol. Soc. Wash. 50: 27, 28; Terra typica: Sonora, Mexico

Distribution: Southern Arizona to western Mexico (Nayarit and Jalisco)
Museum: BMNH, CIT, USNM
Remarks: Together with nominate *ridgwayi* considered as a subspecies of *Glaucidium brasilianum*, but DNA evidence and vocalisations suggest they are distinct

Wing length: ♂ 85–90 mm, ♀ 91–97 mm
Tail length: ♂ 59–64 mm, ♀ 63–67 mm
Tarsus length: ?
Length of bill (cere): ?
Body mass: 60–70 g

Illustration: K. E. Karalus in Karalus and Eckert 1974: Pl. 48; L. Malick in Breeden et al. 1983: 245; S. Webb in Howell and Webb 1995: Pl. 25 (6, upper bird); F. Weick in König et al. 1999: Pl. 49; D. Sibley in Sibley 2000: 283 (adult and juvenile)
Photograph: D. M. Jones in *Living Bird* 1997(Spring): 13; P. Johnsgard in Johnsgard 2002: 21 (*ridgwayi*?)
Literature: Sharpe 1875a: 205–207; Sharpe 1875b: 35–59; Ridgway 1914: 782, 798–803; van Rossem 1937: 27, 28; Sprunt 1955: 196–198; Blake 1963: 216, 217; Karalus and Eckert 1974: 222, 223; Breeden et al. 1983: 244; Howell and Webb 1995: 363; del Hoyo et al. 1999: 217; König et al. 1999: 372, 373; Sibley 2000: 283; Johnsgard 2002: 149–153

- *Glaucidium ridgwayi ridgwayi* Sharpe 1875
  Glaucidium ridgwayi Sharpe 1875, Ibis: 55; Terra typica: Central America, ex Ridgway (1873) Proc. Boston Soc. Nat. Hist. 16: 93 = Mexico

Synonym:
- *Glaucidium brasilianum saturatum* Brodkorb 1941, Occ. Pap. Mus. Zool. U. Michigan 450: 1–4; Terra typica: South Mexico (Chiapas) and Guatemala

Museum: BMNH (holotype), LSUMZ, MZUS, SMNSt, SMTD, USNM, ZFMK, ZMH

Wing length: ♂ 81–108 mm, ♀ 89–113 mm
Tail length: ♂ 52.5–66 mm, ♀ 56–79 mm
Tarsus length: 20.5 mm
Length of bill (cere): 9.5–13 mm
Body mass: ♂ 46–79 g, ♀ 64–102 g

Illustration: A. Brooks in Grosvenor and Wetmore 1937: 22; K. E. Karalus in Karalus and Eckert 1974: Pl. 47; L. Malick in Breeden et al. 1983: 245; A. Cameron in Voous 1988: 159; S. Webb in Howell and Webb 1995: Pl. 25 (6, lower bird); C. Byers in del Hoyo et al. 1999: Pl. 15; F. Weick in König et al. 1999: Pl. 49; D. Sibley in Sibley 2000: 283
Photograph: S. Grossman in Grossman and Hamlet 1965: near 144; E. Gohier in *Living Bird* 1984(Winter): 26; G. Lasley in internet: Owl Pages, photo gallery 2003
Literature: Sharpe 1875a: 205–207; Sharpe 1875b: 35–59; Ridgway 1914: 782, 798–803; Grosvenor and Wetmore 1937: 23; Bent 1961 (repr.): 435; Blake 1963: 216, 217; Land 1970: 139; Karalus and Eckert 1974: 216–222; Breeden et al. 1983: 244; Howell and Webb 1995: 363; del Hoyo et al. 1999: 217; König et al. 1999: 372, 373; Sibley 2000: 283; Johnsgard 2002: 149–153

Glaucidium brasilianum (Gmelin) 1788
Ferruginous Pygmy Owl · Brasil-Sperlingskauz · Chevêchette brune · Caburé Común (Mochuelo Caburé)

Length: 170–200 mm
Body mass: 50–110 g

Distribution: From northern South America east of the Andes (eastern Colombia, Ecuador, the Guianas and northern Brazil) south through the Amazon to eastern Bolivia, eastern Paraguay, Uruguay and central Argentina to La Pampa and Buenos Aires province

Habitat: Tropical and subtropical lowlands and foothills, with mostly humid primary and secondary forest, clearings, forest edges. Riverine woodland, plantations, suburban areas such as parks and large gardens. Usually below 1 500 m, sometimes up to around 2 300 m

- ***Glaucidium brasilianum medianum*** Todd 1916

Glaucidium brasilianum medianum Todd 1916, Proc. Biol. Soc. Wash. 29: 98; **Terra typica:** Bonda, Santa Marta, Columbia

Distribution: Northern Colombia
Museum: AMNH

Wing length:	94 mm
Tail length:	59 mm
Tarsus length:	?
Length of bill (cere):	?
Body mass:	?

Illustration: ?
Photograph: ?
Literature: Meyer de Schauensee 1964: 118; Eck and Busse 1973: 122, 123; Hilty and Brown 1986: 229, 230; del Hoyo et al. 1999: 217; König et al. 1999: 370–372

- ***Glaucidium brasilianum phaloenoides*** (Daudin) 1800

Strix phaloenoides Daudin 1800, Traite d Orn. 2: 206; **Terra typica:** Trinidad

Synonym:
- *Glaucidium brasilianum margaritae* Phelps and Phelps 1951, Proc. Biol. Soc. Wash. 64: 65–72; **Terra typica:** Margarita Island

Distribution: Trinidad and the Guianas, also Margarita Island (off Venezuela)
Museum: AMNH, BMNH, ZSBS

Wing length:	99–104 mm
Tail length:	65–67.5 mm
Tarsus length:	19 mm
Length of bill (cere):	?
Body mass:	70 g

Illustration: S. Webb in Hilty 2003: Pl. 24 (rufous and greybrown)
Photograph: ?
Literature: Sharpe 1875a: 203–205; Land 1970: 129, 130; del Hoyo et al. 1999: 217; König et al. 1999: 370–372; Hilty 2003: 362

- ***Glaucidium brasilianum duidae*** Chapman 1929

Glaucidium brasilianum duidae Chapman 1929, Am. Mus. Novit. 380: 8; **Terra typica:** Mount Duida, 4700 feet, Venezuela

Distribution: Endemic to Mount Duida, Venezuela
Museum: AMNH (holotype)

Wing length:	95–101 mm
Tail length:	55–62 mm
Tarsus length:	?
Length of bill (cere):	?
Body mass:	?

Illustration: F. Weick in König et al. 1999: Pl. 49
Photograph: ?
Literature: Chapman 1929: 1–27; Chapman 1931: 1–135; Kelso 1934d: 65; Eck and Busse 1973: 12, 123; del Hoyo et al. 1999: 217; König et al. 1999: 370–372; Hilty 2003: 362

- ***Glaucidium brasilianum olivaceum*** Chapman 1939

Glaucidium brasilianum olivaceum Chapman 1939, Am. Mus. Novit. 1051: 6; **Terra typica:** Mount Auyan-Tepui, SE Venezuela

Distribution: Endemic to Mount Auyan-Tepui, southeast Venezuela
Museum: AMNH (holotype)

Wing length:	94–99 mm
Tail length:	59–63 mm
Tarsus length:	?
Length of bill (cere):	?
Body mass:	?

Illustration: ?
Photograph: ?
Literature: Chapman 1939: 6; Hilty 2003: 362

- ***Glaucidium brasilianum ucayalae*** Chapman 1929

Glaucidium brasilianum ucayalae Chapman 1929, Am. Mus. Novit. 380: 9; **Terra typica:** Sarayacu, Rio Ucayali, Peru

Distribution: Amazonian Colombia, Venezuela and Brazil, south to southern Peru and northern Bolivia

Museum: AMNH (holotype), FMNH, SMTD, ZFMK, ZMH

Wing length: 98–106 mm
Tail length: 58–65 mm
Tarsus length: ?
Length of bill (cere): ?
Body mass: ?

Illustration: F. Weick in König et al. 1999: Pl. 49
Photograph: ?
Literature: Chapman 1929: 1–27; Kelso 1934d: 66; Hilty and Brown 1986: 229, 230; König et al. 1999: 370–372; Hilty 2003: 362

- ***Glaucidium brasilianum brasilianum*** (Gmelin) 1788
Strix brasiliana Gmelin 1788, Syst. Nat. 1(1): 289;
Terra typica: Brazil, Ceará ex Hellmayr (1929) FMNH-Publ. Zool. Ser. 12: 407

Synonym:
- *Strix ferox* Vieillot 1817, Nouv. Dict. Hist. Nat. 7: 22
- *Strix ferruginea* Maximilian (Wied-Neuwied) 1820, Reise Bras.: 105, footnote
- *Strix infuscata* Temminck 1821, Man. Orn., ed. 2, 1: 97
- *Glaucidium ferox rufus* Bertoni 1901, An. Cient. Paraguayos 1(1): 179; **Terra typica:** Paraguay

Distribution: Northeast Brazil, south to eastern Paraguay, northeast Argentina and northern Uruguay
Museum: AMNH, BMNH, FNSF, LSUMZ, MZUS, SMNSt, SMTD, ÜMB, USNM, ZFMK, ZMH, ZSBS

Wing length: 91.5–106 mm
Tail length: 53–73 mm
Tarsus length: 18–23 mm
Length of bill (cere): ?
Body mass: ♂ 46–74 g, ♀ 62–95 g

Illustration: T. Boyer in Boyer and Hume 1991: 101; C. Byers in del Hoyo et al. 1999: Pl. 15; F. Weick in König et al. 1999: Pl. 49; P. J. Greenfield in Ridgely and Greenfield 2001: II: Pl. 35(11)
Photograph: P. Pinney in Everett 1977: 32; E. Hosking in Hosking and Flegg 1982: 52; C. König in internet: Owl Pages, photo gallery 2003 (Misiones)
Literature: Sharpe 1875a: 200–202; Sharpe 1875b: 35–59; Ridgway 1914: 779–782; Chapman 1929: 10; Eck and Busse 1973: 122, 123; Voous 1988: 157–160; König 1991b: 51–56; Boyer and Hume 1991: 101; del Hoyo et al. 1999: 217; König et al. 1999: 370–372; Ridgely and Greenfield 2001: I: 309, II: 217

- ***Glaucidium brasilianum stranicki*** König and Wink 1995
Glaucidium brasilianum stranicki König and Wink 1995, J. Orn. 136: 461–468; **Terra typica:** Parador de Montana, Santa Rosa de la Calamuchita, 750 m, Córdoba, Argentina

Distribution: Central Argentina and southern Uruguay
Museum: MACN, MLP, SMNSt (holotype)

Wing length: ♂ 95–102 mm, ♀ 102–106 mm
Tail length: ♂ 66–70.5 mm, ♀ 73.5–75 mm
Tarsus length: ?
Length of bill (cere): ?
Body mass: ♂ 76 g, ♀ 87–93 g

Illustration: F. Weick in König et al. 1999: Pl. 49
Photograph: C. König in König 1991b: Abb. 24
Literature: König 1991b: 51–56; König and Wink 1995: 461–468; Heidrich et al. 1995: 1–47; König et al. 1999: 370–372

❖

Glaucidium tucumanum Chapman 1922

Tucuman Pygmy Owl · Tucuman-(Chaco-)Sperlingskauz · Chevêchette de Tucuman · Caburé de Chaco
Glaucidium tucumanum **Chapman 1922, Am. Mus. Novit. 31: 5; Terra typica: Rosario de Lerma, 4 800 feet, Salta, Argentina (see Figure 13)**

Length: 160–175 mm
Body mass: 55–60 g

Synonym:
- *Glaucidium brasilianum pallens* Brodkorb 1938, Occ. Pap. Mus. Zool. U. Michigan 394: 3; **Terra typica:** Puerto Casado, Paraguay

Distribution: Chaco of Bolivia, Paraguay and Argentina, south to Tucumán, northern Córdoba and Santiago del Estero
Habitat: Semi-open, dry thorn forest, thorny scrub, giant cacti and semi-arid and arid bushy areas, from 500 m up to 1 500 m, locally to 1 800 m above sea-level
Museum: AMNH (holotype), FNSF, MZUM (holotype = *G. b. pallens*), SMNSt, ZFMK, ZSBS

Wing length: ♂ 90–95.5 mm, ♀ 94–99.5 mm
Tail length: 60.5–68.5 mm
Tarsus length: 16 mm
Length of bill (cere): ?
Body mass: ♂ 52–56 g, ♀ 60 g

Illustration: C. Byers in del Hoyo et al. 1999: Pl. 15; F. Weick in König et al. 1999: Pl. 49
Photograph: C. König in internet: Owl Pages, photo gallery 2003 (Salta)
Literature: Chapman 1922: 5; Kelso 1934d: 66; Brodkorb 1938: 3; C. König 1991b: 15–75; Heidrich et al. 1995: 1–47; del Hoyo et al. 1999: 217; König et al. 1999: 373–375

Glaucidium palmarum **Nelson 1901**
Colima Pygmy Owl · Colima-Zwergkauz · Chevêchette de Colima · Tecolotito de Colima
Glaucidium palmarum **Nelson 1901, Auk 18: 46;**
Terra typica: **Arroyo de Juan Sánchez, Nayarit**
(see Figure 16)

Length: 130–150 mm
Body mass: 45 g

Synonym:
- *Glaucidium minutissimum oberholseri* R.T. Moore 1937, Proc. Biol. Soc. Wash. 50: 65, 98, 103, 105; Terra typica: Southern Sonora to Southern Sinaloa
- *Glaucidium minutissimum griscomi* R.T. Moore 1947, ibid. 60: 31, 33, 141, 146; Terra typica: Southwest Morelos and northeast Guerrero

Distribution: Endemic to western and central Mexico, from Sonora to Oaxaca
Habitat: Tropical semi-deciduous forest, thorny woods, also swampy woods. From sea-level up to around 1 500 m
Museum: DMNH, LSUMZ, MLZ, USNM (holotype), WFVZ

Wing length: ♂ 81–85 mm, ♀ 84–87.5 mm
Tail length: ♂ 51–54 mm, ♀ 53–56 mm
Tarsus length: 21 mm
Length of bill (cere): 9 mm
Body mass: 42.5–47.5 g

Illustration: S. Webb in Robbins and Howell 1995: Frontispiece; S. Webb in Howell and Webb 1995: Pl. 25; C. Byers in del Hoyo et al. 1999: Pl. 15; F. Weick in König et al. 1999: Pl. 47; F. Weick in *Gefiederte Welt* 2004(10): 314
Photograph: Lockshaw in internet: Owl Pages, photo gallery 2003 (not typical plumage! = error?)
Literature: Nelson 1901a: 46; Robbins and Howell 1995: 1–25; Howell and Webb 1995: 361; Heidrich et al. 1995: 1–47; del Hoyo et al. 1999: 215, 216; König et al. 1999: 359, 360; Johnsgard 2002: 145–148

Glaucidium sanchezi **Lowery and Newman 1949**
Tamaulipas Pygmy Owl · Tamaulipas-Zwergkauz · Chevêchette des Tamaulipas · Tecolotito Tamaulipeco
Glaucidium minutissimum sanchezi **Lowery and Newman 1949, Occ. Pap. Mus. Zool. LSU 22: 1–5;**
Terra typica: **Lano de Garzas, near Cerro Coneja, San Luis Potosi, Mexico, 6 800 feet**
(see Figure 16)

Length: 130–160 mm
Body mass: 51.5–55.5 g

Distribution: Endemic to northeast Mexico: southern Tamaulipas and southeast San Luis Potosi, also extreme northern Hidalgo
Habitat: Subtropical, humid evergreen and semi-deciduous, pine and cloud forest, from 900 up to 2 100 m
Museum: BMNH, DMNH, LSUMZ (holotype)

Wing length: ♂ 86–88.5 mm, ♀ 88.5–93.5 mm
Tail length: ♂ 51–55.5 mm, ♀ 54.5–57 mm
Tarsus length: ?
Length of bill (cere): 11 mm
Body mass: 51.5–55.5 g

Illustration: S. Webb in Robbins and Howell 1995: Frontispiece; S. Webb in Howell and Webb 1995: Pl. 25; C. Byers in del Hoyo et al. 1999: Pl. 15; F. Weick in König et al. 1999: Pl. 47; F. Weick in *Gefiederte Welt*: 2004(10): 314
Photograph: ?
Literature: Blake 1963: 216; Howell and Robbins 1995: 7–25; Robbins and Howell 1995: 1–6; Howell and

Webb 1995: 362; Heidrich et al. 1995: 1–47; del Hoyo et al. 1999: 216; König et al. 1999: 358, 359; Johnsgard 2002: 145–148

Glaucidium griseiceps Sharpe 1875
Central American Pygmy Owl · Graukopf-Zwergkauz · Chevêchette à tête grise · Tecolotito Centroamericano
***Glaucidium griseiceps* Sharpe 1875, Ibis: 41, Pl. 2;**
Terra typica: **Veragua, Panama and Chisec and Choctum, Guatemala**
(see Figure 15)

Length: 140–160 mm
Body mass: 50 g

Synonym:
- *Glaucidium minutissimum rarum* Griscom 1931, Proc. New England Zool. Cl. 12: 41; Terra typica: Permé, Caribbean slope of eastern Panama
- *Glaucidium minutissimum occultum* R.T. Moore 1947, Proc. Biol. Soc. Wash. 60: 31, 33, 141–146; Terra typica: Southern Mexico: southeast Veracruz, northern Oaxaca and Chiapas

Distribution: Southeast Mexico, Guatemala, Belize and Honduras, to Costa Rica, Panama and northwest South America
Habitat: Humid, evergreen forest and humid bushland, also secondary growth. Semi-open country and old plantations. From sea-level up to around 1 300 m
Museum: AMNH, BMNH(lectotype), CMNH, FMNH, FNSF, LSUMZ, MLZ, MZUS, SMNSt, SMTD, USNM, ZMH

Wing length: 84.5–90 mm
Tail length: 44.5–49.5 mm
Tarsus length: 17.8 mm
Length of bill (cere): 10–10.5 mm
Body mass: 49.5–57 g

Illustration: J. G. Keulemans in Sharpe 1875b: Pl. 2; D. Gardner in Stiles and Skutch 1991: Pl. 20; S. Webb in Robbins and Howell 1995: Frontispiece; S. Webb in Howell and Webb 1995: Pl. 25; C. Byers in del Hoyo et al. 1999: Pl. 15; F. Weick in König et al. 1999: Pl. 47; P. J. Greenfield in Ridgely and Greenfield 2001: II: Pl. 35(9); F. Weick in *Gefiederte Welt* 2004(10): 313
Photograph: R. and N. Bowers in internet: Owl Pages, photo gallery 2004
Literature: Sharpe 1875a: 196–198; Sharpe 1875b: 35–59; Ridgway 1914: 781, 782, 795, 796; Kelso 1934d: 64, 65; Land 1970: 139; Stiles and Skutch 1991: 193; Robbins and Howell 1995: 1–25; Howell and Webb 1995: 362; del Hoyo et al. 1999: 216; König et al. 1999: 360, 361; Ridgely and Greenfield 2001: I: 307, 308, II: 216; Johnsgard 2002: 145–148

***Glaucidium parkeri* Robbins and Howell 1995**
Subtropical Pygmy Owl · Parker-Zwergkauz · Chevêchette de Parker · Tecolotito de Parker
***Glaucidium parkeri* Robbins and Howell 1995, Wilson Bull. 107(1): 2;** Terra typica: **Zamora-Chinchipe, Ecuador**
(see Figure 16)

Length: 140 mm
Body mass: 62 g

Distribution: Eastern slope of Andes in Ecuador and Peru (possibly also in northern and southwest Colombia and south to northern Bolivia)
Habitat: Subtropical evergreen forest and cloud forest, also in outlying ridges with subcanopy. From 1 450 m up to 1 975 m above sea-level
Museum: AMNH, ANSP (holotype)

Wing length: 89.5–97 mm
Tail length: 46.5–53.5 mm
Tarsus length: ?
Length of bill (cere): ?
Body mass: 59–64 g

Illustration: S. Webb in Robbins and Howell 1995: Frontispiece; C. Byers in del Hoyo et al. 1999: Pl. 15; F. Weick in König et al. 1999: Pl. 47; P. J. Greenfield in Ridgely and Greenfield 2001: II: Pl. 35(10); F. Weick in *Gefiederte Welt* 2004(10): 314
Photograph: ?
Literature: Robbins and Howell 1995: 1–25; del Hoyo et al. 1999: 216; König et al. 1999: 363, 364; Ridgely and Greenfield 2001: I: 308, II: 216, 217; Weick 2003-2005, *Gefiederte Welt* 2004(10): 314

Glaucidium hardyi Vielliard 1989
Amazonian Pygmy Owl · Amazonas-Zwergkauz · Chevêchette d'Amazonie · Mochuelo Amazónico
Glaucidium hardyi Vielliard 1989, Rev. Bras. Zool. 6: 685–693; Terra typica: 20 km SW of Presidente Medici Rondônia, Brazil (see Figure 15)

Length: 140–150 mm
Body mass: 60 g

Distribution: Southeast Venezuela (Bolivar), east through the Guianas to northern Brazil (Pará), south to southeast Peru, northern and eastern Bolivia and the southern Mato Grosso
Habitat: Humid evergreen rainforest (upper storey), forest edges. From sea-level up to about 850 m
Museum: CMNH, FMNH, LSUMZ, ZFMK, ZUEC (holotype)

Wing length: 90.5–94 mm
Tail length: 45–53 mm
Tarsus length: ?
Length of bill (cere): ?
Body mass: 51.8–63 g

Illustration: S. Webb in Robbins and Howell 1995: Frontispiece; C. Byers in del Hoyo et al. 1999: Pl. 15; F. Weick in König et al. 1999: Pl. 47; F. Weick in *Gefiederte Welt* 2004(10): 313
Photograph: J. P. O'Neill in Burton et al. 1992: 175
Literature: Robbins and Howell 1995: 3; Howell and Robbins 1995: 7–25; König 1991b: 15–76; Heidrich et al. 1995: 1–47; del Hoyo et al. 1999: 216; König et al. 1999: 362, 363; Hilty 2003: 362; Weick 2003–2005, *Gefiederte Welt* 2004(10): 313

Glaucidium minutissimum (Wied) 1830
Pernambuco Pygmy Owl · Pernambuco-Zwergkauz · Chevêchette de Pernambuco · Mochuelo de Pernambuco
Strix minutissima zu Wied-Neuwied 1830, Beitr. Naturgesch. Brasil. 3(1): 242; Terra typica: Interior of State Bahia, Brazil (see Figure 15)

Length: 140–150 mm
Body mass: 51 g

Synonym:
- *Strix pumila* Temminck 1821, Pl. color. livr. 7: Pl. 39; Terra typica: Paraguay and Brazil; name invalid
- *Glaucidium mooreorum* da Silva et al. 2003, Ararajuba 10(2): 122–130, Dec. 2002; Terra typica: Caatinga Forest, near Pernambuco; name invalid and synonym by Wied's name from 1830!

Distribution: Eastern Brazil: from Bahia north to the Caa-tinga Forest near Pernambuco. Range non-overlapping with that of *Glaucidium sicki* König and Weick (new name was needed) from southeast Brazil!
Habitat: Tropical, humid evergreen forest and edges to open bush canopy. About 140 m above sea-level
Museum: AMNH, FMNH, MPEG, MZUSP, UFPE (holotype *mooreorum*)

Wing length: 87–90 mm
Tail length: 50.5–53.5 mm
Tarsus length: 17.8 mm?
Length of bill (cere): 10.2–11.2 mm
Body mass: 51 g

Illustration: C. C. Tofte in *Ararajuba* 2002, 10(2): Frontispiece (*G. mooreorum* = error)
Photograph: Skins in *Ararajuba* 2002, 10(2): 125
Literature: Wied 1830: 242, 243, 246; Sharpe 1875b: 35–59; Cardoso da Silva et al. 2002: 123–130; König and Weick 2005: 1–10

Glaucidium sicki König and Weick 2005
Sick's Pygmy Owl · Sick-Zwergkauz · Chevêchette naine · Mochuelo Minimo (eneno) de Sick (Caburé miudinho de Sick)
Glaucidium sicki König and Weick 2005, Stuttgarter Beitr. Naturkd. Ser. A 268: 1–10; Terra typica: Subtropical rainforest eastern Peru (State Santa Catarina) (see Figure 15)

Synonym:
- *Glaucidium minutissimum* zu Wied-Neuwied 1830, Beitr. Naturgesch. Brasil. 3, 1: 242, 243, 246; Terra typica: Interior of State Bahia, Brazil

Distribution: Southeast Brazil: from Misiones north to southern Bahia, also adjacent Paraguay
Habitat: Tropical, humid evergreen forest and forest edges. From sea-level up to 500 (800) m
Museum: AMNH, ANSP, BMNH, FNSF, Sao Paulo, SMNSt, USNM, ZSBS

Wing length: 85–91 mm
Tail length: 49–54 mm
Tarsus length: 18 mm
Length of bill (cere): ?
Body mass: 50 g

Illustration: J. G. Keulemans in Sharpe 1875b: Pl. 1; A. Cameron in Voous 1988: 155; T. Boyer in Boyer and Hume 1991: 99; S. Webb in Robbins and Howell 1995: Frontispiece; C. Byers in del Hoyo et al. 1999: Pl. 15; F. Weick in König et al. 1999: Pl. 47; F. Weick in *Gefiederte Welt* 2004(10): 313
Photograph: ? Schultz in Burton et al. 1992: 174 (possibly another taxon of Pygmy Owls!)
Literature: Sharpe 1875b: 35–59; Eck and Busse 1973: 121; Voous 1988: 153–155; König 1991b: 15–76; Boyer and Hume 1991: 99; Heidrich et al. 1995: 1–47; Howell and Robbins 1995: 7–25; Robbins and Howell 1995: 1–6; Sick 1984,1: 331; Ornithologia Brasileira. Rio de Janeiro; del Hoyo et al. 1999: 217; König et al. 1999: 361, 362; König and Weick 2005: 1–10

Glaucidium tephronotum Sharpe 1875
Red-chested Owlet · Rotbrust-Sperlingskauz · Chevêchette à pieds jaunes · Mochuelo Pechirrojo
(see Plate 7)

Length: 170–180 (200) mm
Body mass: 80–100 g

Distribution: West Africa, from Liberia, Ivory Coast, Ghana, southern Camaroon and the Congo Basin to Uganda and western Kenya (Mount Elgon, Kakamega, Nandi and Mau Forest)
Habitat: Dense primary forest and forest scrub mosaic. Also in logged forest, forest edges and clearings. From lowlands up to about 2 150 m
Remarks: Relationships from this species uncertain!

- ***Glaucidium tephronotum tephronotum*** Sharpe 1875
Glaucidium tephronotum Sharpe 1875, Ibis: 260; Terra typica: South America = error, West Africa ex Chapin (1921) Auk 38: 456, 457

Distribution: Liberia, Ivory Coast and Ghana
Museum: BMNH, SMNSt, ZSBS

Wing length: 99–109 mm
Tail length: 67–76 mm
Tarsus length: 19–24 mm
Length of bill: 15–17 mm
Body mass: ♂ 80, 85 and 95 g, ♀ 75–103 g

Illustration: J. G. Keulemans in Sharpe 1875b: Pl. 13; H. Grönvold in Bannerman 1953: 540 (b/w); T. Boyer in Boyer and Hume 1991: 102; P. Hayman in Kemp and Kemp 1988: 295; C. Byers in del Hoyo et al. 1999: Pl. 16; F. Weick in König et al. 1999: Pl. 45; N. Borrow in Borrow and Demey 2001: Pl. 5
Photograph: M. Fodgen in Burton et al. 1992: 172; B. Schmidt in internet: Owl Pages, photo gallery 2003
Literature: Sharpe 1875b: 260; Sharpe 1875a: 211, 212; Bannerman 1953: 540; Fry et al. 1988: 140, 141; Boyer and Hume 1991: 102, 103; Kemp and Kemp 1998: 294, 295; del Hoyo et al. 1999: 221; König et al. 1999: 353, 354; Borrow and Demey 2001: 497

- ***Glaucidium tephronotum pycrafti*** Bates 1911
Glaucidium pycrafti Bates 1911, Bull. Brit. Ornith. Cl. 27: 85, Pl. 7; Terra typica: Bitye, Cameroon

Distribution: Cameroon
Museum: BMNH, SMNSt, ZFMK

Wing length: 104–109 mm
Tail length: ?
Tarsus length: ?
Length of bill: ?
Body mass: ?

Illustration: H. Grönvold in Bates 1911: Pl. 7; C. Byers in del Hoyo et al. 1999: Pl. 16; F. Weick in König et al. 1999: Pl. 45; N. Borrow in Borrow and Demey 2001: Pl. 59
Photograph: ?
Literature: Bates 1911: 85; Bannerman 1953: 540; Fry et al. 1988: 140, 141; del Hoyo et al. 1999: 221; König et al. 1999: 353, 354; Borrow and Demey 2001: 497

- ***Glaucidium tephronotum medje*** Chapin 1932
Glaucidium tephronotum medje Chapin 1932, Am. Mus. Novit. 570: 3; Terra typica: Medje, Ituri Forest, Congo

Synonym:
- *Glaucidium tephronotum lukolelae* Chapin 1932, ibid.: 4; Terra typica: Lukolela, Middle Congo River

Glaucidium tephronotum pycrafti · Red-chested Owlet · Rotbrust-Sperlingskauz

- *Glaucidium tephronotum kivuense* Verheyen 1946, Bull. Mus. R. Hist. Nat. Belg. 22: 1; Terra typica: Eastern Zaire: Forêt de Kambatule, Bwanendeke, Kivu

Distribution: Congo Basin, eastern Zaire and southwest Uganda
Museum: AMNH (holotype), MRAC

Wing length:	113–127 mm
Tail length:	81–86.5 mm
Tarsus length:	?
Length of bill:	?
Body mass:	?

Illustration: M. Woodcock in Fry et al. 1988: Pl. 8
Photograph: Zimmerman in Zimmerman 1972: 267 (b/w)
Literature: Chapin 1932: 3, 4; Verheyen 1946: 1; Zimmerman 1972: 291, 292; Eck and Busse 1973: 124, 125; Fry et al. 1988: 141, 142; del Hoyo et al. 1999: 221; König et al. 1999: 353, 354

- ***Glaucidium tephronotum elgonense*** Granvik 1934
Glaucidium tephronotum elgonense Granvik 1934, Rev. Zool. Bot. Afr. 25: 41; Terra typica: Mount Elgon

Distribution: Eastern Uganda and western Kenya: Mount Elgon, Kagamega, Nandi and Mau Forest
Museum: RMTB?

Wing length:	♂ 127 mm ($n = 1$)
Tail length:	?
Tarsus length:	?
Length of bill:	?
Body mass:	♂ 80–95 g, ♀ 82?–103 g

Illustration: D. Zimmerman in Zimmerman et al. 1996: Pl. 56; J. Gale in Stevenson and Fanshawe 2002: Pl. 102
Photograph: ?
Literature: Granvik 1934: 41; Eck and Busse 1973: 124, 125; Fry et al. 1988: 141, 142; Zimmerman et al. 1996: 445; del Hoyo et al. 1999: 221; König et al. 1999: 353, 354; Stevenson and Fanshawe 2002: 204

Glaucidium brodiei (Burton) 1836
Collared Owlet · Wachtelkauz · Chevêchette à collier · Mochuelo Acollarado

Length:	145–170 mm
Body mass:	53–63 g

Distribution: Himalayas from northern Pakistan, east to China and Taiwan, south through Malaysia to Sumatra and Borneo
Habitat: Submontane and montane open forest, forest edges, woodland and scrub. Mainly from 1 350 m up to 2 750 m, but in China near cultivated land down to around 700 m
Remarks: Often considered to be closely related to *G. passerinum*, but relationships uncertain

- ***Glaucidium brodiei brodiei*** (Burton) 1836
Noctua Brodiei Burton 1836, Proc. Zool. Soc. Lond. 1835: 152; Terra typica: Himalayas, restricted to Simla, ex Baker (1927) Fauna 4: 450, Western Himalayas

Synonym:
- *Noctua tubiger* Hodgson 1836, As. Res. 19: 175; Terra typica: Nepal

- *Glaucidium brodiei garoense* Koelz 1952, J. Zool. Soc. India 4: 45; Terra typica: Tura Mountains, Garo Mountains, Assam

Distribution: From northern Pakistan through the Himalayas to southeast Tibet, northern Indochina, southern, central and eastern China, Hainan and south to Malaysia
Museum: BMNH, MHNP, SMNSt, SMTD, ÜMB, USNM, ZFMK, ZMH, ZSBS

Wing length: ♂ 80–90 mm, ♀ 93–101 mm
Tail length: 56–71 mm
Tarsus length: 20–23 mm
Length of bill: 13–15 mm
Body mass: ♂ 52 and 53 g, ♀ 63 g

Illustration: H. Grönvold in Baker 1927: Pl. 6 and 7; T. Boyer in Boyer and Hume 1991: 104; D. Cole in Grimmett et al. 1998: Pl. 30; C. Byers in del Hoyo et al. 1999: Pl. 14; F. Weick in König et al. 1999: Pl. 50; H. Burn in Robson 2000: Pl. 21; L. McQueen in Rasmussen and Anderton 2005: Pl. 79
Photograph: B. v. Elegem in del Hoyo et al. 1999: 104; P. P. Katsura in internet: Owl Pages, photo gallery 2004
Literature: Sharpe 1875a: 212–214; Baker 1927: 450; Delacour and Jabouille 1931: 140, 141; Ali and Ripley 1969: 285, 286; Puget and Hüe 1970: 86, 87; Eck and Busse 1973: 124; Voous 1988: 161–164; Smythies and Hughes 1984: 320, 321; Boyer and Hume 1991: 104; Lekagul and Round 1991: 178; Grimmett et al. 1998: 436; del Hoyo et al. 1999: 210; König et al. 1999: 375, 376; Rasmussen and Anderton 2005: 244

- **Glaucidium brodiei pardalotum** (Swinhoe) 1863
Athene pardalotum Swinhoe 1863, Ibis: 216;
Terra typica: Interior of Formosa
(see Plate 7)

Distribution: Taiwan
Museum: MNBHU, SMNSt, SMTD

Wing length: 88–91 mm
Tail length: 63.5 mm
Tarsus length: 18.5–20 mm
Length of bill: ?
Body mass: ?

Illustration: F. Berille in Etchécopar and Hüe 1978: Pl. 18; A. Cameron in Voous 1988: 163; F. Weick in König et al. 1999: Pl. 50
Photograph: ?
Literature: Swinhoe 1863: 216; Sharpe 1875a: 214, 215; Eck and Busse 1973: 124; Etchécopar and Hüe 1978: 482, 483; Voous 1988: 161–164; del Hoyo et al. 1999: 210; König et al. 1999: 375, 376

- **Glaucidium brodiei sylvaticum** (Bonaparte) 1850
Strix sylvatica "Müll." Bonaparte 1850, Consp. Gen. Av. 1: 40;
Terra typica: Sumatra

Synonym:
- *Glaucidium brodiei peritum* Peters 1940, Checklist Bds. World 4: 133; new name for *sylvaticum*, invalid

Distribution: Sumatra
Museum: NNML

Wing length: 95 mm
Tail length: 53 mm
Tarsus length: 20 mm
Length of bill: 16 mm
Body mass: ?

Illustration: K. Phillipps in McKinnon and Phillipps 1993: Pl. 39?
Literature: Sharpe 1875a: 215; Eck and Busse 1973: 124; McKinnon and Phillipps 1993: 194; del Hoyo et al. 1999: 210; König et al. 1999: 375, 376

- **Glaucidium brodiei borneense** Sharpe 1893
Glaucidium borneense Sharpe 1893, Bull. Brit. Ornith. Cl. 1: 55;
Terra typica: Mount Kalulong, Sarawak, Borneo

Distribution: Borneo
Museum: BMNH

Wing length: ?
Tail length: ?
Tarsus length: ?
Length of bill: ?
Body mass: ?

Illustration: Hughes in Smythies 1960: Pl. 15 (subspecies cannot be identified!)
Photograph: ?
Literature: Sharpe 1893: 55; Smythies 1960: 282; Eck and Busse 1973: 124; McKinnon and Phillipps 1993: 194

Subgenus
Taenioglaux **Kaup 1848**
Taenioglaux Kaup 1848, Isis v. Oken, col. 769.
Type by *Strix radiata* Tickell

Synonym:
- *Smithiglaux* Bonaparte 1854, Rev. Mag. Zool.: 544. Type by *Athene capense* A. Smith

Glaucidium radiatum **(Tickell) 1833**
Jungle Owlet · Dschungelkauz · Chevêchette de la jungle · Mochuelo de Jungle
(see Plate 7)

Length:	200 mm
Body mass:	88–114 g

Distribution: The Himalayas from Himachal Pradesh east to Bhutan, Bangladesh and extreme western Myanmar, south through India, Sri Lanka
Habitat: Sparse and dense deciduous forest. Scrub, terai, bhabar, duns and duars. Teak and bamboo facies on foothills, sometimes near human settlements. From lowlands up to about 900 m, in Nepal to 2 000 m above sea-level

- ***Glaucidium radiatum radiatum*** (Tickell) 1833
Strix Radiata Tickell 1833, J. As. Soc. Beng. 2: 572;
Terra typica: Jungles of Borahbum and Dholbhum

Glaucidium (Taenioglaux) radiatum · Jungle Owlet · Dschungelkauz

Synonym:
- *Glaucidium radiatum principum* Koelz 1950, Am. Mus. Novit. 1452: 3; Terra typica: Oria, Siriohi/Raiputane, western India (holotype AMNH)

Distribution: Himalayas from Hmachal Pradesh east to Bhutan, Bangladesh and extreme western Myanmar, south through India, with the exception of the southwest. Also Sri Lanka
Museum: AMNH, FNSF, MHNP, SMNSt, SMTD, ÜMB, ZFMK, ZMH, ZSBS

Wing length:	124–136 mm
Tail length:	63–84 mm
Tarsus length:	24–29 mm
Length of bill:	17–21 mm
Body mass:	88–114 g

Illustration: G.M. Henry in Henry 1955: Pl. 16; H. Grönvold in Whistler 1963: 349 (b/w); T. Boyer in Boyer and Hume 1991: 105; D. Cole in Grimmett et al. 1998: Pl. 30; C. Byers in del Hoyo et al. 1999: Pl. 16; F. Weick in König et al. 1999: Pl. 50; H. Burn in Robson 2000: Pl. 21; L. McQueen in Rasmussen and Anderton 2005: Pl. 79
Photograph: W. Scherzinger 1968: 204–206; W. Scherzinger in *Der Falke* 1969(9): Backcover; R. Saldino in internet: Owl Pages, photo gallery 2003
Literature: Sharpe 1875a: 217, 218; Baker 1927: 448, 449; Koelz 1950: 3; Henry 1955: 200; Whistler 1963: 348–350; Scherzinger 1968: 204–206; Ali and Ripley 1969: 286–288; Hoppe 1969: 319; Eck and Busse 1973: 127, 128; Boyer and Hume 1991: 105, 106; Grimmett et al. 1998: 436, 437; del Hoyo et al. 1999: 222; König et al. 1999: 376, 377; Robson 2000: 293; Rasmussen and Anderton 2005: 245

- ***Glaucidium radiatum malabaricum*** (Blyth) 1846
Athene malabaricus Blyth 1846, J. As. Soc. Beng. 15: 280;
Terra typica: Malabar Coast and Travancore

Distribution: Southwest India: Malabar Coast from southern Konkan through Goa, western Mysore to Kerala
Museum: FNSF, MZUS, SMNSt

Wing length:	120–134 mm
Tail length:	62–70 mm
Tarsus length:	(20) 22–27 mm
Length of bill:	19–22 mm
Body mass:	?

Illustration: D. Cowen in Ali and Ripley 1968: Pl. 13; D. Cole in Grimmett et al. 1998: Pl. 30; C. Byers in del Hoyo et al. 1999: Pl. 16; F. Weick in König et al. 1999: Pl. 50; L. McQueen in Rasmussen and Anderton 2005:Pl. 79
Photograph: ?
Literature: Sharpe 1875a: 218, 219; Baker 1927: 449, 450; Ali and Ripley 1969: 288; Eck and Busse 1973: 127, 128; Grimmett et al. 1998: 436, 437; del Hoyo et al. 1999: 222; König et al. 1999: 376, 377; Rasmussen and Anderton 2005: 245

Glaucidium castanonotum **(Blyth) 1852**
Chestnut-backed Owlet · Kastanienrückenkauz · Chevêchette à dos marron · Mochuelo de Ceilán
Athene castanonota **Blyth 1852, Cat. Birds Mus. As. Soc. 1849, 1:39; Terra typica: Ceylon**

Length: 190 mm
Body mass: ~100 g

Distribution: Endemic to Sri Lanka
Habitat: Dense forest in humid zones and rubber plantations. Usually in canopy of tall trees on steep hillsides. From lowlands up to 1 950 m
Museum: FNSF, MNBHU, ÜMB, ZFMK
Remarks: Often included in *G. radiatum*, but monotypic

Wing length: 122–137 mm
Tail length: 56–70 mm
Tarsus length: 25–28 mm
Length of bill: 17–18 mm
Body mass: ~100 g

Illustration: J. G. Keulemans in Legge 1878a: Pl. 4; D. Cole in Grimmett et al. 1998: Pl. 30; C. Byers in del Hoyo et al. 1999: Pl. 16; F. Weick in König et al. 1999. Pl. 50; L. McQueen in Rasmussen and Anderton 2005:Pl. 79
Photograph: ?
Literature: Baker 1927: 447, 448; Henry 1955: 201, 202; Ali and Ripley 1969: 288, 289; Eck and Busse 1973: 127, 128; Grimmett et al. 1998: 437; del Hoyo et al. 1999: 222; König et al. 1999: 377, 378; Rasmussen and Anderton 2005: 245

Glaucidium cuculoides **(Vigors) 1831**
Asian Barred Owlet · Kuckuckskauz · Chevêchette barrée (ou cuculoide) · Mochuelo Cuco

Length: 220–250 mm
Body mass: ♂ 150–175 g, ♀ -240 g

Distribution: Western Himalayas of northeast Pakistan and Kashmir, east to Nepal, Bhutan and Myanmar. Across south and southeast China south to Hainan and southeast Asia
Habitat: Open pine and oak forest, montane and submontane, also tropical and subtropical evergreen jungle on foothills. Sometimes near human habitations. From lowlands up to 1 800 m, locally to 2 700 m

- ***Glaucidium cuculoides cuculoides*** (Vigors) 1831
  Noctua cuculoides Vigors 1831, Proc. Comm. Zool. Soc. London: 8; Terra typica: Himalayas, Simla-Almora district

Distribution: Himalayas from northeast Pakistan and Kashmir east to western Sikkim
Museum: SMNSt, SMTD, ÜMB, ZFMK, ZSBS

Wing length: 143–162 mm
Tail length: 75–96 mm
Tarsus length: 24–30 mm
Length of bill: 19–22 mm
Body mass: ♂ 150–176 g, ♀ -210 g

Illustration: G. M. Henry in Ali 1949: Pl. 83; T. Boyer in Boyer and Hume 1991: 106; D. Cole in Grimmett et al. 1998: Pl. 30; C. Byers in del Hoyo et al. 1999: Pl. 16; F. Weick in König et al. 1999: Pl. 52; L. McQueen in Rasmussen and Anderton 2005:Pl. 79
Photograph: Zool. Soc. London in Everett 1977: 41; E. Hosking in Hosking and Flegg 1982: 43; E. Anders in *Gefiederte Welt* 1987(10): 275 (adult and juvenile); H. Reinhard in Burton et al. 1992: 176; B. Marcot in internet: Owl Pages, photo gallery 2003
Literature: Baker 1927: 444, 445; Ali 1949: 157–159; Ali and Ripley 1969: 289–291; Eck and Busse 1973: 126, 127; Anders 1987: 275–277; Voous 1988: 165–169; Boyer and Hume 1991: 105, 106; Grimmett et al. 1998: 436; del Hoyo et al. 1999: 221; König et al. 1999: 381, 382; Rasmussen and Anderton 2005: 244, 245

- ***Glaucidium cuculoides rufescens*** Baker 1926
  Glaucidium cuculoides rufescens Baker 1926, Bull. Brit. Ornith. Cl. 47: 59; Terra typica: Noong-zai-ban, Manipur

Synonym:
- *Glaucidium cuculoides austerum* Ripley 1948, Zoologica 33: 200; Terra typica: Tezu Hills, Mishmi (cotype ZMH)
- *Glaucidium cuculoides deignani* Ripley 1948, ibid.: 200; Terra typica: Ban Nong kho, Southeast Thailand
- *Glaucidium cuculoides delacouri* Ripley 1948, ibid.: 201; Terra typica: Bac Tan Tray, North Indochina

Remarks: *austerum*, *deignani* and *delacouri* included in *rufescens*, but differences in plumage and measurements uncertain

Distribution: Eastern Sikkim, Bhutan, Assam, northeast India, Bangladesh, north and northwest Myanmar, southeast Thailand, southern and northern Indochina, northern Laos and North Vietnam
Museum: BMNH, FNSF, ZFMK, ZMH

Wing length:	♂ 141–154 mm, ♀ 156–162 mm
Tail length:	78–94 mm
Tarsus length:	25–29 mm
Length of bill:	20–22 mm
Body mass:	150–176 g

Illustration: Hughes in Smythies and Hughes 1984: Pl. 18; Hughes (same Pl.) in Ali and Ripley 1970: Pl. 46; D. Cole in Grimmett et al. 1998: Pl. 30; F. Weick in König et al. 1999: Pl. 52; L. B. McQueen in Rasmussen and Anderton 2005: Pl. 79
Photograph: Photos from Hosking, Reinhard etc. see nominate, probably this subspecies!
Literature: Baker 1926: 59; Baker 1927: 445, 446; Delacour and Jabouille 1931: 137, 138; Ali and Ripley 1969: 291, 292; Eck and Busse 1973: 126, 127; Grimmett et al. 1998: 436, 437; del Hoyo et al. 1999: 221; König et al. 1999: 381, 382; Rasmussen and Anderton 2005: 244, 245

- ***Glaucidium cuculoides bruegeli*** (Parrot) 1908
  Athene cuculoides bruegeli Parrot 1908, Verh. Ornith. Ges. Bayern 8: 104; Terra typica: Bankok, Siam

Synonym:
- *Glaucidium cuculoides fulvescens* Baker 1926, Bull. Brit. Ornith. Cl. 47: 60; Terra typica: Kolidoo, Tenasserim

Distribution: Tenasserim, Thailand (not the southeast), southern Laos, Cambodia and South Vietnam
Museum: BMNH, MZUS, SMTD, USNM

Wing length:	♂ 131–158 mm, ♀ 138–161 mm
Tail length:	86 mm
Tarsus length:	?
Length of bill:	?
Body mass:	?

Illustration: W. Mongkol in Lekagul and Round 1991: Pl. 60; F. Weick in König et al. 1999: Pl. 52; H. Burn in Robson 2000: Pl. 21 (*bruegeli* and *deignani*)
Photograph: ?
Literature: Parrot 1908: 104–107; Baker 1927: 446, 447; Delacour and Jabouille 1931: 128; Deignan 1945: 179, 180; Ripley 1948a: 200, 201; del Hoyo et al. 1999: 221; König et al. 1999: 381, 382

- ***Glaucidium cuculoides whitelyi*** (Blyth) 1867
  Athene Whitelyi Blyth 1867, Ibis: 313; Terra typica: Japan, error = China (see Plate 7)

Distribution: West-central and southeast China, NE Vietnam and probably extreme SE Tibet
Museum: SMNKa, SMNSt, SMTD, ZFMK, ZMH

Wing length:	154–168 mm
Tail length:	93–114 mm
Tarsus length:	32 mm
Length of bill:	?
Body mass:	170–240 g

Illustration: J. H. Dick in Meyer de Schauensee 1984: Pl. 48; C. Byers in del Hoyo et al. 1999: Pl. 16(?); F. Weick in König et al. 1999: Pl. 52
Photograph: ?
Literature: Sharpe 1875a: 222, 223; Delacour and Jabouille 1931: 139; Ripley 1948a: 200, 201; Etchécopar and Hüe 1978: 484, 485; Meyer de Schauensee 1984: 269, 270; Voous 1988: 165–169; del Hoyo et al. 1999. 221; König et al. 1999: 381, 382

- ***Glaucidium cuculoides persimile*** Hartert 1910
  Glaucidium cuculoides persimile Hartert 1910, Novit. Zool. 17: 205; Terra typica: Five Finger Mountains, Hainan

Distribution: Endemic to Hainan Island
Museum: AMNH, BMNH, MNBHU, SMTD

Wing length: 160 mm
Tail length: 98 mm
Tarsus length: ?
Length of bill: ?
Body mass: ?

Illustration: A. Cameron in Voous 1988: 167; F. Weick in König et al. 1999: Pl. 52
Photograph: ?
Literature: Hartert 1910: 205; Etchécopar and Hüe 1978: 484, 485; del Hoyo et al. 1999: 221; König et al. 1999: 382, 383

❖

Glaucidium castanopterum (Horsfield) 1821
Javan Owlet · Trillerkauz · Chevêchette spadicée · Mochuelo de Java
Strix castanoptera Horsfield 1821, Trans. Linn. Soc. London 13(1): 140; Terra typica: Java

Length: 240 mm
Body mass: ?

Distribution: Endemic to Java and Bali
Habitat: Fragments of forest (primary and secondary), gardens, villages. Lowlands and up to 900 m, locally to 2 000 m
Museum: FNSF, MNBHU, MZUS, SMNSt, ZSBS
Remarks: Often regarded as an isolated subspecies of *cuculoides*, but of an older taxon (1821 vs. 1831), and specifically distinct and monotypic

Wing length: 144–150 mm
Tail length: 75–96 mm
Tarsus length: 21.5–23 mm
Length of bill: ?
Body mass: ?

Illustration: K. Phillipps in McKinnon and Phillipps 1993: Pl. 39; C. Byers in del Hoyo et al. 1999: Pl. 16; F. Weick in König et al. 1999: Pl. 52
Photograph: ?
Literature: Sharpe 1875a: 216, 217; McKinnon 1990: 181, 182; McKinnon and Phillipps 1993: 194; del Hoyo et al. 1999: 221; König et al. 1999: 382, 383

❖

Glaucidium sjoestedti Reichenow 1893
Sjoestedt's Owlet · Prachtkauz · Chevêchette à queue barrée · Mochuelo del Congo
Glaucidium sjöstedti Reichenow 1893, Orn. Monatsb. 1: 65; Terra typica: Mount Cameroon

Length: 250–280 mm
Body mass: 140 g

Glaucidium (Taenioglaux) sjoestedti
Sjoestedt's Owlet
Prachtkauz

Distribution: Cameroon, Gabon, northern Congo, southern Central African Republic, northwest and central Zaire
Habitat: Tropical lowland primary forest, distribution uncommon and local. Somewhat higher altitudes (Mount Cameroon)
Museum: BMNH, MHNP, MNBHU (holotype), MRAC, SMNSt

Wing length:	♂ 152–165 mm, ♀ 167 and 168 mm
Tail length:	80–110 mm
Tarsus length:	26–30 mm
Length of bill:	21–24 mm
Body mass:	139 g (Dunning)

Illustration: A. Ekblom in *Svensk Vet. AK. Handl.* 1895, 27(1): Pl. 2; H. Grönvold in Bannerman 1953: 541 (b/w); M. Woodcock in Fry et al. 1988: Pl. 9; T. Boyer in Boyer and Hume 1991, 107; P. Hayman in Kemp and Kemp 1998: 301; C. Byers in del Hoyo et al. 1999: Pl. 16; F. Weick in König et al. 1999: Pl. 52; N. Borrow in Borrow and Demey 2001: Pl. 59
Photograph: Bowden and Andrews 1994: 13–18; Swanepoel, Transvaal Mus. in Kemp and Kemp 1998: 300, 301; B. Schmidt in internet: Owl Pages, photo gallery 2003 (4 photographs)
Literature: Bannerman 1953: 541, 542; Eck and Busse 1973: 125, 126; Fry et al. 1988: 143, 144; Boyer and Hume 1991: 107; Bowden and Andrews 1994: 13–18; Kemp and Kemp 1998: 300, 301; del Hoyo et al. 1999: 221; König et al. 1999: 378, 379; Borrow and Demey 2001: 498

❖

Glaucidium castaneum Neumann 1893

Chestnut Owlet · Kastanienkauz · Chevêchette marron (ou châtaine) · Mochuelo Castano
***Glaucidium castaneum* Neumann 1893, Orn. Monatsb. 1: 62; Terra typica: Andundi, Zaire**

Length:	210 mm
Body mass:	~100 g

Distribution: Nothern Zaire: Semliki Valley and southwest Uganda: Bwamba Forest
Museum: IRSNB, MNBHU (holotype), MRAC
Remarks: Often regarded as conspecific with *G. capense*, but is a distinct species

Wing length:	128–139 mm
Tail length:	74.5–78.5 mm
Tarsus length:	?
Length of bill:	?
Body mass:	~100 g

Illustration: Reichenow 1902: Pl. 3; M. Woodcock in Fry et al. 1988: Pl. 9; C. Byers in del Hoyo et al. 1999: Pl. 16; F. Weick in König et al. 1999: Pl. 51; N. Borrow in Borrow and Demey 2001: Pl. 59; J. Gale in Stevenson and Fanshawe 2002: Pl. 102
Photograph: ?
Literature: Neumann 1893: 62; Eck and Busse 1973: 128, 129; Fry et al. 1988: 141, 142; Kemp and Kemp 1998: 296; del Hoyo et al. 1999: 223; König et al. 1999: 380, 381; Borrow and Demey 2001: 498; Stevenson and Fanshawe 2002: 204

❖

Glaucidium etchecopari Érard and Roux 1983

Chestnut-barred Owlet · Etchécoparkauz · Chevêchette de Etchécopar · Mochuelo de Etchécopar
***Glaucidium capense etchecopari* Érard and Roux 1983, Ois. Rev. France Orn. 53: 97–104; Terra typica: Liberia**
(see Plate 7)

Length:	200–210 mm
Body mass:	80–120 g

Distribution: Liberia and Ivory Coast, patchy distribution
Habitat: Old secondary forest and heavily logged forest
Museum: RMTB?

Wing length:	123–132 mm
Tail length:	?
Tarsus length:	?
Length of bill:	?
Body mass:	♂ 83 g, ♀ 93 and 119 g

Illustration: P. Hayman in Kemp and Kemp 1998: 297; F. Weick in König et al. 1999: Pl. 51
Photograph: ?
Literature: Érard and Roux 1983: 97–104; Thiollay 1985: 1–59; Fry et al. 1988: 141, 142; Gatter 1997: 146; Kemp and Kemp 1998: 296, 297; del Hoyo et al. 1999: 222, 223 (subspecies. *castaneum*); König et al. 1999: 379, 380; Borrow and Demey 2001: 498

Glaucidium capense (A. Smith) 1834
African Barred Owlet · Kapkauz · Chevêchette du Cap · Mochuelo de El Cabo

Length: 200–220 mm
Body mass: 90–140 g

Distribution: From eastern Kenya through central African woodland to Angola and Namibia, Mozambique south to the eastern Cape. Mafia Island
Habitat: Woodland and riparian trees in open country. Lowland and montane forest, secondary growth. Coastal forest (subspecies *scheffleri*). From sea-level up to about 1 200 m
Remarks: Recent studies suggest that *scheffleri* and *ngamiense* should both be regarded as distinct species, but further research is needed. Here considered as conspecific with *capense*

- ***Glaucidium (capense) scheffleri*** Neumann 1911
Gaucidium (sic) *scheffleri* Neumann 1911, Orn. Monatsb. 19: 184;
Terra typica: Kibwezi, southern Kenya
(see Plate 7)

Distribution: Extreme southern Somalia, eastern Kenya to northeast Tanzania
Museum: MNBHU (holotype), RMTB, SMNKa, ZFMK, ZSBS

Wing length: 132–140 mm
Tail length: 74–89 mm
Tarsus length: ?
Length of bill: ?
Body mass: ♂ 83–100 g, ♀ 93–113 g

Illustration: D. Zimmerman in Zimmerman et al. 1996: Pl. 56; P. Hayman in Kemp and Kemp 1998: 297; C. Byers in del Hoyo et al. 1999: Pl. 16; F. Weick in König et al. 1999: Pl. 51; J. Gale in Stevenson and Fanshawe 2002: Pl. 102
Photograph: ?
Literature: Neumann 1911: 184; Eck and Busse 1973: 128, 129; Prigogine 1983: 886–895; Fry et al. 1988: 141, 142; Zimmerman et al. 1996: 445; Kemp and Kemp 1998: 296, 297; del Hoyo et al. 1999: 222, 223; König et al. 1999: 379, 380; Stevenson and Fan-shawe 2002: 204

- ***Glaucidium capense ngamiense*** (Roberts) 1932
Smithiglaux capensis ngamiensis Roberts 1932, Ann. Transvaal Mus. 15: 26; Terra typica: Maun, Ngamiland

Synonym:
- *Glaucidium rufum* Gunning and Roberts 1911, Ann. Transvaal Mus. 3: 111; Terra typica: Boror, Mozambique (invalid name)
- *Glaucidium capense robertsi* Peters 1940, Check-list Bds. World 4: 132; Terra typica: Lake Tanganyika to lower Zambesi Valley
- *Glaucidium scheffleri clanceyi* Prigogine 1985, Gerfaut 75: 131–139; Terra typica: Zaire

Distribution: Central Tanzania, southeast Zaire, to southern Angola, northern Botswana, eastern Transvaal and south-central Mozambique, Mafia Island
Museum: IRSNB, RMTB = MRAC, MZUS, ZMA

Wing length: ♂ 131–143 mm, ♀ 131–147 mm
Tail length: 77–93 mm
Tarsus length: 24–28 mm
Length of bill: 18–20 mm
Body mass: ♂ 81–132 g, ♀ 93–139 g

Illustration: G. Arnott in Steyn 1984: Pl. 23; P. Hayman in Kemp and Kemp 1998: 297; C. Byers in del Hoyo et al. 1999: Pl. 16; F. Weick in König et al. 1999: Pl. 51
Photograph: Zool. Soc. London in Everett 1977: 41; P. S. Ginn in König and Ertel 1979: Pl. 146; P. Steyn in Steyn 1984: 265 (b/w juvenile); P. S. Ginn in Ginn et al. 1989: 339; P. Barichievy in Ginn et al. 1989: 339; H. Reinhard in Burton et al. 1992: 174; L. Hes in Kemp and Kemp 1998: 296; P. Chadwick in Kemp and Kemp 1998: 297
Literature: König and Ertel 1979: 205; Steyn 1984: 264–266; Prigogine 1983: 886–895; Prigogine 1985: 131–139; Fry et al. 1988: 141, 142; Ginn et al. 1989: 339; Kemp and Kemp 1998: 296, 297; del Hoyo et al. 1999: 222, 223; König et al. 1999: 379, 380

- ***Glaucidium capense capense*** (A. Smith) 1834
Noctua capensis A. Smith 1834, S. Afr. Q. J. 2: 313;
Terra typica: South Africa = eastern Cape
(see Figure 17)

Distribution: From Mozambique south to the eastern Cape
Museum: MHNP, MNBHU, SMTD, ÜMB

Wing length: 136–150 mm
Tail length: 101 mm
Tarsus length: 25.5 mm
Length of bill: ?
Body mass: 120 and 122 g

Illustration: T. J. Ford in Smith 1839: Pl. 33; M. Woodcock in Fry et al. 1988: Pl. 9; T. Boyer in Boyer and Hume 1991: 103; C. Byers in del Hoyo et al. 1999: Pl. 16; F. Weick in König et al. 1999: 379, 380
Photograph: ?
Literature: Sharpe 1875a: 223; Carlyon and Meakin 1985: 44, 45; Carlyon 1985: 22, 23; Fry et al. 1988: 141, 142; Ginn et al. 1989: 339; Kemp and Kemp 1998: 296, 297; del Hoyo et al. 1999: 222, 223; König et al. 1999: 379, 380

Glaucidium albertinum **Prigogine 1983**
Albertine Owlet · Albertseekauz · Chevêchette du Graben · Mochuelo de Alberto
Glaucidium albertinum **Prigogine 1983, Rev. Zool. Afric. 97(4): 886–895; Terra typica: Musangakye, Zaire**
(see Figure 17)

Length:	210 mm
Body mass:	73 g ($n = 1$)

Distribution: Albertine Rift in eastern Zaire and northern Rwanda
Habitat: Montane forest with dense undergrowth, from about 1 100 m up to 1 700 m above sea-level
Museum: RMTB = MRAC (holotype)

Wing length:	126–138 mm
Tail length:	61–70 mm
Tarsus length:	27 mm
Length of bill:	18.5 mm
Body mass:	73 g ($n = 1$)

Illustration: M. Woodcock in Fry et al. 1988: 141 (b/w); P. Hayman in Kemp and Kemp 1998: 329; C. Byers in del Hoyo et al. 1999: Pl. 16; F. Weick in König et al. 1999: Pl. 51; J. Gale in Stevenson and Fan-shawe 2002: Pl. 102
Photograph: Only from skins in Prigogine 1983: 888, Fig. 1 (b/w)
Literature: Prigogine 1983: 886–895; Fry et al. 1988: 141; Kemp and Kemp 1998: 328, 329; del Hoyo et al. 1999: 223; König et al. 1999: 381; Stevenson and Fanshawe 2002: 204

Genus

Xenoglaux **O'Neill and Graves 1977**

Xenoglaux O'Neill and Graves 1977, Auk 94(3): 410

Xenoglaux loweryi **O'Neill and Graves 1977**

Long-whiskered Owlet · Lowery-Zwergkauz · Chevêchette de Lowery · Mochuelo Peludo (de Lowery)

Xenoglaux loweryi O'Neill and Graves 1977, Auk 94(3): 411; Terra typica: Rio Mayo Valley, northern Peru (see Figure 17)

Length:	130–140 mm
Body mass:	~50 g

Distribution: Northern Peru: Rio Mayo Valley, northwest San Martin, eastern Andes northwest of Rioja

Habitat: Humid cloud forest with epiphytes and dense undergrowth, 1 900–2 200 m above sea-level

Museum: LSUMZ (holotype)

Wing length:	100–105 mm
Tail length:	50.5–55.5 mm
Tarsus length:	17.3–17.8 mm
Length of bill (cere):	9.6 a. 9.8 mm
Body mass:	46–51 g

Illustration: J. P. O'Neill in O'Neill and Graves 1977: Frontispiece; J. Fjeldsa in Fjeldsa and Krabbe 1990: Pl. 25; T. Boyer in Boyer and Hume 1991: 108; H. Burn in del Hoyo et al. 1999: Pl. 17; F. Weick in König et al. 1999: Pl. 52

Photograph: J. P. O'Neill in O'Neill and Graves 1977: 410; J. P. O'Neill in Burton et al. 1992: 175; J. P. O'Neill in del Hoyo et al. 1999: 84

Literature: O'Neill and Graves 1977: 409–416; Fjeldsa and Krabbe 1990: 228; Boyer and Hume 1991: 108; del Hoyo et al. 1999: 225; König et al. 1999: 383, 384

Xenoglaux loweryi · Long-whiskered Owlet · Lowery-Zwergkauz

Micrathene

Genus
Micrathene Coues 1866

Micrathene Coues 1866, Proc. Acad. Nat. Sci. Phila.: 51.
Type by *Athene whitneyi* Cooper

Synonym:
- *Micropallas* Coues 1889, Auk 4: 71

Micrathene whitneyi (Cooper) 1861

Elf Owl · Elfenkauz · Chevêchette des saguaros (ou Chouette elfe) · Tecolotito de los Saguaros
(see Plate 7 and Figure 17)

Length: 130–140 mm
Body mass: 36–44 g

Distribution: From southwestern United States (southern Arizona, New Mexico, southern Texas) to central Mexico, Baja California and Socorro Island

Micrathene whitneyi · Elf Owl · Elfenkauz

Habitat: Semi-open and cactus desert, riparian woodland, dry oak woodland, semi-arid wooded canyons, thorny woodland, semi-open bushland and mesquite. From sea-level up to 2 000 m

- ***Micrathene whitneyi whitneyi*** (Cooper) 1861
  Athene whitneyi Cooper 1861, Proc. Calif. Acad. Sci. 2: 118;
  Terra typica: Fort Mojave, Arizona

Distribution: Southwest United States (southern Nevada, southeast California, southern Arizona, southwest New Mexico and southwest Texas) south to northwest Mexico (Sonora). Winter: migrating to Mexico (one specimen mist-nettet on Tres Marias Island, Mexican migrant?)

Museum: AMNH, BMNH, Cal. Geol. Survey Coll. (holotype), FNSF, LACM, MHNP, MNBHU, USNM

Wing length: 102–115 mm
Tail length: 45–53 mm
Tarsus length: 20 mm
Length of bill (cere): 8–9.5 mm
Body mass: ♂ 36–44 g, ♀ 41–48 g

Illustration: A. Brooks in Grosvenor and Wetmore 1937: 22; W. Weber in Sprunt 1955: unpaged colour plate; K. E. Karalus in Karalus and Eckert 1974: Pl. 38; L. Malick in Breeden et al. 1983: 245; A. Cameron in Voous 1988: 171; B. Coleman in Shaw 1989: 95; T. Boyer in Boyer and Hume 1991: 110; H. Burn in del Hoyo et al. 1999: Pl. 17; F. Weick in König et al. 1999: Pl. 52; D. Sibley in Sibley 2000: 279 (adult and juvenile)

Photograph: R. Kinne in Everett 1977: 33; K. Fink in Campbell and Lack 1985: Pl. 443; K. Fink in Burton et al. 1992: 165; K. Fink in Johnsgard 2002: Pl. 22; R. and N. Bowers in internet: Owl Pages, photo gallery 2004

Literature: Ridgway 1914: 807–809; Sprunt 1955: 198; Bent 1961 (repr.): 438–444; Blake 1963: 217; Northern 1965: 358; Ligon 1968: 1–70; Eck and Busse 1973: 129, 130; Karalus and Eckert 1974: 177–183; Breeden et al. 1983: 244; Howell and Webb 1995: 363, 364; del Hoyo et al. 1999: 225; König et al. 1999: 384, 385; Sibley 2000: 279; Johnsgard 2002: 154–160

- ***Micrathene whitneyi sanfordi*** (Ridgway) 1914
  Micropallas whitneyi sanfordi Ridgway 1914, Bull. USN 50(6): XVII and 807 (809); Terra typica: Mirafloras, Lower California

Distribution: Baja California and parts of the Mexican mainland
Museum: USNM (holotype), MNBHU
Remarks: Sympatric with nominate, although taxon very doubtful! Probably best to unite with *M. w. whitneyi*

Wing length: 99–109.5 mm
Tail length: 45.5–53 mm
Tarsus length: ?
Length of bill (cere): 8–9.5 mm
Body mass: ?

Illustration: H. Quintscher in Eck and Busse 1973: Abb. 35
Photograph: D. Lockshaw in internet: Owl Pages, photo gallery 2002 (These photos probably show subspecies *idonea*!); G. Lasley in internet: Owl Pages, photo gallery 2002; R. and N Bowers in internet: Owl Pages, photo gallery 2005
Literature: Ridgway 1914: 810, 811; Eck and Busse 1973: 129, 130; Voous 1988: 170–175; del Hoyo et al. 1999: 225; König et al. 1999: 384, 385; Johnsgard 2002: 154–160

- ***Micrathene whitneyi idonea*** (Ridgway) 1914
  Micropallas whitneyi idonea Ridgway 1914, US Nat. Mus. Bull. 50(6): XVII, 807, 810; Terra typica: Hidalgo, Texas

Remarks: Validity of this subspecies is also questionable!

Distribution: Southern Texas, south to central Mexico: Puebla and Guanajuato
Museum: USNM, AMNH (holotype)

Wing length: 106.5–110.5 mm
Tail length: 49–50 mm
Tarsus length: ?
Length of bill (cere): 8.5–9 mm
Body mass: ?

Illustration: K. E. Karalus in Karalus and Eckert 1974: Pl. 39; F. Weick in König et al. 1999: Pl. 52
Photograph: See photograph *sanfordi*
Literature: Ridgway 1914: 810; Eck and Busse 1973: 129, 130; Karalus and Eckert 1974: 183, 184; del Hoyo et al. 1999: 225; König et al. 1999: 384, 385; Johnsgard 2002: 154–160

- ***Micrathene whitneyi graysoni*** Ridgway 1886
  Micrathene whitneyi graysoni Ridgway 1886, Auk: 333; Terra typica: Socorro Island

Synonym:
- *Micropallas socorroensis* "Ridgway" Sharpe 1899, Handlist 1: 299; Terra typica: Socorro Island = lapsus!

Distribution: Socorro Island
Museum: USNM (holotype)
Remarks: Probably extinct!

Wing length: ♂ 106.5 mm, ♀ 102–104 mm
Tail length: ♂ 51.5 mm, ♀ 44.5–49 mm
Tarsus length: ?
Length of bill (cere): 8.5–9 mm
Body mass: ?

Illustration: F. Weick in König et al. 1999: Pl. 52
Photograph: ?
Literature: Ridgway 1886: 333; Ridgway 1914: 810–812; del Hoyo et al. 1999: 225; König et al. 1999: 384, 385; Johnsgard 2002: 154–160

Athene

Genus
Athene Boie 1822

Athene Boie 1822, Isis v. Oken 1, col. 549.
Type by *Athene noctua* = *Strix noctua* Scopoli

Synonym:
- *Speotyto* Gloger 1842, Gemein. Handb. und Hilfsb. 1841: 226. Type by *Strix cunicularia* Molina
- *Heteroglaux* Hume 1873, Str. Feath. 1: 467. Type by *Heteroglaux blewitti* Hume

Athene noctua (Scopoli) 1769
Little Owl · Steinkauz · Chouette chevêche ou Chevêche d'Athéne · Mochuelo Común
(see Plate 3)

Length:	200–240 mm
Body mass:	160–200 g

Distribution: Britain (introduced; not Scotland), Eurasia from Iberia north to Denmark, southern Sweden and Latvia, east to Asia Minor, Levant, Arabia, central and eastern Asia to China and Manchuria, south to North Africa and the Red Sea Coast to Somalia and Eritrea. Introduced in New Zealand

Habitat: Wide variety of semi-open habits, from steppe and semi-arid desert to farmland and open woodland, also urban habitations. Extending to boreal and tropical areas. From sea-level up to montane regions to 3 000 m, sometimes to around 4 600 m. In Europe normally below 700 m

- ***Athene noctua vidalii*** A.E. Brehm 1857
Athene Vidalii A.E. Brehm 1857, Allg. Deutsche Naturh. Zeit. 3: 440; Terra typica: Mountains of Spain. Type Murcia, southeast Spain

Synonym:
- *Athene noctua mira* Witherby 1920, Brit. Bds. 13: 283; Terra typica: Northern Limburg, The Netherlands
- *Athene noctua Grüni* von Jordans and Steinbacher 1942, Annal. Naturhist. Mus. Wien 1941(52): 234; Terra typica: Lagos, southern Portugal
- *Athene noctua cantabriensis* Harrison 1957, Bull. Brit. Ornith. Cl. 77: 2; Terra typica: Laredo, Santander, northwest Spain

Distribution: Western Europe from Holland and Belgium south through France, Iberian Peninsula, sympatric with *noctua* in western Germany, introduced in England and Wales, straggling occasionally to Scotland and Ireland. North through Denmark to Poland and the Baltic States. Sympatric with indigena in Ukraine and Russia

Museum: BMNH, NNML, SMNKa, SMTD, ZFMK, ZMA, ZSBS

Wing length:	♂ 154–169 mm, ♀ 160–172 mm
Tail length:	68–83 mm
Tarsus length:	33–36.5 mm
Length of bill:	19–22 mm
Body mass:	160–206 g

Illustration: J. G. Walter in Creaman, ?, D'après Nature 1634: Fol. 7; J.F. Naumann in Naumann 1822: Pl. 48; H. C. Richter in Gould 1862–1873: Pl. 37; J. G. Keulemans in Hennicke 1905: unpaged; P. A. Robert in Géroudet 1979: 360; G. Pettersson in Pettersson 1984: 47; H. Delin in Cramp et al. 1985: Pl. 48; A. Cameron in Voous 1988. 183; A. Thorburn in Thorburn 1990: Pl. 27; H. Burn in del Hoyo et al. 1999: Pl. 17; F. Weick in König et al. 1999: Pl. 54

Photograph: Schönn in Schönn 1987: unpaged; E. Hosking in Hosking and Flegg 1982: 62, 63, 84; W. Layer in Mebs 1987: 53; H. Reinhard in *Gefiederte Welt* 1989(10): 309; O. Alamany in del Hoyo et al. 1999: 131; M. Wilkes in del Hoyo et al. 1999: 135; numerous photographs in Mebs and Scherzinger 2000 (by H. D. Brandl, J. Diedrich, F. Hortig, Stengel, Volmer, Wothe)

Literature: Naumann 1822: 493–500, Nachtr. 189, 190; Hartert 1912–1921: 999 and 1003; Hartert 1923: 389–391; Dementiev and Gladkov 1951: 440, 441; Harrison 1957: 2, 3; Vaurie 1965: 607, 608, 610; Mikkola 1983: 126–135; Cramp et al. 1985: 514–525; Mebs 1987: 52–59; Voous 1988: 181–187; Schönn et al. 1991: 237 pp; del Hoyo et al. 1999: 225, 226; König et al. 1999: 389–391; Mebs and Scherzinger 2000: 311–333

- ***Athene noctua noctua*** (Scopoli) 1769
Strix noctua Scopoli 1769, Annus I Hist.-Nat.: 22; Terra typica: Carnibia = Krain, Slowenia

Synonym:
- *Athene Chiaradiae* Giglioli 1900, Avicula 4: 57; Sacile, Udine (leucistic specimen)
- *Athene Sarda* Kleinschmidt 1907, Falco 3: 65; Terra typica: Sardinia (holotype ZFMK)

- *Athene noctua salentina* Trischitta 1939, (pamphlet): Alcune nuove ...: 2; Terra typica: Southern Italy
- *Athene noctua daciae* Keve and Kohl 1961, Bull. Brit. Ornith. Cl. 81: 51; Terra typica: Reghin, eastern Transylvania, Romania

Distribution: Sardinia, Corsica, mainland Italy, southeast Austria, northwest Yugoslavia, southern Czech Republic, Hungary and Romania north and west of Carpathia
Museum: NHMWien, NNML, SMNKa, SMTD, ZFMK, ZMA, ZMH
Remarks: Sympatric with *vidalii* in southern France, Switzerland, Germany etc.

Wing length: 152–169 mm
Tail length: 73–83 mm
Tarsus length: 29.5–33.5 mm
Length of bill (cere): 13–15.5 g
Body mass: ♂ 105–210 g, ♀ 120–215 g

Illustration: O. Kleinschmidt in Kleinschmidt 1907a: Pl. 2 and 3; P. Barruel in Sutter and Barruel 1958: Pl. 25; O. Kleinschmidt in Kleinschmidt 1958: Pl. 34; H. Delin in Cramp et al. 1985: Pl. 48 (adult and juvenile); F. Weick in Schönn et al. 1991: Pl. 1; L. Jonsson in Jonsson 1992: 325; H. Burn in del Hoyo et al. 1999, Pl. 17; F. Weick in König et al. 1999: Pl. 54; Mullarney andZetterström in Svensson et al. 1999: 215; W. Daunnicht in Mebs and Scherzinger 2000: 15; F. Weick in Weick 2004: Mai
Photograph: S. Dalton in Everett 1977: 125; E. Hosking in Hosking and Flegg 1982: 134; numerous photographs in Schönn et al. 1991 (by Alamany, Ille, Scherzinger, Schönn, etc.); E. Hosking in Burton et al. 1992: 161; numerous photographs in Mebs and Scherzinger 2000 (by Brandl, J. Diedrich, F. Hortig, Stengel, Volmer and Wothe); N. Blake in internet: Owl Pages, photo gallery 2004
Literature: Sharpe 1875a: 133, 134; Kleinschmidt 1907a: 8 pp; Hartert 1912–1921: 999–1002; Hartert 1923: 389, 390; Dementiev and Gladkov 1951: 440–444; Vaurie 1965: 607, 608, 610; Eck and Busse 1973:148–151; Glutz von Blotzheim and Bauer 1980: 501–532; Cramp et al. 1985: 514–525; Bezzel 1985: 648–651; Mebs 1987: 52–59; Schönn and Schönn 1987: 24 pp; Voous 1988: 181–187; Schönn et al. 1991: 237 pp; del Hoyo et al. 1999: 225, 226; König et al. 1999. 389–391; Mebs and Scherzinger 2000: 311–333

- ***Athene noctua indigena*** C.L. Brehm 1855
  Athene indigena C.L. Brehm 1855, Der Vollst. Vogelfang: 37; Terra typica: Greece, wander to Egypt, type Attica

Synonym:
- *Athene glaux kessleri* Semenov 1899, Zap. Imp. Akad. Nauk. Classe Sci. Phys. Math. Ser. 8, 6: 14; Terra typica: Crimea
- *Carine noctua caucasica* Zarudny and Loudon 1904, Orn. Jahrb.: 56; Terra typica: Surround Baku, eastern Caucasus

Distribution: Albania, southeast Yugoslavia, southern and eastern Roumania, southern Ukraine, southern Russia, Caucasus and southwest Siberia, south to Crete, Turkey (except the southeast) and the Middle East south to Haifa
Museum: AMNH, NNML, SMTD, ZFMK, ZMA, ZMH, ZSBS
Remarks: Sympatric with *vidalii* and *noctua*

Wing length: 158–174 mm
Tail length: 75–89 mm
Tarsus length: 32–34.5 mm
Length of bill (cere): 14–15.5 mm
Body mass: ♂ 162–175 g, ♀ 130–207 g

Illustration: H. Delin in Cramp et al. 1985: Pl. 48; F. Weick in König et al. 1999: Pl. 54
Photograph: W. Lange in *Gefiederte Welt* 1992(8): 278, 279; A. Ganz in Shirihai 1996: Pl. 67
Literature: Hartert 1912–1921: 1002, 1003; Hartert 1923: 390; Dementiev and Gladkov 1951: 441, 444, 445; Vaurie 1965: 611; Cramp et al. 1985: 514–525; Voous 1988: 181–187; del Hoyo et al. 1999: 225, 226; König et al. 1999: 389–391

- ***Athene noctua glaux*** (Savigny) 1809
  Noctua glaux Savigny 1809, Descript. Egypte, Hist. Nat. 1, Syst. Oiseaux: 105; Terra typica: Egypt

Synonym:
- *Strix Athene ruficolor* Kleinschmidt 1907, Falco 3: 65; Terra typica: Marakech, southern Morocco
- *Strix saharae* Kleinschmidt 1909, Falco 5: 19; Terra typica: Mouleina near Biskra, southern Algeria
- *Athene noctua solitudinis* Hartert 1924, Novit. Zool. 31: 18; Terra typica: Mount Todera, Air, southern Sahara

Distribution: North Africa, from Morocco southwest to Mauretania, south to the Sahara and east to Egypt. Sometimes Arabian Peninsula
Museum: MNBHU, MZUS, SMTD, ZFMK (Kleinschmidt coll.), ZMH (*saharae*), ZSBS

Wing length:	146–165 mm
Tail length:	69–79 mm
Tarsus length:	31–34.5 mm
Length of bill (cere):	12.5–15 mm
Body mass:	?

Illustration: O. Kleinschmidt in Kleinschmidt 1907a: Pl. 3 (light and dark); H. Delin in Cramp et al. 1985: Pl. 48; M. Woodcock in Fry et al. 1988: Pl. 8; H. Burn in del Hoyo et al. 1999: Pl. 17; F. Weick in König et al. 1999: Pl. 54
Photograph: Eshbol in Shirihai 1996: pl 67; P. Doherty in Shirihai 1996: Pl. 67
Literature: Sharpe 1875a: 135–137; Hartert 1912–1921: 1003, 1004; Hartert 1923: 390; Vaurie 1965: 608, 609; Etchécopar and Hüe 1967: 341, 342; Mikkola 1983: 126–135; Cramp et al. 1985: 514–525; Fry et al. 1988: 144, 145; Shirihai 1996: 316–318; Kemp and Kemp 1998: 302, 303; del Hoyo et al. 1999: 225, 226; König et al. 1999: 389–391

- ***Athene (noctua) lilith*** Hartert 1913
Athene (o. *Carine*) *noctua lilith* Hartert 1913, Die Vögel der paläarktischen Fauna 1912–1921: 1006;
Terra typica: Deir ez Zor, eastern Syria

Distribution: Cyprus and interior Middle East, from southeast Turkey south to southern Sinai
Museum: AMNH, ZFMK, ZSBS
Remarks: Intergrades with *indigena* and *bactriana*. Probably specifically distinct from *noctua* (vocalisation)

Wing length:	152–164 (168) mm
Tail length:	71–78 mm
Tarsus length:	29–31 mm
Length of bill (cere):	14–15 mm
Body mass:	?

Illustration: A. Cameron in Voous 1988: 183; H. Burn in del Hoyo et al. 1999: Pl. 17; F. Weick in König et al. 1999: Pl. 54; Mullarney and Zetterström in Svensson et al. 1999: 215
Photograph: Shirihai in Shirihai 1996: Pl. 67
Literature: Hartert 1912–1921: 1006; Hartert 1923: 390; Vaurie 1965: 609; Etchécopar and Hüe 1967: 341, 342; Mikkola 1983: 126–135; Cramp et al. 1985: 514–525; Voous 1988: 181–187; Shirihai 1996: 316–318; del Hoyo et al. 1999: 225, 226; König et al. 1999: 389–391; Svensson et al. 1999: 214

- ***Athene noctua spilogastra*** (Heuglin) 1869
Athene spilogastra von Heuglin 1863, J. Orn.: 14; Nomen nudum!
Noctua spilogastra von Heuglin 1869, Orn. Nordost Afr. 1: 119, Pl. 4; **Terra typica:** Ethiopian coastland = Massana, Eritrea

Distribution: Red Sea Coast from eastern Sudan to northern Ethiopia
Museum: SMNSt (holotype), ZFMK

Wing length:	146, 147 and 160 mm?
Tail length:	73–76 mm
Tarsus length:	26–32 mm
Length of bill (cere):	?
Body mass:	?

Illustration: von Heuglin in von Heuglin 1863: Pl. 4; F. Weick in König et al. 1999: Pl. 54
Photograph: ?
Literature: von Heuglin 1869: 119; Sharpe 1875a: 138; Mackworth-Praed and Grant 1957: 653; Fry et al. 1988: 144, 145; del Hoyo et al. 1999: 225, 226; König et al. 1999: 389–391

- ***Athene noctua somaliensis*** Reichenow 1905
Athene spilogaster somaliensis Reichenow 1905, Vog. Afr. 3: 822; **Terra typica:** Aurowana, northern Somaliland

Museum: MNBHU (holotype?), SMTD, ZFMK, ZMH
Remarks: Subspecific distinction based mainly on colour, which varies greatly both individually and geographically, thus all African forms might be considered as variations of subspecies *glaux*!?

Wing length:	129–144 mm
Tail length:	?
Tarsus length:	?
Length of bill (cere):	?
Body mass:	?

Illustration: O. Kleinschmidt in Kleinschmidt 1907a: Pl. 3; F. Weick in König et al. 1999: Pl. 54
Photograph: ?

Literature: Reichenow 1905: 822; Mackworth-Praed and Grant 1957: 654; Fry et al. 1988: 144, 145; Kemp and Kemp 1998: 302, 303; del Hoyo et al. 1999: 225, 226; König et al. 1999: 389–391

- ***Athene noctua bactriana*** Blyth 1847
  Athene bactrianus Blyth 1847, in Hutton, J. As. Soc. Beng. 16(2): 776; Terra typica: Old Kandahar, Afghanistan

Distribution: From southeast Azerbaijan, eastern Iraq, Iran and Afghanistan east through Central Asia to Lake Balkash
Museum: SMTD, ZFMK, ZSBS
Remarks: Sympatric with *lilith* and *indigena*

Wing length: 156–177 mm
Tail length: 84–87 mm
Tarsus length: 32 mm
Length of bill (cere): 18–20 mm
Body mass: ♂ 118–172 g, ♀ 165–260 g

Illustration: J. G. Keulemans in Sharpe 1891: Pl. 3; O. Kleinschmidt in Kleinschmidt 1907a: Pl. 3; F. Weick in König et al. 1999: Pl. 54; L. McQueen in Rasmussen and Anderton 2005: Pl. 79
Photograph: ?
Literature: Sharpe 1891: 14; Hartert 1912–1921: 1005, 1006; Hartert 1923: 390; Baker 1927: 442, 443; Dementiev and Gladkov 1951: 441, 445, 446; Vaurie 1965: 609, 610; Ali and Ripley 1969: 298; Cramp et al. 1985: 514–525; del Hoyo et al. 1999: 225, 226; König et al. 1999: 389–391; Rasmussen and Anderton 2005: 245, 246

- ***Athene noctua orientalis*** Severtzov 1873
  Athene orientalis Severtzov 1873, Izv. Imp. O. Liub. Est. Ant. Etn. 8, 1872, 2: 115; Terra typica: Turkestan, but no type design., lectotype from Issyk-Kul, Tian-Shan (by Dementiev (1931) Alauda 2(3): 258)

Distribution: Extreme northwest China and adjacent Siberia (up to 4 200 m!)
Museum: ZFMK, ZMM

Wing length: ♂ 152–156 mm, ♀ 166–169 mm (Tien-Shan), ♂ 161–175 mm, ♀ 170–181 mm (Tibet)
Tail length: ?
Tarsus length: ?
Length of bill (cere): ?
Body mass: ?

Illustration: D. Cole in Grimmett et al. 1998: Pl. 30
Photograph: ?
Literature: Dementiev and Gladkov 1951: 441, 446, 447; Vaurie 1965: 612; Etchécopar and Hüe 1978: 479, 480; Grimmett et al. 1998: 437; del Hoyo et al. 1999: 225, 226; König et al. 1999: 289–291

- ***Athene noctua ludlowi*** Baker 1926
  Athene noctua ludlowi Baker 1926, Bull. Brit. Ornith. Cl. 47: 58; Terra typica: Dochen, 15000 feet, near Rham Tso Lake, S Tibet

Distribution: Southern and central China, southern and eastern Tibet, southern and northern Himalayas
Museum: BMNH (holotype)

Wing length: 169–173 mm (Baker), -182 mm?
Tail length: 88-96 mm, -100 mm?
Tarsus length: 31 and 32 mm
Length of bill (cere): 18–20 mm
Body mass: ?

Illustration: L. McQueen in Rasmussen and Anderton 2005: Pl. 79
Photograph: ?
Literature: Hartert 1923: 390, 391; Baker 1927: 443; Vaurie 1965: 612; Ali and Ripley 1969: 399; Etchécopar and Hüe 1978: 479, 480; del Hoyo et al. 1999: 225, 226; König et al. 1999: 389–391; Rasmussen and Anderton 2005: 245, 246

- ***Athene noctua plumipes*** Swinhoe 1870
  Athene plumipes Swinhoe 1870, Proc. Zool. Soc. Lond.: 448; Terra typica: Near Shato, Nankou, China

Synonym:
- *Athene noctua impasta* Bangs and Peters 1928, Bull. Mus. Comp. Zool. 68: 330; Terra typica: South of Lake Kokonor, eastern Tsinghai

Distribution: Western, central and northern China, Inner Mongolia, Outer Mongolia north to southwest Transbaikalia and west to south-eastern Russian Altai
Museum: BMNH, MCZ (holotype *impasta*), ZFMK, ZMH

Wing length: ♂ 158–170 mm, ♀ 167–178.5 mm
Tail length: 86 mm
Tarsus length: 29 mm
Length of bill (cere): ?
Body mass: ?

Illustration: ?
Photograph: ?
Literature: Swinhoe 1870b: 448, 449; Hartert 1912–1921, 1006; Hartert 1923: 391; Dementiev and Gladkov 1951: 441, 447; Vaurie 1965: 612, 613; Etchécopar and Hüe 1978: 479, 489; Meyer de Schauensee 1984: 270; del Hoyo et al. 1999: 225, 226; König et al. 1999: 389–391

Athene brama (Temminck) 1821
Spotted Owlet · Brahma-Kauz · Chouette (ou Chevêche) brame · Mochuelo Brahmán
(see Plate 3 and Figure 18)

Length: 190–210 mm
Body mass: ~115 g

Distribution: Southern Asia, from Iran and southern Afghanistan through most of the Indian subcontinent (except Sri Lanka), to Vietnam. Also southeast Asia (except peninsular Thailand and Malaysia)
Habitat: Open forest, agricultural fields, semi-deserts, mango groves. Also within villages, ruins and human cultivations. From sea-level locally up to around 1 500 m

- ***Athene brama albida*** **Koelz 1950**
  Athene brama albida Koelz 1950, Am. Mus. Novit. 1452: 2; Terra typica: Saadatabad, Kerman, Iran

Distribution: Southern Iran and southern Pakistan
Museum: AMNH, SMNSt
Remarks: Slightly different from *indica*, probably a synonym (Vaurie 1965; Ali and Ripley 1969)

Wing length: 154–167 mm
Tail length: 74–82.5 mm
Tarsus length: ?
Length of bill (cere): ?
Body mass: ?

Illustration: ?
Photograph: ?
Literature: Koelz 1950: 2; Vaurie 1965: 613; del Hoyo et al. 1999: 226; König et al. 1999: 391, 392

- ***Athene brama indica*** **(Franklin) 1831**
  Noctua Indica Franklin 1831, Proc. Comm. Zool. Soc. London: 115; Terra typica: Unit. Provinces, India

Synonym:
- *Noctua Tarayensis* Hodgson 1836, As. Res. 19: 175; Terra typica: Nepal Terai

Distribution: Northern and central Indian Subcontinent (not Assam and Sri Lanka)
Museum: BMNH, MNBHU, SMTD, ZFMK, ZMH

Wing length: ♂ 153–169 mm, ♀ 159–171 mm
Tail length: 74–84 mm
Tarsus length: 27 and 28 mm
Length of bill (cere): 18–23 mm
Body mass: 110 and 114 g

Illustration: D. Cole in Grimmett et al. 1998: Pl. 30; S. McQueen in *Living Bird* 1998(Spring): 26; H. Burn in del Hoyo et al. 1999: Pl. 17; F. Weick in König et al. 1999: Pl. 54; F. Weick in *Gefiederte Welt* 2003(10): 304; L. McQueen in Rasmussen and Anderton 2005: Pl. 79
Photograph: G. Ziesler in del Hoyo et al. 1999: 125
Literature: Baker 1927: 440; Whistler 1963: 347; Vaurie 1965: 613; Ali and Ripley 1969: 299–301; Grimmett et al. 1998: 437; del Hoyo et al. 1999: 226; König et al. 1999: 391, 392; Rasmussen and Anderton 2005: 246, 247

- ***Athene brama ultra*** **Ripley 1948**
  Athene brama ultra Ripley 1948, Proc. Biol. Soc. Wash. 61: 100; Terra typica: Lakhimpur District, northeast Assam

Distribution: Northeast Assam
Museum: USNM (holotype), SMNSt (Assam!)
Remarks: Different vocalisations, but little morphological differentation from *indica*!

Wing length: 164–167 mm
Tail length: 83–93 mm
Tarsus length: ?
Length of bill (cere): 15 a. 16 mm
Body mass: ?

Illustration: F. Weick in König et al. 1999: Pl. 54
Photograph: ?
Literature: Ripley 1948b: 100; Eck and Busse 1973: 151, 152; König et al. 1999: 391, 392

- ***Athene brama brama*** (Temminck) 1821
  Strix brama Temminck 1821, Pl. col. livr. 12: Pl. 68;
  Terra typica: Pondicherry and western coast of India

Synonym:
- *Carine brama fryi* Baker 1919, Bull. Brit. Ornith. Cl. 40: 60; Terra typica: Rameswaram, Madras, India

Museum: BMNH, MZUS, SMNSt, ZFMK, Zool. Inst. Heidelberg, ZSBS

Wing length: ♂ 141–158 mm, ♀ 151–163 mm
Tail length: 66–74 (77) mm
Tarsus length: 27 and 28 mm
Length of bill: 19–22 mm
Body mass: ~115 g

Illustration: Temminck 1821e: Pl. 68; E. Traviès in Buffon 1851 and *Catal. Sotheby's* 1996: 77 no. 137; A. Cameron in Voous 1988: 191; T. Boyer in Boyer and Hume 1991: 126; D. Cole in Grimmett et al. 1998: Pl. 30; H. Burn in del Hoyo et al. 1999: Pl. 17; F. Weick in König et al. 1999: Pl. 54; L. McQueen in Rasmussen and Anderton 2005: Pl. 79
Photograph: E. Hosking in Hosking and Flegg 1982: 135; H. Reinhard in Burton et al. 1992: 158; L. Körner in internet: Owl Pages, photo gallery 2003; R. Saldino in internet: Owl Pages, photo gallery 2003
Literature: Sharpe 1875a: 138, 139; Hartert 1912–1921: 1006, 1007; Baker 1927: 439, 440; Ali and Ripley 1969: 302; Voous 1988: 189–192; Boyer and Hume 1991: 126; Grimmett et al. 1998: 437; del Hoyo et al. 1999: 226; König et al. 1999: 391, 392; Rasmussen and Anderton 2005: 246, 247

- ***Athene brama pulchra*** Hume 1873
  Athene pulchra Anonymus = Hume 1873, Str. Feath. 1: 469;
  Terra typica: Pegu, Burma

Synonym:
- *Athene brama mayri* Deignan 1941, Auk 58: 396; Terra typica: Ban Mak Khaeng, eastern Thailand (holotype USNM)

Distribution: Myanmar, Thailand (except the southern half of the peninsula), southern Laos, Cambodia
Museum: MNBHU, USNM, ZFMK
Remarks: Subspecies *mayri*, which differs only slightly in winglength, is considered a synonym

Wing length: 134–163 mm
Tail length: 65–74 mm
Tarsus length: 26–28 mm
Lenght of bill: 20–22 mm
Body mass: ?

Illustration: J. G. Keulemans in Sharpe 1891: Pl. 21; F. Weick in König et al. 1999: Pl. 54; T. Worfolk in Robson 2000: Pl. 21
Photograph: ?
Literature: Delacour and Jabouille 1931: 136; Deignan 1945: 181, 182; Smythies and Hughes 1984: 319; del Hoyo et al. 1999: 226; König et al. 1999: 391, 392; Robson 2000: 293

Athene blewitti (Hume) 1873

Forest Owlet · Blewitt-Kauz · Chevêche forestière (Chouette des forêts) · Mochuelo de Blewitt
Heteroglaux Blewitti **Anonymus=Hume 1873, Str. Feath. 1: 468;** Terra typica: **Busnah, Phooljan, India**
(see Figure 18)

Length: 220–230 mm
Body mass: ♂ 241 g ($n = 1$)

Remarks: Spurious use of the generic name *Heteroglaux*

Distribution: West-central and east-central India (Akrani Range, northwest Maharashtra and eastern Madhya Pradesh)
Habitat: Dry to dense moist lowland deciduous forest, at 200–500 m above sea-level
Museum: AMNH (1), BMNH (holotype), MCZ (2), NMI (1)

Wing length: 145–154 mm
Tail length: 63–72 mm
Tarsus length: 33–37 mm
Length of bill: 20 and 21 mm
Body mass: ♂ 241 g ($n = 1$)

Illustration: J. G. Keulemans in Sharpe 1891: Pl. 22 also in *Living Bird* 1998(Spring): 25; T. Boyer in Boyer and Hume 1991: 127; D. Cole in Grimmett et al. 1998: Pl. 30; S. McQueen in *Living Bird* 1998 (Spring): 25; L. McQueen in *Forktail* 1998(11):

Frontispiece; H. Burn in del Hoyo et al. 1999: Pl. 17; F. Weick in König et al. 1999: Pl. 53; F. Weick in *Gefiederte Welt* 2003(10):304; L. McQueen in Rasmussen and Anderton 2005: Pl. 79

Photograph: Rasmussen and Abbott in *Living Bird* 1998 (Spring): 28; F. Ishtiaq in del Hoyo et al. 1999: 151

Literature: Baker 1927: 441; Ali and Ripley 1969: 302, 303; Eck and Busse 1973: 152, 153; Ripley 1976: 1–4; King and Rasmussen 1998: 41–49; Gallagher 1998: 24–28; Rasmussen and Collar 1999: 11–21; del Hoyo et al. 1999: 226, 227; König et al. 1999: 387, 388; Weick 2003, *Gefiederte Welt*: 2003(10): 304; Rasmussen and Anderton 2005: 247

Athene cunicularia (Molina) 1782

Burrowing Owl · Präriekauz, Kaninchenkauz · Chouette des terriers · Lechuza vizcachera

Length:	190–250 mm
Body mass:	120–250 g

Athene cunicularia hypugaea · Burrowing Owl · Präriekauz

Distribution: Probably more than one species from western North America south to Central America and patchily southeast to the Atlantic Coast, also Hispaniola, locally in Cuba and other Caribbean Islands. Locally in northwest South America and the Andean region, on eastern and western (not wooded) slopes. Eastern South America from Brazil south to Patagonia and Tierra del Fuego. Extinct on some small islands. Northern birds winter migrants to Mexico, the Caribbean and, rarely, to Honduras and Panama

Habitat: Dry, open country, treeless plains, grassland and prairie. Also farmland, semi-desert and desert. Montane slopes with ravines and scattered bushes. From sea-level up to 4 500 m. Often associated with burrowing animals

Remarks: The subspecific (specific) status of some taxa (e.g. *brachyptera*, *minor*, *punensis*, *intermedia* etc.) is uncertain and needs re-evalution with modern systematic methods! The number of subspecies is too high (Voous 1988), all taxa in need of critical reevaluation

- ***Athene cunicularia hypugaea*** (Bonaparte)
  Strix hypugaea Bonaparte 1825, Am. Orn. 1: 72 and Pl. 7; Terra typica: Western United States, plains of Platte River (see Plate 3)

Synonym:

- *Speotyto cunicularia becki* Rothschild and Hartert 1902, Novit. Zool. 9: 405; **Terra typica:** Guadalupe Island and Baja California (holotype BMNH)

Distribution: From southern Canada and the western Great Plains, south to El Salvador

Museum: AMNH, ANSP, BMNH, SMTD, USNM

Wing length:	162.5–181 mm
Tail length:	71.5–86 mm
Tarsus length:	40–48.5 mm
Length of bill (cere):	13–15 mm
Body mass:	♂ ∅ 146.5 g, ♀ ∅ 156 g

Illustration: L. Bonaparte 1825: Pl. 7, Fig. 2; J. Audubon in Audubon 1827–1838: Pl. 244; R. Ridgway in Merriam and Fisher 1893: Pl. unpaged; A. Brooks in Grosvenor and Wetmore 1937: 20; A. Brooks in Sprunt 1955: unpaged plate; K. E. Karalus in Karalus and Eckert 1974: Pl. 36; L. Malick in Breeden et al. 1983: 247 (adult and juvenile);

A. Cameron in Voous 1988: 194; T. Boyer in Boyer and Hume 1991: 129; F. Weick in König et al. 1999: Pl. 53 (adult and juvenile); D. Sibley in Sibley 2000: 278 (adult and juvenile); L. A. Fuertes in Johnsgard 2002: Pl. 5; R. Bateman in Dean 2004: 64

Photograph: Grossman in Grossman and Hamlet 1965: near 141; S. Grossman in Everett 1977: 65 (juvenile); E. Hosking in Hosking and Flegg 1982: 25, 156; H. Rosenfeld in *Living Bird* 1983(Winter): Backcover; K. Rudloff in *Der Falke* 1984(6): 215; Burgess in Burton et al. 1992: 164; R. Curtis in del Hoyo et al. 1999: 111; S. J. Krasemann in del Hoyo et al. 1999: 122; D. Sept in del Hoyo et al. 1999: 138 (adult and juvenile); C. Glatzer in internet: Owl Pages, photo gallery 2004; K. Hildreth in internet: Owl Pages, photo gallery 2004; G. Lasley in internet: Owl Pages, photo gallery 2004

Literature: Sharpe 1875a: 142–147; Ridgway 1914: 812, 820; Grosvenor and Wetmore 1937: 21; Sprunt 1955: 200–203; Bent 1961 (repr.): 384–396; Grossman and Hamlet 1965: 448, 449; Eck and Busse 1973: 153–155; Karalus and Eckert 1974: 163–172; Breeden et al. 1983: 246; Minnemann1984, *Der Falke* 1984(6): 214; Voous 1988: 193–199; Boyer and Hume 1991: 127, 128; del Hoyo et al. 1999: 227, 228; König et al. 1999: 386, 387; Sibley 2000: 278; Johnsgard 2002: 161–169

- ***Athene cunicularia rostrata*** (Townsend) 1890
  Speotyto rostrata C.H. Townsend 1890, Proc. US Nat. Mus. 13: 133; Terra typica: Clarion Island, Revillagigedo Group, off western Mexico

Distribution: Clarion Island, in Revillagigedo Islands, off western Mexico
Museum: BMNH, MNBHU, USNM (holotype)
Remarks: Plumage identical to that of *hypugaea*

Wing length: 160–169 mm
Tail length: 70–79.5 mm
Tarsus length: 44.5–49 mm
Length of bill (cere): 15.5–17 mm
Body mass: ?

Illustration: ?
Photograph: ?
Literature: Ridgway 1914: 820; del Hoyo et al. 1999: 227, 228; König et al. 1999: 386, 387; Johnsgard 2002: 161–169

- ***Athene cunicularia floridana*** (Ridgway) 1874
  Speotyto cunicularia var. *floridana* Ridgway 1874, Am. Sportsman 4: 216; Terra typica: 16 miles east of Sarasota Bay, Florida

Synonym:
- *Speotyto bahamensis* Maynard 1899, App. to Catal. West Ind. Bds.: 33; Terra typica: New Providence, The Bahamas
- *Speotyto cunicularia cavicola* Bangs 1900, Auk 17: 287; new name for *bahamensis*
- *Speotyto cunicularia guantanamensis* Garrido 2001; Terra typica: Southern Cuba

Distribution: Florida, extreme south of Georgia, The Bahamas, Cuba (locally) and Isla de Juvendad (Isle of Pines)
Museum: MNBHU, SMTD, USNM (holotype)

Wing length: 154–170 mm
Tail length: 70–80.5 mm
Tarsus length: 41–46.5 mm
Length of bill (cere): 14–15.5 mm
Body mass: ♂ ~148.8 g, ♀ ~149.7 g

Illustration: K. E. Karalus in Karalus and Eckert 1974: Pl. 37; F. Weick in König et al. 1999: Pl. 53; D. Sibley in Sibley 2000: 278; K. Williams in Raffaele et al. 2003: Pl. 43
Photograph: A. M. Bailey in Bent 1961: Pl. 87–89; J. Templeton in MacKenzie 1986: 131
Literature: Ridgway 1914: 820–823; Sprunt 1955: 200–203; Sprunt 1955: 200–203; Bent 1961 (repr.): 396–401; Karalus and Eckert 1974: 174–176; Bond 1986: 121; del Hoyo et al. 1999: 227, 228; König et al. 1999: 386, 387; Sibley 2000: 278; Raffaele et al. 2003: 100

- ***Athene cunicularia troglodytes*** (Wetmore and Swales) 1931
  Speotyto dominicensis Cory 1886, Auk 3: 471; Terra typica: Haiti. Invalid name
  Speotyto cunicularia troglodytes Wetmore and Swales 1931, US Nat. Mus. Bull. 155: 41 and 239; new name for *dominicensis*

Distribution: Hispaniola, including Gonave and Beata Islands
Museum: BMNH, SMNSt, USNM (holotype)

Wing length: 145–165.5 mm
Tail length: 64.5–76.5 mm
Tarsus length: 38–45.5 mm
Length of bill (cere): 14–15.5 mm
Body mass: ?

Illustration: F. Weick in König et al. 1999: Pl. 53
Photograph: ?
Literature: Cory 1886: 471; Ridgway 1914: 823, 824; Wetmore and Swales 1931b: 41, 239; Kelso 1934d: 61; Bond 1986: 121; del Hoyo et al. 1999: 227, 228; König et al. 1999: 386, 387; Raffaele et al. 2003: 100

- ***Athene cunicularia amaura*** (Lawrence) 1878†
Speotyto amaura Lawrence 1878, Proc. US Nat. Mus. 1: 234; Terra typica: Antigua, West Indies

Distribution: Extinct. Formerly Nevis and Antigua Islands, Lesser Antilles (St. Christopher Island?)
Museum: USNM (holotype)

Wing length: 145–154 mm
Tail length: 70–75.5 mm
Tarsus length: 39.5–41.5 mm
Length of bill (cere): 14–15.5 mm
Body mass: ?

Illustration: ?
Photograph: ?
Literature: Ridgway 1914: 825; Eck and Busse 1973: 153–155; del Hoyo et al. 1999: 227, 228; König et al. 1999: 386, 387

- ***Athene cunicularia guadeloupensis*** (Ridgway) 1874†
Speotyto cunicularia guadeloupensis Ridgway 1874, in Baird et al., A history of North American Birds 3: 90; Terra typica: Guadeloupe Island, West Indies

Distribution: Extinct. Formerly Guadeloupe Island, Lesser Antilles
Museum: Mus. Boston Soc. Nat. Hist. (holotype), USNM

Wing length: 152–162.5 mm
Tail length: 75.5–86.5 mm
Tarsus length: 42.5–46.5 mm
Length of bill (cere): 15 a. 15.5 mm
Body mass: ?

Illustration: Line drawings of primaries in Sharpe, 1875: 148, but from St. Nevis birds!, so possibly subspecies *amaura*
Photograph: ?
Literature: Ridgway 1874b: 90; Sharpe 1975: 147–149; Ridgway 1914: 824, 825

- ***Athene cunicularia brachyptera*** (Richmond) 1896
Speotyto brachyptera Richmond 1896, Proc. US Nat. Mus. 18: 663; Terra typica: Margarita Island, off northern Venezuela

Synonym:
- *Speotyto cunicularia arubensis* Cory 1915, Field Mus. Nat. Hist. Publ. Orn. 1: 299; Terra typica: Aruba Island, Dutch Antilles (holotype FMNH, similar in size and plumage to *brachyptera*)
- *Speotyto cunicularia apurensis* Gilliard 1940, Am. Mus. Novit. 1071: 5; Terra typica: Upper Apure Valley, Venezuela (holotype AMNH, also small and dark birds, very similar to *brachyptera*)

Distribution: Margarita Island, off northern Venezuela, Aruba Island, Dutch Antilles off northern Venezuela and northern and central Venezuela
Museum: AMNH, FMNH, SMNSt, USNM (holotype)

Wing length: 142.5–155 mm
Tail length: 63.5–74 mm
Tarsus length: 39.5–43 mm
Length of bill (cere): 13.5–14.5 mm
Body mass: 155 g (Hilty)

Illustration: S. Webb in Hilty 2003: Pl. 24 (*apurensis*)
Photograph: K. Rudloff in *Gefiederte Welt*: 1989(6): 181
Literature: Stone 1899: 302–304; Cory 1915: 663; Gilliard 1940: 5; Eck and Busse 1973: 153–155; Voous 1988: 193–199; del Hoyo et al. 1999: 227, 228; König et al. 1999: 386, 387; Hilty 2003: 363

- ***Athene cunicularia minor*** (Cory) 1918
Speotyto cunicularia minor Cory 1918, Field Mus. Nat. Hist. Publ. Zool. Ser. 13(2): 40; Terra typica: Brazil and probably adjacent parts of British Guiana and Surinam

Remarks: Subspecific status uncertain, very similar in size to *brachyptera*

Distribution: Southern Guiana, Surinam and extreme northern Brazil (Roraima)
Museum: FMNH (holotype), MZUS

Wing length: 142 and 143 mm
Tail length: ?
Tarsus length: ?
Length of bill (cere): ?
Body mass: ?

Illustration: ?
Photograph: ?
Literature: Cory 1918: 40; von Boetticher 1929: 386–392; del Hoyo et al. 1999: 227, 228; König et al. 1999: 386, 387; Hilty 2003: 363

- ***Athene cunicularia carrikeri*** (Stone) 1922
  Speotyto cunicularia carrikeri Stone 1922, Auk 39: 84; Terra typica: Palmas, Boyaca, Columbia

Distribution: Eastern Colombia
Museum: ANSP (holotype)
Remarks: Plumage extremely light. Very different to the geographically nearest subspecies *tolimae* – a very dark bird

Wing length: 173 mm
Tail length: 78 mm
Tarsus length: 46 mm
Length of bill (cere): 14 mm
Body mass: ?

Illustration: ?
Photograph: ?
Literature: Stone 1922: 84; Kelso 1934d: 61; Meyer de Schauensee 1964: 118; Hilty and Brown 1986: 230; Voous 1988: 193–199; del Hoyo et al. 1999: 227, 228; König et al. 1999: 386, 387

- ***Athene cunicularia tolimae*** (Stone) 1899
  Speotyto cunicularia tolimae Stone 1899, Proc. Acad. Nat. Sci. Phila.: 303; Terra typica: Plains of Tolima, Colombia

Distribution: Western Colombia, upper Magdalena Valley (arid tropical zone!) see remarks for subspecies *carrikeri*. Small and dark birds, similar to subspecies *brachyptera*
Museum: ANSP (holotype)

Wing length: 152.5 mm
Tail length: ?
Tarsus length: ?
Length of bill (cere): ?
Body mass: ?

Illustration: ?
Photograph: ?
Literature: Stone 1899: 302–304; Meyer de Schauensee 1964: 118; Hilty and Brown 1986: 230; Fjeldsa and Krabbe 1990: 230, 231; del Hoyo et al. 1999: 227, 228; König et al. 1999: 386, 387

- ***Athene cunicularia pichinchae*** (von Boetticher) 1929
  Speotyto cunicularia pichinchae von Boetticher 1929, Senckenbergiana 11: 391; Terra typica: Mount Pichincha, Ecuador

Distribution: Western Ecuador, north to Quito
Museum: FNSF (holotype), ZMH

Wing length: 175 and 177 mm
Tail length: ?
Tarsus length: ?
Length of bill (cere): ?
Body mass: ?

Illustration: F. Weick in König et al. 1999: Pl. 53; P. J. Greenfield in Ridgely and Greenfield 2001: II: Pl. 35(14)
Photograph: Only b/w photograph from feathers in *Senckenbergiana* 1929(11)
Literature: von Boetticher 1929: 386–392; Fjeldsa and Krabbe 1990: 230, 231; König et al. 1999: 386, 387; Ridgely and Greenfield 2001: I: 310, II: 218

- ***Athene cunicularia punensis*** (Chapman) 1914
  Speotyto cunicularia punensis Chapman 1914, Bull. Am. Mus. Nat. Hist. 35: 318; Terra typica: Puna Island, off Ecuador

Distribution: Arid littoral of western Ecuador, from Bahia de Caraques to northwest Peru. But Chapman (1926, p 250), regarded all Ecuadorian specimens of *A. cunicularia* with the exception Puna birds as a subspecies of *A. c. nanodes*
Museum: AMNH (holotype), MNBHU

Wing length: 180 and 183 mm
Tail length: ?
Tarsus length: ?
Length of bill (cere): ?
Body mass: ?

Illustration: H. Burn in del Hoyo et al. 1999: Pl. 17
Photograph: ?
Literature: Chapman 1914: 318 and 1926: 250; Fjeldsa and Krabbe 1990: 230, 231; del Hoyo et al. 1999: 227, 228; König et al. 1999: 386, 387; Ridgely and Greenfield 2001: I: 310, II: 218

- ***Athene cunicularia nanodes*** (Berlepsch and Stolzmann) 1892
Speotyto cunicularia nanodes von Berlepsch and Stolzmann 1892, Proc. Zool. Soc. Lond.: 388; Terra typica: Lima, Peru

Synonym:
- *Speotyto cunicularia intermedia* Cory 1915, Field Mus. Nat. Hist. Publ. Orn. 1: 300; Terra typica: Pacasmayo, western Peru (holotype FMNH, skin also ZFMK)

Distribution: Southwest and west Peru
Remarks: Plumage very similar to that of subspecies *grallaria*, but sometimes "bleached" (Stone 1899, pp 302–304)
Museum: BMNH, FMNH, SMTD, USNM, ZFMK

Wing length: 164–173 mm
Tail length: 77–83 mm
Tarsus length: 40 and 41 mm
Length of bill (cere): (17) 21–23 mm
Body mass: ?

Illustration: M. Koepcke in Koepcke 1964: 67 (b/w)
Photograph: ?
Literature: von Berlepsch and Stolzmann 1892: 387, 388; Stone 1899: 302–304; Koepcke 1964: 67; Fjeldsa and Krabbe 1990: 230, 231; König et al. 1999: 386, 387

- ***Athene cunicularia juninensis*** (Berlepsch and Stolzmann) 1902
Speotyto cunicularia juninensis von Berlepsch and Stolzmann 1902, Proc. Zool. Soc. Lond. 2: 41; Terra typica: Andes, central Peru and western Bolivia

Distribution: Central Peru and western Bolivia to northwest Argentina
Museum: BMNH, ZFMK

Wing length: ♂ 193–200 mm, ♀ 213 mm
Tail length: ♂ 93–100 mm, ♀ 110 mm
Tarsus length: ♂ 41–43 mm, ♀ 50 mm
Length of bill: 21–23 mm
Body mass: ?

Illustration: J. Fjeldsa in Fjeldsa and Krabbe 1990: Pl. 25 (adult and juvenile); F. Weick in König et al. 1999: Pl. 53
Photograph: ?
Literature: von Berlepsch and Stolzmann 1902: 41; von Boetticher 1929: 390–392; Fjeldsa and Krabbe 1990: 230, 231; del Hoyo et al. 1999: 227, 228; König et al. 1999: 386, 387

- ***Athene cunicularia boliviana*** (L. Kelso) 1939
Speotyto cunicularia boliviana L. Kelso 1939, Biol. Leaflet 11: unpaged; Terra typica: Warnes, Santa Cruz, Bolivia, 400 m

Distribution: Bolivia, sympatric with subspecies *cunicularia* in Tucumán
Museum: USNM

Wing length: ?
Tail length: ?
Tarsus length: ?
Length of bill: ?
Body mass: ?

Illustration: ?
Photograph: ?
Literature: Kelso 1939: Biol. Leaflet, 11: unpaged

- ***Athene cunicularia grallaria*** (Temminck) 1822
Strix grallaria Temminck 1822, Pl. col. livr. 25: Pl. 146; Terra typica: Brazil = Faxina, Sao Paulo (Hellmayr 1929)

Synonym:
- *Speotyto cunicularia beckeri* Cory 1915, Field Mus. Nat. Hist. Publ. Orn. 1: 299; Terra typica: Sao Marcello, Rio Preto, Bahia, Brazil (holotype FMNH)

Distribution: Central and eastern Brazil, probably sympatric with subspecies *cunicularia*!
Museum: FMNH, NNML, SMNKa, SMNSt, SMTD, USNM, ZFMK
Remarks: Very similar to the nominate in plumage and size!

Wing length: 168–198 mm
Tail length: 109–114 mm
Tarsus length: 37–50 mm
Length of bill: ?
Body mass: ?

Illustration: Temminck 1822: Pl. 146; H. Burn in del Hoyo et al. 1999: Pl. 17; F. Weick in König et al. 1999: Pl. 53

Photograph: K. W. Fink in Campbell 1974: 117; G. Ziesler in del Hoyo et al. 1999: 103
Literature: von Boetticher 1929: 390–392; del Hoyo et al. 1999: 227, 228; König et al. 1999: 386, 387

- ***Athene cunicularia partridgei*** **Olrog 1976**
Athene cunicularia partridgei Olrog 1976, Neotropica 22(68): 180; Terra typica: Corrientes, northeast Argentina

Distribution: Northeast Argentina, sympatric with subspecies *juninensis*
Museum: IML

Wing length: ? (similar in size to subspecies *grallaria* and *juninensis*!)

Illustration: J. Fjeldsa in Fjeldsa and Krabbe 1990: Pl. 25
Photograph: ?
Literature: Olrog 1976: 180; Fjeldsa and Krabbe 1990: 230, 231

- ***Athene cunicularia cunicularia*** **(Molina) 1782**
Strix cunicularia Molina 1782, Sagg. Stor. Nat. Chili: 263; Terra typica: Chile

Distribution: Southern Bolivia and southern Brazil, south to Tierra del Fuego. Also Paraguay and Uruguay. Probably sympatric with subspecies *boliviana* (southern Bolivia) and *grallaria* (southern Brazil)
Museum: MZUS, SMNSt, SMTD, ZFMK, ZMH

Wing length: 179–200 mm
Tail length: 78–114 mm
Tarsus length: 46–60 mm
Length of bill (cere): 15.5–16 mm
Body mass: 170–240 g

Illustration: H. Burn in del Hoyo et al. 1999: Pl. 17; F. Weick in König et al. 1999: Pl. 53; A. Burke in Jaramillo et al. 2003: Pl. 59 (adult, juvenile, flight)
Photograph: T. Daskam 1977: Agenda, unpaged (Directorio); C. König in König 1983, Frontispiece
Literature: Sharpe 1875a: 146, 147; Ridgway 1914: 812, 813; Eck and Busse 1973: 153–155; Daskam 1977: unpaged (Directorio); Voous 1988: 193–199; del Hoyo et al. 1999: 227, 228; König et al. 1999: 386, 387; Jaramillo et al. 2003: 146

Tribus / Tribe
Aegolini

Genus
Aegolius Kaup 1829

Aegolius Kaup 1829, Skizz. Entwickl.-Gesch. Europ. Thierw.: 34. Type by *Strix tengmalmi* Gmelin = *Strix funereus* von Linné 1758

Synonym:
- *Nyctale* Brehm 1828, Isis v. Oken, col. 1271 = nomen nudum. Type by *Strix tengmalmi* Gmelin
- *Scotophilus* Swainson 1837, Class. Birds II: 217. Type by *Strix tengmalmi* Gmelin
- *Gisella* Bonaparte 1854, Rev. Mag. Zool. II, 6: 541. Type by *Nyctale harrisii* Cassin
- *Cryptoglaux* Richmond 1901, Auk 18: 193. Type by *Strix tengmalmi* Gmelin
- *Microscops* Buturlin 1910, Nascha Ochota 4: 10. Type by *Strix acadica* Gmelin

Aegolius funereus (Linné) 1758
Boreal Owl, Tengmalm's Owl · Raufußkauz · Chouette de Tengmalm · Lechuza de Tengmalm
(see Plate 3)

Length: 230–260 mm
Body mass: 100–200 g

Distribution: In the north, distribution matches that of the coniferous belt of the Holarctic. South of the latter, isolated populations exist locally in Europe, Pyrenees, Carpathians, the Balkans, Alps, Black Forest etc. Also in the Caucasus, Tien Shan, Himalayas etc. In North America, distribution is largely confined to the forested areas of the Rocky Mountains and the northern coniferous belt. East of the Rockies south to New Mexico, west of the Rockies from Alaska to Oregon

Habitat: Coniferous forest with old trees, also mixed woodland. Winters in mature forest. Occurs from lowlands up to 1 200 m (central and southern Germany) to 2 000 m in the Alps and Asia. In the North American Rockies up to 3 000 m

- ***Aegolius funereus funereus*** (Linné) 1758
  Strix funerea von Linné 1758, Syst. Nat. ed. 10, 1: 93;
  Terra typica: Europe = Sweden

Synonym:
- *Strix tengmalmi* Gmelin 1788, Syst. Nat. 1: Pl. 1
- *Cryptoglaux tengmalmi transvolgensis* Buturlin 1910, Nascha Ochota 4: 11; Terra typica: Govern. of Kazan, Perm and Orenburg

Distribution: Europe locally in Pyrenees, Carpathians, Balkans, Alps, Black Forest etc. From northern Scandinavia south to the Pyrenees and Alps, east to Greece and from the Baltic Republics to Russia, north of the Caspian Sea, excluding the Caucasus

Museum: BMNH, FNSF, MNBHU, NNML, SMTD, ZFMK, ZMA, ZMH, ZSBS

Wing length: ♂ 162–176 mm, ♀ 164–182 mm
Tail length: ♂ 89–102 mm, ♀ 95–114 mm
Tarsus length: 21–24 mm
Length of bill: 19.5–21.5 mm
Body mass: 98–215 g

Illustration: Rudbeck 1695–1710 in Krook 1988 (repr.): Pl. 91 and 95; J. F. Naumann in Naumann 1822: Pl. 48; E. Lear in Gould 1832–1837, vol. 1: 49; H. C. Richter in Gould 1862–1873 (repr.) unpaged; J. G. Keulemans in Hennicke 1905: unpaged; P. Barruel in Sutter and Barruel 1958: Pl. 24; L. Binder in Wüst 1970: 245; P. Robert in Géroudet 1979: 361; H. Delin in Cramp et al. 1985: Pl. 48; A. Cameron in Voous 1988: 287; A. Thorburn in Thorburn 1990: Pl. 27; T. Boyer in Boyer and Hume 1991: 164; H. Burn in del Hoyo et al. 1999: Pl. 17; F. Weick in König et al. 1999: Pl. 55; Mullarney and Zetterström in Svensson et al. 1999: 211; F. Weick in Weick 2004: November; F. Weick in Kröher and Weick 2004: 185, 224

Photograph: H. Schrempp in *Der Falke* 1974(2): Frontcover; H. Schrempp in *Der Falke* 1980(9): Frontcover; A. Leinonen in Mebs 1987: 61; G. Sauer in Mebs 1987: 63; F. Adam in Mebs 1987: 65; Holynki in *Der Falke* 1999(10): 301; numerous photographs in Mebs and Scherzinger 2000: 87, 288, 289, 305 (by Hautala, Hedvall, Höfer, Scherzinger and Wothe)

Literature: Naumann 1822: 500–507, Nachtr. 190, 191; Sharpe 1875a: 284–286; Hartert 1912–1921: 995–998; Dementiev and Gladkov 1951: 435–438; Vaurie 1965: 604, 605; März 1968: 48 pp; Eck and Busse 1973: 184–186; Glutz von Blotzheim and Bauer 1980: 533–578; Mikkola 1983: 256–286; Cramp et al. 1985: 606–620; Bezzel 1985: 663–667; Mebs 1987: 60–67; Voous 1988: 284–292; Boyer and Hume 1991: 163–165; del Hoyo et al. 1999: 228; König et al. 1999: 392–395; Svensson et al. 1999: 211; Mebs and Scherzinger 2000: 288–310

- ***Aegolius funereus pallens*** (Schalow) 1908
  Nyctale tengmalmi pallens Schalow 1908, J. Orn. 56: 109; Terra typica: Kashka Su, Tian Shan

Synonym:
- *Cryptoglaux tengmalmi sibirica* Buturlin 1910, Nascha Ochota 11: 11; Terra typica: Khanka Lake, southern Ussuriland

Distribution: Western Siberia, Tian Shan, southern Siberia south through northeast China (Heilongjiang) to the Russian Far East (including Sakhalin)
Museum: MNBHU, ZFMK
Remarks: Similar to the nominate, but slightly paler and greyer

Wing length: ♂ 162–176 mm, ♀ 164–182 mm
Tail length: ?
Tarsus length: ?
Length of bill: ?
Body mass: ?

Illustration: L. B. McQueen in Rasmussen and Anderton 2005: Pl. 79
Photograph: ?
Literature: Dementiev and Gladkov 1951: 436, 440; Vaurie 1965: 605, 606; Voous 1988: 284–292; Cramp et al. 1985: 616; del Hoyo et al. 1999: 228, 229; König et al. 1999: 392–395; Rasmussen and Anderton 2005: 247

- ***Aegolius funereus magnus*** Buturlin 1907
  Nyctala magna Buturlin 1907, Psovaya i Ruzheinaya Okhota 13(6): 87; Terra typica: Kolyma region and Kamchatka

Synonym:
- *Nyctala jakutorum* Buturlin 1908, J. Orn. 56: 287; Terra typica: Yakutia

Distribution: Northeast Siberia, from Kolyma to Kamchatka, accidental in Alaska
Museum: ZFMK, ZMA, ZMM
Remarks: Subspecies intergrades in the west with *pallens*, intermediates have been named *jakutorum*! This subspecies is distinctly paler than *pallens*, with larger white spots

Wing length: ♂ 172–188 mm, ♀ 180–192 mm
Tail length: ?
Tarsus length: ?
Length of bill: ?
Body mass: 109 g ($n = 1$)

Illustration: H. Burn in del Hoyo et al. 1999: Pl. 17; F. Weick in König et al. 1999: Pl. 55
Photograph: ?
Literature: Hartert 1912–1921: 999; Kelso 1934d: 58; Dementiev and Gladkov 1951: 436, 439, 440; Vaurie 1965: 606; Cramp et al. 1985. 616; Voous 1988: 284–292; del Hoyo et al. 1999: 228, 229; König et al. 1999: 392–395

- ***Aegolius funereus caucasicus*** (Buturlin) 1907
  Nyctala Caucasica Buturlin 1907, Psovaya i Ruzheinaya Okhota 13(6): 87; Terra typica: Kislovodsk, northern Caucasus

Synonym:
- *Aegolius tengmalmi beickianus* Stresemann 1928, Orn. Monatsb. 36: 41; Terra typica: Northeast Tsinghai, western China (holotype MNBHU)
- *Aegolius funerea juniperi* Koelz 1939, Proc. Biol. Soc. Wash. 52: 80; Terra typica: Northern Punjab

Distribution: Northern Caucasus south to Transcaucasia, northwest Himalayas, northwest India and mountains of western China
Museum: ZMM, MNBHU (holotype *beickianus*)

Wing length: (151) 154–166 mm
Tail length: 75–92 mm
Tarsus length: ?
Length of bill: ?
Body mass: ?

Illustration: ?

Photograph: ?

Literature: Hartert 1912–1921: 998; Stresemann 1928: 41; Dementiev and Gladkov 1951: 436, 438; Vaurie 1965: 606, 607; Ali and Ripley 1969: 317; Cramp et al. 1985: 616; del Hoyo et al. 1999: 228, 229; König et al. 1999: 392–395

- ***Aegolius funereus richardsoni*** (Bonaparte) 1938
  Nyctale Richardsoni Bonaparte 1938, Geogr. and Comp. List: 7; new name for *Strix tengmalmi* Audubon, Bds. Am. folio ed.: Pl. 380; **Terra typica:** Bangor, Maine

Distribution: North America, from Alaska to western United States: Washington, Idaho, Montana, Wyoming, Colorado and locally New Mexico. East through northern Ontario, central Quebec and Labrador, south to southern British Columbia, central Alberta, southern Quebec and New Brunswick

Museum: AMNH, ANSP, BMNH, FNSF, SMTD, USNM, ZFMK

Wing length: ♂ 163–171.5 mm, ♀ 171.5–189 mm
Tail length: 95.5–107 mm
Tarsus length: ?
Length of bill (cere): 13.5–16 mm
Body mass: ♂ 93–139 g, ♀ 132–215 g

Illustration: J. Audubon in Audubon 1827–1838: Pl. 242 (380?); L. A. Fuertes in Gilbert Pearson 1936: Pl. 55; L. A. Fuertes in Forbush and May 1955: Pl. 47; J. F. Landsdowne in Landsdowne and Livingston 1967: Pl. 21; K. E. Karalus in Karalus and Eckert. 1974:Pl. 11; G. Miksch-Sutton in *Living Bird* 1980(1): 5; L. Malick in Breeden et al. 1983: 247 (adult and juvenile); F. Weick in König et al. 1999: Pl. 55; D. Sibley in Sibley 2000: 277 (adult and juvenile); L. A. Fuertes in Johnsgard 2002: Pl. 9

Photograph: B. and P. Wood in MacKenzie 1986: 113; E. T. Jones in MacKenzie 1986: 119; P. and G. Haywood in Johnsgard 2002: Pl. 31 and 32; A. Cook in internet: Owl Pages, photo gallery 2004 (2 photographs); D. Aspery in internet: Owl Pages, photo gallery 2004

Literature: Ridgway 1914: 624–627; Forbush and May 1955: 277, 278; Sprunt 1955: 212–214; Bent 1961 (repr.): 220–228; Landsdowne and Livingston 1967: 21; Karalus and Eckert 1974: 66–71; Breeden et al. 1983: 246; del Hoyo et al. 1999: 228; König et al. 1999: 392–395; Johnsgard 2002: 224–233

Aegolius acadicus (Gmelin) 1788

Northern Saw-whet Owl · Sägekauz · Chouette Scie (Petit Nyctale) · Tecolotito Cabezón (Tecolote abetero Norteno) (see Plate 3)

Length: 180–210 mm
Body mass: 55–125 g

Distribution: Occurs south of a line from Newfoundland and the mouth of the St. Lawrence River in the east, to Queen Charlotte Island in the west. In the east extending south to North Carolina, in the west south to Arizona and in the higher mountains as far as central Mexico (Oaxaca)

Habitat: Dense woodland, often moist in swampy areas. Locally in more open woodland. In winter with a wide range of habitats, varying in vegetation and altitude. Breeds generally above 1 350 m, in Mexico up to 2 500 m

- ***Aegolius acadicus acadicus*** (Gmelin) 1788
  Strix acadica Gmelin 1788, Syst. Nat. 1(1): 296;
  Terra typica: North America = Nova Scotia, ex Latham

Synonym:
- *Nyctale acadica scotaea* Osgood 1901 North Am. Fauna 21: 43; **Terra typica:** Masset, Graham Island, Queen Charlotte Islands (Biol. Survey = holotype, not different to nominate)
- *Aegolius acadicus brodkorbi* Briggs 1954, Proc. Biol. Soc. Wash. 67: 180; **Terra typica:** Amalepac Oaxaca

Remarks: Holotype *brodkorbi* is from a single juvenile specimen, may be a hybrid of *acadica* × *ridgwayi*, range overlaps

Distribution: From southern Alaska, British Columbia east to the Gulf of St. Lawrence and south to the southern United States (California, Arizona, New Mexico) and Florida. Also patchily in the highlands of Mexico, northeast Sonora, central Michoacan, east to Pueblo, Hidalgo and central Oaxaca. Isolated population in southeast Coahuila, southwest Nuevo León and northern San Luis Potosi

Museum: AMNH, ANSP, BMNH, FNSF, MSUZ, SMTD, USNM, ZFMK

Wing length: ♂ 125–141 mm, ♀ 134–146 mm
Tail length: 65–73 mm
Tarsus length: 21.5 mm
Length of bill (cere): 11–14 mm
Body mass: ♂ 54–96 g, ♀ 65–124 g

Illustration: J. J. Audubon in Audubon 1827–1838: Pl. 243; J. Cassin in Cassin 1854: Pl. 11; L. A. Fuertes in Gilbert Pearson 1936: Pl. 55; A. L. Fuertes in Forbush and May 1955: Pl. 47; J. F. Landsdowne in Landsdowne and Livingston 1968: Pl. 33; K. E. Karalus in Karalus and Eckert 1974: Pl. 9; E. A. Gilbert in A Thought for Christmas 1975; L. Malick in Breeden et al. 1983: 247; A. Cameron in Voous 1988: 295; T. Boyer in Boyer and Hume 1991: 166; H. Burn in del Hoyo et al. 1999: Pl. 17; F. Weick in König et al. 1999: Pl. 55 (adult and juvenile); D. Sibley in Sibley 2000: 277 (adult and juvenile); L. A. Fuertes in Johnsgard et al. 2002: Pl. 10

Photograph: J. T. McKeen in Everett 1977: 39; G. Manzaros in *Living Bird* 1983(Winter): 13; Anonymus in MacKenzie 1986: 122; R. B. Ranford in MacKenzie 1986: 127; B. Coleman in Shaw 1989: 95; Austing in Burton et al. 1992: 168; Haeg in *SCRO Mag.* 1997(1): 36; P. Johnsgard in Johnsgard 2002: Pl. 30; A. G. Nelson in Johnsgard 2002: Pl. 33; F. Bednar in internet: Owl Pages, photo gallery 2002; J. S. Huy in internet: Owl Pages, photo gallery 2004; J. Hobbs in internet: Owl Pages, photo gallery 2004; J. Noll Ueblacker in internet: Owl Pages, photo gallery 2004

Literature: Ridgway 1914: 627–633; Forbush and May 1955: 278, 279; Bent 1961 (repr.): 228–241; Landsdowne and Livingston 1968: 33; Karalus and Eckert 1974: 56–64; Breeden et al. 1983: 246; Voous 1988: 293–298; Boyer and Hume 1991: 166, 167; Howell and Webb 1995: 370; del Hoyo et al. 1999: 228, 229; König et al. 1999: 395–397; Sibley 2000: 277; Johnsgard 2002: 234–241

- ***Aegolius acadicus brooksi*** (Fleming)
  Cryptoglaux acadica brooksi Fleming 1916, Auk 33: 422;
  Terra typica: Graham Island, Queen Charlotte Islands

Distribution: Queen Charlotte Islands, British Columbia
Museum: AMNH, Skidegate Mus.

Wing length:	♀ 140.5 mm ($n = 3$)
Tail length:	?
Tarsus length:	?
Length of bill:	?
Body mass:	62.7–118.6 g

Illustration: K. E. Karalus in Karalus and Eckert 1974: Pl. 10; H. Burn in del Hoyo et al. 1999: Pl. 17; F. Weick in König et al. 1999: Pl. 55

Photograph: R. Steinberg 1997a: 28 (skins and mounted specimen); S. Sealy in *Kauzbrief* 2004(16): 12 (b/w)

Literature: Fleming 1916: 420–423; Bent 1961 (repr.): 242; Karalus and Eckert 1974: 64, 65; Steinberg 1997a: 24–30; del Hoyo et al. 1999. 228, 229; König et al. 1999: 395–397

Aegolius ridgwayi **(Alfaro) 1905**
Unspotted Saw-whet Owl · Ridgwaykauz · Chouette de Ridgway (Nyctale immaculée) · Tecolote abetero Sureno
Cryptoglaux ridgwayi **Alfaro 1905, Proc. Biol. Soc. Wash. 18: 217;**
Terra typica: Cerro de la Candelaria, Costa Rica

Length:	180–200 mm
Body mass:	80 g

Synonym:
- *Cryptoglaux rostrata* Griscom 1930, Am. Mus. Novit. 438: 1; Terra typica: Sacapulas, 4 500 feet, Rio Negro Valley, Guatemala (holotype AMNH, probably a hybrid of *acadicus* × *ridgwayi*?)
- *Aegolius ridgwayi tacanensis* Moore 1947, Proc. Biol. Soc. Wash. 60: 141; Terra typica: Vulcan Tacaná, Chiapas, southern Mexico (holotype MLZ, probably a hybrid of *acadicus* × *ridgwayi*?)

Distribution: From southern Mexico through Guatemala, Honduras and eastern Salvador to Costa Rica and western Panama

Museum: AMNH, MLZ, USNM (holotype)

Remarks: The *rostratus* "question" needs further studies: large bill, colour of forehead as in *acadicus*. Validity of subspecies uncertain. It has been suggested that *rostratus* and *tacanensis* are perhaps hybrids and thus synonyms

Wing length:	133–146 mm
Tail length:	64 mm
Tarsus length:	25 mm
Length of bill (cere):	13–20 mm
Body mass:	80 g (Stiles and Skutch 1991)

Illustration: J. A. Gwynne in Ridgely and Gwynne 1989: 191(b/w); D. Gardner in Stiles and Skutch 1991: Pl. 20; T. Boyer in Boyer and Hume 1991: 168; S. Webb in Howell and Webb 1995: Pl. 25; H. Burn in del Hoyo et al. 1999: Pl. 17; F. Weick in König et al. 1999: Pl. 55

Photograph: ?
Literature: Alfaro 1905: 217; Ridgway 1914: 633, 634 (descr. young male); Griscom 1930a: 438 (descr. *rostrata/ridgwayi-acadica*); Kelso 1934d: 59; Moore 1947: 141; Blake 1963: 224; Land 1970: 142 (*rostrata, tacanensis);* Ridgely and Gwynne 1989: 192; Stiles and Skutch 1991: 196; Boyer and Hume 1991: 167, 168; Howell and Webb 1995: 370; del Hoyo et al. 1999: 229; König et al. 1999: 397, 398

Aegolius harrisii **(Cassin) 1849**
Buff-fronted Owlet · Blaßstirnkauz · Nyctale de Harris (Chouette de Harris) · Mochuelo Canela (Lechuzita acanelada)

Length: 180–230 mm
Body mass: 130 g

Distribution: Lowlands of southeast Brazil to Paraguay and northeast Argentina. At 375–2 000 m from Tucumán to Jujuy, northwest Argentina and patchily in Bolivia (west of Comurapa Santa Cruz and at 3 900 m in Cochabamba), Yurinagui Alto in Junin, Cushi in Pasco, Cajamarca and around Huancabamba in Peru, Zambiza in northeast and Pichincha in northwest Ecuador. Narino, Cauca and "Bogota" Colombia, Mérida, Caracas and Neblina Mountains in Venezuela. Generally rare
Habitat: Fairly open, humid forest, up to the treeline, also in drier areas. In northwest Argentina and Bolivia mainly in ravines with dense wood (*Podocarpus*, *Alnus* and *Polylepsis*)

Aegolius harrisii · Buff-fronted Owlet · Blaßstirnkauz

- ***Aegolius harrisii harrisii*** (Cassin) 1849
Nyctale Harrisii Cassin 1849, Proc. Acad. Nat. Sci. Phila. 4: 157;
Terra typica: South America

Distribution: The Andes, from northwest Venezuela south to northern and central Peru
Museum: ANSP (holotype), BMNH, FMNH (specimen from Mt. Neblina), FNSF, MHH, SMNSt, ÜMB
Remarks: Two specimens collected from Cerro Neblina in 1985 probably different subspecies from nominate *harrisii* (Willard et al. 1991)

Wing length: 142–164 mm
Tail length: 78–89 mm
Tarsus length: 28 mm
Length of bill (cere): 16.5–18.5 mm
Body mass: 130 g

Illustration: H. Quintscher in Eck and Busse 1973: Abb. 42 (b/w); J. Fjeldsa in Fjeldsa and Krabbe 1990: Pl. 25; T. Boyer in Boyer and Hume 1991: 168; H. Burn in del Hoyo et al. 1999: Pl. 17; F. Weick in König et al. 1999: Pl. 55; F. Weick in *Orn. Mitt.* 1999(4): Abb. 12; P. J. Greenfield in Ridgely and Greenfield 2001: II: Pl. 35(13); S. Webb in Hilty 2003: Pl. 24
Photograph: R. Harling in König 1994: 11 (b/w mounted bird)
Literature: Sharpe 1875a: 283, 284; Eck and Busse 1973: 183, 184; Hilty and Brown 1986: 233; Fjeldsa and Krabbe 1990: 231; Willard et al. 1991: 1–80; Boyer and Hume 1991: 168; König 1994: 1–25; König 1999a: 428; König 1999b: 127–138; del Hoyo et al. 1999: 229; König et al. 1999: 398, 399; Ridgely and Greenfield 2001: I: 314, 315, II: 222; Hilty 2003: 367

- ***Aegolius harrisii iheringi*** (Sharpe) 1899
Gisella iheringi Sharpe 1899, Bull. Brit. Ornith. Cl. 8: 40;
Terra typica: Sao Paulo, Brazil

Distribution: Eastern Bolivia, Paraguay, central and eastern Brazil, south to northeast Argentina and northeast Uruguay

Museum: BMNH, FNSF, ZSBS
Remarks: Perhaps a distinct species (vocalisations)

Wing length:	♂ 144 mm, ♀ 167 mm
Tail length:	♂ 70.5 mm, ♀ 83 mm
Tarsus length:	27–30 mm
Length of bill (cere):	16.5–18.5 g
Body mass:	?

Illustration: F. Weick in König 1999b: 137 (b/w line drawing); H. Burn in del Hoyo et al. 1999: Pl. 17
Photograph: ?
Literature: Sharpe 1875a: 7–8; Eck and Busse 1973: 182–184; Fjeldsa and Krabbe 1990: 231; König 1999b: 127–138; del Hoyo et al. 1999: 229; König et al. 1999: 298, 299; Hilty 2003: 367

- ***Aegolius harrisii dabbenei*** Olrog 1979
  Aegolius harrisii dabbenei Olrog 1979, Acta Zool. Lilloana 33: 5–7

Distribution: Northwest Argentina (Tucumán, Salta and Jujuy, probably western Bolivia)
Museum: Instituto M. Lillo, Tucumán (holotype), SMNSt
Remarks: Taxonomic status of this subspecies very doubtful and uncertain – needs further studies

Wing length:	146 mm
Tail length:	82 mm
Tarsus length:	?
Length of bill (cere):	?
Body mass:	?

Illustration: ?
Photograph: ?
Literature: Dabbene 1926: 395 (records from Tucumán); Olrog 1979a: 1–324; Olrog 1979b: 5–7; Maijer and Hohnwald 1997: 223–235; Fjeldsa and Krabbe 1990: 231; König 1999b: 127–138; del Hoyo et al. 1999: 229; König et al. 1999: 398, 399

Tribus / Tribe
Ninoxini

Ninox

Genus
Ninox Hodgson 1837

Ninox **Hodgson 1837, Madras J. Lit. Sci. 5: 23.**
Type by *Ninox nipalensis* Hodgson = *Strix lugubris* Tickell

Synonym:
- *Hieracoglaux* Kaup 1848, Isis v. Oken: 768. Type by *Falco connivens* Latham
- *Spiloglaux* Kaup 1848, ibid.: 768. Type by *Strix novae-Seelandiae* Gmelin
- *Cephelptynx* Kaup 1852, Contrib. Orn.: 105. Type by *Noctua punctulata* Quoy and Gaimard
- *Ctenoglaux* Kaup 1852, ibid.: 109. Type by *Strix hirsuta* Temminck
- *Rhabdoglaux* Bonaparte 1854, Rev. Mag. de Zool.: 544. Type by *Athene strenua* Gould
- *Berneyornis* Mathews 1916, Birds Austr. 5: 305. Type by *Athene strenua* Gould

Ninox rufa (Gould) 1846

Rufous (Hawk) Owl · Rostkauz, Roter Buschkauz · Ninoxe rousse · Ninox Rojizo
(see Plate 8)

Length: ♂ 44–52 mm, ♀ 40–46 mm
Body mass: ♂ 1 150–1 300 g, ♀ 700–1 050 g

Distribution: New Guinea, Aroe (Aru) and Waigeo Islands, northern Australia (west of the Gulf of Carpentaria) and eastern coast of Queensland
Habitat: Tropical rainforest, well wooded savanna, monsoon woodland, gallery forest and along waterways. Lowlands and foothills, in Australia up to 1 200 m, in New Guinea up to 2 000 m from sea-level

- ***Ninox rufa rufa*** (Gould) 1846
  Athene rufa Gould 1846, Proc. Zool. Soc. Lond. 14: 18; Terra typica: Port Essington

Distribution: Northeast Western Australia (Kimberleys) and northern North Territory (Arbhem Land)
Museum: AMNH, BMNH, NMV (HLW coll.)

Wing length: ♂ 374–383 mm, ♀ 347–357 mm
Tail length: ?
Tarsus length: ?
Length of bill: ?
Body mass: ♂ 1 150–1 300 mm, ♀ 700–1 050 mm

Illustration: N. Cayley in Cayley 1961 (repr.): Pl. 5(17); N. Day in Simpson and Day 1986: 153; P. Slater in Slater et al. 1989: 193 (adult and juvenile); T. Boyer in Boyer and Hume 1991: 112; N. Arlott in del Hoyo et al. 1999: Pl. 18; F. Weick in König et al. 1999: Pl. 56 (adult and juvenile)
Photograph: E. Hosking in Hosking and Flegg 1982: 18; J. Calaby in Frith et al. 1983: 302; D. Hollands in Hollands 1991: 4, 10, 12, 15, 19–21, 24; D. Hollands in Burton et al. 1992: 147; P. Kranz in internet: Owl Pages, photo gallery 2003; D. Lewis in internet: Owl Pages, photo gallery 2003; J. and S. Boller in internet: Owl Pages, photo gallery 2004
Literature: Mees 1964a: 6, 7, 53; Eck and Busse 1973: 134; Schodde and Mason 1881; Frith et al. 1983: 302; Simpson and Day 1986. 152; Slater et al. 1989: 192; Boyer and Hume 1991: 112; Hollands 1991: 11–27 and 207; del Hoyo et al. 1999: 231; König et al. 1999: 399, 400

- ***Ninox rufa meesi*** Mason and Schodde 1980
  Ninox rufa meesi Mason and Schodde 1980, Emu 80(3): 141–144; Terra typica: Queensland

Synonym:
- *Ninox rufa marginata* Mees 1964, Zool. Verh. Leiden 65: 8; Terra typica: Cardwell, Queensland (invalid name)

Distribution: Queensland: coastal and subcoastal Cape York to about Endeavour River and Mitchell River
Museum: AMNH, HLW
Remarks: Plumage as in the nominate, but smaller

Wing length: ♂ 313–349 mm, ♀ 306–352 mm
Tail length: ?
Tarsus length: ?
Length of bill: ?
Body mass: ?

Illustration: ?
Photograph: ?
Literature: Mees 1964a: 8, 9, 53; Mason and Schodde 1980: 141–144; Eck and Busse 1973: 134; del Hoyo et al. 1999: 231; König et al. 1999: 399, 400

- ***Ninox rufa queenslandica*** **Mathews 1911**
Ninox rufa queenslandica Mathews 1911, Bull. Brit. Ornith. Cl. 27: 62; Terra typica: The Hollows, Mackay, Queens Park Creek

Distribution: Coastal and subcoastal Queensland, from Endeavour River south to lower Burdekin River and perhaps Water
Remarks: Size as in *meesi*, but darker above, and with darker crossbands below. Probabaly *meesi* is only a pale morph of *queenslandica*!
Museum: AMNH (holotype)

Wing length:	♂ 348 mm ($n = 1$)
Tail length:	?
Tarsus length:	?
Length of bill:	?
Body mass:	?

Illustration: N. Arlott in del Hoyo et al. 1999: Pl. 18; F. Weick in König et al. 1999: Pl. 56
Photograph: ?
Literature: Mees 1964a: 9, 10, 53; del Hoyo et al. 1999: 231; König et al. 1999: 399, 400

- ***Ninox rufa humeralis*** **(Bonaparte) 1850**
Athene humeralis Bonaparte 1850, Consp. Gen. Av. 1: 40; Terra typica: Oceania = Triton Bay, New Guinea

Synonym:
- *Ninox aruensis* Schlegel 1866, Nederl. Tijdschr. Dierk. 3: 329; **Terra typica:** Aru Island
- *Ninox franseni* Schlegel 1866, ibid.: 256; **Terra typica:** Waigeo Island; **Remarks:** Much smaller than *humeralis*; **Wing length:** 260 mm (1 ♀) (holotype NNML!)
- *Ninox undulata* Ramsay 1879, Proc. Linn. Soc. New S. Wales 3: 249; **Terra typica:** South coast of New Guinea

Distribution: New Guinea, also Aroe (Aru) and Waigeo Islands
Museum: AMNH, MZUS, NNML, SMTD, ZMH, ZSBS

Wing length: ♂ 327–347 mm, ♀ 307–330 mm (260 and 270 mm Aru Island)

Illustration: T. Medland in Iredale 1956: Pl. 7; N. Arlott in del Hoyo et al. 1999: Pl. 18; F. Weick in König et al. 1999: Pl. 56
Literature: Iredale 1956: 136; Mees 1964a: 10, 33; Beehler et al. 1986: 131; del Hoyo et al. 1999: 231; König et al. 1999: 399, 400

❖

Ninox strenua **(Gould) 1838**
Powerful Owl · Riesenkauz · Ninoxe puissante (ou géante) · Ninox Robusto
Athene ? strenua **Gould 1838, Syn. Birds Austr. 3: Pl. 47; Terra typica: New South Wales**
(see Plate 8)

Length:	♂ 520–650 mm, ♀ 480–540 mm
Body mass:	1 050–1 700 g

Synonym:
- *Ninox strenua victoriae* Mathews 1912, Aust. Avian Rec. 1: 75; **Terra typica:** Victoria, Australia

Distribution: Coastal and subcoastal southeast Queensland (south from Dawson River), eastern New South Wales and southeast Victoria to extreme southeast of South Australia

Ninox strenua · Powerful Owl · Riesenkauz

Ninox connivens
Barking Owl
Kläfferkauz

Habitat: Open wet and dry forest, densely forested gullies and ravines. Prefers to be near water. Coastal woodland and scrub. Sometimes in cultivated areas and pine plantations. From sea-level up to 1500 m
Museum: AMNH, BMNH, HLW, NMV, SMNSt, SMTD, USNM

Wing length: ♂ 398–427 mm, ♀ 381–410 mm
Tail length: 279.5 mm
Tarsus length: 53–54.5 mm
Length of bill: ?
Body mass: ♂ 1 130–1 700 g, ♀ 1 050–1 600 g

Illustration: Gould in Gould 1838a: Pl. 47; H. C. Richter in Gould 1848–1869: Pl. 35; N. Cayley in Cayley 1961 (repr.): Pl. 5(18); N. Day in Simpson and Day 1986: 153 (adult and juvenile); P. Slater in Slater et al. 1989: 193; T. Boyer in Boyer and Hume 1991: 113; N. Arlott in del Hoyo et al. 1999: Pl. 18; F. Weick in König et al. 1999: Pl. 56 (adult and juvenile)
Photograph: E. Hosking in Hosking and Flegg 1982: 150; Calaby and Lindsay in Frith et al. 1983: 302; I.R. McCann in *Grampian Birds* 1982: 10; D. Hollands in Hollands 1991: 28, 32–44, 47; D. Hollands in Burton et al. 1992: 146; T. and P. Gardner in del Hoyo et al. 1999: 148; T. Quested in internet: Owl Pages, photo gallery 2003; C. Hübner in internet: Owl Pages, photo gallery 2004
Literature: Fleay 1944: 97–112; Mees 1964a: 11–13, 53; Eck and Busse 1973: 133; Frith et al. 1983: 302; Simpson and Day 1986: 152; Slater et al. 1989: 192; Hollands 1991: 28–49 and 207, 208; Boyer and Hume 1991: 112, 113; del Hoyo et al. 1999: 231; König et al. 1999: 401, 402

Ninox connivens **(Latham) 1801**
Barking (or Winking) Owl · Kläfferkauz · Ninoxe (Chouette) aboyense · Ninox Labrador
(see Plate 8)

Length: 380–440 mm
Body mass: 425–510 g

Distribution: Northern Moluccas, New Guinea and north, northwest and northeast Australia, Queensland, southwest Australia and east and southeast Australia
Habitat: Riparian forest and forest edges in lowlands, often near swamps, streams and other wetlands. In New Guinea and Karkas Island up to around 1 000 m

- ***Ninox connivens rufostrigata*** (G.R. Gray) 1861
Athene rufostrigata G.R. Gray 1861, Proc. Zool. Soc. Lond. 1860: 344; **Terra typica:** Eastern Gilolo

Distribution: Northern Moluccas: Morotai, Halmahera, Bacan and Obi
Museum: AMNH, BMNH (holotype), NNML

Wing length: 258–291 mm (1 × 295 mm)
Tail length: 165–198 mm
Tarsus length: 43–51 mm
Length of bill: ?
Body mass: ?

Illustration: Johnstone and Darnell 1997: Fig. 2(8) (line drawing); D. Gardner in Coates and Bishop 1997: Pl. 33; F. Weick in König et al. 1999: Pl. 56
Photograph: ?
Literature: Gray 1861: 344; Sharpe 1875a: 177; Mees 1964a: 36, 37; Eck and Busse 1973: 134, 135; Coates and Bishop 1997: 361; del Hoyo et al. 1999: 231, 232; König et al. 1999: 402, 403

- ***Ninox connivens assimilis*** Salvadori and D'Albertis 1875
Ninox assimilis Salvadori and D'Albertis 1875, Ann. Mus. Civ. Genova 7: 802, 809; **Terra typica:** Mount Epa, New Guinea

Synonym:
- *Ninox albomaculata* Ramsay 1879, Proc. Linn. Soc. New S. Wales 3: 249; Terra typica: Laloki, New Guinea

Distribution: Central and eastern New Guinea, west to the Merauke and Sepik River, also Manam and Karkas Islands
Museum: AMNH, BMNH, MGD (holotype), NNML, SMTD

Wing length:	(244?) 255–277 mm
Tail length:	143 mm
Tarsus length:	36 mm
Length of bill:	30 mm
Body mass:	♂ 380 g, ♀ 430 g

Illustration: N. Arlott in del Hoyo et al. 1999: Pl. 18
Photograph: ?
Literature: Mees 1964a: 36, 56; Rand and Gilliard 1967: 355, 356; Eck and Busse 1973: 134, 135; Beehler et al. 1986: 131, 132; del Hoyo et al. 1999: 231, 232; König et al. 1999: 402, 403

- ***Ninox connivens occidentalis*** Ramsay 1887
  Ninox connivens occidentalis Ramsay 1887, Proc. Linn. Soc. New S. Wales; 1886, 2(1): 1086; Terra typica: Derby, northwest Australia

Synonym:
- *Ninox connivens suboccidentalis* Mathews 1912, Novit. Zool. 18: 255; Terra typica: Port Keats, Northern Territory

Distribution: Western Australia, Northern Territory and northwest Queensland. (holotype AMNH)
Museum: AM (Sydney) (holotype), AMNH, BMNH, HLW, NMV; USNM, WAM
Remarks: Sometimes included in subspecies *peninsularis*

Wing length:	272–306 mm (Northwest Australia and Northern Territory); 280–310 mm (Northwest Queensland)
Tail length:	?
Tarsus length:	?
Length of bill:	?
Body mass:	?

Illustration: F. Knight in Olson 2001: 205
Photograph: ?
Literature: Mees 1964a: 34–36, 56; del Hoyo et al. 1999: 231, 232; König et al. 1999: 402, 403

- ***Ninox connivens peninsularis*** Salvadori 1876
  Ninox peninsularis Salvadori 1876, Ann. Mus. Civ. Genova 7: 992; Terra typica: Cape York

Distribution: Cape York Peninsula, Queensland, Thursday Island, Banks or Moa Island (Torres Strait). Boundary of subspecies *connivens* possibly Endeavour River?
Museum: AMNH, BMNH, HLW
Remarks: Smaller size and slightly darker plumage than nominate. Also smaller size than *occidentalis*

Wing length:	257–288 mm
Tail length:	?
Tarsus length:	?
Length of bill:	?
Body mass:	?

Illustration: ?
Photograph: ?
Literature: Mees 1964a: 34, 56; del Hoyo et al. 1999: 231, 232; König et al. 1999: 402, 403

- ***Ninox connivens connivens*** (Latham) 1801
  Falco connivens Latham 1801, Index Orn. Suppl. XII; Terra typica: "Nova Hollandia"

Synonym:
- *Noctua frontata* Lesson 1830, Traite d Orn.: 106; Terra typica: ? Patrie ignorée
- *Athene? fortis* Gould 1838, Syn. Birds Austr.: Pl. 3 (and descript.); Terra typica: New South Wales
- *Ninox connivens addenda* Mathews 1912, Aust. Avian Rec. 1: 120; Terra typica: Southwest Australia

Distribution: Western Australia, southern Australia, Victoria, New South Wales, Queensland. In western Australia only in the extreme southwest
Museum: AMNH, BMNH, CSIRO, FNSF, HLW, MZUS, NMV, NNML, SMNSt, SMTD, WAM, ZSBS

Wing length:	282–325 mm
Tail length:	178–190.5 mm
Tarsus length:	47–53 mm
Length of bill:	?
Body mass:	♂ 425–510 g, ♀ 390–475 g

Illustration: Gould in Gould 1838a: Pl. 3; H. C. Richter in Gould 1848–1869: Pl. 34; N. Cayley in Cayley 1961 (repr.): Pl. 5(16); N. Day in Simpson and Day 1986: 153 (adult and juvenile); P. Slater in Slater et al. 1989: 193 (adult and juvenile); T. Boyer in Boyer and Hume 1991: 114; N. Arlott in del Hoyo et al. 1999: Pl. 18; F. Weick in König et al. 1999: Pl. 56; M. Oberhofer in Olson 2001: 203; F. Knight in Olson 2001: 205

Photograph: E. Hosking in Hosking and Flegg 1982: 148; J. Calaby and J. Lowe in Frith et al. 1983: 304; D. Hol-lands in Hollands 1991: 4, 60–67 (2 subspecies?); D. Hollands in Burton et al. 1992: 144; D. P. Lewis in internet: Owl Pages, photo gallery 2003

Literature: Sharpe 1875a: 175–177; Fleay 1940: 91–95; Mees 1964a: 31–33; 56; Frith et al. 1983: 304; Simpson and Day 1986: 152; Slater et al. 1989: 192; del Hoyo et al. 1999: 231, 232; König et al. 1999: 402, 403

Ninox rudolfi **A.B. Meyer 1882**
Sumba Boobook · Sumbakauz · Ninoxe (ou Chouette) de Sumba · Ninox de Sumba
Ninox rudolfi **A.B. Meyer 1882, Ibis: 232, Pl. 6;** Terra typica: **Sumba Island**
(see Figure 20)

Length:	300–360 mm
Body mass:	?

Distribution: Sumba Island, Lesser Sunda Islands

Habitat: Evergreen and deciduous primary and secondary forest. Also monsoon and rain forest, open forest and forest edges, sometimes farmland. From sea-level up to 930 m

Museum: BMNH, SMTD (lectotype)

Wing length:	227–243 mm
Tail length:	145 mm
Tarsus length:	~40 mm
Length of bill:	~30 mm
Body mass:	?

Illustration: J. G. Keulemans in Meyer 1882: Pl. 6; D. Gardner in Coates and Bishop 1997: Pl. 33; Johnstone and Darnell 1997: Fig. 2(7); N. Arlott in del Hoyo et al. 1999: Pl. 18; F. Weick in König et al. 1999: Pl. 57; F. Weick in *Gefiederte Welt* 2005(1): 22

Photograph: J. Olsen in Olsen et al. 2002: Fig. 6

Literature: Meyer 1882: 232, 233; Mees 1964a: 12, 13; Eck 1970: 137–140; Eck and Busse 1973: 138, 139; White and Bruce 1986: 254; Coates and Bishop 1997: 361; del Hoyo et al. 1999: 232; König et al. 1999: 403, 404; Olsen et al. 2002: 223–231; Weick 2003–2005, *Gefiederte Welt* 2005(1): 23

Ninox boobook **(Latham) 1801**
Southern Boobook · Boobookkauz (Kuckuckskauz) · Ninoxe coucou · Ninox Australiano

Length:	270–360 mm
Body mass:	145–315 g

Distribution: From Roti, Timor, etc., New Guinea south to all areas (also drier) of Australia. Tasmanian population regarded here as a distinct species

Habitat: Wide variety of habitats; from semi-desert to farmland, woodland, tropical rainforest, suburbs, cultivated country, orchards and parks. From sea-level up to 2 500 m (Timor)

- ***Ninox boobook rotiensis*** Johnstone and Darnell 1997
  Ninox boobook rotiensis Johnstone and Darnell 1997, West Austral. Nat. 21: 161–172; Terra typica: Roti Island, Indonesia

Distribution: Roti Island, Indonesia

Museum: WAM

Wing length:	♂ 188 mm ($n = 1$)
Tail length:	100 mm
Tarsus length:	30 mm
Length of bill:	25 mm
Body mass:	146 g

Illustration: Johnstone and Darnell 1997: Fig. 2(5), 3(3), 4(1)

Photograph: ?

Literature: Johnstone and Darnell 1997: 161–172; del Hoyo et al. 1999: 232; König et al. 1999: 405–407

- ***Ninox boobook fusca*** (Vieillot) 1817
  Strix fusca Vieillot 1817, Nouv. Dict. Hist. Nat. 7: 22;
  Terra typica: Timor (Saint Domingue et Porto Ricco = error)

Synonym:
- *Strix maugei* Temminck 1821, livr. Pl. col. 8, Pl. 46; Terra typica: Timor (Antilles = error)
- *Strix (Athene) guteruhi* S. Müller 1845, Verh. Nat. Gesch. Nederl., Land- en Volkenk.:279; Terra typica: Timor

Distribution: Timor and Semau Islands
Museum: AMNH, NNML (holotype *guteruhi*),SMTD, ZSBS

Wing length:	214–225 mm
Tail length:	108–122 mm
Tarsus length:	31–36 mm
Length of bill:	25–32 mm
Body mass:	♂ 180 g ($n = 1$)

Illustration: J. G. Keulemans in Sharpe 1875a: Pl. 12(1); D. Gardner in Coates and Bishop 1997: Pl. 33; Johnstone and Darnell 1997: Fig. 2(1), 3(1), 4(3); N. Arlott in del Hoyo et al. 1999: Pl. 18; F. Weick in König et al. 1999: Pl. 57
Photograph: ?
Literature: Sharpe 1875a: 172, 173; Mees 1964a: 15; Sayers 1976a: 129, 130; White and Bruce 1986: 253, 254; Coates and Bishop 1997: 362; Johnstone and Darnell 1997: 161–173; del Hoyo et al. 1999: 232; König et al. 1999: 405–407

- ***Ninox boobook plesseni*** Stresemann 1929
  Ninox fusca plesseni Stresemann 1929, Orn. Monatsb. 37: 47;
  Terra typica: Tanglapoi in western Alor

Distribution: Alor Island, Lesser Sunda Island
Museum: MNBHU (holotype, only known from the type!)

Wing length:	♀ 212 mm
Tail length:	?
Tarsus length:	?
Length of bill:	?
Body mass:	?

Illustration: Johnstone and Darnell 1997: Fig. 2(4), 3(2), 4(2)
Photograph: ?
Literature: Stresemann 1929: 47, 48; Mees 1964a: 13–15; Sayers 1976a: 129; White and Bruce 1986: 253, 254; Coates and Bishop 1997: 362; Johnstone and Darnell 1997: 161–173; del Hoyo et al. 1999: 232; König et al. 1999: 405–407

- ***Ninox boobook moae*** Mayr 1943
  Ninox novaeseelandiae moae Mayr 1943, Emu 43: 13;
  Terra typica: Moa Island

Distribution: Romang, Leti and Moa Islands
Museum: AMNH
Remarks: Doubtful subspecies, intermediate between subspecies *fusca* and *cinnamomina*. Hartert identified it as a dark morph of *ocellata*!

Wing length:	♂ 208–221 mm ($n = 6$), ♀ 215–228 mm ($n = 3$)
Tail length:	?
Tarsus length:	?
Length of bill:	?
Body mass:	?

Illustration: ?
Photograph: ?
Literature: Hartert 1904: 190–191; Mayr 1943: 3–17; Mees 1964a: 15, 53; Sayers 1976a: 130; White and Bruce 1986: 253, 254; del Hoyo et al. 1999: 232; König et al. 1999: 405–407

- ***Ninox boobook cinnamomina*** Hartert 1906
  Ninox boobook cinnamomina Hartert 1906, Novit. Zool. 13: 293;
  Terra typica: Tepa, Barbar Island

Distribution: Tepa and Baba (Babber) Islands
Museum: AMNH (6)

Wing length:	♂ 210–215 mm, ♀ 212–215 mm
Tail length:	115–123 mm
Tarsus length:	34–38 mm
Length of bill:	27–28.5 mm
Body mass:	?

Illustration: D. Gardner in Coates and Bishop 1997: Pl. 33; Johnstone and Darnell 1997: Fig. 2(3); F. Weick in König et al. 1999: Pl. 57
Photograph: ?
Literature: Hartert 1906: 288–302; Mees 1964a: 15, 16, 53; Sayers 1976a: 130; White and Bruce 1986: 253,

254; Coates and Bishop 1997: 362; Johnstone and Darnell; 1997: 161–173; del Hoyo et al. 1999: 232; König et al. 1999: 405–407

- ***Ninox boobook remigialis*** Stresemann 1930
  Ninox novaeseelandiae remigialis Stresemann 1930, Bull. Brit. Ornith. Cl. 50: 61; Terra typica: Kei Islands

Distribution: Kei Islands
Museum: BMNH (holotype)
Remarks: Known only from holotype, similar to *moae*, but barring on primaries and secondaries less pronounced

Wing length:	?
Tail length:	?
Tarsus length:	?
Length of bill:	?
Body mass:	?

Illustration: ?
Photograph: ?
Literature: Stresemann 1930: 61; Mees 1964a: 16 (no measurements!); Sayers 1976a: 130; Johnstone and Darnell 1997: 161–173

- ***Ninox boobook pusilla*** Mayr and Rand 1935
  Ninox novaeseelandiae pusilla Mayr and Rand 1935, Am. Mus. Novit. 814: 3; Terra typica: Dogwa, Oriomo River, Territory of Papua

Length: 250–275 mm

Distribution: Southern New Guinea, opposite Cape York
Museum: AMNH (holotype, five skins)
Remarks: Similar, but smaller and a little darker than the dark morph of *ocellata*

Wing length:	193–205 mm
Tail length:	?
Tarsus length:	35 mm
Length of bill:	25 mm
Body mass:	?

Illustration: ?
Photograph: ?
Literature: Mayr and Rand 1935: 3; Mees 1964a: 16, 53; Rand and Gillmore 1967: 254, 255; Sayers 1976: 130; Beehler et al. 1986: 132

- ***Ninox boobook ocellata*** (Bonaparte) 1850
  Athene ocellata "Hombr. and Jaquin" Bonaparte 1850, Consp. Gen. Av. 1: 42; Terra typica: Raffles Bay, Cobourg Peninsula, Northern Territory, Australia

Synonym:
- *Ninox boobook mixta* Mathews 1912, Novit. Zool. 18: 255; Terra typica: Northwest Australia, Parry's Creek
- *Ninox boobook melvillensis* Mathews 1912, Aust. Avian Rec. 1: 34; Terra typica: Melville Island, Northern Territory (holotype AMNH)
- *Ninox boobook macgillivrayi* Mathews 1913, Aust. Avian Rec. 1: 194; Terra typica: Cape York, Northern Queensland
- *Spiloglaux novaeseelandiae everardi* Mathews 1916, Birds Austr. 5: 332; Terra typica: Everard Ranges, central Australia
- *Ninox ooldeanensis* Cayley 1929, Emu 28: 162; Terra typica: Near Ooldea, South Australia
- *Ninox novaeseelandiae arida* Mayr 1943, Emu 43: 16; Terra typica: Fitzroy River, 3 miles west of Mount Anderson, West Kimberley District
- *S(piloglaux) b(oobook) parocellata* Mathews 1946, Working List of Australian Birds: 55; Terra typica: SW Australia (see Mees 1961: 106)
- *S(piloglaux) o(cellata) carteri* Mathews 1946, ibid.: 55; Terra typica: Mid West Australia (see Mees 1961: 106)
- *Ninox novaeseelandiae rufigaster* Mees 1961, J. Roy. Soc. W. Aust. 44: 106; Terra typica: Perth (holotype WAM)

Distribution: Queensland, Northern Territory (including Melville Island), southwest Australia (ranges ca. 300 miles inland), southern Australia, islands in the Torres Strait, also Sawu Island (west of Timor)
Museum: AMNH, AMO, BMNH, HLW, NMV, NNML, SAM, SMTD, SW, USNM, WAM
Remarks: This subspecies needs further studies. Populations from Melville Island (smaller size) and southwest Australia (*rufigaster*), sometimes classed as separate subspecies, but here, following Schodde and Mason, these are included with *ocellata*. According to Mayr (1943): "The classification of the boobook owls is one of the most difficult taxonomic problems I have ever encountered. The main difficulty is that variation seems not to be correlated with geographic districts, but more or less with rainfall or humidity. And therefore one encounters, not infrequently, indistinguishable populations at widely-separated locali-

ties." Such difficulties occur mainly in the range of subspecies *ocellata* (Mees 1964a)

Wing length: 199–213 mm (Melville Island), 203–227 mm (Sawu Island and Groote Island), 205–240 mm (southern and western Australia and Northern Territories), 218–234 mm (Queensland), 215–246 mm (southwest Australia)
Tail length: 114–141 mm
Tarsus length: 37–46 mm
Length of bill: 26–29 mm
Body mass: ?

Illustration: N. Cayley in Cayley 1961 (repr.): Pl. 5(15); O. Seymor in Mees 1964a: Frontispiece; N. Day in Simpson and Day 1986: 153; P. Slater in Slater et al. 1989: 193; Johnstone and Darnell 1997: Fig. 2(2), 4(4); N. Arlott in del Hoyo et al. 1999: Pl. 18; F. Weick in König et al. 1999: Pl. 57 (light and dark); F. Knight in Olson 2001: 205
Photograph: Serventi in Burton et al. 1992: 143 (light); Lewis in internet: Owl Pages, photo gallery 2002 (dark)
Literature: Sharpe 1875a: 170–172; Mathews 1912a: 34; Mathews 1912b: 255; Mees 1961: 106; Mees 1964a: 17–22, 54; Sayers 1976a: 131–132; Schodde and Mason 1980: pp 136; Simpson and Day 1986: 152; Slater et al. 1989: 192; Johnstone and Darnell 1997: 161–173; del Hoyo et al. 1999: 231; König et al. 1999: 405–407

- ***Ninox (boobook) lurida*** Vis 1887
Ninox boobook var. *lurida* de Vis 1887, Proc. Linn. Soc. New S. Wales 1886, 2(1): 1135; Terra typica: Few miles from Cardwell

Synonym:
- *N(inox) lurida* de Vis 1889, Rep. Sci. Exp. Bellenden-Ker Range: 84; Terra typica: Herbert Gorge, Bellenden-Ker
- *Spiloglaux boweri* Mathews 1913, Aust. Avian Rec. 2: 74; Terra typica: Cairns, North Queensland
- *Ninox yorki* Cayley 1929, Emu 28: 162; Terra typica: Cape York Peninsula (holotype AMS)

Distribution: Northeast Queensland (Cairns), Bartle Frere, Kirrima Range, Murray River, Cardwell and Bellenden-Ker Ranges (between Cooktown and Paluma)
Museum: AMNH, AMS, HLW, NMVB (large and dark specimen, lacking spots)

Wing length: 207–221 mm (244 mm = Cape York Peninsula)
Tail length: ?
Tarsus length: ?
Length of bill: ?
Body mass: ?

Illustration: N. Cayley in Cayley 1961: Pl. 5(14); P. Slater in Slater et al. 1989: 193; N. Arlott in del Hoyo et al. 1999: Pl. 18; F. Weick in König et al. 1999: Pl. 57
Photograph: D. Hollands in Hollands 1991: 58 and backcover
Literature: Cayley 1929: 162; Mees 1964a: 25, 26, 55; Sayers 1976a: 132, 133; Hollands 1991: 59, 209, 210; del Hoyo et al. 1999: 231; König et al. 1999: 405–407; F. Knight in Olson 2001: 205

- ***Ninox boobook boobook*** (Latham) 1801
Strix Boobook Latham 1801, Index Orn. Suppl. XV; Terra typica: New Holland=New South Wales apud Mathews (see Plate 8)

Synonym:
- *Athene marmorata* Gould 1846, Proc. Zool. Soc. Lond.: 18; Terra typica: South Australia
- *Ninox boobook halmaturina* Mathews 1912, Novit. Zool. 18: 254; Terra typica: Kangaroo Island (holotype AMNH); not always separable from nominate *boobook*!
- *Spiloglaux boobook leachi* Mathews 1913, Aust. Avian Rec. 2: 74; Terra typica: Victoria
- *Spiloglaux boobook tregellari* Mathews 1913, ibid.: 74; Terra typica: Frankston, Victoria

Distribution: Victoria, New South Wales, South Queensland, eastern South Australia (Spencer Gulf) and Kangaroo Island off the coast of South Australia
Museum: AMNH, AMS, BMNH, CSIRO, NMV, NNML, SAMB, SMNSt, SMTD, USNM, ZFMK

Wing length: 227–261 mm
Tail length: 160 mm
Tarsus length: 44 mm
Length of bill: ?
Body mass: ♂ 194–360 g, ♀ 170–298 g (1 × 315 g)

Illustration: H. C. Richter in Gould 1848–1869, vol 1: Pl. 32; N. Cayley in Cayley 1961 (repr.): Pl. 5(12); B. Fremlin in Fremlin 1986: 27; N. Day in Simpson and Day 1986: 153 (adult and juvenile); P. Slater in Slater et al. 1989: 193 (adult and juvenile); T. Boyer in Boyer and Hume 1991: 115; N. Arlott in del Hoyo et al. 1999: Pl. 18; F. Weick in König et al. 1999: Pl. 57 (adult and juvenile); F. Knight in Olson 2001: 205 (adult and juvenile)

Photograph: H. Beste in Everett 1977: 76, 77; E. Hosking in Hosking and Flegg 1982: 45; J. Calaby in Frith et al. 1983: 303; D. Hollands in Hollands 1991: 4, 50, 53, 54, 57; H. Beste in Burton et al. 1992: 143; M. Sacchi in del Hoyo et al. 1999: 118; T. Quested in internet: Owl Pages, photo gallery 2002

Literature: Sharpe 1875a: 168–170; Mayr 1943: 3–17; Mees 1964a: 23–25, 54; Eck and Busse 1973: 138, 139; Schodde and Mason 1980: pp 136; Frith et al. 1983: 303; Simpson and Day 1986: 152; Slater et al. 1989: 192; Hollands 1991: 50–59, 209, 210; Boyer and Hume 1991: 114, 115; del Hoyo et al. 1999: 232; König et al. 1999: 405–407

Ninox leucopsis (Gould) 1838

Tasmanian Boobook · Tasman-Boobookkauz

Noctua Maculata **Vigors and Horsfield 1826, Trans. Linn. Soc. London 15: 189; Terra typica: Australia = Tasmania, invalid name**

Athene leucopsis **Gould 1938, Proc. Zool. Soc. Lond. 1837: 99; Terra typica: Tasmania**

Length: 280–300 mm
Body mass: ?

Synonym:
- *Spiloglaux boobook clelandi* Mathews 1913, Aust. Avian Rec. 2: 74; **Terra typica:** Flinder Island
- *Spiloglaux boobook leachi* Mathews 1913, ibid.: 74; **Terra typica:** Victoria = Melbourne (migratory bird)
- *Spiloglaux novaeseelandiae tasmanica* Mathews 1917, Aust. Avian Rec. 3: 70; **Terra typica:** Tasmania

Distribution: Tasmania and Islands in the Bass Strait (migratory to Victoria and New South Wales)

Museum: AMNH (holotypes *clelandi, leachi* and *tasmanica)*, AMS, BMNH, FNSF, HLW, NMV, NNML, SAM, SMTD

Remarks: This taxon is generally closer to *Ninox novaeseelandiae* (Christidis and Norman 1998), but its morphological similarity to *Ninox boobook*, suggests it is a distinct species

Wing length: 198–222 mm
Tail length: 127–132 mm
Tarsus length: 37 and 38 mm
Length of bill: ?
Body mass: ?

Illustration: H. C. Richter in Gould 1848–1869, vol 1: Pl. 32; N. Cayley 1961 (repr.): Pl. 5(13); N. Arlott in del Hoyo et al. 1999: Pl. 18; F. Weick in König et al. 1999: Pl. 57

Photograph: ?

Literature: Gould 1838b: 99; Sharpe 1875a: 174, 175; Mathews 1913b: 74; Mathews 1917: 70; Mees 1964a: 27–30, 55; Sayers 1976a: 133; del Hoyo et al. 1999: 232, 233; König et al. 1999: 405–407

Ninox novaeseelandiae J.F. Gmelin 1788

Morepork, New Zealand Boobook · Neuseeland-Boobook · Ninoxe de Nouvelle-Seelandia · Ninox Maori

Length: 260–290 mm
Body mass: 150–170 g

Distribution: New Zealand, including most offshore islands, Lord Howe Island (subspecies extinct)

Habitat: Forest, farmland and plantations, urban areas. From lower altitudes up to the treeline

- ***Ninox novaeseelandiae albaria*** **Ramsay 1888 †**
  Ninox albaria Ramsay 1888, Tab. List. Austral. Bds.: 36; Terra typica: Lord Howe Island

Distribution: Lord Howe Island (extinct)

Museum: AMNH (seven skins) a light brown plumage with an ocellated undersurface

Wing length: ♂ 209–215 mm, ♀ 218–222 mm
Tail length: ?
Tarsus length: ?
Length of bill: ?
Body mass: ?

Illustration: ?
Photograph: ?
Literature: Ramsay 1888: 36; Mees 1964a: 29, 55; Eck and Busse 1973: 138, 139; König et al. 1999: 404, 405

- ***Ninox novaeseelandiae undulata*** (Latham) 1801
Strix undulata Latham 1801, Index Orn. Suppl. XVII;
Terra typica: Insula Norfolk

Synonym:
- *Ninox boobook royana* Mathews 1912, Aust. Avian Rec. 1: 120; Terra typica: Norfolk Island

Distribution: Norfolk Island, probably extinct. Similar to *albaria*, but darker above and below
Museum: AMNH (holotype *royana*), HLW

Wing length: 196–208 mm
Tail length: ?
Tarsus length: ?
Length of bill: ?
Body mass: ?

Illustration: N. Arlott in del Hoyo et al. 1999: Pl. 18; J. Stuart in Olson 2001: 48 (from 1839)
Photograph: ?
Literature: Mees 1964a: 29, 55; Sayers 1976a: 133, 134; Norman et al. 1998a: 33–36; del Hoyo et al. 1999: 232, 233; König et al. 1999: 404, 405

- ***Ninox novaeseelandiae novaeseelandiae*** (J.F. Gmelin) 1788
(Strix) novae Seelandiae J.F. Gmelin 1788, Syst. Nat. 1(1): 296;
Terra typica: New Zealand (South Island)

Synonym:
- *(Strix) fulva* Latham 1790, Index Orn. 1: 65; Terra typica: Nova Zeelandia
- *Strix novaeseelandiae maculata* Kerr 1792, Anim. Kingdom 1: 538; Terra typica: New Zealand
- *Noctua zelandica* Quoy and Gaimard 1830, Voy. *Astrolabe* 1: 168, Pl. 2(1); Terra typica: New Zealand
- *Noctua venatica* Peal 1848, U.S. Expl. Exp.: 75; Terra typica: New Zealand (North Island)

Distribution: New Zealand, North and South Islands, Taranga or Hen Island, Great Barrier Island, Little Barrier Island, Three Kings Island, Kapiti and Stewart Islands
Museum: AMNH, BMNH, CMC, DMW, MZUS, SMNSt, SMTD, ZFMK

Wing length: 183–203 mm
Tail length: 135–146 mm
Tarsus length: 34–37 mm
Length of bill: ?
Body mass: ♂ 140–156 g, ♀ 170–216 g

Illustration: Quoy and Gaimard 1830: Pl. 2(1) (b/w); J. G. Keulemans in Buller 1882: Pl. 20; G. E. Lodge in Fleming 1982 (repr.): with *Sceloglaux;* E. Power in Falla et al. 1993: Pl. 39; N. Arlott in del Hoyo et al. 1999: Pl. 18; F. Weick in König et al. 1999: Pl. 57
Photograph: G. Moon in Robertson et al. 1985: 257; T. Little in internet: Owl Pages, photo gallery 2002
Literature: Sharpe 1875a: 173, 174; Mees 1964a: 29, 30, 55; Falla et al. 1966: 184–186; Sayers 1976a: 128, 129; Schodde and Mason 1980: pp 136; Robertson et al. 1985: 257; Boyer and Hume 1991: 114, 115; Moon 1992; Falla et al. 1978: 168; Norman et al. 1998a: 33–36; del Hoyo et al. 1999: 232, 233; König et al. 1999: 404, 405

Ninox scutulata (Raffles) 1822

Brown Hawk Owl · Falkenkauz (Schildkauz) · Ninoxe (ou Chouette) hirsute · Ninox Pardo
(see Plate 8)

Length: 270–330 mm
Body mass: 170–230 g

Distribution: Indian Subcontinent to eastern Siberia and Japan. South to the Andamans, Malay Peninsula, Great and Lesser Sundas, Moluccas, Taiwan and the Philippines. One record from Australia
Habitat: Wide variety of distribution: deciduous, evergreen and coniferous forest. Scrub, mangroves, river thickets and rainforest. Also plantations, parks and suburbs. From lowlands up to 1 200–1 500 m

- ***Ninox scutulata florensis*** (Wallace) 1864
Athene florensis Wallace 1864, Proc. Zool. Soc. Lond. 1863: 488;
Terra typica: Flores

Synonym:
- *Ninox macroptera* Blasius 1888, Braunschw. Anz. 11(9): 86; Terra typica: Mindoro (Ornis 1888: 551)
- *Ninox scutulata ussuriensis* Buturlin 1910, Mess. Orn.: 187; Terra typica: Lake Khanka, southern Ussuriland

- *Ninox scutulata yamashinae* Ripley 1953, Tori 13: 49; Terra typica: Fukumoto, Yamato, Amami Oshima, North Ryukyu Island

Distribution: Southeast Siberia, southeast Manchuria, northern China and North Korea. Winter stragglers to Ryukyu Island, Borodinos, Taiwan, and Lanyu, Seven Sisters, Izu, Bonin and Volcano Islands. Migrating to the Philippines and Wallacea (Flores, Luzon, Jolo, Calayan and Mindanao!)
Museum: BMNH (holotype *florensis*), USNM

Wing length:	22–245 mm
Tail length:	134.5–140 mm
Tarsus length:	28–30 mm
Length of bill:	25 mm
Body mass:	?

Illustration: F. Berille in Etchécopar and Hüe 1978: Pl. 18; Y. V. Kostin in Flint et al. 1984: Pl. 13; F. Weick in König et al. 1999: Pl. 58 (named *japonica* = error!)
Photograph: J. Pukinski in Pukinski 1983: 55, 56, 82 (b/w); Knystaustas et al. 1987: 3 colour photographs near 68
Literature: Wallace 1863: 488; Sharpe 1875a: 156–167; Hartert 1912–1921: 992, 993; Hartert 1923: 388, 389; Dementiev and Gladkov 1951: 433–435; Vaurie 1965: 617, 618; du Pont 1971: 176; Nakamura 1975: 37, 38; White and Bruce 1986: 256; Dickinson et al. 1991: 228, 229; Coates and Bishop 1997: 362; del Hoyo et al. 1999: 233 (*ussuriensis*); König et al. 1999: 407, 408 (*ussuriensis* as synonym of *japonica* and ignored *florensis* = error!)

- ***Ninox scutulata japonica*** (Temminck and Schlegel) 1844
Strix hirsuta japonica Temminck and Schlegel 1844, in Siebold's Fauna Japonica, Aves: 28, Pl. 9b;
Terra typica: Japan

Synonym:
- *Ninox scutulata totogo* Momiyama 1931, Amoeba 3(1–2): 68; Terra typica: Botal Tobago, southeast of Taiwan (holotype NNML)

Distribution: Eastern China, central and southern Korea, Japan, Taiwan and Lanyu. Winters in Sundas, Wallacea and the Philippines
Museum: BMNH, FMNH, MZB, NMP, NNML, SMTD

Wing length:	206–219 mm (*totogo*), 213–225 mm (*japonica*)
Tail length:	107–127 mm
Tarsus length:	25.5–30 mm
Length of bill (cere):	14 and 15 mm
Body mass:	?

Illustration: J. Wolf in Temminck and Schlegel 1844: Pl. 9b; T. Miyamoto in Kobayashi 1965: Pl. 24; S. Takano in Massey et al. 1983: 193; A. Cameron in Voous 1988: 179; D. Gardner in Coates and Bishop 1997: Pl. 33
Photograph: ?
Literature: Sharpe 1875a: 156–167; Hartert 1912–1921: 992, 993; Hartert 1923: 388, 389; Vaurie 1965: 618; Kobayashi 1965: 64; Massey et al. 1983: 192; White and Bruce 1986: 256; Voous 1988: 176–180; Coates and Bishop 1997: 362; del Hoyo et al. 1999: 233; König et al. 1999: 407, 408

- ***Ninox scutulata lugubris*** (Tickell) 1833
Strix lugubris Tickell 1833, J. As. Soc. Beng. 2: 572;
Terra typica: Dampara, Dholbhum, Bengal

Distribution: Northern and central India to western Assam
Museum: BMNH, FNSF, ZMH, ZSBS

Wing length:	♂ 207–227 mm, ♀ 218, 219 mm
Tail length:	115–140 mm
Tarsus length:	24–29 mm
Length of bill:	21–24 mm
Body mass:	♀ 186 g ($n = 1$)

Illustration: Hodgson 1837b: 23 and Pl. 14; D. Cole in Grimmett et al. 1998: Pl. 30(?); F. Weick in König et al. 1999: Pl. 58; L. McQueen in Rasmussen and Anderton 2005: Pl. 75
Photograph: H. Denzau in del Hoyo et al. 1999: 119; S. Bhatta Charwa in internet: Owl Pages, photo gallery 2004
Literature: Sharpe 1875a: 156–167; of Tweeddale 1878: 340; Baker 1927: 454, 455; Ali and Ripley 1969: 292–294; Voous 1988: 176–180; Grimmett et al. 1998: 438; del Hoyo et al. 1999: 233; König et al. 1999: 407, 408; Rasmussen and Anderton 2005: 247, 248

- ***Ninox scutulata hirsuta*** (Temminck) 1824
  Strix hirsuta Temminck 1824, Pl. col. livr. 49: Pl. 289;
  Terra typica: Ceylon

Distribution: Southern India and Sri Lanka. Similar to *burmanica*, but darker. Head darker than back
Museum: BMNH, NNML, SMTD

Wing length:	190–212 mm
Tail length:	112–122 mm
Tarsus length:	24–28 mm
Length of bill:	22–24 mm
Body mass:	172–227 g

Illustration: Temminck 1824: Pl. 289; G. M. Henry in Henry 1955: 201 (b/w); D. V. Cowes in Ali and Ripley 1968: Pl. 13; L. McQueen in Rasmussen and Anderton 2005: Pl. 75
Photograph: S. Bhattacharwa in internet: Owl Pages, photo gallery 2003
Literature: Sharpe 1875a: 156–167; Baker 1927: 457; Henry 1955: 201, 202; Ali and Ripley 1969: 295; Eck and Busse 1973: 139–141; Grimmett et al. 1998: 438; del Hoyo et al. 1999: 233; König et al. 1999: 407, 408

- ***Ninox (scutulata) obscura*** (Hume) 1873
  Ninox obscurus Hume 1873, Str. Feath. 1: 11;
  Terra typica: Camorta, Nicobar Islands

Distribution: Andaman and Nicobar Islands
Museum: BMNH, FNSF
Remarks: Probably a distinct species?!

Wing length:	197–220 mm
Tail length:	120–126 mm
Tarsus length:	26.5–28 mm
Length of bill:	22 and 23 mm
Body mass:	?

Illustration: J. G. Keulemans 1874: Pl. 4; D. Cole in Grimmett et al. 1998: Pl. 30; N. Arlott in del Hoyo et al. 1999: Pl. 18; F. Weick in König et al. 1999: Pl. 58; L. McQueen in Rasmussen and Anderton 2005: Pl. 75
Photograph:
Literature: Sharpe 1875a: 159; Baker 1927: 457, 458; Ali and Ripley 1969: 295, 296; Grimmett et al. 1998: 438; del Hoyo et al. 1999: 233; König et al. 1999: 407, 408; Rasmussen and Anderton 2005: 248

- ***Ninox scutulata burmanica*** Hume 1876
  Ninox burmanica Hume 1876, Str. Feath. 4: 285;
  Terra typica: Pegu and Tenasserim

Distribution: Eastern Assam: south of the Brahmaputra River, through Myanmar, Laos, Vietnam to southern China. South to northern Malay Peninsula, Thailand and Indonesia
Museum: BMNH, SMNSt, SMTD, USNM, ZMH

Wing length:	206–222 mm (228)
Tail length:	128–134 mm
Tarsus length:	24 and 25 mm
Length of bill:	21 and 22 mm
Body mass:	?

Illustration: W. Mongkol in Lekagul and Round 1991: Pl. 60; H. Burn in Robson 2000: Pl. 21
Photograph: G. Chuen Hang in *SCRO Mag.* 1999(2): Pl. 60
Literature: Sharpe 1875a: 162, 163; Baker 1927: 455, 456; Delacour and Jabouille 1931: 142, 143; Deignan 1945: 180, 181; Ali and Ripley 1969: 294, 295; Smythies and Hughes 1984: 321; Lekagul and Round 1991: 177; H. H. Chew 1999 in *SCRO Mag.* 1999(2): 53–56; del Hoyo et al. 1999: 233; König et al. 1999: 407, 408; Robson 2000: 294

- ***Ninox scutulata palawanensis*** Ripley and Rabor 1962
  Ninox scutulata palawanensis Ripley and Rabor 1962, Postilla 73: 4; Terra typica: Palawan

Distribution: Palawan Island, the Philippines (resident!)
Museum: YPM (holotype)

Wing length:	195 mm
Tail length:	108 mm
Tarsus length:	?
Length of bill:	?
Body mass:	?

Illustration: ?
Photograph: ?
Literature: Ripley and Rabor 1962: 4; du Pont 1971: 176; Dickinson et al. 1991: 228, 229; del Hoyo et al. 1999: 233; König et al. 1999: 407, 408; Kennedy et al. 2000: 180

- ***Ninox scutulata randi*** Deignan 1951
  Ninox scutulata randi Deignan 1951, Proc. Biol. Soc. Wash. 64: 41; Terra typica: Catagan, base of Mount Malindang, Mindanao

Distribution: Philippines: Luzon, Mariduque, Mindoro, Negros, Cebu, Siquijor, Mindanao, Basilan. Prop. Fuga? (Ripley and Rabor 1958, resident)
Museum: BMNH, FMNH, USNM (holotype), USNM (holotype from Fuga)

Wing length:	228–242 mm
Tail length:	134 mm
Tarsus length:	30–33 mm
Length of bill:	25–26.5 mm
Body mass:	200–220 g ($n = 6$)

Illustration: G. Sandström in du Pont 1971: Pl. 38; F. Weick in König et al. 1999: Pl. 58; A. P. Sutherland in Kennedy et al. 2000: Pl. 35
Photograph: ?
Literature: Deignan 1951: 40, 41; Ripley and Rabor 1958: 40; du Pont 1971: 176; Dickinson et al. 1991: 228, 229; del Hoyo et al. 1999: 233; König et al. 1999: 407, 408; Kennedy et al. 2000: 180

- ***Ninox scutulata scutulata*** (Raffles) 1822
  Strix scutulata Raffles 1822, Trans. Linn. Soc. London 13(2): 280; Terra typica: Sumatra

Synonym:
- *Athene malaccensis* Eyton 1845, Ann. Mag. Nat. Hist. 16: 228; Terra typica: Malacca

Distribution: Southern Malay Peninsula, Riau Archipelago, Sumatra and Banka Islands
Museum: BMNH, FMNH, NNML, SMNSt, USNM, ZFMK, ZSBS

Wing length:	212–228 mm
Tail length:	116–127 mm
Tarsus length:	28–31 mm
Length of bill:	?
Body mass:	172–227 g

Illustration: T. Boyer in Boyer and Hume 1991: 116; K. Phillipps in McKinnon and Phillipps 1993: Pl. 39; F. Weick in König et al. 1999: Pl. 58
Photograph: ?
Literature: Sharpe 1875a: 156–167; Voous 1988: 177–180; Boyer and Hume 1991: 116; McKinnon and Phillipps 1993: 195; del Hoyo et al. 1999: 233; König et al. 1999: 407, 408

- ***Ninox scutulata javanensis*** Stresemann 1928
  Ninox scutulata javanensis Stresemann 1928, Orn. Monatsb. 36: 54; Terra typica: Indramaju, Cheribn, Java

Distribution: Western Java, (Bali?)
Museum: MNBHU (holotype), ZFMK

Wing length:	178–183 mm
Tail length:	?
Tarsus length:	?
Length of bill:	?
Body mass:	?

Illustration: A. Cameron in Voous 1988: 179; F. Weick in König et al. 1999: Pl. 58
Photograph: ?
Literature: Stresemann 1928: 54; Voous 1988: 177–180; McKinnon and Phillipps 1993: 195; del Hoyo et al. 1999: 233; König et al. 1999: 407, 408

- ***Ninox scutulata borneensis*** (Bonaparte) 1850
  Strix hirsuta borneensis «Schlegel» Bonaparte 1850, Consp. Gen. Av. 1: 41; Terra typica: Malaysia, Borneo

Synonym:
- *Ninox labuanensis* Sharpe 1875, Cat. Birds Brit. Mus. 2: 165 (in text); Terra typica: Labuan Island

Distribution: Borneo and northern Natuna Islands
Museum: BMNH (holotype), FNSF, MHNP, NNML, SMTD

Wing length:	176–197 mm
Tail length:	98–106.5 mm
Tarsus length:	25–27 mm
Length of bill:	?
Body mass:	?

Illustration: A. M. Hughes in Smythies 1968: Pl. 15; F. Weick in König et al. 1999: Pl. 58
Photograph: ?
Literature: Sharpe 1875a: 164, 165; Smythies 1968: 282; Voous 1988: 177–180; König et al. 1999: 407, 408

Ninox affinis **Beavan 1867**
Andaman Hawk Owl · Andamanenkauz · Ninoxe des Andaman · Ninox de Andamán

Length: 250–280 mm
Body mass: ?

Distribution: Andaman and Nicobar Islands
Habitat: Forest, open woodland, plantations, mangroves. Sometimes near human settlements
Remarks: Regarded by del Hoyo et al. as monotypic

- ***Ninox affinis affinis*** Beavan 1867
  Ninox affinis "Tytler" Beavan 1867, Ibis: 316;
  Terra typica: Aberdeen Point, Port Blair, Andaman Island

Distribution: South Andaman, Andaman Island
Museum: BMNH (holotype)

Wing length: 167–169 mm
Tail length: 102–106 mm
Tarsus length: 27 and 28 mm
Length of bill: 20 mm
Body mass: ?

Illustration: J.G. Keulemans in Walden 1874: Pl. 5; T. Boyer in Boyer and Hume 1991: 117; D. Cole in Grimmett et al. 1998: Pl. 30; N. Arlott in del Hoyo et al. 1999: Pl. 18; F. Weick in König et al. 1999: Pl. 58; L. McQueen in Rasmussen and Anderton 2005: Pl. 75
Photograph: ?
Literature: Walden 1874: 129–131; Sharpe 1875a: 155; Baker 1927: 456; Ali and Ripley 1969: 296, 297; Eck and Busse 1973: 142; Boyer and Hume 1991: 117; Grimmett et al. 1998: 439; Sankaran 1998: 17–22; del Hoyo et al. 1999: 233; König et al. 1999: 408, 409; Rasmussen and Anderton 2005: 248

- ***Ninox affinis isolata*** Stuart Baker 1926
  Ninox scutulata isolata Stuart Baker 1926, Bull. Brit. Ornith. Cl. 47: 60; Terra typica: Camorta and Car Nicobar Islands

Synonym:
- *Ninox affinis rexpimenti* Abdulali 1979, J. Bombay Nat. Hist. Soc. 75: 744–772; Terra typica: Great Nicobar Island

Distribution: Nicobar Islands: Great Nicobar, Camorta, Trinkat, Car Nicobar
Museum: BMNH (holotype *isolata*), BNHS (holotype *rexpimenti*) not conspecific with *N. scutulata*!

Wing length: 185–205 mm
Tail length: 118–130 mm
Tarsus length: 27 and 28 mm
Length of bill: 22 mm
Body mass: ?

Illustration: ?
Photograph: ?
Literature: Sharpe 1875a: 159; Stuart Baker 1926: 60; Stuart Baker 1927: 456, 457; Ali and Ripley 1969: 297; Abdulali 1979: 744–772; del Hoyo et al. 1999: 233; König et al. 1999: 408, 409; Rasmussen and Anderton 2005: 247, 248

Ninox superciliaris **(Vieillot) 1817**
White-browed Hawk Owl · Madagaskar-Kauz · Ninoxe à sourcils blancs · Ninox Malgache
Strix superciliaris **Vieillot 1817, Nouv. Dict. Hist. Nat. 7: 33;**
Terra typica: Madagascar, fide Gurney (1969) Ibis: 453
(see Plate 8)

Length: 230–300 mm
Body mass: 236 g

Distribution: Northeast, southwest and southern Madagascar
Habitat: Evergreen rainforest, gallery forest, clearings. Also thorny scrubs in semi-arid areas, wooded savanna and deciduous dry woodland. From sea-level up to 800 m
Museum: BMNH, FNSF, MNBHU, MHNP (holotype), NNML, SMNSt
Remarks: Sometimes regarded as rather distinct from entire genus *Ninox*, and included in genera *Athene* or *Strix*, but differs greatly from *Athene* in its long tail, other wing formula and blackish irides

Wing length: 180–193 mm
Tail length: 88–102 mm
Tarsus length: 37 and 38 mm
Length of bill: ?
Body mass: 236 g ($n = 1$)

Illustration: J. G. Keulemans in Milne-Edwards and Grandidier 1876: Atlas 1, Pl. 39 and line draw. bill,

Ninox superciliaris · Madagascar Hawk Owl · Madagaskar-Kauz

foot, wing, tail; K. Bretagnolle in Langrand 1990: Pl. 26 (irides); T. Boyer in Boyer and Hume 1991: 117 (irides); J. Lewington in Kemp and Kemp 1998: 289; N. Arlott in del Hoyo et al. 1999: Pl. 18; F. Weick in König et al. 1999: Pl. 58 (light and dark morph)

Photograph: D. Richards and O. Langrand in Kemp and Kemp 1998: 288, 289; P. Morris in Morris and Hawkins 1998: 168a, b; J. Sargatal in del Hoyo et al. 1999: 104; R. Seitre in del Hoyo et al. 1999: 107; T. Quested in internet: Owl Pages, photo gallery 2003 (Benty Reserve 1995)

Literature: Sharpe 1875a: 181, 182; Eck and Busse 1973: 143; Langrand 1990: 227, 228; Boyer and Hume 1991: 117, 118; Kemp and Kemp 1998: 298, 299; Morris and Hawkins 1998: 202; del Hoyo et al. 1999: 233; König et al. 1999: 409, 410

❖

Ninox philippensis Bonaparte 1855

Philippine Hawk Owl · Philippinenkauz · Ninoxe des Philippines · Ninox Filipino

Length:	150–180 mm
Body mass:	125 g

Distribution: Endemic to the Philippines, occurring on most islands. Mindoro Hawk Owl treated as a distinct species, *Ninox mindorensis*, different in vocalisations (Morris 1999, personal communication to C. König). It is not worth arranging different groups of subspecies (as in Delacour and Mayr 1945), because of the high variability between subspecies

Habitat: Primary forest, secondary growth, remnants, gallery forest and forest edges. From lowlands to higher altitudes. Roosting in daytime mostly in denser and darker parts of the forest

- ***Ninox philippensis philippensis*** Bonaparte 1855
  Ninox philippensis Bonaparte 1855, Compt. Rend. Acad. Sci. Paris 41: 655; **Terra typica:** Philippines

Distribution: Chiefly northeast Philippine Islands: Luzon, Polillo, Marinduque, Catanduanes, Samar, Leyte. Buad, Biliran (sight recorded)

Museum: BMNH, FNSF, MHNP ? (holotype not located!), SMTD, ÜMB, ZFMK

Remarks: Smallest subspecies, white below, striped with rufous-cinnamon

Wing length:	158–169 mm
Tail length:	75–85 mm
Tarsus length:	27.5–29 mm
Length of bill:	20–23 mm
Body mass:	125 g ($n = 1$)

Illustration: J. Smit in *Trans. Zool. Soc. London* 1877(9): Pl. 25; G. Sandström in du Pont 1971: Pl. 38; H. Quintscher in Eck and Busse 1973: Abb. 37 (b/w); T. Boyer in Boyer and Hume 1991: 118; N. Arlott in del Hoyo et al. 1999: Pl. 19; F. Weick in König et al. 1999: Pl. 59; A. P. Sutherland in Kennedy et al. 2000: Pl. 35

Photograph: ?

Literature: Sharpe 1875a: 167, 168; Delacour and Mayr 1945: 108, 109; du Pont 1971: 174; Eck and Busse 1973: 141, 142; Boyer and Hume 1991: 118; Dickinson et al. 1991: 229, 230; del Hoyo et al. 1999: 235; König et al. 1999: 410, 411; Kennedy et al. 2000: 180, 181

- ***Ninox philippensis spilocephala*** Tweeddale 1879
  Ninox spilocephala of Tweeddale 1879, Proc. Zool. Soc. Lond. 1878: 939; Terra typica: Zamboanga, Mindanao

Distribution: Chiefly southeast Philippines: Basilan, Mindanao, Siargao and Dinagat
Museum: BMNH (sintype), FNSF, MNBHU, NNML, SMTD, USNM
Remarks: Very variable in colour and size! Below striped or variegated

Wing length:	164–177 mm (1 × 190 mm?)
Tail length:	71–81 mm
Tarsus length:	28–31.5 mm
Length of bill:	22–24 mm
Body mass:	?

Illustration: H. Quintscher in Eck and Busse 1973: Abb. 37 (b/w); N. Arlott in del Hoyo et al. 1999: Pl. 19; F. Weick in König et al. 1999: Pl. 59; A. P. Sutherland in Kennedy et al. 2000: Pl. 35
Photograph: ?
Literature: of Tweeddale 1878: 939; Delacour and Mayr 1945: 108, 109; du Pont 1971: 175; Eck and Busse 1973: 141, 142; Dickinson et al. 1991: 229, 230; del Hoyo et al. 1999: 235; König et al. 1999: 410, 411; Kennedy et al. 2000: 180, 181

- ***Ninox philippensis reyi*** Oustalet 1880
  Ninox Reyi Oustalet 1880, Bull. Assoc. Sci. France 2(2): 206; Terra typica: Sulu Archipelago

Synonym:
- *Ninox everetti* Sharpe 1897, Bull. Brit. Ornith. Cl. 6: 47; Terra typica: Siasi, Sulu Archipelago

Distribution: Distributed throughout Sulu Archipelago: Bongao, Jolo, Siasi, Sanga Sanga, Sibutu, Tawitawi
Museum: BMNH (holotype *everetti*), MHNP (holotype *reyi*?)
Remarks: Together with *spilonota*, the largest subspecies of *philippensis*. Head and back barred, below barred (not belly)

Wing length:	194 mm
Tail length:	?
Tarsus length:	?
Length of bill:	?
Body mass:	?

Illustration: ?
Photograph: ?
Literature: Sharpe 1897: 47; Delacour and Mayr 1945: 109; du Pont 1971: 175; Eck and Busse 1973: 141, 142; Dickinson et al. 1991: 229, 230; König et al. 1999: 410, 411; Kennedy et al. 2000: 180, 181

- ***Ninox philippensis centralis*** Mayr 1945
  Ninox philippensis centralis Mayr 1945, Zoologica 30(12): 108; Terra typica: Siquijor

Distribution: Distributed throughout the central part of the Philippines: Guimaras, Negros, Siquijor, Panay, Bohol, Boracay, Carabo and Semirara
Museum: USNM (holotype), ZMH
Remarks: Longtailed, much larger than subspecies *proxima*, stripes below ill defined, white margins edged with ochra

Wing length:	181–191 mm
Tail length:	89–93 mm
Tarsus length:	?
Length of bill:	?
Body mass:	?

Illustration: ?
Photograph: ?
Literature: Mayr in Delacour and Mayr 1945: 108; du Pont 1971: 175; Dickinson et al. 1991: 230; del Hoyo et al. 1999: 235; König et al. 1999: 410, 411; Kennedy et al. 2000: 180, 181

- ***Ninox philippensis spilonota*** Bourns and Worcester 1894
  Ninox spilonotus Bourns and Worcester 1894, Occ. Pap. Minn. Acad. Sci. 1: 8; Terra typica: Cebu, Sibuyan, Tablas and Mindoro (sic)

Distribution: Heterogenous geographical distribution: Sibuyan, Cebu, Tablas, Camiguin Sur
Museum: AMNH, BMNH, USNM (sintype)
Remarks: Largest subspecies, coarsely barred below, less barred further down on flanks. Back barred, but less further down. Less white scapulars

Wing length:	188–194 mm
Tail length:	96–101 mm
Tarsus length:	?
Length of bill:	?
Body mass:	?

Illustration: N. Arlott in del Hoyo et al. 1999: Pl. 19; F. Weick in König et al. 1999: Pl. 59
Photograph: ?
Literature: Mayr in Delacour and Mayr 1945: 109; du Pont 1971: 176; Dickinson et al. 1991: 230; del Hoyo et al. 1999: 235; König et al. 1999: 410, 411; Kennedy et al. 2000: 180, 181

- ***Ninox philippensis proxima*** **Mayr 1945**
Ninox philippensis proxima Mayr 1945, Zoologica 30(12): 109; Terra typica: Masbate

Synonym:
- *Ninox philippensis ticaoensis* du Pont 1972, Nemouria 6: 6; Terra typica: Sitio Calpi, Danao, Sao Jacinto, Ticao

Distribution: Masbate and Ticao, two neighbouring islands
Museum: USNM (holotype *proxima*), DMNH (holotype *ticaoensis*)
Remarks: Similar plumage, but larger than nominate *philippensis*. Coarser and darker stripes below. Tail less distinctly barred

Wing length:	175 mm
Tail length:	79 and 82 mm
Tarsus length:	?
Length of bill:	?
Body mass:	?

Illustration: ?
Photograph: ?
Literature: Mayr in Delacour and Mayr 1945: 108; du Pont 1972: 6; Dickinson et al. 1991: 230; del Hoyo et al. 1999: 235; König et al. 1999: 410, 411; Kennedy et al. 2000: 180, 181

❖

Ninox mindorensis Ogilvie-Grant 1896

Mindoro Hawk Owl · Mindorokauz · Ninoxe de Mindoro · Ninox Mindoro
Ninox mindorensis Ogilvie-Grant 1896, Ibis: 463; Terra typica: Lowlands of Mindoro

Length:	200 mm
Body mass:	100–118 g

Synonym:
- *Ninox plateni* "Blasius" Hartlaub 1899, Abh. Naturwiss. Ver. Bremen 16: 271; Terra typica: Mindoro

Distribution: Philippines: Mindoro
Habitat: Forest and wooded lowland areas
Museum: AMNH (Rothsch. = holotype *plateni*), BMNH (holotype *mindorensis*), MNBHU, SNMB
Remarks: Different from *Ninox philippensis* in vocalisations (P. Morris 1999, personal communication to C. König). Small size, vermiculated, or barred above and below

Wing length:	♂ 159–175 mm, ♀ 157–171 mm
Tail length:	77–88 mm
Tarsus length:	27–31.5 mm
Length of bill:	21–23.5 mm
Body mass:	♂ 108 and 118 g, ♀ 100 and 105 g

Illustration: H. Quintscher in Eck and Busse 1973: Abb. 37 (b/w); F. Weick in König 1999a: Pl. 59; A. P. Sutherland in Kennedy et al. 2000: Pl. 35
Photograph: P. Morris in del Hoyo et al. 1999: 140
Literature: Mayr in Delacour and Mayr 1945: 108, 109; Ripley and Rabor 1958: 41; du Pont 1971: 175; Dickinson et al. 1991: 230; del Hoyo et al. 1999: 235; König et al. 1999: 411, 412; Kennedy et al. 2000: 180, 181

❖

Ninox sumbaensis Olsen, Wink, Sauer-Gürth and Trost 2002

Little Sumba Hawk Owl · Kleiner Sumbakauz · Petite Ninoxe de Sumba
Ninox sumbaensis Olsen, Wink, Sauer-Gürth and Trost 2002, Emu 102(3): 225; Terra typica: Sumba Island, Indonesia
(see Plate 8 and Figure 20)

Length:	230 mm
Body mass:	90 g ($n = 1$)

Distribution: Sumba Island, Lesser Sundas, Indonesia
Habitat: Small patches of primary and secondary woodland. About 600 m from sea-level
Museum: Univers. Heidelberg (IPG-20415)

Wing length:	176 mm
Tail length:	99.5 mm
Tarsus length:	?
Length of bill:	?
Body mass:	90 g ($n = 1$)

Illustration: F. Weick in *Gefiederte Welt* 2005(1): 22; F. Weick in Owls (unpubl.): Pl. 67
Photograph: J. Olsen in Olsen et al. 2002: Frontcover, 226–230
Literature: Coates and Bishop 1997: 361 (*Otus* spec.?); King and Yong 2001: 91–93; Olsen et al. 2002: 223–231; Weick 2003–2005, *Gefiederte Welt* 2005(1): 23; Bird Life International (2004) Threatened Birds of the World (CD-ROM)

❖

Ninox dubiosa **sp. nov.**
Dubious Hawk Owl · Vielfleckfalkenkauz
Ninox dubiosa **sp. nov.**
(see Figure 20)

Length:	280 mm
Body mass:	?

Terra typica: ?

Distribution: unknown
Habitat: unknown
Museum: FNSF (labelled only with no. 25238, no other data!)
Remarks: While working at the FNSF on *Ninox* owls for the colour plates in König et al. (1999), I discovered this dubious skin between many other *Ninox* skins, labelled only with the inscription "*Ninox* spec. no. 25238". Subsequent mensural analyses of a series of all *Ninox* species confirmed the distinctness of this specimen

Wing length:	195 and 200 mm
Tail length:	120 mm
Tarsus length:	26 mm
Length of bill:	18 mm
Body mass:	?

Illustration: F. Weick in *Gefiederte Welt* 2005(1): 22; F. Weick in König et al. (unpublished plate for a new edition): Pl. 67
Photograph: Only photograph from skin in photo collection Weick
Literature: Weick 2003–2005, *Gefiederte Welt* 2005(1): 23

❖

Ninox burhani **Indrawan and Somadikarta 2004**
Togian Hawk Owl · Togian-Falkenkauz
Ninox burhani **Indrawan and Somadikarta 2004, Bull. Brit. Ornith. Cl. 124(3): 162;** Terra typica: **Benteng Village, Togian Island, Indonesia**

Length:	~250 mm
Body mass:	100 g

Distribution: Togian, Malenge and Batudaka Islands, probably also Walea Kodi and Walea Bahi, Togian Archipelago, off central Sulawesi, Indonesia
Habitat: Disturbed lowland and hill forest, mixed gardens, sago swamps, sometimes near villages. From sea-level up to 400 m
Museum: MZB (holotype and paratype)

Wing length:	183.5 mm
Tail length:	98 mm
Tarsus length:	29 mm
Length of bill:	22.5 mm
Body mass:	100 g

Illustration: ?
Photograph: M. Indrawan in Indrawan and Somadikarta 2004: 161, Fig. 2 and 3; M. Indrawan in *World Birdwatch* 26(4): 2
Literature: Indrawan and Somadikarta 2004: 160–171; Anonymus 2004: New Owl from islands where no owls were known. *World Birdwatch* 26(4): 2

❖

Ninox ochracea **(Schlegel) 1866**
Ochre-bellied Hawk Owl · Ockerbauchkauz · Ninoxe ocrée · Ninox Ocráceo
Noctua ochracea **Schlegel 1866, Nederl. Tijdschr. Dierk. 1865, 3: 183;** Terra typica: **Negri-lama, Gulf of Tomini, Celebes**
(see Plate 8 and Figure 19)

Length:	250–260 mm
Body mass:	?

Synonym:
- *Ninox perversa* Stresemann 1938, Orn. Monatsb. 46: 149; new name for *ochracea*, but invalid name

Distribution: Sulawesi and Butung Island
Habitat: Primary and tall secondary lowland forest, also riverine forest. Lowlands, sometime up to 1 000 m (possibly 1 780 m)
Museum: BMNH, MNBHU, MZB, NNML (holotype), SMTD

Wing length:	179.5–196 mm
Tail length:	92–104.5 mm
Tarsus length:	28–31 mm
Length of bill:	22–25 mm
Body mass:	?

Illustration: J. G. Keulemans in Sharpe 1875a: Pl. 11(2); T. Boyer in Boyer and Hume 1991: 118; K. Phillipps in Holmes and Phillipps 1996: Pl. 10; D. Gardner in Coates and Bishop 1997: Pl. 33; N. Arlott in del Hoyo et al. 1999: Pl. 19; F. Weick in König et al. 1999: Pl. 59; J. Lewington in Rasmussen 1999: Frontispiece; F. Weick in *Gefiederte Welt* 2005(1): 21
Photograph: ?
Literature: Sharpe 1875a: 167; Stresemann 1938: 149; Stresemann 1940: 427, 428; Mees 1964a: 7 (footnote); Eck and Busse 1873: 144, 145; White and Bruce 1986: 256, 257; Boyer and Hume 1991: 118; Holmes and Phillipps 1996: 362; Coates and Bishop 1997: 362; del Hoyo et al. 1999: 235; König et al. 1999: 412; Rasmussen 1999: 457–464; Indrawan and Somadikarta 2004: 160–171; Weick 2003–2005, *Gefiederte Welt* 2005(1): 21, 22

❖

Ninox ios Rasmussen 1999
Cinnabar Hawk Owl · Zinnoberkauz · Ninoxe cinabre
Ninox ios Rasmussen 1999, Wilson Bulletin 111(4): 458; Terra typica: Northern Sulawesi, Indonesia
(see Figure 19)

Length:	220 mm
Body mass:	78 g

Distribution: Northern Sulawesi, Indonesia
Habitat: Wooded valleys, around 1 120 m above sea-level
Museum: NNML (holotype)

Wing length:	172 mm
Tail length:	97 mm
Tarsus length:	22.5 mm
Length of bill:	18 mm
Body mass:	78 g ($n = 1$)

Illustration: J. Lewington in Rasmussen 1999: Frontispiece; F. Weick in *Gefiederte Welt* 2005(1): 21; F. Weick in König et al. unpublished: Pl. 65
Photograph: Skin photograph (fide Blok) in collection Weick
Literature: Rozendaal and Dekker 1989: 85–109; Rasmussen 1999: 457–464; Weick 2003–2005, *Gefiederte Welt* 2005(1): 21, 22

❖

Ninox squamipila (Bonaparte) 1850
Moluccan Hawk Owl · Molukkenkauz · Ninoxe des Moluques · Ninox Moluqueno

Length:	260–360 mm
Body mass:	210 g

Distribution: Endemic to Moluccas and Tanimbar Islands
Habitat: Primary and tall secondary forest, forest edges. Prefers dense vegetation. From sea-level up to mountain forests at 1 400 m (Seram) and 1 750 m (Buru)

- *Ninox squamipila hypogramma* (G.R. Gray) 1861
  Athene hypogramma G.R. Gray 1861, Proc. Zool. Soc. Lond. 1860: 344; Terra typica: Batjan and Halmahera

Distribution: Moluccas: Halmahera, Ternate and Bacan
Museum: BMNH, MNBHU, NNML

Wing length:	220–241 mm
Tail length:	140–157 mm
Tarsus length:	25.5 mm
Length of bill:	?
Body mass:	?

Illustration: J. G. Keulemans in Sharpe 1875a: Pl. 10; N. Arlott in del Hoyo et al. 1999: Pl. 19; F. Weick in König et al. 1999: Pl. 61
Photograph: ?
Literature: Gray 1861: 344; Lister 1888: 525–527; Sharpe 1875a: 183, 184; White and Bruce 1986: 255; Boyer and Hume 1991: 119; Coates and Bishop 1997: 363; del Hoyo et al. 1999: 235; König et al. 1999: 416; Rasmussen 1999: 457–464

- ***Ninox squamipila hantu*** (Wallace) 1863
  Athene hantu Wallace 1863, Proc. Zool. Soc. Lond.: 22;
  Terra typica: Buru

Distribution: Moluccas: Buru Island
Museum: BMNH (holotype)

Wing length: 190–212 (231?) mm
Tail length: 127–147 mm
Tarsus length: 33 mm
Length of bill: ?
Body mass: 140 g (Novit. Zool. 21)

Illustration: J. G. Keulemans in Sharpe 1875a: Pl. 11(1); N. Arlott in del Hoyo et al. 1999: Pl. 19; F. Weick in König et al. 1999: Pl. 61 (two morphs); J. Lewington in Rasmussen 1999: Frontispiece
Photograph: ?
Literature: Wallace 1863: 22; Sharpe 1875a: 185, 186; Lister 1888: 525–527; Hartert 1914a: 383; Eck and Busse 1973: 135–138; White and Bruce 1986: 255; Coates and Bishop 1997: 363; del Hoyo et al. 1999: 235; König et al. 1999: 416; Rasmussen 1999: 457–464

- ***Ninox squamipila squamipila*** (Bonaparte) 1850
  Athene squamipila Bonaparte 1850, Consp. Gen. Av. 1: 41;
  Terra typica: Ceram

Distribution: Moluccas: Seram Island
Museum: AMNH, BMNH, MHNP (holotype?), MNBHU, SMNSt

Wing length: 190–212 mm
Tail length: 135 mm
Tarsus length: 32 mm
Length of bill: ?
Body mass: 210 g (Novit. Zool. 21)

Illustration: J. G. Keulemans in Sharpe 1875a: Pl. 12(2)2; T. Boyer in Boyer and Hume 1991: 119; D. Gardener in Coates and Bishop 1997: Pl. 33; N. Arlott in del Hoyo et al. 1999: Pl. 19; F. Weick in König et al. 1999: Pl. 61
Photograph: ?
Literature: Sharpe 1875a: 184; Lister 1888: 525–527; Hartert 1914a: 79; White and Bruce 1986: 255; Boyer and Hume 1991: 119; Coates and Bishop 1997: 363; del Hoyo et al. 1999: 235; König et al. 1999: 416; Rasmussen 1999: 457–464

- ***Ninox squamipila forbesi*** Sclater 1883
  Ninox forbesi Sclater 1883, Proc. Zool. Soc. Lond.: 52 and Pl. 11;
  Terra typica: Lutu, Timor Laut, Tanimbar Islands

Distribution: Tanimbar Islands, Lesser Sundas
Museum: BMNH (holotype)

Wing length: 190–212 mm
Tail length: 135 mm
Tarsus length: ?
Length of bill: ?
Body mass: ?

Illustration: W. Hart in Gould 1875–1888, vol 1: Pl. 6; I. Smit in Sclater 1883: Pl. 11; N. Arlott in del Hoyo et al. 1999: Pl. 19; F. Weick in König et al. 1999: Pl. 61
Photograph: ?
Literature: Sclater 1883: 52; Lister 1888: 525–527; White and Bruce 1986: 255; Coates and Bishop 1997: 363; del Hoyo et al. 1999: 235; König et al. 1999: 416

❖

Ninox natalis Lister 1889

Christmas Hawk Owl · Christmas-Kauz · Ninoxe de Christmas · Ninox de la Christmas
Ninox natalis **Lister 1889, Proc. Zool. Soc. Lond. 1888: 525;**
Terra typica: **Christmas Island, Indian Ocean**

Length: 260–290 mm
Body mass: 130–200 g

Distribution: Christmas Island, Indian Ocean
Habitat: Dense evergreen and deciduous rainforest, less often in secondary forest. Sometimes near human habitations. Resting under dense foliage. Relatively tame and approachable
Museum: BMNH (holotype), NNML
Remarks: Often considered as a subspecies of *Ninox squamipila*, but separable by morphological and plumage details, DNA analyses, vocalisations and geographical isolation

Wing length: 178–183 mm
Tail length: 127 mm
Tarsus length: 40.5 mm
Length of bill: ?
Body mass: 130–200 g

Ninox natalis · *Christmas Hawk Owl* · *Christmas-Kauz*

Illustration: N. Arlott in del Hoyo et al. 1999: Pl. 19; F. Weick in König et al. 1999: Pl. 61; F. Weick in König et al. unpublished: Pl. 65

Photograph: R. Hill in internet: Owl Pages, photo gallery 2002; P. Meek in internet: Owl Pages, photo gallery 2002

Literature: Lister 1888: 525–527; Norman et al. 1998b: 197–208; Higgins 1999, vol 4; Hill and Lill 1998a: 209–220; Hill and Lill 1998b: 221–226; del Hoyo et al. 1999: 235; König et al. 1999: 416, 417

Ninox meeki **Rothschild and Hartert 1914**
Manus Hawk Owl · Manuskauz · Ninoxe de l'Amirauté (ou de Manus) · Ninox de la Manus
Ninox meeki **Rothschild and Hartert 1914, Bull. Brit. Ornith. Cl. 33: 105; Terra typica: Manus Island, Admirality Islands**

Length:	250–300 mm
Body mass:	?

Distribution: Manus Island, Admirality Islands
Museum: AMNH (Rothsch. coll.), BMNH

Wing length:	230–240 mm
Tail length:	120–130 mm
Tarsus length:	37 mm
Length of bill (cere):	17–19 mm
Body mass:	?

Illustration: H. Quintscher in Eck and Busse 1973: Pl. 1; T. Boyer in Boyer and Hume 1991: 121; N. Arlott in del Hoyo et al. 1999: Pl. 19; F. Weick in König et al. 1999: Pl. 61

Photograph: ?

Literature: Rothschild and Hartert 1914b: 105; Hartert 1914b: 289; Coates 1976; Coates 1985; Buckingham et al. 1995; Eastwood 1995: 53–55; Gregory 1995: 112–115; del Hoyo et al. 1999: 236; König et al. 1999: 418

Ninox theomacha **(Bonaparte) 1855**
Jungle Hawk Owl · Einfarbkauz · Ninoxe brune · Ninox Papú
(see Plate 8)

Length:	200–250 mm
Body mass:	?

Distribution: New Guinea, Waigeo and Misool Islands, Entrecasteaux Archipelago, Louisiade Archipelago
Habitat: Lowland forest, forest edges, montane and submontane rainforest, tree groves in open country. From lowlands up to 770–1 250 m, sometimes up to about 2 000 m

- ***Ninox theomacha hoedtii*** (Schlegel) 1871
  Noctua Hoedtii Schlegel 1871, Nederl. Tijdschr. Dierk. 4: 3; Terra typica: Misool

Distribution: Waigeo and Misool Islands
Museum: NNML (holotype), FNSF

Wing length:	178 mm
Tail length:	99 mm
Tarsus length:	28 mm
Length of bill:	26.5 mm
Body mass:	?

Illustration: F. Weick in König et al. 1999: Pl. 60

Photograph: ?

Literature: Sharpe 1875a: 178; Rand and Gilliard 1967: 254; Eck and Busse 1973: 135–138; Beehler et al. 1986: 132; del Hoyo et al. 1999: 236; König et al. 1999: 413, 414

- ***Ninox theomacha theomacha*** (Bonaparte) 1855
  Spiloglaux theomacha Bonaparte 1855, Compt. Rend. Acad. Sci. Paris 41: 654; Terra typica: Triton Bay, New Guinea

Synonym:
- *Ninox terricolor* Ramsay 1880, Proc. Linn. Soc. New S. Wales 4: 466; Terra typica: Goldie River, New Guinea

Distribution: New Guinea
Museum: BMNH, FNSF, MHNP (holotype), NNML, SMNSt, SMTD, YPM

Wing length:	175–186 mm
Tail length:	93 mm
Tarsus length:	32 mm
Length of bill:	24 mm
Body mass:	?

Illustration: A. E. Gilbert in Rand and Gilliard 1954: Pl. 18; T. Medland in Iredale 1956: Pl. 7; D. Zimmerman in Beehler et al. 1986: 132; T. Boyer in Boyer and Hume 1991: 119; N. Arlott in del Hoyo et al. 1999: Pl. 19; F. Weick in König et al. 1999: Pl. 60
Photograph: Mayr and Gilliard 1954: Pl. 18; Zool. Soc. London in Everett 1977: 41, left bot.; Lindgren in Burton et al. 1992: 148; D. Hadden in del Hoyo et al. 1999: 91
Literature: Rand and Gilliard 1967: 254; Eck and Busse 1973: 135–138; Beehler et al. 1986: 132; Boyer and Hume 1991: 119; del Hoyo et al. 1999: 236; König et al. 1999: 413, 414

- ***Ninox theomacha goldii*** Gurney 1883
  Ninox goldii Gurney 1883 Ibis: 171; Terra typica: Southeast New Guinea, error = Fergusson Island, fide Rothschild and Hartert (1918) Novit. Zool. 25: 325

Synonym:
- *Ninox goodenoviensis* de Vis 1890, Ann. Rept. Brit. New Guinea 1888/1889: 58; Terra typica: Goodenough Island, D'Entrecasteaux Archipelago

Distribution: D'Entrecasteaux Archipelago: Goudenough, Fergusson and Normanby Islands
Museum: BMNH (holotype), FNSF, SMTD

Wing length:	215–227 mm
Tail length:	?
Tarsus length:	?
Length of bill:	?
Body mass:	?

Illustration: T. Medland in Iredale 1956: Pl. 7; H. Quintscher in Eck and Busse 1973: Abb. 36 (b/w line drawing); D. Zimmerman in Beehler et al. 1986: 132; N. Arlott in del Hoyo et al. 1999: Pl. 19; F. Weick in König et al. 1999: Pl. 60
Photograph: ?
Literature: Gurney 1883: 171; Rothschild and Hartert 1918: 325; Rand and Gilliard 1967: 254; Eck and Busse 1973: 135–138; Beehler et al. 1986: 132; del Hoyo et al. 1999: 236; König et al. 1999: 413, 414

- ***Ninox theomacha rosseliana*** Tristram 1889
  Ninox rosseliana Tristram 1889, Ibis: 557;
  Terra typica: Rossel Island, Louisiade Archipelago

Distribution: Louisiade Archipelago: Tagula and Rossel Islands
Museum: BMNH (holotype)
Remarks: Relationship of species *theomacha* is uncertain. Included as a subspecies, maybe a distinct species, but further research is needed!

Wing length:	206 mm
Tail length:	127 mm
Tarsus length:	33 mm
Length of bill:	?
Body mass:	?

Illustration: F. Weick in König et al. 1999: Pl. 60
Photograph: ?

Literature: Tristram 1889: 557, 558; Rand and Gilliard 1967: 254; Eck and Busse 1973: 135–138; Beehler et al. 1986: 132; del Hoyo et al. 1999: 236; König et al. 1999: 413, 414

Ninox punctulata **(Quoy and Gaimard) 1830**
Speckled Hawk Owl · Pünktchenkauz · Ninoxe pointillée · Ninox Punteado
Noctua punctulata **Quoy and Gaimard 1830, Voy. "*Astralobe*", Zool. 1: 165, Atlas Ois. Pl. 1(f1);** Terra typica: **Celebes** **(see Figure 21)**

Length:	260 mm
Body mass:	151 g

Distribution: Sulawesi, with Kabaena, Muna and Butung Islands

Habitat: Primary lowland and hill forest, also tall secondary forest, forest edges. Woodland with streams, cultivated areas near habitations. From lowlands up to 1 100 m
Museum: BMNH, FNSF, MZB, NNML, SMTD
Remarks: Relationship to other *Ninox* members uncertain

Wing length:	157–177 mm
Tail length:	76 mm
Tarsus length:	33 mm
Length of bill:	28 mm
Body mass:	♂ 151 g ($n = 1$)

Illustration: Quoy and Gaimard 1830: Atlas Ois. Pl. 1 (♀ 1); T. Boyer in Boyer and Hume 1991: 120; K. Phillipps in Holmes and Phillipps 1996: Pl. 10; D. Gardner in Coates and Bishop 1997: Pl. 33; N. Arlott in del Hoyo et al. 1999: Pl. 19; F. Weick in König et al. 1999: Pl. 60 (2 variants)
Photograph: ?
Literature: Stresemann 1940: 428, 429; White and Bruce 1986: 255; Boyer and Hume 1991: 120; Andrew 1993; Holmes and Phillipps 1996: 36; Catterall 1997; Coates and Bishop 1997: 363; del Hoyo et al. 1999: 236; König et al. 1999: 414, 415

Ninox odiosa Sclater 1877
Russet or New Britain Hawk Owl · Neubritannienkauz · Ninoxe odieuse · Ninox de Nueva Bretana
Ninox odiosa Sclater 1877, Proc. Zool. Soc. Lond.: 108;
Terra typica: New Britain

Length:	200–230 mm
Body mass:	209 g

Distribution: New Britain Island in the Bismarck Archipelago
Habitat: Forested lowlands and hills, cultivated areas, plantations and settlements. Up to around 800 m above sea-level
Museum: BMNH (holotype), FNSF, MNBHU, SMTD, ZMH, ZSBS
Remarks: Relationship uncertain, but possibly related to *Ninox jacquinoti*. Widely distributed and quite common, but little studied and in need of further investigation

Wing length:	♂ 170 mm, ♀ 181–187 mm
Tail length:	?
Tarsus length:	?
Length of bill:	?
Body mass:	♀ 209 g ($n = 1$)

Illustration: W. Hart in Gould 1875–1888, vol 1: Pl. 5; T. Boyer in Boyer and Hume 1991: 122; N. Arlott in del Hoyo et al. 1999: Pl. 19; F. Weick in König et al. 1999: Pl. 60
Photograph: ?
Literature: Sclater 1877: 108; Dahl 1899: 163; Eck 1971: 173–218; Eck and Busse 1973: 144; Boyer and Hume 1991: 122; del Hoyo et al. 1999: 237; König et al. 1999: 415

Ninox variegata (Quoy and Gaimard) 1830
Bismarck or New Ireland Hawk Owl · Neuirlandkauz, Bismarckkauz · Ninoxe de Nouvelle Irlande · Ninox de las Bismarck

Length:	250–300 mm
Body mass:	?

Distribution: Endemic to Bismarck Archipelago: New Ireland, New Britain and New Hanover
Habitat: Forested lowlands, hills and lower mountains, up to 1 000 m from sea-level

- *Ninox variegata variegata* (Quoy and Gaimard) 1830
  Noctua variegata Quoy and Gaimard 1830, Voy. "*Astrolabe*", Zool. 1: 166, Atlas Ois. Pl. 1(f2);
  Terra typica: Carteret Harbor, New Ireland

Synonym:
- *Ninox solomonis* Sharpe 1876, Proc. Zool. Soc. Lond.: 673, Pl. 62; Terra typica: Solomon Island, error = South New Ireland (Mayr [1933] *Ibis:* 552)
- *Ninox novae britanniae* Ramsay 1877, Proc. Linn. Soc. New S. Wales 2: 105; Terra typica: New Britain
- *Ninox novaebritanniae novaehibernicae* Mathews 1926, Bull. Brit. Ornith. Cl. 46: 131; New name for *Ninox variegata* (invalid)

Distribution: Bismarck Archipelago: New Britain and New Ireland
Museum: BMNH, MNBHU

Wing length: 192–210 mm
Tail length: 117 and 118 mm
Tarsus length: 30.5–33 mm
Length of bill (cere): 17 mm
Body mass: ?

Illustration: Quoy and Gaimard 1830, Atlas Ois.: Pl. 1, f2 (b/w); J. Smit in Sharpe 1876a: Pl. 57 (*N. solomonis);* H. Quintscher in Eck and Busse 1973: Abb. 36 (b/w); T. Boyer in Boyer and Hume 1991: 121; N. Arlott in del Hoyo et al. 1999: Pl. 19; F. Weick in König et al. 1999: Pl. 61
Photograph: ?
Literature: Sharpe 1875a: 185; Sharpe 1876a: 673; Hartert 1925a: 121; Eck and Busse 1973: 135–138; Boyer and Hume 1991: 121; del Hoyo et al. 1999: 236, 237; König et al. 1999: 418, 419

- *Ninox variegata superior* Hartert 1925
  Ninox Variegata superior Hartert 1925, Novit. Zool. 32: 121; Terra typica: New Hanover, Bismarck Archipelago

Distribution: New Hanover, Bismarck Archipelago
Museum: AMNH (Rothsch. coll.), BMNH (Cayley Webster coll.)

Wing length: 211–224 mm
Tail length: ?
Tarsus length: ?
Length of bill (cere): 18 mm
Body mass: ?

Illustration: F. Weick in König et al. 1999: Pl. 61
Photograph: ?
Literature: Hartert 1925a: 121; Eck and Busse 1973: 135–138; Coates 1985; König et al. 1999: 418, 419

❖

Ninox jacquinoti (Bonaparte) 1850
Solomon Hawk Owl · Salomonenkauz, Jacquinotkauz · Ninoxe de Jacquinot · Ninox de la Solomón

Length: 250–300 mm
Body mass: 174 g

Distribution: Solomon Archipelago
Habitat: Primary and tall secondary forest. Lowlands and foothills, up to 1 500 m above sea-level

- *Ninox jacquinoti eichhorni* (Hartert) 1929
  Spiloglaux jacquinoti eichhorni Hartert 1929, Am. Mus. Novit. 364: 7; Terra typica: Bougainville, Choiseul and Buka Island, Solomon Archipelago

Distribution: Solomon Archipelago: Bougainville, Choiseul and Buka Islands
Museum: AMNH (holotype), BMNH

Wing length: 185–197 mm
Tail length: 96–105 mm
Tarsus length: ?
Length of bill: ?
Body mass: ?

Illustration: ?
Photograph: ?
Literature: Hartert 1929b: 6, 7; Eck and Busse 1973: 145, 146; del Hoyo et al. 1999: 237; König et al. 1999: 412, 413

- *Ninox jacquinoti jacquinoti* (Bonaparte) 1850
  Athene jacquinoti Bonaparte 1850, Consp. Gen. Av. 1: 42; Terra typica: Oceania = St. George Island, Solomon Islands, ex Hombron and Jacquinot (1853) Voy. Pôle Sud, Zool. 3: 51

Synonym:
- *Athene taeniata* Hombron and Jacquinot 1853, Voy. Pôle Sud, Zool. 3: 50, 51

Distribution: Solomon Archipelago, Ysabel and St. George Island
Museum: AMNH, BMNH, MHNP (holotype), SMTD

Wing length: 195–208 mm
Tail length: 106–112 mm
Tarsus length: 42 mm
Length of bill: ?
Body mass: ?

Illustration: Hombron and Jacquinot 1853: Pl. 3; H. Quintscher in Eck and Busse 1973: Abb. 38; T. Boyer in Boyer and Hume 1991: 122; N. Arlott in del Hoyo et al. 1999: Pl. 19; F. Weick in König et al. 1999: Pl. 59
Photograph: D. Hadden in Duncan 2003: 284
Literature: Sharpe 1875a: 186 (*taeniata*); Sharpe 1875b: 259; Hartert 1929b: 6, 7; Mayr 1935: 2, 3;

Mayr 1945b; Eck and Busse 1973: 145, 146; Boyer and Hume 1991: 122; del Hoyo et al. 1999: 237; König et al. 1999: 412, 413

- ***Ninox jacquinoti granti*** Sharpe 1888
  Ninox granti Sharpe 1888, Proc. Zool. Soc. Lond.: 183; Terra typica: Guadalcanal, Solomon Archipelago

Distribution: Guadalcanal, Solomon Archipelago
Museum: BMNH (holotype), SMTD

Wing length: 178, 180 and 183 mm
Tail length: 91.5, 104 mm
Tarsus length: 33–35.5 mm
Length of bill: 24 mm
Body mass: ?

Illustration: N. Arlott in del Hoyo et al. 1999: Pl. 19; F. Weick in König et al. 1999: Pl. 59
Photograph: ?
Literature: Sharpe 1888c: 183, 184; Lister 1888: 527; Eck and Busse 1973: 145, 146; del Hoyo et al. 1999: 237; König et al. 1999: 412, 413

- ***Ninox jacquinoti mono*** Mayr 1935
  Ninox jacquinoti mono Mayr 1935, Am. Mus. Novit. 820: 2; Terra typica: Mono Island, Solomon Archipelago

Distribution: Mono Island, Solomon Archipelago
Museum: AMNH (holotype)

Wing length: 190–196 mm
Tail length: ?
Tarsus length: ?
Length of bill: ?
Body mass: ?

Illustration: N. Arlott in del Hoyo et al. 1999: Pl. 19
Photograph: ?
Literature: Mayr 1935: 2, 3; Eck and Busse 1973: 145, 146; del Hoyo et al. 1999: 237; König et al. 1999: 412, 413

- ***Ninox jacquinoti floridae*** Mayr 1935
  Ninox jacquinoti floridae Mayr 1935, Am. Mus. Novit. 820: 2; Terra typica: Florida Island, Solomon Archipelago

Distribution: Florida Island, Solomon Archipelago
Museum: AMNH (holotype)

Wing length: 218–225 (228) mm
Tail length: 120 mm
Tarsus length: ?
Length of bill: ?
Body mass: ?

Illustration: N. Arlott in del Hoyo et al. 1999: Pl. 19
Photograph: ?
Literature: Mayr 1935: 2; Eck and Busse 1973: 145, 146; del Hoyo et al. 1999: 237; König et al. 1999: 412, 413

- ***Ninox jacquinoti malaitae*** Mayr 1931
  Ninox jacquinoti malaitae Mayr 1931, Am. Mus. Novit. 504: 14, 15; Terra typica: Malaita Island, Solomon Archipelago

Distribution: Malaita Island, Solomon Archipelago
Museum: AMNH (holotype)

Wing length: 164 and 165 mm
Tail length: 84 and 93 mm
Tarsus length: 30 and 32 mm
Length of bill (cere): 15 and 16 mm
Body mass: 174 g ($n = 1$)

Illustration: N. Arlott in del Hoyo et al. 1999: Pl. 19
Photograph: ?
Literature: Mayr 1931: 15, 16; Eck and Busse 1973: 145, 146; del Hoyo et al. 1999: 237; König et al. 1999: 412, 413

- ***Ninox jacquinoti roseoaxillaris*** (Hartert) 1929
  Spiloglaux roseoaxillaris Hartert 1929, Am. Mus. Novit. 364: 6; Terra typica: Bauro, San Cristobal, Solomon Archipelago

Distribution: Bauro and San Cristobal Island, Solomon Archipelago
Museum: AMNH (Beck coll.)
Remarks: Relationships uncertain – regarded by some as related to *Ninox odiosa*

Wing length: 157 mm
Tail length: 87 mm
Tarsus length: 30 mm
Length of bill (cere): 17 mm
Body mass: ?

Illustration: N. Arlott in del Hoyo et al. 1999: Pl. 19
Photograph: ?
Literature: Hartert 1929b: 6, 7; Eck and Busse 1973: 145, 146; del Hoyo et al. 1999: 237; König et al. 1999: 412, 413

Genus
Uroglaux **Mayr 1937**
Uroglaux Mayr 1937, Am. Mus. Novit. 939: 6. Type by *Athene dimorpha* Salvadori 1874

Uroglaux dimorpha (Salvadori) 1874

Papuan Hawk Owl · Rundflügelkauz · Chouette ou Ninoxe papoue · Ninox Hálcon
Athene dimorpha Salvadori, 1874, Ann. Mus. Civ. Genova, 6: 308;
Terra typica: Sorong, New Guinea

Length:	300–340 mm
Body mass:	?

Distribution: Northwest New Guinea: Irian Jaya. Southeast New Guinea: Papua New Guinea. Yapen Island. Records exist now from central New Guinea, but possibly occurs throughout the whole area
Habitat: Rainforest, forest edges and clearings. Gallery forest in savanna. Up to about 1 500 m
Museum: BMNH, MGD (holotype), MHNP, MNBHU, NNML, ZFMK

Remarks: Formerly regarded as closely allied to genus *Ninox*, but with rounded instead of pointed wings. Relationship to genus *Sceloglaux* questionable; both are probably relict species!

Wing length:	200–225 mm
Tail length:	145–156 mm
Tarsus length:	32 and 33 mm
Length of bill:	30 mm
Body mass:	?

Illustration: W. Hart in Gould 1875–1888, vol 1: Pl. 7; T. Medland in Iredale 1956: Pl. 7; Grossman and Hamlet 1965: 439 (b/w); D. Zimmerman in Beehler et al. 1986: 131 (b/w line drawing); T. Boyer in Boyer and Hume 1991: 111; J. Lewington in del Hoyo et al. 1999: Pl. 20; F. Weick in König et al. 1999: Pl. 62
Photograph: E. Lindgren in Burton et al. 1992: 155; E. Lindgren in del Hoyo et al. 1999: 78
Literature: Grossman and Hamlet 1965: 439; Rand and Gilliard 1967: 256, 257; Eck and Busse 1973: 130, 131; Beehler et al. 1986: 131; Boyer and Hume 1991: 111; del Hoyo et al. 1999: 239; König et al. 1999: 419

Uroglaux dimorpha
Papuan Hawk Owl
Rundflügelkauz

Genus
Sceloglaux **Kaup 1848**
Sceloglaux Kaup 1848, Isis v. Oken, col. 768. Type by *Athene albifacies* G.R. Gray 1844

Wing length:	264 mm
Tail length:	165 mm
Tarsus length:	67.5 mm
Length of bill:	?
Body mass:	~600 g

Sceloglaux albifacies **(G.R. Gray) 1844**
Laughing Owl · Lachkauz, Weißwangenkauz · Chouette rieuse (ou à joue blanche) · Ninox Reidor
Athene albifacies G.R. Gray 1844, Zool. Voy. "Erebus and Terror", Bds.: 2 and Pl. 1; Terra typica: Waikouaiti, South Island, New Zealand

Length:	350–400 mm
Body mass:	~600 g

Synonym:
- *Sceloglaux rufifacies* Buller 1904, Ibis: 639; Terra typica: Wairarapa district, North Island, New Zealand

Distribution: New Zealand: North Island (southern half), South Island (east of the Southern Alps) and Stewart Island (now extinct!)

Habitat: Areas of lower rainfall: rocky valleys and cliffs, open country and temperate woodland. Scrub, forest edges

Museum: BMNH, MZUS, NHMWien, ÜMB

Illustration: W. Hart in Gould 1869, Suppl.: Pl. 2; J. G. Keulemans in Buller 1888: Pl. 20; J. G. Keulemans in Buller 1904: Pl. 8 (*rufifacies*); G. E. Lodge in Fleming 1982: Pl. ?; E. Power in Falla et al. 1993: Pl. 39; F. Weick in König et al. 1999: Pl. 62; J. Lewington in del Hoyo et al. 1999: Pl. 20

Remarks: The wings are too short, and the tail and tail projection too long in this image! The measurements of length, given in del Hoyo et al. (1999) at up to 470 mm, are also too large. Most skins exceed 400 mm in length

Photograph: H. C. C. Wright (1889–1910), A. Turnbull Library, in internet: Owl Pages, photo gallery 2002

Remarks: This photo of a live bird shows the true proportions of wings, tail and tail projection!

Literature: Buller 1888: vol 1; Buller 1904: 639; Falla et al. 1966: 186 187; Williams and Harrison 1972: 4–19; Eck and Busse 1973: 147, 148; Falla et al. 1978: 172, 173; del Hoyo et al. 1999: 239; König et al. 1999: 420

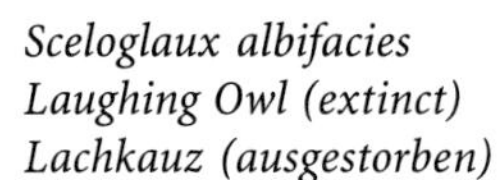

Sceloglaux albifacies
Laughing Owl (extinct)
Lachkauz (ausgestorben)

Subfamilia / Subfamily
Asioninae · Eared Owls and Allies · Ohreulen und Verwandte

Genus
***Pseudoscops* Kaup 1848**
Pseudoscops Kaup 1848, Isis v. Oken, col. 769. Type by *Ephialtes grammicus* Gosse 1847

Remarks: Olson unites *grammicus* and *clamator* in the genus *Pseudoscops* but, as shown by molecular analyses (Wink and Heidrich), *clamator* (and also *stygius*) belong to the genus *Asio*. So the present species is treated in the monotypic genus *Pseudoscops*

***Pseudoscops grammicus* (Gosse) 1847**
Jamaican Owl · Jamaikaeule · Hibou de la Jamaique · Lechuza Jamaicano, Búho Jamaicano
Ephialtes grammicus Gosse 1847, Birds Jamaica: 19 and note; Terra typica: Tait-Shafton, Jamaica

Length: 270–330 (350?) mm
Body mass: ?

Distribution: Endemic to Jamaica, Greater Antilles
Habitat: Open woodland, semi-open country with groups of trees, forest edges and clearings, parkland, gardens. Mainly in coastal areas and lowlands, but also at higher elevations
Museum: BMNH, NHMWien, ÜMB, USNM

Wing length:	197–229 mm
Tail length:	96–131 mm
Tarsus length:	39 mm
Length of bill (cere):	19–22 mm
Body mass:	?

Illustration: Gosse 1847: Pl. 4; Grossman and Hamlet 1965: 463 (b/w); T. Boyer in Boyer and Hume 1991: 161; J. Lewington in del Hoyo et al. 1999: Pl. 20; F. Weick in König et al. 1999: Pl. 25; K. Williams in Raffaele et al. 2003: Pl. 43
Photograph: J. H. Carmichael in Audubon Mag. and Russel 1977: 180; Y. Rey-Millet in Downer and Sutton 1990: Pl. 20; R. and N. Bowers in internet: Owl Pages, photo gallery 2003 (2 photographs)
Literature: Gosse 1847: 19; Sharpe 1875a: 242–244; Ridgway 1914: 674, 675; Eck and Busse 1973: 182; Bond 1986 (repr.): 123; Downer and Sutton 1990: 71; Boyer and Hume 1991: 161; Olson 1995: 35–39; del Hoyo et al. 1999: 239; König et al. 1999: 288; Raffaele et al. 2003: 100

Pseudoscops grammicus · Jamaican Owl · Jamaikaeule

Genus
***Asio* Brisson 1760**
***Asio* Brisson 1760, Ornith. 1: 28. Type by *Strix otus* Linné 1758**

Synonym:
- *Rhinoptynx* Kaup 1851, Arch. Naturgesch. 17(1): 107. Type by *Otus mexicanus* Cuvier = *Bubo clamator* Vieillot

***Asio clamator* (Vieillot) 1807**
Striped Owl · Streifenohreule, Schreieule · Hibou strié · Búho cornudo Cariblanco, Búho Gritón
(see Plate 9)

Length: 305–380 mm
Body mass: 320–556 g

Distribution: From southern Mexico, through Middle America, locally in Colombia, Venezuela, the Guianas, eastern Peru, Bolivia, Brazil, Paraguay, northern Argentina, Uruguay and the Caribbean Islands
Habitat: Variable habitats: tropical forest, semi-open woodland, semi-open grassland with groups of trees, bushes or scrub. Thicket hedges, marshes, savanna, riparian woodland, clearings, suburban and agricultural areas, also plantations. Absent from dense forest

- ***Asio clamator forbesi* Lowery and Dalquest 1951**
  Rhinoptynx clamator forbesi Lowery and Dalquest 1951, Univ. Kansas Publ. Mus. Nat. Hist. 3(4): 576, 577;
  Terra typica: Presidio, Vera Cruz

Synonym:
- *Otus mexicanus* Cuvier 1829, Regne Anim. 1: 341; Terra typica: Mexico. Invalid name?

Distribution: From South Mexico to Costa Rica and Panama
Museum: UKMNH (holotype)

Wing length: ♂ 228–244 mm, ♀ 244–273 mm
Tail length: ♂ 127–130 mm, ♀ 132–150 mm
Tarsus length: ?
Length of bill (cere): 21 mm
Body mass: ♂ 335–347 g, ♀ 400–502 g

Illustration: D. E. Tibbitts in Blake 1963: 221 (b/w); J. A. Gwynne in Ridgely and Gwynne 1989: Pl. 12; D. Gardner in Stiles and Skutch 1991: Pl. 20; S. Webb in Howell and Webb 1995: Pl. 26
Photograph: K. W. Fink in Burton et al. 1992: 133; K. W. Fink in Johnsgard 2002: Pl. 42
Literature: Bangs 1907: 31, 32; Ridgway 1914: 670–673; Lowery and Dalquest 1951: 576, 577; Blake 1963: 221; Eck and Busse 1973. 177; Voous 1988: 266–269; Ridgely and Gwynne 1989: 191, 192; Stiles and Skutch 1991: 195, 196; Boyer and Hume 1991: 151, 152; Howell and Webb 1995: 368, 369; del Hoyo et al. 1999: 239; König et al. 1999: 428, 429

- ***Asio clamator clamator* (Vieillot) 1807**
  Bubo clamator Vieillot 1807, Ois. Am. Sept. 1: 52 and Pl. 20;
  Terra typica: Cayenne

Distribution: Colombia and Venezuela, the Guianas, south to eastern Peru and central and northeast Brazil
Museum: BMNH, FNSF, MZUS, SMNSt, SMTD, USNM, ZFMK, ZMH, ZSBS

Wing length: 236–277 mm
Tail length: 127–165 mm
Tarsus length: 58 mm
Length of bill (cere): 22 mm
Body mass: ♂ 335–385 g, ♀ 400–556 g

Illustration: P. Barruel in Haverschmidt 1968: Pl. 13; G. Tudor in Hilty and Brown 1986: Pl. 9; T. Boyer in Boyer and Hume 1991: 151; J. Lewington in del Hoyo et al. 1999: Pl. 20; F. Weick in König et al. 1999: Pl. 64; P. J. Greenfield in Ridgely and Greenfield 2001: II: Pl. 36(10); S. Webb in Hilty 2003: Pl. 24
Photograph: L. C. Marigo in del Hoyo et al. 1999: 82; L. Koerner in internet: Owl Pages, photo gallery 2002; J. C. Motta jr. in internet: Owl Pages, photo gallery 2004 (juvenile)
Literature: Bangs 1907: 31, 32; Ridgway 1914: 670–673; Haverschmidt 1968: 162; Land 1970: 141; Eck and Busse 1973: 177; Hilty and Brown 1986: 232; Voous 1988: 266–269; Boyer and Hume 1991: 151, 152; del Hoyo et al. 1999: 239; König et al. 1999: 428, 429; Ridgely and Greenfield 2001: I: 313, 314, II: 221; Hilty 2003: 366

- ***Asio clamator oberi*** (E.H. Kelso) 1936
  Rhinoptynx clamator oberi E.H. Kelso 1936, Auk 53: 82;
  Terra typica: Island of Tobago

Distribution: Tobago and northeast Trinidad
Museum: USNM (holotype)
Remarks: Status of subspecies doubtful. Very similar to the nominate *clamator*, being only slightly larger in size

Wing length:	284 mm
Tail length:	154 mm
Tarsus length:	?
Length of bill (cere):	22 mm
Body mass:	?

Illustration: ?
Photograph: ?
Literature: Kelso 1936: 82; Herklots 1961: 131 (merged with *clamator*); Eck and Busse 1973: 177; del Hoyo et al. 1999: 239; König et al. 1999: 428, 429 (merged with *clamator*)

- ***Asio clamator midas*** (Schlegel) 1862
  Otus midas "Lichtenstein" Schlegel 1862, Mus. Pais-Bas 2, Oti: 2, note; Terra typica: Montevideo, Uruguay

Synonym:
- *(Strix) americana* Gmelin 1788, Syst. Nat. 1(1): 288; Terra typica: South America. Invalid name
- *Strix maculata* Vieillot 1817, Nouv. Dict. Hist. Nat. 7: 45; Terra typica: Paraguay
- *Rhinoptynx clamator mogenseni* L. and E. H. Kelso 1935, Auk 52: 451; Terra typica: Argentina (Tucumán) and Uruguay

Distribution: Eastern Bolivia, southern Brazil, south to northern Argentina and Uruguay
Museum: FNSF, NNML (holotype), SMNSt, ZFMK
Remarks: Often placed in monotypic genus *Rhinoptynx*, but certainly related to genus *Asio* (confirmed by molecular analyses). Requires further studies!

Wing length:	267–294 mm
Tail length:	144–150 mm
Tarsus length:	?
Length of bill:	?
Body mass:	350–500 g

Illustration: S. Frisch in Frisch 1981: 121; A. Cameron in Voous 1988: 266; J. Lewington in del Hoyo et al. 1999. Pl. 20; F. Weick in König et al. 1999: Pl. 64 (adult and juvenile)
Photograph: K. Kussmann in Eck and Busse 1973: Abb. 20a; W. Wozniak in Eck and Busse 1973: Abb. 20b; J.C. Motta jr. in internet: Owl Pages, photo gallery 2003
Literature: Eck and Busse 1973: 177; Belton 1984, Bull. Am. Mus. Nat. Hist.: 178; Voous 1988: 266–269; del Hoyo et al. 1999: 239; König et al. 1999: 428, 429

Asio stygius (Wagler) 1832
Stygian Owl · Styx-Eule, Dunkle Ohreule · Hibou obscur, Hibou maître bois · Búho Negruzco
(see Plate 9)

Length:	380–460 mm
Body mass:	675 g

Distribution: Mexico, along Pacific slope (Durango, Guerrero and Chiapas), Atlantic slope from Vera Cruz locally to Guatemala. Cozumel Island and Belize. South through Nicaragua and Honduras to northern South America and patchily south to northeast Argentina. Cuba and Hispaniola
Habitat: Montane pine and pine-oak forest. Also evergreen and deciduous forest and humid woodland. Parks and open areas with groups of trees. From sea-level up to about 3 100 m
Remarks: Placed by some in genus *Rhinoptynx*, but closely related to *Asio otus*

- ***Asio stygius robustus*** L. Kelso 1934
  Asio stygius robustus L. Kelso 1934, Auk 51: 522;
  Terra typica: Mirador, Veracruz, Mexico

Synonym:
- *Asio stygius lambi* Moore 1937, Proc. Biol. Soc. Wash. 50: 103; Terra typica: Babizos, 6 000 feet, Northeast Sinaloa (holotype M(oore)LZ)

Distribution: Western and southern Mexico, along Pacific slope (Durango, Guerrero and Chiapas), Atlantic slope: Vera Cruz patchily to Guatemala, northwest Venezuela, Colombia and Ecuador
Museum: AMNH, FNSF, MLZ, SMNSt, MCZB, USNM (holotype)

Asio stygius robustus · Stygian Owl · Styx-Eule

Wing length: ♂ 292–305 mm, ♀ 340–349 mm
Tail length: ♂ 157 mm, ♀ 169 and 171 mm
Tarsus length: ?
Length of bill (cere): 21–22 mm
Body mass: ♂ 591 g, ♀ 675 g

Illustration: G. Tudor in Hilty and Brown 1986: Pl. 9 (b/w); J. Fjeldsa in Fjeldsa and Krabbe 1990: Pl. 25; S. Webb in Howell and Webb 1995: Pl. 26; F. Weick in König et al. 1999: Pl. 63; P. J. Greenfield in Ridgely and Greenfield 2001: II: Pl. 36(12); S. Webb in Hilty 2003: Pl. 25 (subspecies *stygius*, below with buff in place of white ground colour)
Photograph: J. Culbertson in Johnsgard 2002: Pl. 41; R. and N. Bowers in internet: Owl Pages, photo gallery 2003; J. C. Motta jr. in internet: Owl Pages, photo gallery 2004 (juvenile)
Literature: Kelso 1934b: 522, 523; Hilty and Brown 1986: 232; Fjeldsa and Krabbe 1990: 229; Voous 1988: 262–265; Howell and Webb 1995: 368; del Hoyo et al. 1999: 240; König et al. 1999: 423, 424; Johnsgard 2002: 211, 212; Ridgely and Greenfield 2001: I: 314, II: 221, 222; Hilty 2003: 366, 367

- ***Asio stygius siguapa*** (d'Orbigny) 1839
  Otus siguapa d'Orbigny 1839, in La Sagra's Hist. fis. pol. y nat Isla de Cuba 3, Aves: 40 and Pl. 2; Terra typica: Cuba

Synonym:
- *Asio noctipetens* Riley 1916, Smiths. Misc. Coll. 66(15):1; Terra typica: Constanza, 4 000 feet, Santa Domingo

Distribution: Cuba, Isle of Pines, Hispaniola and Gonave Island (Hispaniola population rare or extinct)
Museum: USNM, ANSP
Remarks: Very similar to *siguapa*

Wing length: 291–305 mm
Tail length: 157–171 mm
Tarsus length: 45–47 mm
Length of bill (cere): 21–22 mm
Body mass: ?

Illustration: d'Orbigny 1839: Pl. 2; A. Brooks in Wetmore and Swales 1931a: Pl. 19; L. Poole in Bond 1985 (repr.): 122 (b/w); A. Cameron in Voous 1988: 262; K. Williams in Raffaele et al. 2003: Pl. 43
Photograph: ?
Literature: Riley 1916: 1; Wetmore and Swales 1931a: 231–247; Kelso 1934d: 38; Bond 1942: 308, 309; Voous 1988: 262–265; del Hoyo et al. 1999: 240; König et al. 1999: 423, 424; Raffaele et al. 2003: 100

- ***Asio stygius stygius*** (Wagler) 1832
  Nyctalops stygius Wagler 1832, Isis v. Oken, col. 1221; Terra typica: Brazil or South Africa = Minas Geraës

Distribution: Northern Brazil, south to eastern Bolivia, northeast Argentina and southeast Brazil
Museum: BMNH, FNSF, ÜMB, ZSBS

Wing length: 324–348 mm
Tail length: 164.5–170 mm
Tarsus length: 48 mm
Length of bill (cere): 22 mm
Body mass: 632–675 g

Illustration: T. Boyer in Boyer and Hume 1991: 155; J. Lewington in del Hoyo et al. 1999: Pl. 20; F. Weick in König et al. 1999:Pl. 63; S. Webb in Hilty 2003: Pl. 25
Photograph: E. Endrigo in del Hoyo et al. 1999: 82; J. C. Motta jr. in internet: Owl Pages, photo gallery 2004
Literature: Eck and Busse 1973: 176, 177; Voous 1988: 262–265; Fjeldsa and Krabbe 1990: 229; Boyer and Hume 1991: 155, 156; del Hoyo et al. 1999: 240; König et al. 1999: 423, 424; Narosky and Yzurieta 2003: 144

- ***Asio stygius barberoi*** W. Bertoni 1930
  Asio stygius var. *barberoi* W. Bertoni 1930, Rev. Soc. Cient. Paraguay 2(6): 243, Pl. 22; Terra typica: Monte Sociedad, Paraguayan Chaco

Distribution: Paraguay, northern Argentina (Tucumán, Santiogo del Estero, Chaco, Formosa and Misiones)
Museum: MHN Paraguay
Remarks: Similar to subspecies *stygius*, but about 10% larger (different plumage below, similar to *robustus*)

Wing length:	356–380 mm
Tail length:	198 mm ($n = 1$)
Tarsus length:	?
Length of bill:	?
Body mass:	?

Illustration: W. Bertoni in Bertoni 1930: Pl. 22
Photograph: ?
Literature: Bertoni 1930: 243; Kelso 1934d: 38; Voous 1988: 262–265; Fjeldsa and Krabbe 1990: 239; del Hoyo et al. 1999: 240

❖

Asio otus **(Linné) 1758**
Northern Long-eared Owl · Waldohreule · Hibou moyen-duc · Búho Chico
(see Plate 9)

Length:	350–400 mm
Body mass:	210–430 g

Distribution: Europe and northern Asia, north to treeline, east to Okhotsk and Japan (Hokkaido), south to Iran, Turkestan, the Himalayas, Kansu and Sichuan, the Canary and Atorae Islands and northwest Africa. North America from British Columbia, Canada, south to northern Mexico and the southern United States
Habitat: Rather open landscapes, patchy woodland, hedges, groups of trees or small woods. Deciduous, mixed and coniferous forests with clearings and forest edges. Semi-open taiga forest, parks and cemeteries with old trees, gardens, marshland and farmland with groups of trees

- ***Asio otus otus*** (Linné) 1758
  Strix Otus von Linné 1758, Syst. Nat. ed. 10, 1: 92; Terra typica: Europe = Sweden

Synonym:
- *Strix deminuta* Pallas 1773, Reise versch. Prov. Russ. Reichs 2: 706; Terra typica: Ural
- *Otus albicollis* and *italicus* Daudin 1800, Traite d Orn. 2: 213; Terra typica: Europe and Italy
- *Asio otus turcmenica* Zarudny and Bilkewitch 1918, Isvestia Zakaspiiskago Muzeia: 16; Terra typica: Tedzhen and Murgab

Remarks: For further synonyms see Hartert 1912–1921: 984, 985

Distribution: Eurasia, from British Isles and Iberia, east to Sea of Okhotsk, Japan (Hokkaido). South to Mediterranean Islands, Middle East, northern Pakistan. Isolated population in east-central China. Azores, northwest Africa (Morocco to northwest Tunisia)
Museum: BMNH, FNSF, MNBHU, SMNKa, SMNSt, SMTD, ÜMB, ZFMK, ZMA, ZMH, ZSBS

Wing length:	♂ 280–315 mm, ♀ 285–319 mm
Tail length:	130–149 mm
Tarsus length:	37–42 mm
Length of bill:	26–32 mm
Body mass:	♂ 210–330 g, ♀ 230–430 g

Illustration: Rudbeck 1693–1710 in Krook 1988 (repr.): Pl. 93; C. Atkinson (ca. 1785) in *Catal. Christie's* 1989: 24, no. 37; T. Bewick in Bewick 1809: 84; J. F. Naumann 1822: Pl. 45; H. C. Richter in Gould 1832–1837: Pl. ?; H. C. Richter in Gould 1863, vol 1(4): Pl. 31; L. Binder in Wüst 1970: 242; P. A. Robert in Géroudet 1979: 342; C. Tunnicliffe in Cusa 1984: 38, 39; H. Delin in Cramp et al. 1985: Pl. 51 (adult and juvenile); A. Cameron in Voous 1988:

255; T. Boyer in Boyer and Hume 1991: 153; L. Jonsson in Jonsson 1992: 315; J. Lewington in del Hoyo et al. 1999: Pl. 20; F. Weick in König et al. 1999: Pl. 63 (adult and juvenile); Mullarney and Zetterström in Svensson et al. 1999: 213 (adult and juvenile); F. Weick in Weick 2004: Februar

Photograph: F. Belle and C. Vienne in Everett 1977: 73; E. Hosking in Hosking and Flegg 1982: 65, 130; numerous photographs in Mebs 1987: 36, 38, 40, 41 (by Reinhard, Sauer, Danegger, Schendel); numerous photographs in Mebs and Scherzinger 2000: 9, 30, 63 and 247–265 (by Brandl, Fürst, Limbrunner, Nill, Stengel and Wothe); G. Laez and J. P. Delobello in Kemp and Kemp 1998: 276

Literature: Sharpe 1875a: 227–229; Hartert 1912–1921: 984–986; Dementiev and Gladkov 1951: 422–426; Kleinschmidt 1958: 30; Vaurie 1965: 593, 594; Ali and Ripley 1969: 313, 314; Wüst 1970: 241; Eck and Busse 1973: 177–179; Glutz von Blotzheim and Bauer 1980: 386–421; Mikkola 1983: 216–232; Cramp et al. 1985: 572–588; Bezzel 1985: 657–660; Mebs 1987: 36–43; Fry et al. 1988: 150–152; Voous 1988: 252–261; Boyer and Hume 1991: 152–155; del Hoyo et al. 1999: 240, 241; König et al. 1999: 424–426; Svensson et al. 1999: 212; Mebs and Scherzinger 2000: 247–269

- *Asio otus canariensis* Madarász 1901
  Asio canariensis von Madarász 1901, Orn. Monatsb. 9: 54;
  Terra typica: Tafira, Gran Canaria, Canary Islands

Distribution: Canary Islands: Gran Canaria, Tenerife, La Palma

Museum: BMNH, NNML, ZFMK, ZMH

Remarks: Much darker plumage above than nominate, some specimens almost black (Hartert 1912–1921)

Wing length:	♂ 252–276 mm, ♀ 268–284 mm
Tail length:	120–142 mm
Tarsus length:	36.5–39.5 mm
Length of bill:	25–28.5 mm
Body mass:	?

Illustration: H. Delin in Cramp et al. 1985: Pl. 51

Photograph: ?

Literature: Hartert 1912–1921: 986, 987; Vaurie 1965: 593; Cramp et al. 1985: 572–588; Voous 1988: 252–261; del Hoyo et al. 1999: 240, 241; König et al. 1999: 424–426

- *Asio otus wilsonianus* (Lesson) 1830
  Otus Wilsonianus Lesson 1830, Traite d Orn. livr. 2: 110;
  Terra typica: USA = Pennsylvania ex Wilson

Distribution: South-central and southeast Canada: Manitoba east to Nova Scotia, south in United States to northern Oklahoma and Virginia

Museum: AMNH, SMTD, ÜMB, USNM, ZFMK, ZMH

Wing length:	264–305 mm
Tail length:	122–160 mm
Tarsus length:	42 mm
Length of bill (cere):	15–18.5 mm
Body mass:	♂ 223–304 g, ♀ 284–409 g

Illustration: J. Audubon in Audubon 1827–1838: Pl. 241 (irides orange!); R. Ridgway in Baird et al. 1860: 18; R. Ridgway in Merriam and Fisher 1893: Pl. 20; L. A. Fuertes in Forbush and May 1955: Pl. 45; A. Brooks in Sprunt 1955: unpaged colour plate; J. F. Landsdowne in Landsdowne and Livingston 1966: Pl. 20 (probably subspecies *tuftsi*); K. E. Karalus in Karalus and Eckert 1974: Pl. 12; A. Wilson in Wilson 1975 (repr.) (1766–1813): 51/4; L.A. Fuertes in McCracken-Peck 1982: 63; L. Malick in Breeden et al. 1983: 239; J. Lewis in del Hoyo et al. 1999: Pl. 20; F. Weick in König et al. 1999: Pl. 63 (adult and juvenile); D. Sibley in Sibley 2000: 272

Photograph: P. and P. Wood in MacKenzie 1986: 120; G. K. Peck in MacKenzie 1986: 121; P. Johnsgard in Johnsgard 2002: Pl. 27; D. Brinzal in internet: Owl Pages, photo gallery 2004; J. Hobbs in internet: Owl Pages, photo gallery 2004; D. Roberson in internet: Owl Pages, photo gallery 2004

Literature: Ridgway 1914: 654–658; Forbush and May 1955: 274, 275; Sprunt 1955: 208–210; Bent 1961 (repr.): 153–165; Karalus and Eckert 1974: 72–80; Breeden et al. 1983: 238; Howell and Webb. 1995: 367, 368; del Hoyo et al. 1999: 240, 241; König et al. 1999: 424–426; Sibley 2000: 272; Johnsgard 2002: 202–210

- *Asio otus tuftsi* Godfrey 1948
  Asio otus tuftsi Godfrey 1948 Can. Field-Nat. 61(6) 1947: 196;
  Terra typica: Last Mountain Lake, Saskatchewan, Canada

Distribution: Western Canada: southern Yukon, southern British Columbia, east to Saskatchewan. South

to Mexico: north-western Baja California, Nuevo León and southern United States: west Texas
Museum: AMNH, USNM
Remarks: Difference to nominate doubtful. Paler, greyer and with pale tawny brown facial disc

Wing length: ~292–294 mm
Tail length: ~144–151 mm
Tarsus length: ?
Length of bill (cere): ~17 mm
Body mass: ?

Illustration: J. F. Landsdowne in Landsdowne and Livingston 1966: Pl. 20 (from Simcoe, Ontario!); K. E. Karalus in Karalus and Eckert 1974: Frontispiece
Photograph: ?
Literature: Landsdowne and Livingston 1966: text Pl. 20; Eck and Busse 1973: 177–179; Karalus and Eckert 1974: 80–82; del Hoyo et al. 1999: 240, 241; Johnsgard 2002: 202–210

Asio abyssinicus **(Guérin Méneville) 1843**
Abyssinian Long-eared Owl · Äthiopienohreule · Hibou d'Abyssinie · Búho Abisinio
(see Plate 9)

Length: 400–440 mm
Body mass: 245–400 g

Distribution: Highlands of Ethiopia and Eritrea. Ruwenzori and Mitumba Mountains in eastern Zaire, western Uganda and Mount Kenya
Habitat: Open moorland and grassland with patches of cedar and oak forest. Giant heath, humid forested valleys and gorges. In high mountains from about 2 800 m up to 3 900 m above sea-level
Remarks: Often regarded as a subspecies of *Asio otus*, but considered specifically distinct on the grounds of vocal differences

- ***Asio abyssinicus abyssinicus*** (Guérin Méneville) 1843
Otus abyssinicus Guérin Méneville 1843, Rev. Zool.: 321;
Terra typica: Eritrea

Distribution: Highlands of Ethiopia and Eritrea
Museum: BMNH, FNSF, MHNP, MNBHU, MZUS, SMNSt, ZFMK

Wing length: 327–360 mm
Tail length: 182–190 mm
Tarsus length: 53–55 mm
Length of bill (cere): 18–22 mm
Body mass: 245–400 g

Illustration: O. Kleinschmidt in von Erlanger 1904: Pl. 18; M. Woodcock in Fry et al. 1988: Pl. 7; T. Boyer in Boyer and Hume 1991: 156; P. Hayman in Kemp and Kemp 1998: 273; J. Lewington in del Hoyo et al. 1999: Pl. 20; F. Weick in König et al. 1999. Pl. 63
Photograph: P. Smitterberg in Kemp and Kemp 1998: 272; G. Ekström in del Hoyo et al. 1999: 83
Literature: Sharpe 1875a: 227–229; von Erlanger 1904: 231–233; Mackworth-Praed and Grant 1957: 646; Fry et al. 1988: 150–152; Voous 1988: 252–261; Boyer and Hume 1991: 156; Kemp and Kemp 1998: 272, 273; del Hoyo et al. 1999: 241; König et al. 1999: 426, 427

- ***Asio abyssinicus graueri*** Sassi 1912
Asio abyssinicus graueri Sassi 1912, Anz. Akad. Wiss. Wien 49: 122;
Terra typica: West of Lake Tanganyika

Distribution: East Africa: Ruwenzori and Mitumba Mountains, eastern Zaire (Mount Kabobo), western Uganda and Mount Kenya
Museum: NHM Wien
Remarks: Smaller, greyer, more blackish pattern, but sometimes considered as inseparable from nominate

Wing length: 309–342 mm
Tail length: ?
Tarsus length: ?
Length of bill: ?
Body mass: ?

Illustration: M. Woodcock in Fry et al. 1988: Pl. 7; D. Zimmerman in Zimmerman et al. 1996: Pl. 55; J. Lewington in del Hoyo et al. 1999: Pl. 20; F. Weick in König et al. 1999: Pl. 63; J. Gale in Stevenson and Fanshawe 2002: Pl. 99
Photograph: ?
Literature: Fry et al. 1988: 150–152; Zimmerman et al. 1996: 446; del Hoyo et al. 1999: 241; König et al. 1999: 426, 427; Stevenson and Fanshawe 2003: 198

Asio madagascariensis (A. Smith) 1834
Madagascar Long-eared Owl · Madagaskar-Ohreule · Hibou malagache · Búho Malgache
Otus Madagascariensis A. Smith 1834, S. Afr. Q. J., 2: 316;
Terra typica: **Madagascar**
(see Plate 9)

Length: ♂ 360–400 mm, ♀ –510 mm
Body mass: ?

Synonym:
- *Asio Chauvini* Lamberton 1928, Bull. Acad. Malgache 10(1927): 40 and Pl.; Terra typica: Brickaville District, Madagascar (based on aberrant plumage)

Distribution: Endemic to Madagascar
Museum: BMNH, FNSF, MHNP, MNBHU, MZUS (juvenile), SMTD, ZMH
Remarks: Incomprehensibly, with the exception of Langrand (1990), Morris and Hawkins (1998) and König et al. (1999), the measurements of bodylength in the owl literature are much too small! *Asio madagascariensis* is the largest owl in Madagascar and also the largest and most powerful member of the genus *Asio*! (see literature)
Remarks: Specifically distinct from *Asio otus* by geographic isolation, larger size (allopatric), different eartufts and plumage

Wing length: ♂ 260–310 mm, ♀ 274–340 mm
Tail length: ♂ 122–165 mm, ♀ 155–195 mm
Tarsus length: 42–48 mm
Length of bill: 40.5 mm
Body mass: ?

Illustration: J. G. Keulemans in Milne-Edwards and Grandidier 1876: Pl. 38; Lamberton 1928: 40 and Pl.; V. Bretagnolle in Langrand 1990: Pl. 26; T. Boyer in Boyer and Hume 1991: 156; J. Lewington in Kemp and Kemp 1998: 275; J. Lewington in del Hoyo et al. 1999. Pl. 20; F. Weick in König et al. 1999. Pl. 63; F. Weick in König et al. unpublished: Pl. 66 (juvenile)
Photograph: F. Hawkins in Morris and Hawkins 1998: Pl. 169a; P. Morris in Morris and Hawkins 1998: Pl. 169b; O. Langrand in Kemp and Kemp 1998: 275 (juvenile)
Literature: Milne-Edwards and Grandidier 1876: 112–118; Lamberton 1928: 40; Sharpe 1875a: 232, 233 (length type = 445 mm); Grossman and Hamlet 1965: 460 (length 330 mm); Langrand 1990: 228, 229 (length 400–500 mm); Boyer and Hume 1991: 156, 157 (length 320 mm) Kemp and Kemp 1998: 274, 275 (length 350 mm); Morris and Hawkins 1998: 204 (length 400–500 mm); del Hoyo et al. 1999: 241 (length 310–360 mm); König et al. 1999: 427, 428 (length 400–500 mm)

Asio flammeus (Pontoppidan) 1763
Short-eared Owl · Sumpfohreule · Hibou des marais · Búho Campestre, Lechuza Campestre
(see Plate 9)

Length: 340–420 mm
Body mass: 200–450 g

Distribution: North America, from western and northern Alaska and the Bering Strait, east to Labrador, south to California and North Carolina. Hispaniola, Cuba and some other Caribbean Islands, Galapagos Islands, Juan Fernandez and Hawaiian Islands, Falkland Islands, locally in South America. Greenland, British Isles, Atlantic coast of southwest France and northwest Spain. From Iberian Peninsula (locally) and Norway, east through central Europe and central Asia to northeast Siberia, Kamchatka, Sakhalin and northern China. Also on some islands in the Bering Sea, the Ponapé and Caroline Islands

Asio f. flammeus · Short-eared Owl · Sumpfohreule

Habitat: Open country with bushes or scattered trees: tundra, marshes, humid grassland, savannas, moorland and swampy areas. Also shrub steppe and tundra, heathland and dry, stony areas, sand dunes. Large clearings in woodland, forest edges. Páramo and puna above the treeline in the Andes. Also in extensively cultivated landscapes. From sea-level to uplands, and in the Andes up to 4000 m

Remarks: The taxonomy of this species requires further study, especially with regard to the status of the taxon *galapagoensis* and some South American subspecies

- ***Asio flammeus flammeus*** (Pontoppidan) 1763
  Strix Flammea Pontoppidan 1763, Dansk Atlas 1: 617 and Pl. 25; **Terra typica:** Sweden

Synonym:
- *Strix accipitrina* Pallas 1771, Reise Russ. Prov. Russ. Reichs: 455; **Terra typica:** Caspian Sea
- *Noctua minor* Gmelin 1771, N. Comm. Petrop. 15: 447 and Pl. 12; **Terra typica:** in desertis Tanain reperti
- *Strix brachyotis* J.R. Forster 1772, Philos. Trans. LXII: 384; **Terra typica:** Severn River, Keawatin
- *Strix arctica* Sparrman 1788, Mus. Carls. Fasc. 51: Pl. 51; **Terra typica:** Northern Sweden
- *Otus palustris* Bechstein 1791, Gemein. Naturgesch. Deutschland II: 344; **Terra typica:** Hessen and Pomeriana
- *Otus leucopsis* C.L. Brehm 1855, Vogelfang: 413; **Terra typica:** Sarepta
- *Asio accipitrinus Mc Ilhennyi* Stone 1900, Proc. Acad. Nat. Sci. Phila. 1899: 478; **Terra typica:** Point Barrow, Alaska
- *Asio accipitrinus pallidus* Zarudny and Loudon 1906, Orn. Monatsb. 14: 151; **Terra typica:** Western Siberia, eastern Orenburg, Turgui and Turkestan

Distribution: From Iceland and the British Isles locally through Europe and Asia, east to Kamchatka and Kommandeur Islands, south to Spain, northwest Africa (Morocco and Tunisia), Caucasus, northeast Mongolia and northern China. In North America from western and northern Alaska and the Bering Strait east to Labrador, south to California and South Carolina

Museum: BMNH, MNBHU, NNML, SMNKa, SMNSt, SMTD, USNM, ZFMK, ZMA

Wing length:	♂ 281–326 mm, ♀ 309–335 mm
Tail length:	134–154 mm
Tarsus length:	43–48 mm
Length of bill:	27.5–30 mm
Body mass:	♂ 206–396 g, ♀ 260–475 g

Illustration: Rudbeck (1693–1710) in Krook1991 (repr.): Pl. 92; Pontoppidan 1763: Pl. 25; Gmelin 1771: Pl. 12; C. Atkinson (~1785) in *Catal. Christie's* 1989: 28; J. F. Naumann 1822: Pl. 45; J. J. Audubon in Audubon 1827–1838 (repr.) Pl. 244; H. C. Richter in Gould 1863, vol 1: Abb. 32; J. G. Keulemans in Dresser 1871–1896: Pl. 304; R. Ridgway in Merriam and Fisher 1893: Pl. 21; J. G. Keulemans in Hennicke 1905: unpaged; Krause in *J. Orn.* 1910: Pl. 9 (*pallidus*); L. A. Fuertes in Gilbert Pearson 1936: Pl. 56; O. Kleinschmidt in Kleinschmidt 1958: 30; P. Barruel in Sutter and Barruel 1958: 61; L. Binder in Wüst 1970: Abb. 132; K. E. Karalus in Karalus and Eckert 1974: Pl. 14; P. A. Robert in Géroudet 1979: 324; L. Malick in Breeden et al. 1983: 239; G. Pettersson in Pettersson 1984: 65–67; H. Delin in Cramp et al. 1985: Pl. 52 (adult and juvenile); A. Cameron in Voous 1988: 270; T. Boyer in Boyer and Hume 1991: 158; L. Jonsson in Jonsson 1992: 313; J. Lewington in del Hoyo et al. 1999: Pl. 20; F. Weick in König et al. 1999: Pl. 64 (adult and juvenile); L. A. Fuertes in Johnsgard 2002: Pl. 8; R. Bateman in Dean 2004: 42, 43; F. Weick in Kröher and Weick 2004: 209

Photograph: Quedens in *Vogelkosmos* 1966(2): 47; E. Hosking in Hosking and Flegg 1982: 44, 85, 89; Hautala in Mikkola 1983: Pl. 67; R. B. Ranford in MacKenzie 1986: 123, 133; Storsberg and Zeininger in Mebs 1987: 83, 85–87; B. Coleman in Shaw 1989: 86, 89, 90; D. R. Franz in del Hoyo et al. 1999: 96; P. Smith in del Hoyo et al. 1999: 128; B. S. Speak in del Hoyo et al. 1999: 132; numerous photographs in Mebs and Scherzinger 2000: 112, 270–285 (by Giel, Hecker, Hlasek, Limbrunner, Paltanavicius, Reinhard, Scherzinger, Wothe, Zeininger); P. Johnsgard in Johnsgard 2002: Pl. 28; D. Baccus in internet: Owl Pages, photo gallery 2004; A. Cook in internet: Owl Pages, photo gallery 2004; P. Moore in internet: Owl Pages, photo gallery 2004

Literature: Sharpe 1875a: 234–238; Hartert 1912–1921: 987–990; Hartert 1923: 388; Ridgway 1914: 661–667; Dementiev and Gladkov 1951: 427–433;

Kleinschmidt 1958: 31; Gerber 1960; Vaurie 1965: 594–596; Scherzinger 1968: 270–273; Eck and Busse 1973: 179–181; Karalus and Eckert 1974: 83–95; Glutz von Blotzheim and Bauer 1980: 421–452; Mikkola 1983: 233–250; Breeden et al. 1983: 238; Cramp et al. 1985: 588-601; Bezzel 1985: 660–663; Mebs 1987: 82–87; Voous 1988: 270–278; Boyer and Hume 1991: 157–159; del Hoyo et al. 1999: 241, 242; König et al. 1999: 429–431; Svensson et al. 1999: 213; Mebs and Scherzinger 2000: 270–287; Sibley 2000: 273; Johnsgard 2002: 213–220

- ***Asio flammeus ponapensis*** Mayr 1933
  Asio flammeus ponapensis Mayr 1933, Am. Mus. Novit. 609: 1; Terra typica: Ponapé Island, eastern Caroline Island, Micronesia

Distribution: Ponapé = Pohnpei Island, eastern Caroline Islands, Micronesia
Museum: AMNH (holotype)
Remarks: Endemic, but very rare (Pratt et al. 1987). Distinctly smaller and shorter winged than nominate, but less is known about this subspecies!

Wing length:	285–290 mm
Tail length:	?
Tarsus length:	?
Length of bill:	?
Body mass:	?

Illustration: ?
Photograph: ?
Literature: Mayr 1933: 1–3; Pratt et al. 1987: 215, 216; Voous 1988: 270–278; del Hoyo et al. 1999: 241, 242; König et al. 1999: 429–431

- ***Asio flammeus sandwichensis*** (Bloxam) 1827
  Strix Sandwichensis Bloxam 1827, Voy. «Blonde» 1826: 250, Hawaiian Islands

Distribution: Endemic to Hawaiian Islands, but uncommon (Pratt et al. 1987)
Museum: MNBHU, SMTD

Wing length:	285–304 mm
Tail length:	152–157 mm
Tarsus length:	42 and 43 mm
Length of bill:	?
Body mass:	?

Illustration: H. D. Pratt in Pratt et al. 1987: Pl. 11; J. Lewington in del Hoyo et al. 1999: Pl. 20
Photograph: ?
Literature: Sharpe 1875a: 238, 239; Pratt et al. 1987: 215, 216; Voous 1988: 270–278; del Hoyo et al. 1999: 241, 242; König et al. 1999: 429–431

- ***Asio flammeus domingensis*** (P.L.S. Müller) 1776
  Strix domingensis P.L.S. Müller 1776, Naturhist. Suppl.: 70; Terra typica: Hispaniola

Synonym:
- *Asio portoricensis* Ridgway 1882, Proc. US Nat. Mus. 4: 366; Terra typica: North side of Puerto Rico

Distribution: Puerto Rico, Hispaniola, Cuba (resident), Greater Antilles (uncommon in Puerto Rico, rare on Cayman Island). Rare visitor to Florida (Sibley 2000)
Museum: ANSP, USNM

Wing length:	274–281 mm (*portoricensis*), 294–297 mm (*domingensis*)
Tail length:	130–132 mm
Tarsus length:	52.5–57.5 mm
Length of bill:	28–30 mm
Body mass:	?

Illustration: D. Sibley in Sibley 2000: 273; K. Williams in Raffaele et al. 2003: Pl. 43
Photograph: ?
Literature: Ridgway 1914: 667–670; Wetmore and Swales 1931a: 244, 245; Kelso 1934d: 39; Bond 1986: 122; del Hoyo et al. 1999: 241, 242; König et al. 1999: 429–431; Raffaele et al. 2003: 100

- ***Asio flammeus pallidicaudas*** Friedmann 1949
  Asio flammeus pallidicaudas Friedmann 1949, Smiths. Misc. Coll. 111(9): 2; Terra typica: Cantaura, Anzoategui, northern Venezuela

Distribution: Northern Venezuela, Gayana
Museum: USNM (holotype)
Remarks: Slightly different in plumage (tail- and wing-banding) from *bogotensis*

Wing length:	304 mm
Tail length:	140 mm
Tarsus length:	52 mm
Length of bill:	30 mm
Body mass:	350 g

Illustration: S. Webb in Hilty 2003: Pl. 24
Photograph: ?
Literature: Friedmann 1949a: 2, 3; Meyer de Schauensee and Phelps 1978: 120; del Hoyo et al. 1999: 241, 242; König et al. 1999: 429–431; Hilty 2003: 367

- ***Asio flammeus bogotensis*** Chapman 1915
Asio flammeus bogotensis Chapman 1915, Bull. Am. Mus. Nat. Hist. 34: 370; **Terra typica:** Savanna of Bogotá, Colombia

Synonym:
- *Asio galapagoensis aequatorialis* Chubb 1916, Bull. Brit. Ornith. Cl. 36: 46; **Terra typica:** Mount Pichincha, Ecuador

Distribution: Colombia, Ecuador and northwest Peru
Museum: AMNH, BMNH
Remarks: Plumage darker than nominate and with rusty wash

Wing length:	303–310 mm
Tail length:	134–140 mm
Tarsus length:	?
Length of bill:	?
Body mass:	?

Illustration: J. Fjeldsa in Fjeldsa and Krabbe 1990: Pl. 26; J. Lewington in del Hoyo et al. 1999: Pl. 20; P. J. Greenfield in Ridgely and Greenfield 2001: II: Pl. 36(11)
Photograph: ?
Literature: Chapman 1915: 370; Chubb 1916b: 46; Hilty and Brown 1986: 232, 233; Voous 1988: 270–278; del Hoyo et al. 1999: 241, 242; König et al. 1999: 429–431; Ridgely and Greenfield 2001: I: 314, II: 222

- ***Asio flammeus suinda*** (Vieillot) 1817
Strix suinda Vieillot 1817, Nouv. Dict. Hist. Nat. 7: 34; **Terra typica:** Paraguay and Rio de la Plata

Distribution: Southern Peru, west-central Bolivia, Paraguay and southeast Brazil, south to Tierra del Fuego
Museum: MNBHU (holotype), SMTD, SMNSt, ZFMK, ZMH
Remarks: Stronger feet and beak and darker plumage than nominate, but lighter than *bogotensis*

Wing length:	310–323 mm
Tail length:	141–155 mm
Tarsus length:	?
Length of bill (cere):	18.5 and 19 mm
Body mass:	?

Illustration: J. Fjeldsa in Fjeldsa and Krabbe 1990: Pl. 26; F. Weick in König et al. 1999: Pl. 64; P. Burke in Jaramillo et al. 2003: Pl. 58
Photograph: B. Davidow in del Hoyo et al. 1999: 108
Literature: Kelso 1934d: 39; Fjeldsa and Krabbe 1990: 229; del Hoyo et al. 1999: 241, 242; König et al. 1999: 429–431; Jaramillo et al. 2003: 144; Narosky and Yzurieta 2003: 144

- ***Asio flammeus sanfordi*** Bangs 1919
Asio flammeus sanfordi Bangs 1919, Proc. New Engl. Zool. Cl. 6: 97; **Terra typica:** Sea Lion Island, Falkland Islands

Distribution: Falkland Islands
Museum: BMNH (Darwin's collection) but no skin examined!
Remarks: Lighter and smaller than *suinda* (Voous 1988)

Wing length:	?
Tail length:	?
Tarsus length:	?
Length of bill:	?
Body mass:	?

Illustration: ?
Photograph: Only a photograph from skin by H. Taylor in Steinheimer 2004: 307, Fig. 2
Literature: Bangs 1919: 95–98; Voous 1988: 270–278; del Hoyo et al. 1999: 241, 242; König et al. 1999: 429–431; Yaramillo et al. 2003: 144

- ***Asio (flammeus) galapagoensis*** (Gould) 1837
Otus (Brachyotus) galapagoensis Gould 1837, Proc. Zool. Soc. Lond.: 10; **Terra typica:** Galapagos Archipelago

Distribution: Galapagos Archipelago
Museum: BMNH, FNSF, MNBHU, USNM
Remarks: Possibly specifically distinct, but a cline in plumage: *suinda-bobotensis-galapagoensis* suggests close relationship

Wing length:	278–288 mm
Tail length:	136–143 mm
Tarsus length:	?
Length of bill (cere):	19–20.5 mm
Body mass:	?

Illustration: J. Gould in Gray 1841: unpaged plate and in Rice 2004: 255; J. Lewington in del Hoyo et al. 1999: Pl. 20; F. Weick in König et al. 1999: Pl. 64

Photograph: H. D. Dossenbach in Dossenbach 1972: 160; E. Hosking in Hosking and Flegg 1982: 131; C. König in König 1983: 188; P. Pagni in Pölking 1986: 13; Survival Anglia Ltd. in Bellamy 1989: 56; Gillsater in Burton et al. 1992: 139; C. König in internet: Owl Pages, photo gallery 2004

Literature: Sharpe 1875a: 238, 239; Abs et al. 1965: 49–56; Voous 1988: 270–278; del Hoyo et al. 1999: 241, 242; König et al. 1999: 429–431

Asio capensis (A. Smith) 1834

Marsh Owl · Kapohreule · Hibou du Cap ou Hibou marais · Búho Moro
(see Plate 9)

Length:	290–380 mm
Body mass:	225–485 g

Distribution: Isolated populations in northwest Africa (Morocco, Algeria), in western Africa from Senegal to Chad and Cameroon, also southern Sudan and Ethiopian Highlands and from southern Congo south to the Cape. Madagascar

Habitat: Open country, moist open grassland and coastal marshes, with or without trees and bushes. Inland marshes, moor and montane grassland and savanna. From sea-level up to around 1 500 m in Madagascar and 3 000 m in Ethiopia

- ***Asio capensis tingitanus*** (Loche) 1867
  Phasmoptynx Capensis tingitanus Loche 1867, Expl. Scient. Albérie, Ois. 1: 99; Terra typica: Harrach-Bache, near Algier

Synonym:
- *Asio nisuella* subsp. *maroccanus* Reichenow 1901, Vög. Afr. 1: 660; Terra typica: Morocco

Distribution: Northern and central Morocco (formerly also northern Algeria and Lake Chad), accidental in Portugal, Spain and the Canary Islands

Museum: BMNH, MNBHU, MZUS, NNML, SMTD, ZFMK, ZMA

Wing length:	284–312 mm
Tail length:	132–153 mm
Tarsus length:	53–60 mm
Length of bill:	29–34 mm
Body mass:	310–350 g

Illustration: J. G. Keulemans in Dresser 1871–1896: Pl. 305 (probably subspecies *capensis*); J. Bree in Bree 1875/1876, last pl. (yellow irides) ex Hartert 1912–1921: 991; P. Barruel in Etchécopar and Hüe 1967: Pl. 11; H. Delin in Cramp et al. 1985: Pl. 52 (adult and juvenile); L. Jonsson in Jonsson 1992: 312; J. Lewis in del Hoyo et al. 1999: Pl. 20; F. Weick in König et al. 1999: Pl. 64; Mullarn and Zetterström in Svensson et al. 1999: 213

Photograph: all examined photographs shows subspecies *capensis*

Literature: Hartert 1912–1921: 990–992; Dementiev and Gladkov 1951: 427 (map); Bannerman 1953: 531, 531 (*capensis*?); Vaurie 1965: 596; Etchécopar and Hüe 1967: 346, 347; Mikkola 1983: 251–255; Cramp et al. 1985: 601–606; Fry et al. 1988: 153–155; Voous 1988: 279–283; Boyer and Hume 1991: 159–161; Jonsson 1992: 312; del Hoyo et al. 1999: 242; König et al. 1999: 431, 432; Svensson et al. 1999: 212

- ***Asio capensis capensis*** (A. Smith) 1834
  Otus capensis A. Smith 1834, S. Afr. Q. J. 2(4): 316; Terra typica: South Africa

Synonym:
- *Strix (Brachyotus) helveola* Lichtenstein 1842, Vert. Samml. Säugeth. and Vögel Kaffernl.: 11; Terra typica: Northeast Cape Colony
- *Asio tingitanus andrewsmithi* W. Sclater 1922, Bull. Brit. Ornith. Cl. 42: 24; new name for *Otus capensis*

Distribution: Isolated areas in West Africa, from Senegal to Chad(?) and Cameroon, also from southern Sudan and Ethiopian Highlands and from southern Congo south to the Cape

Museum: BMNH, FNSF, MZUS, SMNKa, SMTD, ÜMB, ZMA, ZMH, ZSBS

Wing length:	285–330 mm
Tail length:	132–150 (163) mm
Tarsus length:	51–56.5 mm
Length of bill:	28–33 mm
Body mass:	♂ 243–340 g, ♀ 305–355 (376) g

Illustration: T. J. Ford in Smith 1839: Pl. 67; McLachlan and Liversidge 1963: Pl. 11; G. Arnott in Steyn 1982: Pl. 23; M. Woodcock in Fry et al. 1988: Pl. 7;

A. Cameron in Voous 1988: 282; T. Boyer in Boyer and Hume 1991: 160; D. Zimmerman in Zimmerman et al. 1996: Pl. 55; P. Hayman in Kemp and Kemp 1998: 269; F. Weick in König et al. 1999: Pl. 64; N. Borrow in Borrow and Demey 2001: Pl. 58; J. Gale in Stevenson and Fanshawe 2002: Pl. 99

Photograph: E. Hosking in Hosking and Flegg 1982: 46, 132; P. Steyn in Steyn 1982: 257 (b/w); J. F. Reynolds in Mikkola 1983: 70, 71; P. L. Ginn in Ginn et al. 1989: 334; McIlleron in Ginn et al. 1989: 334; Newman in Burton et al. 1992: 140; S. Carlyon in Kemp and Kemp 1998: 268; L. Hes in Kemp and Kemp 1998: 268; J. Laurie in Kemp and Kemp 1998: 269

Literature: Sharpe 1875a: 239–241; Hartert 1912–1921: 990, 991; Bannerman 1953: 531, 532; McLachlan and Liversidge 1963: 192; Steyn 1982: 254–258; Fry et al. 1988: 153–155; Voous 1988: 279–283; Ginn et al. 1989: 334; Boyer and Hume 1991: 159–161; Zimmerman et al. 1996: 446; Kemp and Kemp 1998: 268, 269; del Hoyo et al. 1999: 242; König et al. 1999: 431, 432; Borrow and Demey 2001: 499; Stevenson and Fanshawe 2002: 198

- ***Asio capensis hova*** Stresemann 1922
  Asio helveola hova Stresemann 1922, Orn. Monatsb. 30: 64; new name to replace *Otus capensis major* Schlegel

Synonym:
- *Otus capensis major* Schlegel 1873, Mus. Pays-Bas 2, Fev. Ois. de proje: 3; Terra typica: Bombetok Bay, Madagascar; invalid name

Distribution: Madagascar (endemic)
Museum: MNBHU, NNML (holotype)
Remarks: Larger and darker than subspecies *capensis*, bill and talons more powerful. Probably specifically distinct due to geographical isolation, but this needs further investigation

Wing length:	322–380 mm
Tail length:	176–186 mm
Tarsus length:	57–70 mm
Length of bill:	?
Body mass:	♂ 485 g ($n = 1$)

Illustration: J.G. Keulemans in Milne-Edwards and Grandidier 1876: Atlas 1: Pl. 37; V. Bretagnolle in Langrand 1990: Pl. 26; F. Weick in König et al. 1999: Pl. 64

Photograph: P. Morris in Morris and Hawkins 1998: Pl. 170a; A. Greensmith in Morris and Hawkins 1998: Pl. 170b

Literature: Hartert 1912–1921: 992; Stresemann 1922: 64; Eck and Busse 1973: 179–181; Langrand 1990: 229; Kemp and Kemp 1998: 268, 269; Morris and Hawkins 1998: 204; del Hoyo et al. 1999: 242; König et al. 1999: 431, 432

Genus

Nesasio **J.L. Peters 1937**

Nesasio J.L. Peters 1937, Journ. Wash. Acad. Sci. 27: 82.
Type by *Pseudoptynx solomonensis* Hartert

Nesasio solomonensis (Hartert) 1901

Fearful Owl · Salomoneneule · Chouette des Solomones, Hibou redoutable · Búho de las Salomón

Pseudoptynx solomonensis Hartert 1901, Bull. Brit. Ornith. Cl. 12: 25; Terra typica: Ysabel Island, Solomon Islands

Length: 380 mm
Body mass: ?

Distribution: Solomon Archipelago. Bougainville, Choiseul and Santa Isabel. Endemic

Habitat: Primary and tall secondary forest in lowlands and hills. From sea-level up to about 800 m

Museum: AMNH, BMNH

Wing length: 300 mm
Tail length: 170 mm
Tarsus length: 60 mm
Length of bill: ?
Body mass: ?

Illustration: Grossman and Hamlet 1965: 463 (b/w); K. Lilly in Lloyd and Lloyd 1971: 127; T. Boyer in Boyer and Hume 1991: 162; F. Weick 1996: in AMNH (M. LeCroy); J. Lewington in del Hoyo et al. 1999: Pl. 20; F. Weick in König et al. 1999: Pl. 63

Photograph: G. Dutson in del Hoyo et al. 1999: 150

Literature: Hartert 1901: 25; Hartert 1925b: 263; Peters 1937: 81–83; Grossman and Hamlet 1965: 463; Eck and Busse 1973: 181, 182; Webb 1992: 52–57; del Hoyo et al. 1999: 242; König et al. 1999: 422, 423

Nesasio salomonensis
Fearful Owl
Salomoneneule

Part III — Teil III

Owls in Flight — Eulen im Flug

Upward wingstroke
showing wing form
and underwing pattern

Der Flügelschlag aufwärts
zeigt die Flügelformel
und die Unterflügelzeichnung

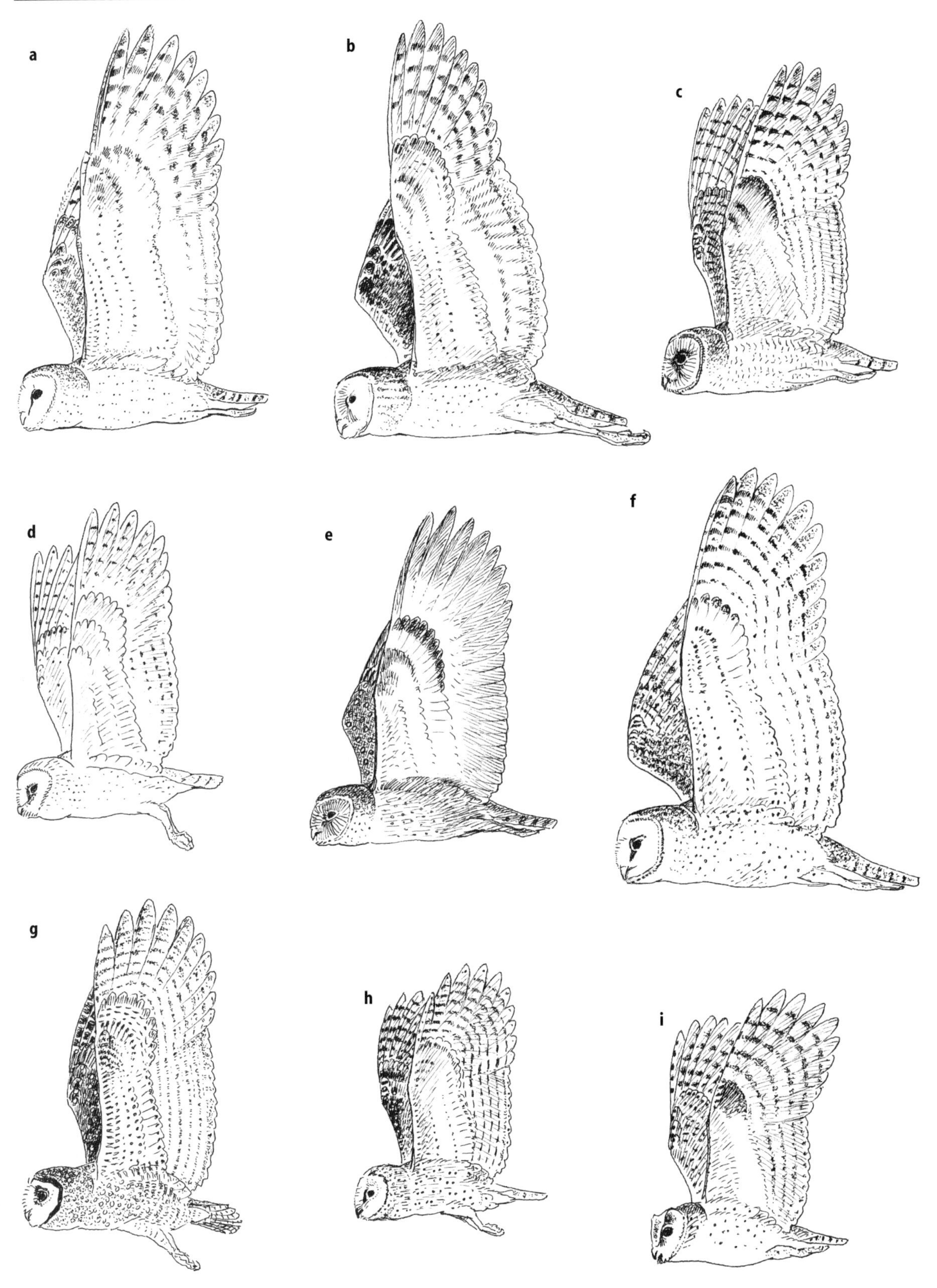

Plate 1. a *Tyto alba*; **b** *Tyto longimembris*; **c** *Tyto glaucops*; **d** *Tyto soumagnei*; **e** *Tyto nigrobrunnea*; **f** *Tyto n. novaehollandiae*; **g** *Tyto multipunctata*; **h** *Tyto prigoginei*; **i** *Phodilus badius*

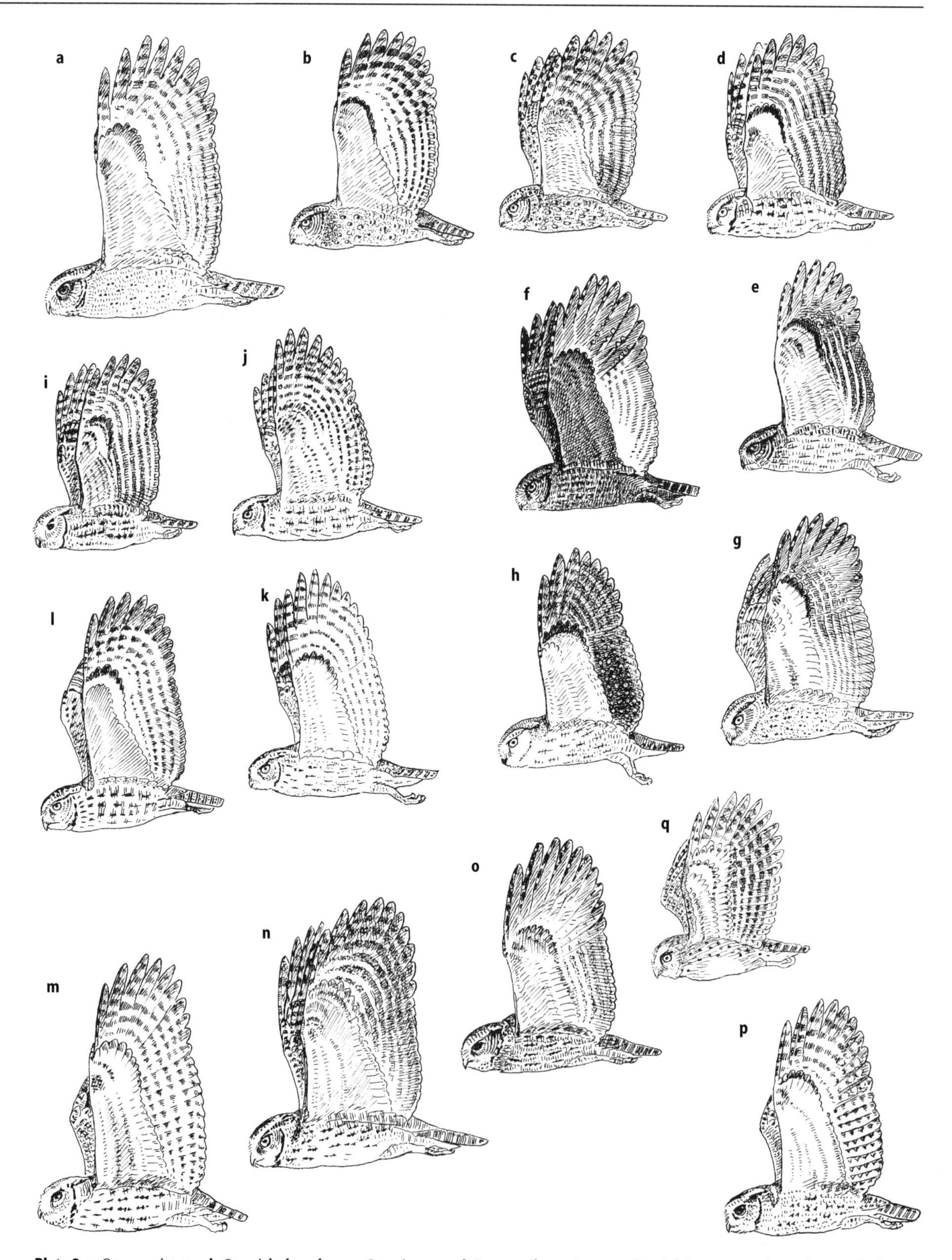

Plate 2. a *Otus sagittatus*; **b** *Otus i. holerythrus*; **c** *Otus ireneae*; **d** *Otus rutilus*; **e** *Otus pauliani*; **f** *Otus capnodes*; **g** *Otus moheliensis*; **h** *Otus pembaensis*; **i** *Otus flammeolus*; **j** *Otus scops*; **k** *Otus brucei*; **l** *Otus insularis*; **m** *Otus semitorques*; **n** *Otus megalotis*; **o** *Otus mentawi*; **p** *Otus lempiji*; **q** *Otus thilohoffmanni*

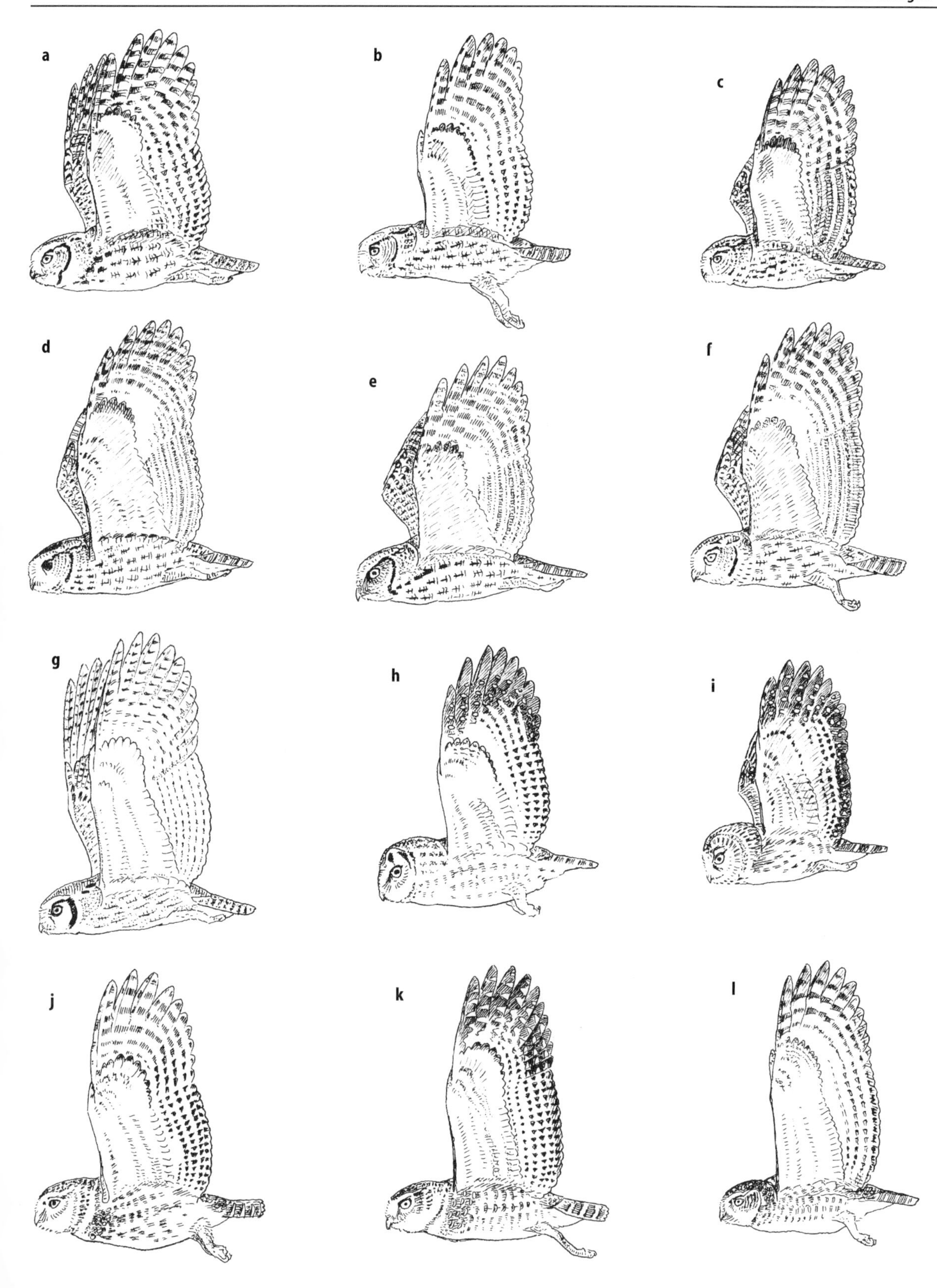

Plate 3. **a** *Otus kennicotti*; **b** *Otus asio*; **c** *Otus trichopsis*; **d** *Otus atricapillus*; **e** *Otus hoyi*; **f** *Otus choliba*; **g** *Ptilopsis granti*; **h** *Aegolius funereus*; **i** *Aegolius acadicus*; **j** *Athene noctua*; **k** *Athene c. hypugaea*; **l** *Athene brama*

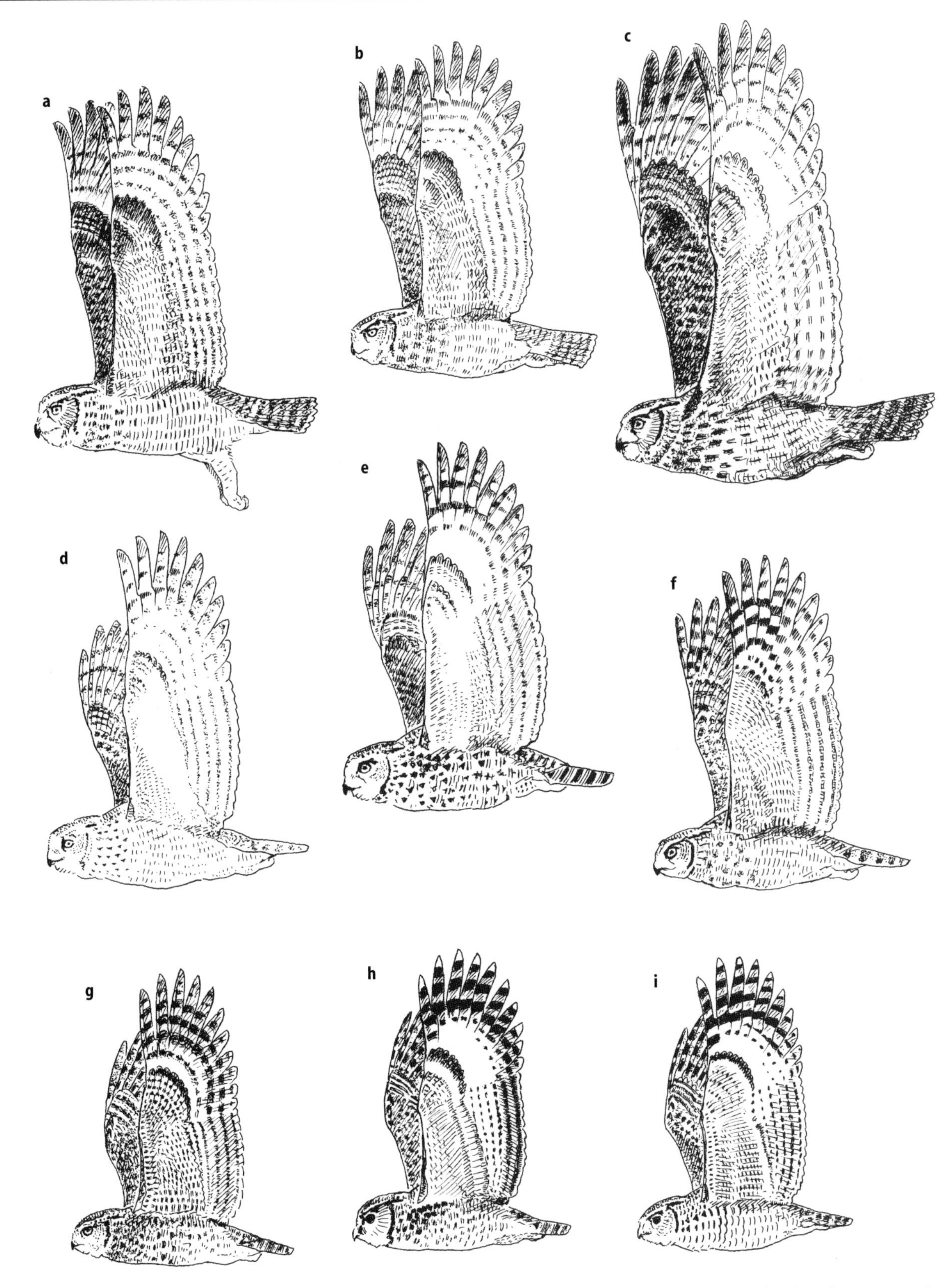

Plate 4. **a** *Bubo virginianus*; **b** *Bubo magellanicus*; **c** *Bubo bubo*; **d** *Bubo ascalaphus*; **e** *Bubo capensis*; **f** *Bubo africanus*; **g** *Bubo leucostictus*; **h** *Bubo vosseleri*; **i** *Bubo poensis*

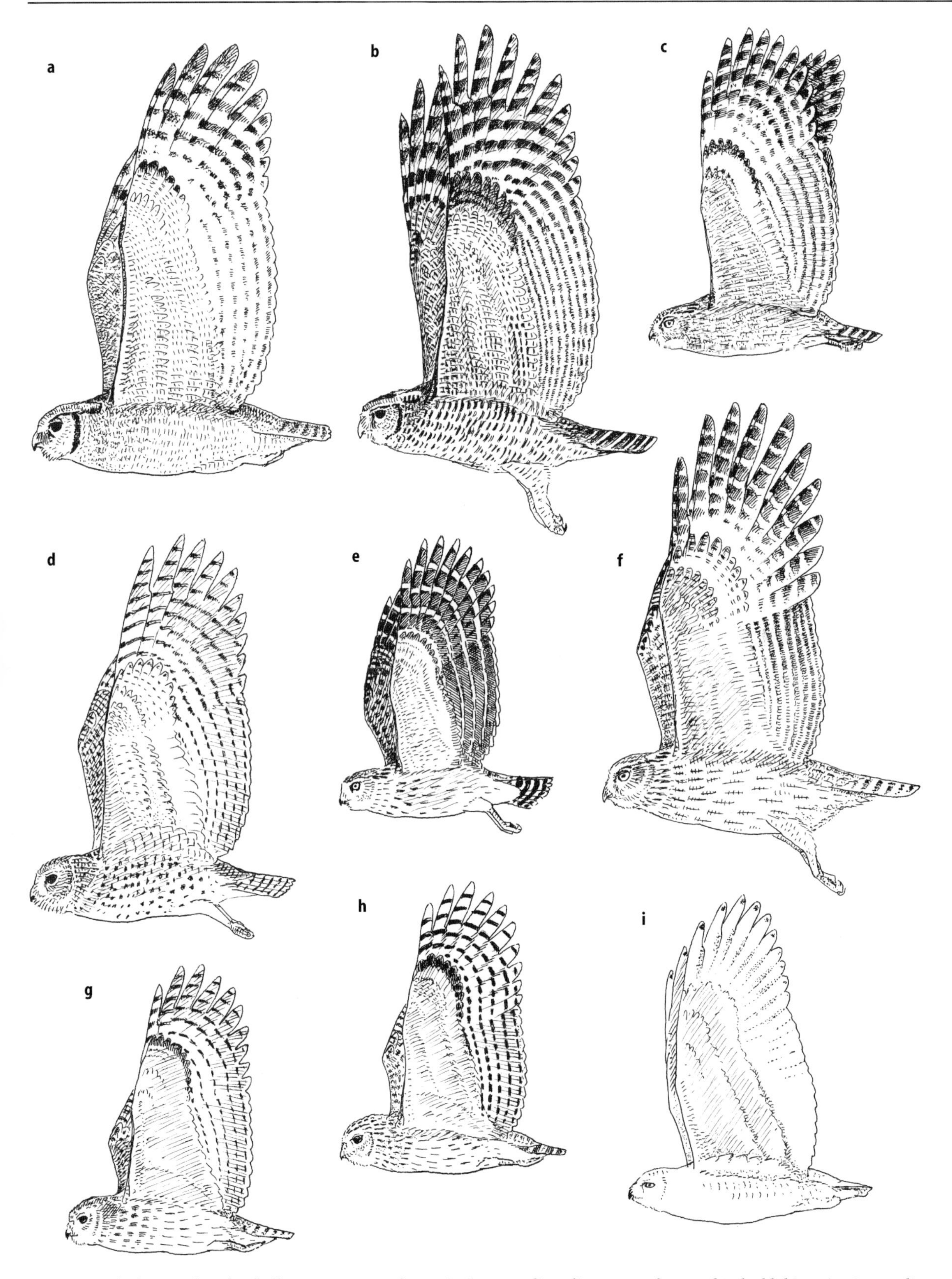

Plate 5. a *Bubo lacteus*; **b** *Bubo shelleyi*; **c** *Ketupa zeylonensis*; **d** *Scotopelia peli*; **e** *Ketupa ketupu*; **f** *Bubo blakistoni*; **g** *Scotopelis ussheri*; **h** *Scotopelia bouvieri*; **i** *Nyctea scandiaca*

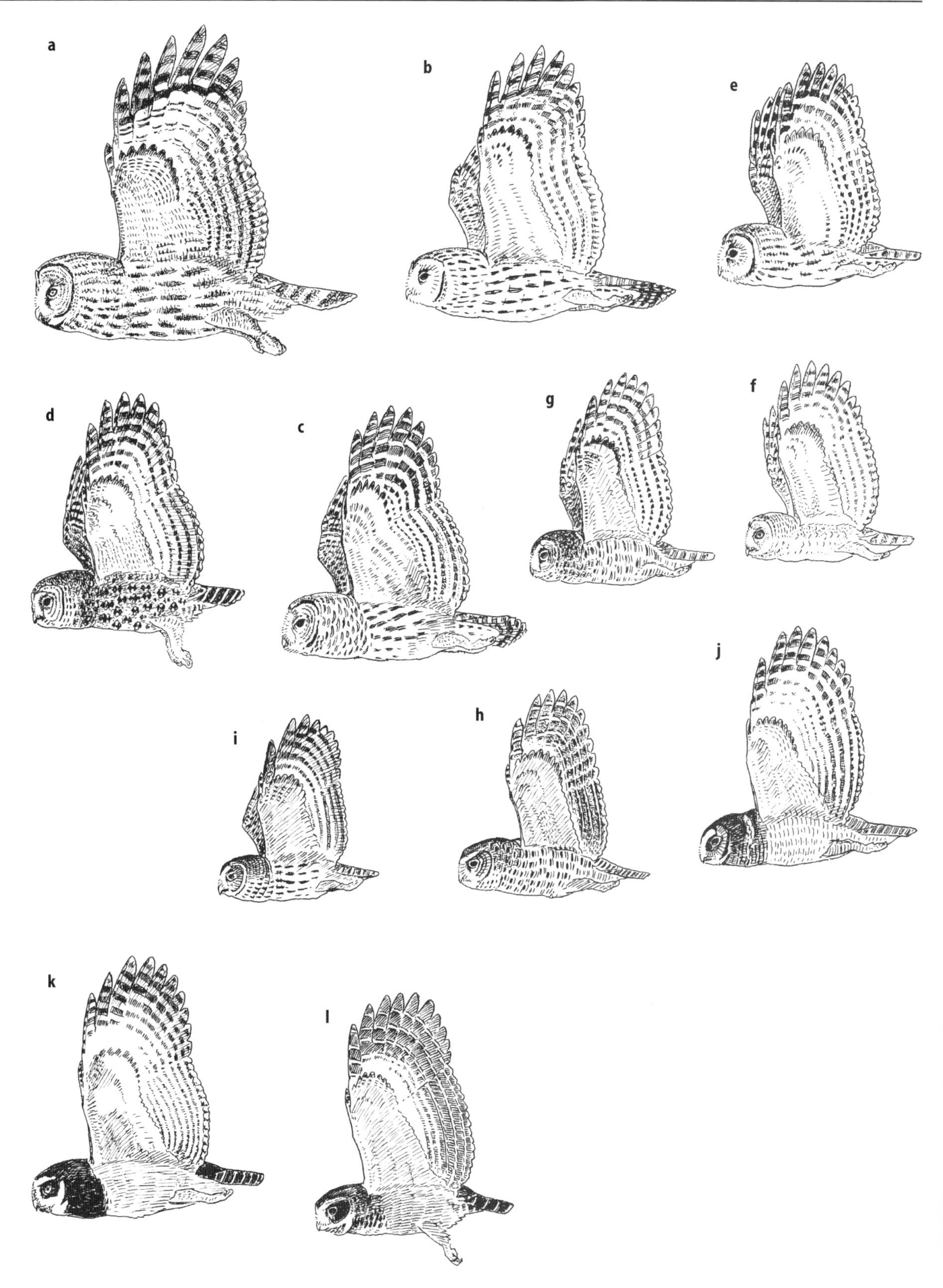

Plate 6. a *Strix n. lapponica*; **b** *Strix u. liturata*; **c** *Strix v. varia*; **d** *Strix occidentalis*; **e** *Strix aluco*; **f** *Strix butleri*; **g** *Strix woodfordi*; **h** *Strix rufipes*; **i** *Strix virgata*; **j** *Strix leptogrammica*; **k** *Pulsatrix perspicillata*; **l** *Pulsatrix koeniswaldiana*

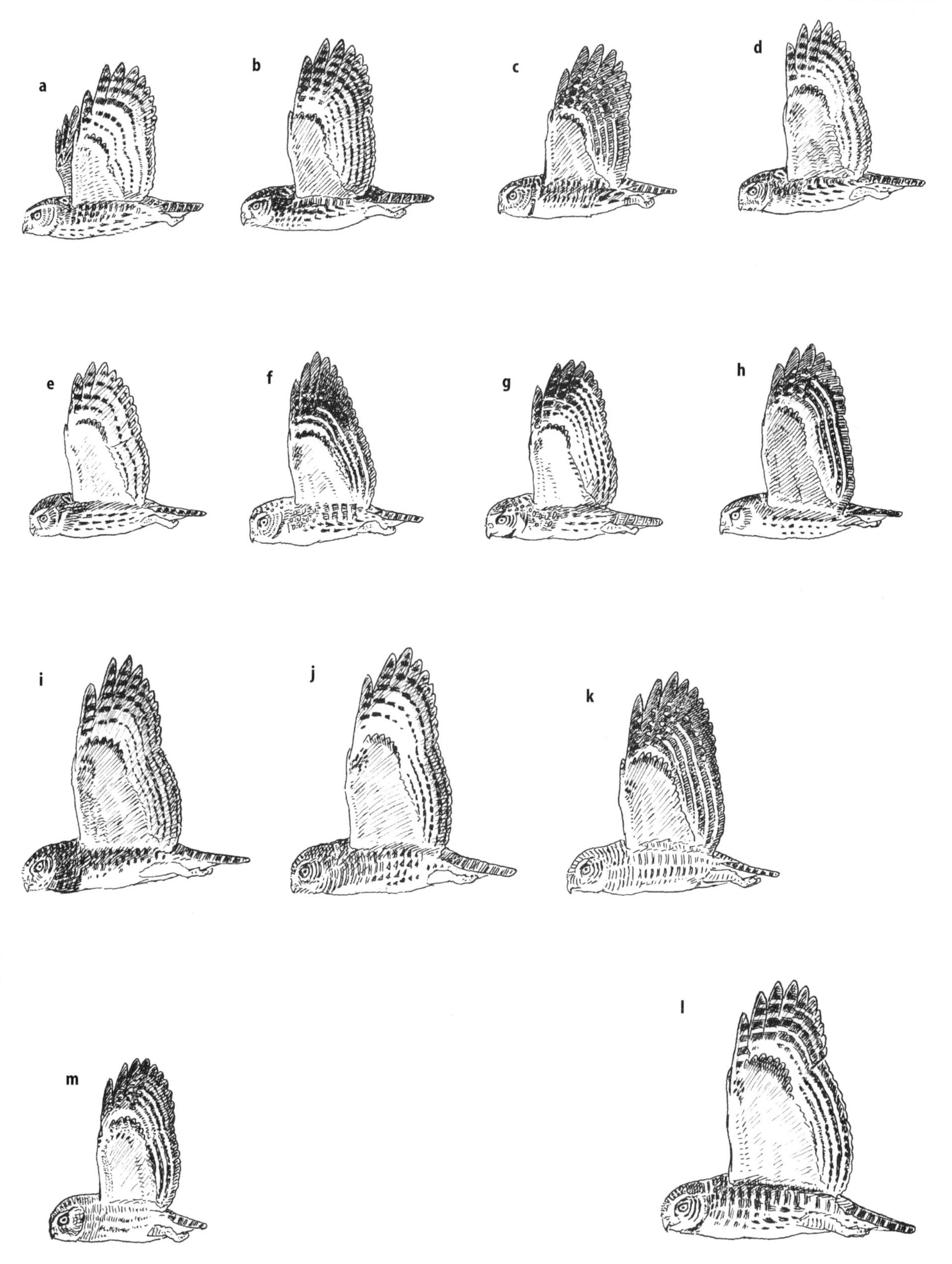

Plate 7. a *Glaucidium passerinum*; **b** *Glaucidium perlatum*; **c** *Glaucidium b. pardalotum*; **d** *Glaucidium californicum*; **e** *Glaucidium ridgwayi*; **f** *Glaucidium costaricanum*; **g** *Glaucidium bolivianum*; **h** *Glaucidium tephronotum*; **i** *Glaucidium etchecopari*; **j** *Glaucidium scheffleri*; **k** *Glaucidium radiatum*; **l** *Glaucidium c. whitelyi*; **m** *Micrathene whitneyi*

Plate 8. a *Ninox rufa*; **b** *Ninox strenua*; **c** *Ninox connivens*; **d** *Ninox b. boobook*; **e** *Ninox scutulata*; **f** *Ninox theomacha*; **g** *Ninox superciliaris*; **h** *Ninox sumbaensis*; **i** *Ninox ochracea*; **j** *Surnia ulula*

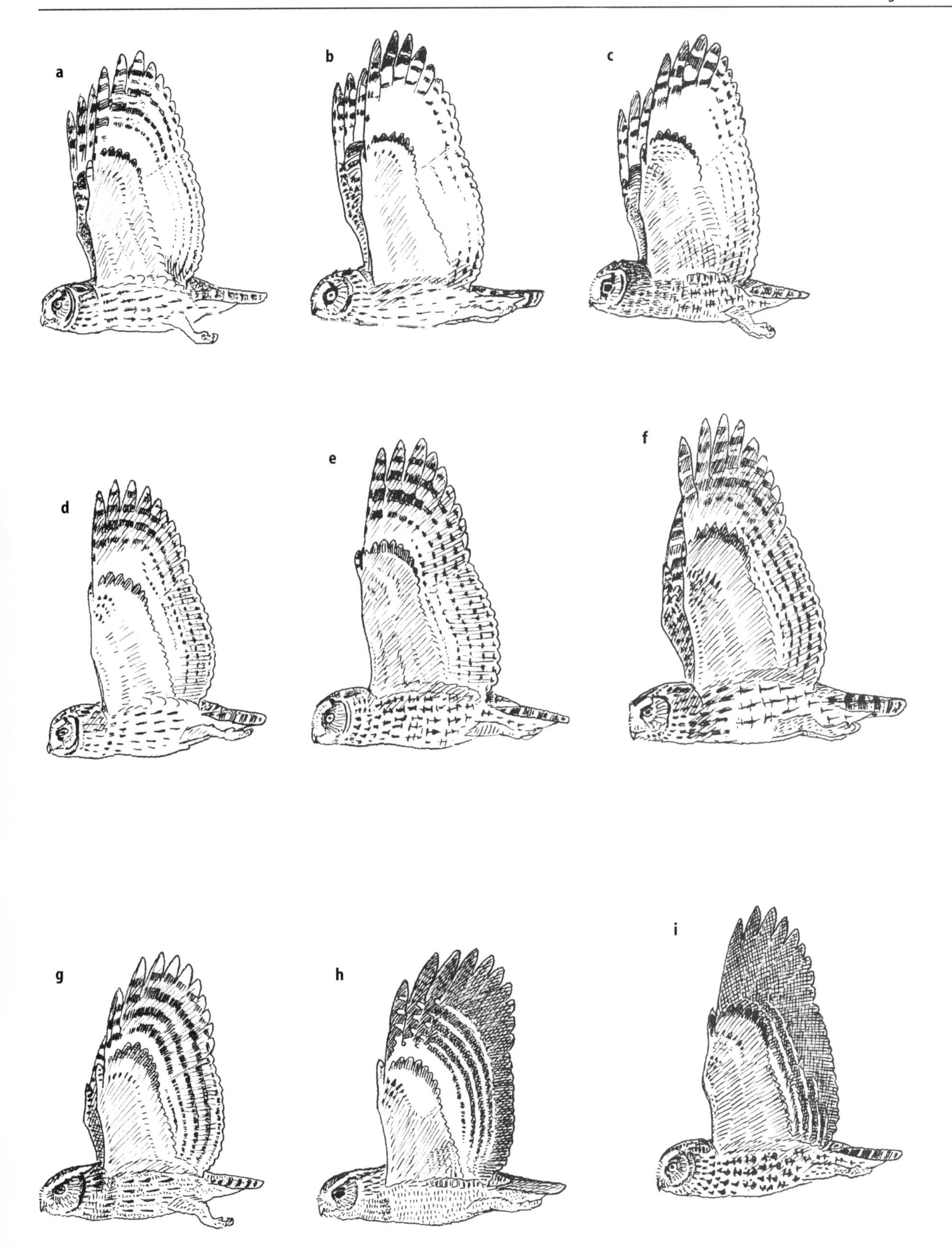

Plate 9. a *Asio o. otus*; **b** *Asio f. flammeus*; **c** *Asio capensis*; **d** *Asio clamator*; **e** *Asio abyssinicus*; **f** *Asio madagascariensis*; **g** *Jubula letti*; **h** *Lophostrix cristata*; **i** *Asio stygius*

Part IV Teil IV

Wing Formula and Topography
Schwingenformel und Topographie

Shortfall of Primary Tips — Projektionen zur Handschwingenspitze

Topography of an Owl — Topographie einer Eule

Table 1. Shortfall of primary tips. The shortfall of the ten primary tips to the wingpoint of the longest primary can be used to determine the wing formula and the shape of the wing. The shortfall (distance) of each of the primaries p10 to p1 to the top of the wing is measured in millimetres (p10 is the outermost primary; see also Plate 10 for the positions and terms used in describing the owl wing). The longest primary tip results in 0 (no distance); "*miss*" indicates that the primary was missing on the examined wing (moult). Synonyms are given in brackets.

The data in the table mainly concern species of the tribe Otini. In addition, some examples of other genera and species are given. The order of species follows the systematic list.

Tabelle 1. Projektionen zur Handschwingenspitze. Die Projektionen der zehn Handschwingen zur Handschwingenspitze sind kennzeichnend für die Schwingenformel und die Flügelform. Gemessen wird dabei die Projektion (Überstand) jeder der Handschwingen p10 bis p1 zur Flügelspitze in mm (p10 kennzeichnet die äußerste Handschwinge; zur Lage und zu den Bezeichnungen des Eulenflügels siehe auch Tafel 10). Für die längste Handschwinge ergibt sich dabei jeweils 0 (keine Projektion); „*miss*" bedeutet, dass die Handschwinge am untersuchten Flügel (Mauser) fehlte. Synonyme erscheinen in Klammern.

Die Angaben dieser Tabelle betreffen in erster Linie Arten vom Tribus Otini. Beispiele für einige andere Arten und Gattungen sind ebenfalls gegeben. Die Reihenfolge der Arten folgt der systematischen Liste.

Wing length (mm)	Species/taxon	p10	p9	p8	p7	p6	p5	p4	p3	p2	p1
212	*Tyto soumagnei*	14?	2	0	3	7	15	23	30	38	48
347	*Tyto alba pratincola*	2	0	10	28	45	67	93	111	123	138
185	*Otus sagittatus*	57	27	11	2	0	0	4	12	21	32
130	*Otus r. rufescens*	31	15	3	1	0	1	5	9	15	21
128	*Otus r. malayensis*	31	13	2	0	0	0	2	7	13	20
128	(*Otus r. burbidgei*)	35	17	6	0	0	1	4	8	11	18
137	(*Otus r. mantis*)	38	17	5	0	0	0	5	10	14	20
129	*Otus thilohoffmanni*	37	19	6	1	0	1	6	13	18	24
134	*Otus i. icterorhynchus*	36	15	5	2	0	2	6	11	16	19
142	*Otus i. holerythrus* ♂	42	19	5	0	2	5	11	18	24	29
143	*Otus i. holerythrus* ♀	34	16	4	0	0	4	10	17	23	28
119	*Otus irenae*	42	21	9	3	0	0	miss	12	21	miss
147	*Otus spiloc. spilocephalus*	45	20	7	2	0	5	12	20	27	34
142	*Otus spiloc. huttoni*	41	18	3	1	0	5	13	20	25	30
161	*Otus spiloc. latouchi*	48	21	8	2	0	3	11	20	29	35
149	*Otus spiloc. hambroecki*	46	19	7	1	0	2	9	18	25	31
144	*Otus spiloc. siamensis*	39	19	6	0	0	0	7	15	20	26
149	(*Otus spiloc. rupchandi*)	41	17	4	0	0	2	10	19	26	33
134	*Otus spiloc. luciae*	41	18	7	2	0	0	5	12	17	23
150	*Otus spiloc. vandewateri*	47	20	4	1	0	1	6	14	22	29
142	*Otus stresemanni*	38	17	5	0	0	0	5	14	21	29
148	*Otus angelinae*	44	22	6	1	0	2	7	17	24	31
140	*Otus balli*	42	17	6	1	0	1	6	14	21	28
158	*Otus alfredi*	48	23	8	3	0	2	10	16	25	33
135	*Otus mirus*	40	17	3	0	1	5	11	18	23	28
147	*Otus longicornis*	43	17	4	0	0	2	6	12	17	25
133	*Otus mindorensis*	37	12	2	0	1	3	9	15	20	25
142	*Otus hartlaubi*	30	11	1	0	2	8	18	25	30	35
153	*Otus rutilus*	42	18	5	0	1	5	12	19	26	33
159	*Otus madagascariensis*	40	17	5	0	1	6	12	21	27	34

Table 1. *Continued*

Wing length (mm)	Species/taxon	p10	p9	p8	p7	p6	p5	p4	p3	p2	p1
173	*Otus mayottensis*	47	19	5	0	1	6	13	21	28	36
164	*Otus moheliensis*	40	15	6	0	0	6	16	23	30	37
164	*Otus capnodes*	41	16	4	0	1	5	12	20	27	35
144	*Otus pauliani*	37	16	1	0	3	11	15	19	24	miss
152	*Otus pembaensis*	32	9	0	0	1	6	12	18	25	32
157	*Otus pembaensis*	33	11	2	0	2	7	13	20	28	35
127	*Otus f. flammeolus*	26	5	0	0	2	10	17	23	27	32
137	(*Otus f. idahoensis*)	23	6	0	0	5	15	24	30	34	37
141	(*Otus f. rarus*)	27	9	1	0	4	16	25	31	36	41
162	*Otus s. scops*	12	0	0	0	14	21	30	36	42	48
159	*Otus s. pulchellus*	12	0	0	4	13	22	28	36	42	48
168	*Otus scops cyprius*	14	0	0	5	15	25	32	39	45	46
150	*Otus s. turanicus*	11	0	0	2	12	20	27	35	40	46
152	(*Otus s. erlangeri*)	14	1	0	6	15	23	31	37	41	45
155	(*Otus s. longipennis*)	11	0	0	5	14	23	29	36	42	48
153	(*Otus scops sibiricus*)	13	0	0	4	12	20	26	32	38	45
168	*Otus brucei brucei*	21	5	0	1	7	17	25	31	37	46
161	*Otus b. obsoletus*	20	3	0	2	6	16	25	31	38	43
166	*Otus b. semenowi*	19	2	0	0	3	15	24	31	35	39
160	*Otus b. exiguus*	21	3	0	0	6	15	23	29	36	40
136	*Otus (b.) pamelae*	17	3	0	0	5	12	20	27	31	35
132	*Otus (b.) socotranus*	21	6	0	0	3	11	18	23	27	32
142	*Otus s. senegalensis*	24	5	1	0	1	7	16	23	27	31
132	(*Otus s. pusillus*)	20	5	0	1	4	12	18	23	27	30
129	(*Otus s. hendersoni*)	20	5	1	0	3	10	18	23	29	34
128	(*Otus s. graueri*)	18	5	1	0	2	8	16	21	25	29
130	*Otus s. feae*	20	6	0	0	3	8	14	19	22	26
128	*Otus s. nivosus*	20	4	0	0	2	7	13	17	22	27
143	*Otus sunia sunia*	28	8	0	0	5	15	22	28	32	38
122	*Otus sunia leggei*	28	10	2	0	1	6	12	17	21	25
144	*Otus s. modestus*	32	9	0	0	2	9	14	20	27	31
145	(*Otus s. nicobaricus*)	30	7	1	0	2	8	15	21	28	35
151	(*Otus s. distans*)	31	9	0	0	6	16	24	29	35	40
146	(*Otus s. khasiensis*)	27	7	0	0	4	12	22	24	34	38
138	*Otus s. rufipennis*	28	7	0	0	2	7	19	23	27	32
145	*Otus s. malayanus*	26	7	0	0	5	16	23	33	36	40
148	*Otus s. stictonotus*	26	5	0	0	6	16	25	31	37	43
150	*Otus s. japonicus*	28	8	1	0	6	17	26	32	38	44
164	*Otus e. elegans*	30	13	1	0	8	21	28	35	41	45
176	*Otus e. elegans*	28	7	0	0	9	23	33	39	46	51

Table 1. *Continued*

Wing length (mm)	Species/taxon	p10	p9	p8	p7	p6	p5	p4	p3	p2	p1
151	(*Otus e. interpositus*)	31	9	0	0	7	19	25	33	38	42
174	*Otus e. calayensis*	39	13	2	0	5	14	24	31	38	42
180	*Otus magicus morotensis*	40	16	4	0	1	7	14	22	30	37
175	*Otus magicus leucospilus*	36	12	2	0	0	7	14	22	29	37
182	*Otus magicus magicus*	35	12	4	0	0	6	14	22	31	41
189	*Otus magicus magicus* ♀	44	15	1	0	0	5	11	21	32	41
178	*Otus magicus bouruensis*	43	16	4	0	0	2	10	18	27	30
162	*Otus magicus albiventris*	37	11	3	0	1	8	16	24	31	37
159	*Otus (m.) tempestatis*	39	12	3	0	1.5	6.5	14	23	30	35
167	*Otus (m.) sulaensis*	37	15	4	0	1	6	15	21	29	34
175	*Otus beccarii*	39	14	3	0	2	5	12	21	39	56
169	*Otus mant. mantananensis*	40	13	2	0	1	6	14	22	29	39
158	*Otus mant. romblonis*	40	14	3	0	2	6	14	19	24	35
181	*Otus mant. cuyensis*	40	14	4	0	0	4	12	19	26	36
157	*Otus mant. sibutuensis*	39	14	3	0	1	5	12	17	23	33
151	*Otus man. manadensis*	36	12	2	0	1	7	15	23	30	37
156	*Otus man. manadensis* ♀	40	15	3	0	0	2	9	17	25	32
127	*Otus (m.) siaoensis*	29	14	6	0	3	8	15	20	27	37?
149	*Otus man. mendeni*	39	15	4	0	0	4	10	19	24	30
176	*Otus (m.) kalidupae*	37	12	2	0	1	7	14	22	28	33
162	*Otus collari*	39	15	5	0	2	10	19	28	34	40
166	*Otus insularis*	37	13	3	0	1	8	15	22	27	32
161	*Otus alius*	37	13	2	0	3	10	20	28	34	39
145	*Otus umbra*	29	9	1	0	4	12	19	25	30	39
163	*Otus enganensis*	44	13	2	0	3	12	20	27	33	40
161	*Otus mentawi*	45	17	6	0	0	2	9	17	25	33
172	*Otus b. solokensis*	46	19	6	0	0	2	10	22	30	38
171	*Otus b. brookii*	43	17	3	0	0	2	10	20	miss	miss
140	*Otus l. lempiji*	39	16	7	2	0	3	9	16	21	25
155	*Ozus (l.) cnephaeus*	41	20	7	2	0	4	12	19	26	32
156	*Otus l. hypnodes*	40	14	2	0	0	3	10	18	26	32
145	*Otus l. lemurum*	44	18	6	0	0	3	12	19	25	30
143	*Otus l. kangeana*	38	18	6	0	0	2	8	15	20	25
174	*Otus lettia lettia*	43	16	4	0	0	5	15	25	31	36
168	*Otus l. erythrocampe*	45	20	7	1	0	2	11	20	28	34
178	*Otus lettia glabripes*	50	22	7	1	0	3	15	24	31	39
164	*Otus lettia umbratilis*	40	17	6	0	0	4	13	21	29	35
150	*Otus b. bakkamoena*	38	15	3	0	0	6	15	20	25	31
180	*Otus b. plumipes*	36	13	3	0	0	8	20	27	33	51!
170	*Otus b. deserticolor*	35	14	2	0	0	6	16	24	31	36

Table 1. *Continued*

Wing length (mm)	Species/taxon	p10	p9	p8	p7	p6	p5	p4	p3	p2	p1
161	*Otus b. gangeticus*	40	16	6	0	0	3	12	22	29	36
185	*Otus s. ussuriensis*	47	17	5	0	0	10	23	32	38	45
184	*Otus s. semitorques*	43	18	4	0	0	10	25	36	42	48
181	*Otus s. pryeri*	43	18	4	0	0	8	17	24	33	40
193	*Otus m. megalotis*	60	30	12	2	0	2	13	19	25	33
176	*Otus m. everetti*	42	18	6	1	0	2	7	15	miss	30
149	*Otus m. nigrorum*	40	20	8	1	0	1	6	13	21	26
161	*Otus fuliginosus*	44	22	8	3	0	1	6	16	20	30
215	*Otus silvicola*	48	18	6	1	0	8	19	32	41	48
212	*Otus silvicola*	56	23	7	1	0	9	20	30	40	50
188	*Otus k. kennicotti*	39	11	0	0	2	11	22	31	38	43
176	(*Otus k. saturatus*)	41	12	2	0	1	9	18	27	32	36
191	(*Otus k. brewsteri*)	59	28	4	0	2	10	22	31	39	44
170	*Otus k. bendirei*	42	12	1	0	1	9	29	29	34	37
177	*Otus k. aikeni*	37	11	0	0	6	14	24	29	35	41
176	(*Otus k. gilmani*)	38	12	1	0	3	11	23	30	40	44
167	*Otus k. suttoni*	37	10	1	0	2	7	16	27	31	35
179	*Otus seductus*	47	21	6	2	0	3	11	22	31	38
177	*Otus cooperi*	41	17	5	2	0	4	11	19	27	34
155	*Otus lambi* ♂	46	18	6	0	0	5	15	23	30	36
166	*Otus lambi* ♀	55	27	10	2	0	3	11	21	27	32
162	*Otus asio asio* ♀	36	10	1	0	0	6	15	23	29	34
174	(*Otus a. naevius*)	43	13	2	0	2	10	24	32	39	45
182	*Otus a. maxwelliae*	41	11	2	0	1	10	22	32	38	44
168	(*Otus a. swenki*)	43	15	4	0	2	13	23	29	35	38
170	*Otus a. hasbroucki*	38	13	1	0	1	11	22	30	35	42
157	*Otus a. floridanus*	41	17	3	0	0	5	16	23	29	36
170	*Otus a. mccalli* ♀	42	15	3	0	1	10	22	29	35	39
151	*Otus t. aspersus*	33	9	1	0	0	7	17	21	31	37
158	*Otus t. trichopsis*	31	9	1	0	0	2	9	15	23	31
154	(*Otus t. ridgwayi*) ♂	36	13	2	0	1	10	20	26	34	39
146	(*Otus t. guerrensis*)	48	26	3	0	0	4	11	20	28	35
151	*Otus t. mesamericanus*	43	14	2	0	3	9	miss	25	28	34
140	(*Otus t. pumilus*) ♂	34	12	2	0	0	2	10	17	24	30
142	(*Otus t. pumilus*) ♀	39	15	4	0	0	3	10	16	24	41
168	*Otus c. luctisonus*	46	17	4	0	0	3	12	19	26	37
166	*Otus c. margaritae*	38	22	3	1	0	3	13	20	26	32
176	*Otus c. crucigerus*	51	21	5	1	0	3	12	20	26	36
172	(*Otus c. portoricens*)	45	18	5	3	0	0	6	16	21	36
182	(*Otus c. caucae*)	49	20	4	0	0	4	10	19	26	36

Table 1. *Continued*

Wing length (mm)	Species/taxon	p10	p9	p8	p7	p6	p5	p4	p3	p2	p1
157	*Otus c. decussatus*	41	15	4	0	0	3	11	20	27	33
174	*Otus c. choliba*	48	18	5	0	0	4	12	20	27	37
160	*Otus c. wetmorei*	38	21	2	0	0	2	12	19	25	30
180	*Otus c. uruguaiensis*	48	20	5	2	0	6	18	30	37	44!
173	*Otus koepckeae* ♂	40	15	3	0	0	4	14	23	32	40
179	*Otus koepckeae* ♀	46	19	4	0	1	3	11	21	29	38
170	*Otus roboratus* ♂	51	17	3	0	0	2	8	15	23	31
175	*Otus roboratus* ♀	47	23	7	0	0	4	9	20	26	38
140	*Otus pazificus*	41	18	5	0	0	3	10	20	24	32
150	*Otus pazificus*	38	17	4	1	0	2	11	18	25	30
180	*Otus clarkii* ♂?	44	17	5	1	0	1	7	16	25	35
190	*Otus clarkii* ♀	50	24	7	1	0	5	10	17	26	38
133	*Otus barbarus* ♂	37	14	3	0	0	5	11	18	24	31
140	*Otus barbarus* ♀?	39	16	5	0	0	4	10	17	21	29
212	*Otus i. ingens* ♀	57	23	9	2	0	0	7	18	30	42
202	(*Otus i. minimus*)	52	24	8	1	0	2	12	25	35	44
189	*Otus colombianus*	59	23	7	0	0	2	9	17	26	37
193	*Otus colombianus*	51	20	7	1	0	2	8	17	25	34
175	*Otus w. watsonii*	44	20	6	1	0	3	12	23	31	45
181	*Otus watsonii usta*	43	18	7	0	0	3	10	19	27	36
175	(*Otus w. morelia*)	52	25	9	2	0	2	7	15	21	29
175	*Otus atricapillus*	49	19	5	1	0	3	11	21	33	39
179	(*Otus a. argentinus*)	49	28	7	0	0	3	13	23	31	39
200	*Otus sanctae-catarinae*	56	25	7	0	0	3	10	19	26	37
170	*Otus hoyi*	48	24	7	0	1	6	15	23	40	47
157	*Otus g. hastatus*	45	21	8	2	0	3	10	20	30	37
157	*Otus g. cassini*	44	18	5	1	0	2	8	18	24	30
177	*Otus g. guatemalae* ♀	52	26	10	3	0	4	13	21	28	33
160	*Otus g. guatemalae* ♂	42	17	4	1	0	4	12	20	27	33
168	(*Otus g. thompsoni*)	50	22	7	3	0	3	21	33	37	41
171	(*Otus g. fuscus*) ♂	50	23	9	2	0	3	11	19	27	34
168	*Otus vermiculatus*	52	22	6	1	0	2	9	17	26	33
174	*Otus n. napensis* ♀	43	17	6	0	0	6	15	27	33	42
169	*Otus n. helleri* ♀	44	15	3	0	0	4	14	23	30	37
161	*Otus roraimae*	47	23	6	2	0	2	9	17	23	33
155	*Otus n. nudipes*	52	26	11	4	0	0	5	12	18	23
171	*Otus n. nudipes*	57	27	12	5	0	miss	9	20	30	38
157	*Otus n. newtoni*	45	19	7	2	0	1	7	16	23	30
164	*Otus n. newtoni*	51	25	10	2	0	1	4	miss	20	26
208	*Otus a. albogularis*	63	30	14	2	0	2	9	18	30	41

Table 1. *Continued*

Wing length (mm)	Species/taxon	p10	p9	p8	p7	p6	p5	p4	p3	p2	p1
201	*Otus a. meridensis*	55	24	8	3	0	3	8	17	28	39
202	*Otus a. macabrum*	60	23	8	1	0	3	10	18	29	41
207	*Otus a. remotus*	51	22	6	1	0	4	10	21	32	41
159	*Pyrroglaux podarginus*	42	18	4	0	0	1	4	11	19	miss
163	*Pyrroglaux podarginus*	50	25	8	3	0	0	3	8	18	28
147	*Gymnoglaux l. lawrencii*	49	22	8	4	0	1	3	9	18	26
154	*Gymnoglaux l. lawrencii*	50	23	8	1	0	1	6	15	23	32
153	*Gymnoglaux l. exsul*	52	25	10	1	0	1	6	13	22	31
191	*Ptilopsis leucotis*	18	0	0	1	7	16	24	33	39	46
197	*Ptilopsis leucotis*	17	1	0	1	9	18	27	36	44	51
197	*Ptilopsis granti* ♂	16	1	0	3	10	19	28	35	43	50
201	*Ptilopsis granti* ♀	22	1	0	4	10	19	28	37	43	49
233	*Mimizuku gurneyi*	60	26	9	1	0	0	6	16	26	38
242	*Mimizuku gurneyi*	miss	26	3	0	0	3	12	18	28	39
428	*Nyctea scandiaca*	40	8	3	0	28	45	60	78	93	108
376	*Bubo v. virginianus*	39	8	0	11	32	49	63	78	92	105
240	*Strix virgata squamulata*	58	24	9	2	0	3	11	27	39	55
320	*Strix varia varia*	84	33	9	0	6	18	37	60	76	90
438	*Strix n. nebulosa*	111	36	11	2	0	6				
238	*Jubula letti* ♂	68	30	11	2	0	1	7	21	39	55
276	*Jubula letti* ♀	79	41	15	2	0	3	13	28	43	53
325	*Lophostrix c. stricklandi*	86	42	12	1	0	7	21	39	61	76
305	*Lophostrix cristata*	86	31	11	1	0	miss	12	35	58	74
295	*Lophostrix cristata*	90	38	14	2	0	1	9	35	54	72
329	*Pulsatrix p. saturata*	101	50	21	5	0	5	26	47	63	75
315	*Pulsatrix koeniswaldiana*	100	55	25	7	0	2	5	12	22	46
100	*Glaucidium passerinum*	29	9	1	0	1	8	11	15	19	23
101	*Glaucidium californicum*	23	7	2	0	3	9	12	16	19	22
105	*Xenoglaux loweryi*	49	24	13	6	0	4	5	6	8	11
102	*Micrathene whitneyi*	28	9	2	0	4	9	13	17	20	23
160	*Athene n. noctua*	25	6	2	0	4	15	26	34	40	43
140	*Aegolius a. acadicus*	28	8	1	0	5	12	19	26	32	39
215	*Ninox s. scutulata*	60	15	2	0	12	24	30			
176	*Ninox p. philippensis*	52	22	6	0	1	6	15	25	34	41
170	*Ninox p. spilocephala*	53	22	6	0	1	6	15	25	34	42
164	*Ninox mindorensis*	53	22	7	1.5	0	5	13	23	31	39
185	*Ninox ochracea*	53	20	4	0	1	8	19	32	42	49
179	*Ninox ochracea*	49	17	2	0	0	10	20	31	44	51
172	*Ninox ios*	53	16	2	0	2	11	21	33	39	47
183.5	*Ninox burhani*	52	22	6	0	2	8	21	35	43	51

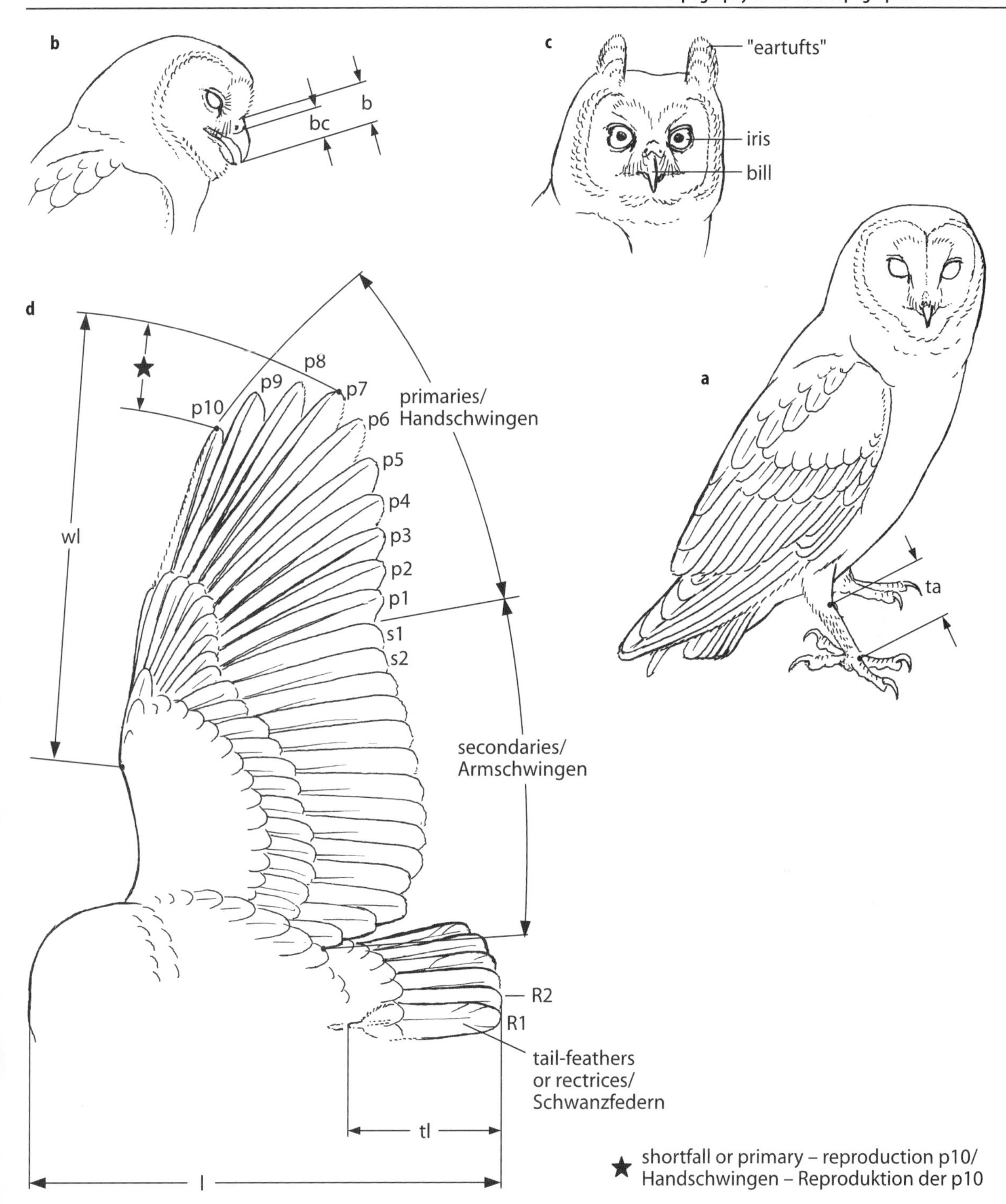

Plate 10. Topography of an owl. a Perched owl; *ta* length of tarsometatarsus. **b, c** Head; *bc* length of bill, from tip to cere; *b* length of bill, from tip to frontal feathers. **d** Wing; *wl* wing length, from bend to wingtip; *d* shortfall (distance) of primary *p10* to wingtip; *p1–p10* primaries, *p10* is the outermost primary; *s1, s2* secondaries; *r1, r2* tail feathers or rectrices; *tl* tail length, from base to tip of central tail feathers; *l* total length

Tafel 10. Topographie einer Eule. a Eule sitzend; *ta* Tarsal- oder Lauflänge **b, c** Kopf; *bc* Schnabellänge von der Spitze bis zur Wachshaut; *b* Schnabellänge von der Spitze bis zur Stirnbefiederung. **d** Flügel; *wl* Flügellänge vom Bug bis zur Spitze der längsten Handschwinge; *d* Projektion (Abstand) einer Handschwingenspitze bis zur Flügelspitze; *p1–p10* Handschwingen, *p10* ist die äußerste Handschwinge; *s1, s2* Armschwingen; *r1, r2* Schwanzfedern; *tl* Schwanzlänge vom Ansatz der zentralen Schwanzfeder bis zur Spitze; *l* Gesamtlänge

Part V Teil V

Owls Described or Rediscovered in the Last 20 Years

Neu beschriebene und wiederentdeckte Eulen der vergangenen 20 Jahre

Figure 1. *Bottom left:* Madagascar Red Owl, *Tyto soumagnei*, an extremely rare small barn owl, rediscovered in 1994; the typical wingbars of *Tyto soumagnei* are especially clear in flight (*top*); *bottom right: Tyto alba hypermetra*, a similar but easily distinguishable species also found in Madagascar (Lagrand 1995; Thorston 1996)

Abbildung 1. Malegasseneule, *Tyto soumagnei*, eine sehr seltene, erst 1994 wiederentdeckte kleine Schleiereule (*links*); *Tyto soumagnei* zeigt vor allem im Flug ihre typische Flügelbänderung (*oben*); *rechts* die ebenfalls auf Madagaskar lebende, gut unterscheidbare *Tyto alba hypermetra* (Lagrand 1995; Thorston 1996)

Figure 2.
Left: The tiny barn owl, *Tyto prigoginei*, rediscovered in 1996 in the Itombwe Forest (Zaire). *Right:* The Bay Owl, *Phodilus badius*, morphologically quite distinct from *Tyto prigoginei* (Butynski et al. 1996)

Abbildung 2.
Die winzige Schleiereule *Tyto prigoginei* (*links*) wurde 1996 im Itombwe Forst (Zaire) wiederentdeckt. Von dieser unterscheidet sich morphologisch die großäugige Maskeneule, *Phodilus badius* (*rechts*) (Butynski et al. 1996)

Figure 3.
Left: Otus alfredi, a scops owl from Flores Island, rediscovered in 1994–1997. Prior to rediscovery, the Flores Scops Owl had been considered as a red morph of *Otus spilogaster* ssp. *stresemanni* (*middle*) or *Otus magicus albiventris* (*right*) (Collar et al. 1994)

Abbildung 3.
Links: *Otus alfredi*, eine Zwergohreule der Insel Flores, 1994–1997 wiederentdeckt. Die Flores-Zwergohreule wurde zwischenzeitlich als rote Morphe *Otus spilocephalus* ssp. *stresemanni* (*Mitte*) bzw. *Otus magicus albiventris* (*rechts*) artlich zugeordnet (Collar et al. 1994)

Figure 4. *Top left:* Serendib Scops Owl, *Otus thilohoffmanni*, a scops owl from southwestern Sri Lanka first decribed in 2004, living sympatrically with the Oriental Scops Owl, *Otus sunia leggei* (*middle*). This new scops owl is most like the Reddish Scops Owl, *Otus rufescens malayensis* (*top right*), but lacks visible eartufts and has yellow eyes (Warakagoda and Rasmussen 2004)

Abbildung 4. *Links*: Serendib-Zwergohreule, *Otus thilohoffmanni*, eine erst 2004 vom südwestlichen Sri Lanka beschriebene Zwergohreule. Sie kommt dort sympatrisch mit der Orient-Zwergohreule, *Otus sunia leggei* (*Mitte*) vor. Am ähnlichsten ist diese Eule der Röteleule, *Otus rufescens malayensis* (*rechts*), hat aber keine sichtbaren Federohren und ist gelbäugig (Warakagoda und Rasmussen 2004)

Figure 5.
Light (*left*) and dark (*right*) morph of the Anjouan Scops Owl, *Otus capnodes*. Rediscovered in 1992 by Safford (Safford 1993)

Abbildung 5.
Helle (*links*) und dunkle (*rechts*) Morphe der Anjouan-Zwergohreule, *Otus capnodes*. Wiederfund durch Safford im Jahre 1992 (Safford 1993)

Figure 6.
Left: Red colour morph of the Madagascar Scops Owl, *Otus rutilus*. *Middle* and *right*: Grey and rufous morphs, respectively, of the Torotoroka Scops Owl, *Otus madagascariensis*, also from Madagascar. The latter two owls have similar plumage, but very distinct vocalizations (Rasmussen et al. 2000)

Abbildung 6.
Links: Rote Farbmorphe der Madagaskar-Zwergohreule, *Otus rutilus*. Graue (*Mitte*) und rote (*rechts*) Morphe der ebenfalls auf Madagaskar verbreiteten Tototoroka-Zwergohreule, *Otus madagascariensis*, sehr ähnlich im Gefieder, jedoch mit deutlich verschiedenem Gesang (Rasmussen et al. 2000)

Figure 7.
Left: Nicobar Scops Owl, *Otus alius*, described by Rasmussen as a new species in 1998. *Middle* and *right:* The very similar *Otus umbra* from Simeulue Island and *Otus enganensis* from Enggano Island (off West Sumatra), respectively (Rasmussen 1998)

Abbildung 7.
Links: Nicobaren-Zwergohreule, *Otus alius*, 1998 als neue Art von Rasmussen beschrieben. *Mitte:* Die sehr ähnlichen *Otus umbra* von der Simeulue-Insel (*Mitte*) und *Otus enganensis* von der Insel Enggano (vor West Sumatra) (*rechts*) (Rasmussen 1998)

Figure 8.
Left: Sangihe Scops Owl, *Otus collari*, from Sangihe Island (between Sulawesi and Mindanao), described as a new species by Lambert and Rasmussen in 1998. *Middle: Otus manadensis* from Sulawesi. *Right:* The tiny *Otus siaoensis* from Siau Island (near Sangihe), often considered as a subspecies of *Otus manadensis* (Lambert and Rasmussen 1998)

Abbildung 8.
Links: Sangihe-Zwergohreule, *Otus collari,* eine 1998 von Lambert und Rasmussen beschriebene neue Art von der Insel Sangihe (zwischen Sulawesi und Mindanao). *Mitte: Otus manadensis* von Sulawesi und *rechts* die winzige *Otus siaoensis* von der Insel Siau (nahe Sangihe), die meist als Subspezies von *Otus manadensis* angesehen wird (Lambert und Rasmussen 1998)

Figure 9.
Mohéli Scops Owl, *Otus moheliensis*, brown morph sitting (*bottom left*) and in flight (*top right*), as well as rufous morph (*top left*). The species was described by Lafontaine and Moulaert in 1999, and is endemic to Mohéli Island (Mwali, Comores). *Bottom right:* Dark colour morph of *Otus mayottensis*, from the neighbouring Mayotte (Maore) Island (Lafontaine and Moulaert 1999)

Abbildung 9.
Mohéli-Zwergohreule, *Otus moheliensis*, braune Farbmorphe sitzend (*links*) und im Flug (*oben rechts*) sowie rostfarbene Farbmorphe (*Mitte*). Von Lafontaine und Moulaert 1999 beschriebene endemisch auf Mohéli (Mwali, Komoren) lebende neue Eulenart. Zum Vergleich dunkle Morphe von *Otus mayottensis*, von der Nachbarinsel Mayotte (Maore) (*rechts*) (Lafontaine und Moulaert 1999)

Figure 10.
Left: Otus cnephaeus, often regarded as subspecies of *Otus lempiji* (*middle*), light morph with light-coloured eyes, but with distinct vocalizations. *Right: Otus lettia*, larger and with lighter plumage, but dark-eyed like *Otus cnephaeus* (Del Hoyo et al. 1999)

Abbildung 10.
Links: Otus cnephaeus, oft als Subspezies zu *Otus lempiji* (*Mitte*), helle und helläugige Morphe, gestellt, aber mit deutlich anderem Gesang. *Rechts: Otus lettia*, größer und heller im Gefieder als *Otus cnephaeus*, ebenfalls dunkeläugig (Del Hoyo et al. 1999)

Figure 11. Tumbes Screech Owl, *Otus pacificus:* greybrown (*top left*) and rufous (*bottom left*) morphs. Described by Hekstra in 1982 as a subspecies of *Otus guatemalae*, later regarded as conspecific with *Otus roboratus* (*top right*), but distinct in both morphology and vocalizations. *Bottom right:* Koepcke's Screech Owl, *Otus koepckeae*, described in 1982 as a subspecies of *Otus choliba* by Hekstra, but since recognised as a distinct species (König et al. 1999)

Abbildung 11. Tumbes-Kreischeule, *Otus pacificus*, graubraune (*links oben*) und rotbraune Morphe (*links unten*), bereits 1982 von Hekstra als Rasse von *Otus guatemalae* beschrieben, später artlich mit *Otus roboratus* (*rechts oben*) vereinigt, jedoch von beiden in Morphologie und Stimme unterscheidbar. Die Koepcke-Kreischeule, *Otus koepckeae* (*rechts unten*) wurde von Hekstra irrtümlich als Rasse von *Otus choliba* beschrieben, jedoch später als eigene Art anerkannt (König et al. 1999)

Figure 12. The Montane Forest-Screech Owl, *Otus hoyi* (*top left*), is distinct from the Variable Screech Owl *Otus atricapillus* (*top middle* and *top right*) in plumage and vocalizations (described by König and Straneck 1989). The dark-eyed (*top middle*) and yellow-eyed (*top right*) *Otus atricapillus* also occupy different habitats from *Otus hoyi*. *Bottom left:* Cinnamon Screech Owl, *Otus petersoni*, described in 1986 and named in honour of R. T. Peterson, a wildlife artist; *bottom right:* Cloudforest Screech Owl, *Otus marshalli*, described in 1981 and named in honour of J. T. Marshall jr., an owl expert (Weske and Terborgh 1981; Fitzpatrick and O'Neill 1986; König and Straneck 1989; König et al. 1994, 1999, 2001)

Abbildung 12. *Oben links:* Bergwald-Kreischeule, *Otus hoyi*, in Gefieder und Stimme von *Otus atricapillus* (*oben Mitte* und *rechts*) gut unterscheidbar (von König und Straneck 1989 beschrieben). Die Schwarzkappen-Kreischeule *Otus atricapillus* (dunkeläugig *oben Mitte* sowie gelbäugig *rechts*) bewohnt auch ein anderes Habitat als *Otus hoyi*. *Links unten:* Zimt-Kreischeule, *Otus petersoni*, eine zu Ehren des Vogelmalers R. T. Peterson benannte, 1986 beschriebene Kreischeule. *Rechts unten:* Nebelwald-Kreischeule, *Otus marshalli*, 1981 beschrieben und zu Ehren des Eulenkenners J. T. Marshall benannt (Weske und Terborgh 1981; Fitzpatrick und O'Neill 1986; König und Straneck 1989; König et al. 1994, 1999, 2001)

Figure 13. Yungas Pygmy Owl, *Glaucidium bolivianum* (*left*), and the Peruvian Pygmy Owl, *Glaucidium peruanum* (*middle*) were described by König in 1991. The slightly larger and greyer Tucuman Pygmy Owl, *Glaucidium tucumanum* (*right*) is often considered as a subspecies of *G. brasilianum*, but has a clearly distinct song (König 1991; König et al. 1999)

Abbildung 13. *Links:* Yungas-Sperlingskauz, *Glaucidium bolivianum*; diese Art, wie auch der Peru-Sperlingskauz, *Glaucidium peruanum* (*Mitte*) wurden von König 1991 beschrieben. Der etwas größere und grauere Tucuman-Sperlingskauz, *Glaucidium tucumanum* (*rechts*) wird häufig als Subspezies von *Glaucidium brasilianum* betrachtet, hat aber einen deutlich anderen Gesang (König 1991; König et al. 1999)

Figure 14. *Left:* Cloudforest Pygmy Owl, *Glaucidium nubicola*, first described in 1999. This small, stout owl is distributed in the western slopes of the Andes, from Colombia to Ecuador. The rufous morph of *Glaucidium costaricanum* (*middle*) and greybrown morph of *Glaucidium jardinii* (*right*), were formerly considered as members of one species (Robbins and Stiles 1999)

Abbildung 14. *Links:* Der Nebelwald-Sperlingskauz, *Glaucidium nubicola*, wurde erst 1999 beschrieben. Der kleine, kompakte Kauz lebt in den westlichen Anden von Kolumbien bis Ecuador. Die rotbraune Morphe von *Glaucidium costaricanum* (*Mitte*) und die graubraune Morphe von *Glaucidium jardinii* (*rechts*) wurden früher als Rassen einer Art betrachtet (Robbins und Stiles 1999)

Figure 15. *Left top* and *bottom:* Amazonian Pygmy Owl, *Glaucidium hardyi*, described in 1989, and, like some other recently described pygmy owls, distinguishable primarily by song. *Bottom right:* Pernambuco Pygmy Owl, *Glaucidium minutissimum*, now distributed only throughout a small region near Pernambuco. The highly similar Sick's Pygmy Owl, *Glaucidium sicki* (*middle*), can be distinguished only by its vocalizations and typical "occipital-face". *Top right:* Central American Pygmy Owl, *Glaucidium griseiceps*, another small owlet similar in plumage that is distinguishable primarily by its voice (Vielliard 1989; Robbins and Howell 1995; König and Weick 2005)

Abbildung 15. *Links oben* und *unten:* Amazonas-Zwergkauz, *Glaucidium hardyi*, 1989 beschrieben und wie einige der neu beschriebenen Sperlingskäuze hauptsächlich am arteigenen Gesang unterscheidbar. *Unten rechts:* Pernambuco-Zwergkauz, *Glaucidium minutissimum*, Restvorkommen nur noch in einem kleinen Gebiet nahe Pernambuco. Der sehr ähnliche Sick-Zwergkauz (*Mitte*) unterscheidet sich im Gesang und einem deutlichen „Occipitalgesicht". *Rechts oben:* Graukopf-Zwergkauz, *Glaucidium griseiceps*, ein weiteres in seinem Federkleid ähnliches Käuzchen, das hauptsächlich durch seinen Gesang unterscheidbar ist (Vielliard 1989; Robbins und Howell 1995; König und Weick 2005)

Figure 16. *Left:* Subtropical Pygmy Owl, *Glaucidium parkeri*, described in 1995 by Robbins and Howell. The Tamaulipas Pygmy Owl, *Glaucidium sanchezi* (*middle:* sitting; *top right:* flying) and the Colima Pygmy Owl, *Glaucidium palmarum* (*right*), despite their great similarity, are distinct species, and can be distinguished by song (Robbins and Howell 1995; Howell and Robbins 1995)

Abbildung 16. *Links:* Parker-Zwergkauz, *Glaucidium parkeri*, 1995 von Robbins and Howell beschrieben. Der sehr ähnliche Tamaulipas-Zwergkauz, *Glaucidium sanchezi* (*Mitte:* sitzend; *oben rechts:* fliegend) und der Colima-Zwergkauz, *Glaucidium palmarum* (*rechts*), sind trotz großer Ähnlichkeit stimmlich unterscheidbare Arten (Robbins und Howell 1995; Howell und Robbins 1995)

Figure 17. Long-whiskered Owlet, *Xenoglaux loweryi*, a tiny owlet discovered in 1977 by O'Neill and Graves in the cloud-forest of North Peru (*top left*). *Top right:* The Elf Owl, *Micrathene whitneyi*, similar in size, but unrelated to *Xenoglaux*. *Bottom right:* Albertine Owlet, *Glaucidium albertinum*, described in 1983 by A. Prigogine, as a new species from the Rift Valley, Central Africa. *Bottom left:* An example of the very rare African Barred Owlet, *Glaucidium c. capense*, rediscovered in 1985 by Carlyon at the Eastern Cape, South Africa (O'Neill and Graves 1997; Prigogine 1983)

Abbildung 17. Lowery-Zwergkauz, *Xenoglaux loweryi*, ein winziges Eulchen, das 1977 von O'Neill und Graves im Nebelwald von Nordperu entdeckt wurde (*links oben*). *Oben Mitte:* Der etwa gleichgroße Elfenkauz, *Micrathene whitneyi*, der aber nicht mit *Xenoglaux* verwandt ist. Der Albertseekauz, *Glaucidium albertinum*, wurde 1983 von A. Prigogine als neue Art des Rift Valley (Zentralafrika) beschrieben (*rechts*). *Unten:* Die Nominatform des Kapkauzes, *Glaucidium c. capense*, die sehr selten ist und erst 1985 durch Carlyon am östlichen Kap (Südafrika) wiederentdeckt wurde (O'Neill und Graves 1997; Prigogine 1983)

Figure 18. *Left* and *middle:* Forest Owlet, *Athene blewitti*, rediscovered in 1998 by King and Rasmussen. *Right:* The similar Spotted Owlet, *Athene brama*, which has a widespread distribution throughout the Indian Subcontinent (King and Rasmussen 1998; Rasmussen and Collar 1999)

Abbildung 18. Blewitt-Kauz, *Athene blewitti*, 1998 wiederentdeckt durch King and Rasmussen (*links und Mitte*); *rechts:* der ähnliche, in Indien weit verbreitete Brahma-Kauz, *Athene brama* (King und Rasmussen 1998; Rasmussen und Collar 1999)

Figure 19. *Left:* Cinnabar Hawk Owl, *Ninox ios*, a hawk owl, described in 1999 from a skin collected in 1985 in North Sulawesi. *Right:* The similar, somewhat larger, Ochre-bellied Hawk Owl, *Ninox ochracea*, also living endemic in Sulawesi (Rasmussen 1999)

Abbildung 19. Zinnoberkauz, *Ninox ios*, erst 1999 nach einem 1985 auf Nord-Sulawesi gesammeltem Balg (*links*) beschrieben wurde. *Rechts:* Der ähnliche, etwas größere Ockerbauchkauz, *Ninox ochracea*, der ebenfalls endemisch auf Sulawesi vorkommt (Rasmussen 1999)

Figure 20.
Left: Little Sumba Hawk Owl, *Ninox sumbaensis*, discovered and described in 2002. Since 1980, ornithologists have reported an unknown owl (thought to be a screech owl) from Sumba Island. *Right:* The larger, dark-eyed Sumba Boobook, *Ninox rudolfi*, also endemic to Sumba. *Middle:* An unknown hawk owl (from the collection of the Senckenberg Museum, Frankfurt, labelled without data, painted and named as Dubious Hawk Owl, *Ninox dubiosa*) (Olsen et al. 2002; Weick 2006)

Abbildung 20.
Links: Ninox sumbaensis, ein kleiner Sumbakauz, der erst 2002 entdeckt und beschrieben wurde. Seit 1980 wurde auf Sumba (Kleine Sunda-Inseln) eine neue Eulenart (meist Zwergohreule) vermutet. *Rechts* der ebenfalls endemisch auf Sumba lebende, größere und dunkeläugige Sumbakauz, *Ninox rudolfi. In der Mitte:* Ein Falkenkauz, dessen Balg mit Etikett ohne Daten sich in der Sammlung des Senckenbergmuseums in Frankfurt befindet, hier als Vielfleck-Falkenkauz, *Ninox dubiosa* sp. nov., abgebildet (Olsen et al. 2002; Weick 2006)

Figure 21.
Left: Togian Hawk Owl, *Ninox burhani*, described by Indrawan and Somadikarta in 2004. The Togian Islands are small islands in the Gulf of Tomin (Sulawesi). This owl resembles the Brown Hawk Owl, *Ninox scutulata*, a migratory visitor of Sulawesi, in plumage but is distinctly smaller in size. There is no similarity between *Ninox burhani* and the three endemic Sulawesi hawk owls, e.g. the Speckled Hawk Owl, *Ninox punctulata* (*right*), which is similar in size, but with different plumage (Indrawan and Somadikarta 2004)

Abbildung 21.
Der Togian-Falkenkauz, *Ninox burhani*, erst 2004 von Indrawan und Somadikarta beschrieben (*links*). Die Togian-Inseln liegen im Golf von Tomin (Sulawesi). Der Kauz ähnelt dem dort auf seinen Wanderungen ebenfalls vorkommenden Schildkauz, *Ninox scutulata*, ist aber erheblich kleiner. Keinerlei Ähnlichkeit besteht mit den drei endemischen Falkenkäuzen Sulawesis, wie z. B. dem gleichgroßen Pünktchenkauz, *Ninox punctulata*, mit deutlich anderem Gefieder (*rechts*) (Indrawan und Somadikarta 2004)

Figure 22. Taxonomy of the American Great Horned Owls (*Bubo* spp.) taking into consideration the mostly parallel variation within the polymorphic subspecies (Weick 1999). **a, b** *Bubo virginianus wapacuthu*; **c** *wapacuthu*, *heterocnemis* and *saturatus*; **d** *saturatus*, *wapacuthu* and *heterocnemis* and smaller: *pallescens* and *mayensis*; **e** *occidentalis*; **k** *pallescens* and *virginianus* (rare); **l** *virginianus*; **m** *pallescens*; **x, y** *Bubo magellanicus*

Abbildung 22. Taxonomie der Amerikanischen Uhus (*Bubo* spp.) unter Berücksichtigung eines größtenteils parallel variierenden Polymorphismus innerhalb der Subspezies (Weick 1999). **a, b** *Bubo virginianus wapacuthu*; **c** *wapacuthu*, *heterocnemis* and *saturatus*; **d** *saturatus*, *wapacuthu* und *heterocnemis* und kleiner: *pallescens* und *mayensis*; **e** *occidentalis*; **k** *pallescens* und *virginianus* (selten); **l** *virginianus*; **m** *pallescens*; **x, y** *Bubo magellanicus*

Figure 23. Taxonomy of the American Great Horned Owls (*Bubo* spp.) taking into consideration the mostly parallel variation within the polymorphic subspecies (Weick 1999). **f** *Bubo saturatus* and *heterocnemis*; **g** *saturatus* and *pacificus* (smaller); **h** *pacificus*, *pallescens*, *nacurutu*, *deserti* and *mayensis*; **i, j** *elachistus* (small!); **n** *pallescens*, *mayensis* and *nacurutu* (rare); **o** *mayensis*, *nacurutu* and *deserti*; **p** *mayensis* and *virginianus* (rare); **q** *B. v. nacurutu* and *B. magellanicus* (smaller); **r** *nigrescens*

Abbildung 23. Taxonomie der Amerikanischen Uhus (*Bubo* spp.) unter Berücksichtigung eines größtenteils parallel variierenden Polymorphismus innerhalb der Subspezies (Weick 1999). **f** *Bubo saturatus* und *heterocnemis*; **g** *saturatus* und *pacificus* (kleiner); **h** *pacificus*, *pallescens*, *nacurutu*, *deserti* und *mayensis*; **i, j** *elachistus* (klein!); **n** *pallescens*, *mayensis* und *nacurutu* (selten); **o** *mayensis*, *nacurutu* und *deserti*; **p** *mayensis* und *virginianus* (selten); **q** *B. v. nacurutu* and *B. magellanicus* (kleiner); **r** *nigrescens*

Part VI Teil VI

References Literaturverzeichnis

References/Literaturverzeichnis

For the most references the owl taxon covered is given.
Bei den meisten Literaturangaben ist zusätzlich das jeweils behandelte Eulen-Taxon angegeben.

Abdulali H (1965) The birds of the Andaman and Nicobar Islands. J Bombay Nat Hist Soc 61:534

Abdulali H (1967) The birds of the Nicobar Islands, with notes to some Andaman birds. J Bombay Nat Hist Soc 64: 139–190

Abdulali H (1972) A catalogue of birds in the collection of the Bombay Natural History Society. J Bombay Nat Hist Soc 11:102–129

Abdulali H (1978) The birds of Great and Car Nicobars. J Bombay Nat Hist Soc 75:749–772

Abdulali H (1979) A catalogue of birds in the collection of the Bombay Natural History Society. J Bombay Nat Hist Soc 75:744–772 (*Ninox affinis rexpimenti*)

Abs M, Curio E, Kramer P, Niethammer J (1965) Zur Ernährungsweise der Eulen auf Galapagos. Deutsche Galapagos Expedition, 1962/63. J Orn 1965(1):49–56

Adam CJG (1989) Eastern Screech Owl in Saskatchewan and adjacent areas. Blue Jay 47(3):164–188

Aharoni J (1931) Orn Monatsb 39(6):171–173 (*Strix butleri*)

Alexander B (1901) Bull Brit Ornith Cl 12:10 (*Glaucidium albiventer*)

Alexander F (1995) Is the Sokoke Scops Owl in the Shinba Hills? Kenya Birds 4(1):32, 33

Alfaro A (1905) Proc Biol Soc Wash 18:217 (*Cryptoglaux ridgwayi*)

Ali S (1949) Indian hill birds. Oxford University Press

Ali S (1953) The birds of Travancore and Cochin. Oxford University Press, Bombay

Ali S (1962) The birds of Sikkim. Oxford University Press, Madras

Ali S (1964) The book of Indian birds, 7th ed. Leaders Press Private Ltd., Bombay

Ali S (1969) The birds of Kerala. Oxford University Press, Madras

Ali S (1977) Field guide to the birds of the Eastern Himalayas. OUP, Dehli

Ali S (1978) Mystery birds of India. 3. Blewitt's Owl or Forest Spotted Owl. Hornbill 4–6

Ali S, Ripley SD (1968) Handbook of the birds of India and Pakistan. 1. Oxford University Press

Ali S, Ripley SD (1969) Handbook of the birds of India and Pakistan. 2 and 3. Oxford University Press

Ali S, Ripley SD (1970) Handbook of the birds of India and Pakistan. 4. Oxford University Press

Ali S, Biswas B, Ripley SD (1996) The birds of Bhutan. Zoological Survey of India, Occas. Paper, 136

Allen GM, Greenway JC jr (1935) A specimen of *Tyto (Heliodilus) soumagnei*. Auk 52:414–417

Allen RP (1961) Birds of the Carribean. Viking Press, NY

Allison (1946) Notes d'Ornith. Musée Hende, Shanghai, I, fasc. 2:12 (*Otus bakkamoena aurorae*)

Amadom D, Bull J (1988) Hawks and owls of the world. Checklist West Found Vertebr Zool

Amadon D (1953) Owls of Sao Thomé. Bull Am Mus Nat Hist 100(4)

Amadon D (1959) Remarks on the subspecies of the Grass Owl *Tyto capensis*. J Bombay Nat Hist Soc 56:344–346

Amadon D, du Pont JE (1970) Notes to Philippine birds. Nemouria 1:1–14

Amadon D, Eckelberry DR (1955) Observations on Mexican birds. Condor 57:65–80

Amadon D, Jewett SG Jr (1946) Notes on Philippine birds. Auk 63:551–558

Anders E (1987) Über den Trillerkauz. Gefiederte Welt 1987(10):275–277

Andrew A, Tremul P (1995) Observation of a Barking Owl and prey. Aust Raptor Assoc News 16(1):26

Andrew P (1993) The birds of Indonesia. Kukila Checklist, 1. Indonesian Ornithological Society Jakarta

Andrew P, Milton GR (1988) A note on the Javan Scops Owl, *Otus angelinae* Finsch. Kukila 3(3–4):79–81

Angell T (1974) Owls. University Wisconsin Press, 80 pp (North American owls)

Anonymus (2004) New owl from islands where no owls were known. World Birdwatch 26(4):2

Araya B, Chester S (1993) The birds of Chile – A field guide. Latour, Santiago

Archiv Vogelschutzwarte Karlsruhe (1999) Eulen verstehen. Bezirksstelle Naturschutz und Landschaftspflege Karlsruhe

Aronson L (1980) Hume's Tawny Owl *Strix butleri* in Israel. Dutch Birding 1:18, 19

Ash JS, Miskell JE (1983) Birds of Somalia, their habitat, status and distribution. Scopus Spec Suppl 1

Ash JS, Miskell JE (1998) Birds of Somalia. Pica Press

Atkinson PW, Koroma AP, Ranft R, Rowe SG, Wilkinson R (1994) The status, identification and vocalication of African fishing owls, with particular reference to the Rufous Fishing Owl *Scotopelia ussheri*. Bull ABC 1:67–72

Atkinson PW, Peet N, Alexander J (1991) The status and conservation of the endemic bird species of Sao Thomé and Principe, W Africa. Bird Conserv Int 1(3):255–282

Audubon JJ (1827–1838) Birds of America. (Faksimile 1981)

Audubon Mag., Russell F (1977) Die Wunder der Vogelwelt. Droemer

Austin OL (1948) The birds of Korea. Bull Mus Comp Zool 101:1–301

Austin OL, Kuroda N (1953) The birds of Japan, their status and distribution. Bull Mus Comp Zool 109:277–637

Austing GR (1957) The Saw Whet Owl. Nat Hist 66:154–158

Baird SF, Cassin J, Lawrence GN (1860) Birds of North America, Atlas

Baker ECS (1919) Bull Brit Ornith Cl 40:60 (*Carine brama fryi*)

Baker ECS (1926) Bull Brit Ornith Cl 47:58 (*Athene noctua ludlowi*), 59 (*Glaucidium cuculoides rufescens*), 60 (*Glaucidium cuculoides fulvescens*), (*Ninox scutulata isolata*)

Baker ECS (1927a) The fauna of British India. Birds 4. London, Striges, pp 383–458

Baker ECS (1927b) Remarks on oriental owls with description of four new races. Bull Brit Ornith Cl 47:58–61

Baker ECS (1935) Bull Brit Ornith Cl 56:36 (*Strix indranee shanensis*)

Baker RH (1951) The avifauna of Micronesia, its origin, evolution and distribution. Univ Kansas Publ Mus Nat Hist 3:1–159

Bangs O (1899) A new Barred Owl from Corpus Christi, Texas. Proc New Engl Zool Cl 1:31, 32 (*Syrnium nebulosum helveolum*)

Bangs O (1900) Auk 17:287 (*Speotyto cunicularia cavicola* = new name for *bahamensis*)

Bangs O (1907) An owl (*Rhinoptynx clamator*), added to the Costa Rica Ornis. Proc Biol Soc Wash 20:31, 32 (*Asio clamator forbesi*)

Bangs O (1908) Auk 25:316 (*Strix varia albogilva*)

Bangs O (1913) Proc New Engl Zool Cl 4:91 (*Gymnasio lawrencei exsul*)

Bangs O (1919) Proc New Engl Zool Cl 6:97 (*Asio flammeus sanfordi*)

Bangs O (1932) In: La Touche's Handb. Bds. E China 2(2):113 (*Bubo bubo inexpectatus*)

Bangs O, Noble GK (1918) Birds of Peru. Auk 40:448–449 (*Otus roboratus*)

Bangs O, Penard TE (1918) Bull Mus Comp Zool 62:51 (*Pulsatrix perspicillata trinitatis*)

Bangs O, Penard TE (1921) Description of six new subspecies of American birds. Proc Biol Soc Wash 34:89–92 (*Otus choliba luctisonus*)

Bangs O, Peters JL (1928) Bull Mus Comp Zool 68:330 (*Athene noctua impasta*)

Bannerman DA (1933) The birds of tropical W Africa. 3. Crown Agents for the Colonies

Bannerman DA (1953) Birds of West- and Equatorial Africa. 1. Oliver and Boyd

Bannerman DA, Bannerman WM (1968) History of the birds of the Cape Verde Islands. Birds of the Atlantic Islands 4, Edinburgh

Barbour Th (1912) Proc Biol Soc Wash XXIV:57 (*Hybris nigrescens noctividus*)

Barlow C, Wacher T, Disley T (1997) A field guide to the birds of the Gambia and Senegal. Pica Press

Barré N, Baran A, Jouanin C (1996) Oiseaux de la Réunion. Les Édit du Pacifique

Barrowclough GF, Guitiérrez RJ (1990) Generic variation and differentation in the Spotted Owl (*Strix occidentalis*). Auk 107:737–744

Barthel PH (1988) *Asio flammeus and Asio otus*. Limicola 2(1):1–21

Barton BS (1799) Fragment of the Natural History of Pennsylvania. 11 (*Strix varia*)

Basilio A (1963) Aves de la Isla deFernando Poo. Edit Coculsa

Batchelor T (1996) Breeding the African Grass Owl (*Tyto capensis*). Tyto 1(5):145, 146

Bates GL (1911) Bull Brit Ornith Cl 27:85, pl 7 (*Glaucidium pycrafti*)

Bates GL (1929) Bull Brit Ornith Cl 49:90 (*Jubula*)

Bates GL (1930) Handbook of the birds of West Africa. J. Bale and Son and Danielson

Bates GL (1937) Description of two new races of Arabian birds. Bull Brit Ornith Cl 57:150–151 (*Otus senegalensis pamelae*)

Bates RSP, Lowther EHN (1952) Breeding birds of Kashmir. Oxford University Press, Bombay

Bauer H-G, Berthold UP (1997) Die Brutvögel Mitteleuropas – Bestand and Gefährdung, 2. Aufl. Aula

Beavan RC (1867) Ibis 316 (*Ninox affinis*)

Bechstein JM (1791) Gemein Naturgesch Deutschland II:344 (*Otus palustris*)

Beck A (1990) Black and White Owl. Peregrine Fund Newsl 19:3

Becking JH (1994) On the biology and voice of the Javan Scops Owl, *Otus angelinae*. Bull Brit Ornith Cl 114:211–224

Beckon WN (1989) An undescribed form of owl in Fiji. Notornis 36(2):114–116

Beehler BM, Finch BW (1985) Species checklist of the birds of New Guinea. Australasian Ornith. Monographs 1. Royal Australasian Ornithological Union, Victoria, 126 pp

Beehler BM, Pratt TK, Zimmerman DA (1986) Birds of New Guinea. Princeton University Press

Belcher C, Smooker GD (1936) Birds of the Colony of Trinidad and Tobago. III. Ibis 13(6):1–35

Bell HL (1970) The Rufous Owl in New Guinea. Emu 70:31

Bellamy P (1989) Wildtiere aus aller Welt. Müller, Erlangen, 200 pp

Belthoff JR, Dufty AM (1995) Locomotor activity levels and dispersal of Western Screech Owl. Anim Behav 50(2): 558–561

Belton W (1984) Birds of Rio Grande do Sul, Brazil. Bull Am Mus Nat Hist 1:178(4)

Bendire CE (1891) Auk 8:140 (*Megascops asio macfarlanei*)

Benson CW (1960) Birds of the Comoro Islands. Ibis 103b:59–63 (*Otus rutilus mayottensis*), (*Otus pauliani*)

Benson CW (1981) Ecological difference between the Grass Owl *Tyto capensis* and the Marsh Owl *Asio capensis*. Bull Brit Ornith Cl 101:372–376

Benson CW, Brooke RK, Dowsett RJ, Stuart-Irwin MP (1971) The birds of Zambia. Collins
Benson CW, Stuart-Irwin MP (1967) The distribution and systematic of *Bubo capensis* Smith. Arnoldia Rh 3(19):1–19
Bent AC (1961) Life histories of North American birds - Owls. Dover (reprint), pp 140–482
Bertoni M, Bertoni W (1901) An Cient Paraguayos 1(1):174 (*Nyctale fasciata*), 175 (*Syrnium koeniswaldianum*)
Bertoni W (1901a) Av Nuevas Paraguay 177, 178 (*Syrnium borellianum*)
Bertoni W (1901b) An Cient Paraguayos 1(1):173 (*Nyctale Bergiana*), 179 (*Glaucidium ferox rufus*)
Bertoni W (1930) Rev Soc Cient Paraguay 2(6):243 (*Asio stygius* var. *baberoi*)
Bewick T (1809) History of British birds. Newcastle
Bezzel E (1985) Kompendium der Vögel Mitteleuropas. Nonpasseriformes, Aula
Bianchi VL (1906) Bull Brit Ornith Cl 16:69 (*Bubo bubo tibetanus*)
Biggs HC, Kemp AC, Mendelsohn HP, Mendelsohn JM (1979) Weights of southern African raptors and owls. Durban Mus Novit 12(17):73–813
Billberg GJ (1828) Syn Faun Scand 1, pt 2, tab A (*Tyto*)
Bird Life International (2004) Threatened Birds of the World (CD-ROM)
Bishop KD (1989) Little known *Tyto* owls of Wallacea. Kukila 4(1–2):37–43
Bishop LB (1931) Proc Biol Soc Wash 44:93–94 (*Bubo virginianus leucomelas*), (*Strix varia brunnescens*), 95 (*Strix varia albescens*)
Biswas B (1961) The birds of Nepal, 3. J Bombay Nat Hist Soc 58:63–134
Black CT (1936) A new striped owl from Tobago. Auk 53:82
Blackhall S (1995) A hawking owl. Aust Raptor Assoc News 16(1):18
Blake ER (1963) Birds of Mexico. University of Chicago Press
Blakiston W, Pryer H (1878) A catalogue of the Birds of Japan. Ibis (4)2:209–250, 246 (*Strix uralensis japonica*)
Blasius AWH (1888a) Braunschweig Anz 9, 11. Jan: 86 (*Ninox macroptera*)
Blasius AWH (1888b) Braunschweig Anz 52, 1. March: 467 (*Syrnium wiebkeni*)
Bloom PH (1983) Notes on the distribution and biology of the Flammulated Owl in California. Western Birds 14(1): 49–52
Bloxham A (1827) Voyage "Blonde" 1826. p 250 (*Strix sandwichensis*)
Blyth E (1845) J As Soc Beng 14(1):185 (*Syrnium nivicolum*)
Blyth E (1846) J As Soc Beng 15:8 (*Ephialtes spilocephalus*), 280 (*Athene malabaricus*)
Blyth E (1847) In Hutton: J As Soc Beng 16(2):776 (*Athene bactrianus*)
Blyth E (1850a) J As Soc Beng 18:801 (*Strix pusilla*)
Blyth E (1850b) J As Soc Beng 18:802 (*Strix parva*)
Blyth E (1852) Cat Birds Mus As Soc 1849(1):39 (*Glaucidium radiatum castanonotum*)
Blyth E (1862) Ibis 388 (*Strix affinis*)
Blyth E (1867) Ibis 313 (*Athene whitelyi*)
Bohmke BW, Macek M (1994) Breeding the Spectacled Owl. AFA Watchbird 21(5):4–7
Boie F (1822) Isis v Oken, 1, col 549 (*Athene*)
Boie F (1826) Isis v Oken, 2, col 970 (*Glaucidium*)
Bonaparte CLJL (1825) Am Orn 1:72 and pl 7 (*Strix hypugaea*)
Bonaparte CLJL (1838) Geogr and Comp Liest 7 (*Strix pratincola*), (*Nyctale richardsoni*)
Bonaparte CLJL (1850) Consp Gen Av I:40 (*Athene humeralis*), 41 (*Strix squamipila*), (*Strix hirsuta borneensis*), 42 (*Athene ocellata*), (*Athene jacquinoti*), 44 (*Scotopelia*), (*Strix peli*), (*Ciccaba myrtha*), 50 (*Strix sylvatica*), 52 (*Syrnium albitarse*), 53 (*Syrnium macabrum*), (*Syrnium squamulatum*)
Bonaparte CLJL (1854) Rev et Mag Zool II(6):253 (*Macabra*), 541 (*Gisella*), 544 (*Rhabdoglaux*), (*Phalaenopsis nana*)
Bonaparte CLJL (1855) Compt Rend Acad Sci Paris 41:654 (*Phalaenopsis jardinii*), 654 (*Spiloglaux theomacha*), 655 (*Ninox philippensis*)
Bond J (1928) On the birds of Dominica, St. Lucia, St. Vincent and Barbados. BWI Proc Acad Nat Soc Phil 80:523–545
Bond J (1942) Notes on the Devil Owl. Auk 59:308, 309
Bond J (1949) Identity of Trinidad barn owls. Auk 66:91
Bond J (1956) Checklist of the birds of the West Indies. Academy of Natural Sciences of Philadelphia, Philadelphia
Bond J (1975) Origin of the Puerto Rican Screech Owl *Otus nudipes*. Ibis 117:244
Bond J (1986) Birds of West Indies. Collins (reprint)
Bond J, Meyer de Schauensee R (1941) Description of new birds from Bolivia. 4. Not Nat 93:1–7 (*Otus guatemalae bolivianus*), (*Otus albigularis remotus*)
Bond J, Meyer de Schauensee R (1943) The birds of Bolivia. II. P Acad Nat Sci Phila 95:167–221
Borrow N, Demey R (2001) The birds of Western Africa. A & C Black
Bosakowski T (1987) Census of barred and spotted owls. In: Nero RW, Clark RJ, Knapton RJ, Hamre RH (eds) Biology and conservation of northern forest owls. Symp. Proc. Gen. Techn. Raptors. RM 142, US Dep. Agric. Forest Service, Fort Collins, pp 307, 308
Boshoff A (1987) More on the distribution and status of the Giant Eagle Owl in the Cape Province. Bee-Eater 38(2):18, 19
Bouchner M, Bárta D (1979) Greifvögel and Eulen. Dausien
Bourns FS, Worcester DC (1894) Occ Papers Minn Acad Sci 1:8 (*Ninox spilonotus*)
Bowden CCR, Andrews M (1994) Mount Kupe and its birds. Bull ABC 1(1):13–18 (*Glaucidium sjoestedti*)
Boyer T, Hume R (1991) Owls of the world. Dragons World
Brack C (1996) Breeding the Forest Eagle Owl (*Bubo nipalensis*) in the Berlin Zoo. Tyto 1(3):75, 76
Bradley (1962) Nat Hist Rennell Is IV:12 (*Tyto alba bellonae*)
Brandt T, Seebaß C (1994) Die Schleiereule. Aula
Brasil (1915) Rev France Orn 4:202 (*Tyto alba lifuensis*)
Brazil MA (1991) The birds of Japan. C. Helm
Brazil MA, Yamamoto S (1989a) Status and distribution of the owls of Japan. Rapt Mod World 389–401
Brazil MA, Yamamoto S (1989b) Behaviour ecology of Blakiston's Fish Owl *Ketupa blakistoni* in Japan. Rapt Mod World 403–410

Bree CR (1875/1876) A history of the birds of Europe not observed in the British Isles, vol 1. G. bell and Sons, London

Breeden LR, et al. (1983) Field guide to the birds of North America. Nat Geog Soc

Bregulla HL (1992) Birds of Vanuatu. A. Nelson, oswestry

Brehm AE (1857) Allg Deutsche Naturhist Zeit 3:440 (*Strix kirchhoffi*), (*Athene vidalii*)

Brehm CL (1828) Isis v Oken 1271 (*Nyctala*)

Brehm CL (1831) Handbuch der Naturgeschichte aller Vögel Deutschlands. Ilmenau, p 106 (*Strix guttata*)

Brehm CL (1855) Der vollständige Vogelfang. 37 (*Athene indigena*), 40 (*Strix splendens*), 43 (*Scops pygmea*), 413 (*Otus leucopsis*)

Brehm CL (1858a) Naumannia 214 (*Strix flammea obscura*)

Brehm CL (1858b) Vogelfang. 215 (*Strix adspersa*)

Bretagnolle V, Attié C (1996) Comments on a possible new species of scops owl *Otus* sp. on Réunion. Bull ABC 3(1):36

Brewster W (1882) Bull Nuttall Ornith Cl 7:31 (*Scops asio bendirei*)

Brewster W (1888) Auk 5:87 (*Megascops aspersus*), 88 (*Megascops vinaceus*), 136 (*Glaucidium gnoma hoskinsii*)

Brewster W (1891) Auk 8:139 (*Megascops asio aikeni*), 141 (*Megascops asio saturatus*)

Brewster W (1902) Bull Mus Comp Zool 41:93 (*Megascops asio xantusi*), 96 (*Bubo virginianus elachistus*)

Briffet C, Sutari SB (1993) The birds of Singapore. Times Edit., Singapore

Briggs MA (1954) Apparent neoteny in the Saw Whet Owl of Mexico and Central America. Proc Biol Soc Wash 67: 179–181 (*Aegolius acadicus brodkorbi*)

Brisson MJ (1760) Ornith 1:28 (*Asio*)

Brodkorb P (1937) Proc Biol Soc Wash 50:33 (*Otus choliba wetmorei*)

Brodkorb P (1938) Occ Pap Mus Zool U Michigan 394:3 (*Glaucidium brasilianum pallens*)

Brodkorb P (1941) Occ Pap Mus Zool U Michigan 450:1–4 (*Glaucidium brasilianum saturatum*)

Brook RV, Oatley TB, Hurly ME, Kuntz WD (1983) The South African distribution and status of the nominate race of the Barred Owl. Ostrich 54(3):173, 174

Brooke RK (1973) Notes on the distribution and food of the Cape Eagle Owl in Rhodesia. Ostrich 44:137–139

Brosse A, Erard C (1986) Les oiseaux des Rég. Forest du NE de Gabon 1

Brown CJ, Rickert BR, Morsbach RJ (1987) The breeding biology of the Asian Scops Owl. Ostrich 58(2):58–64

Brown LH (1970) African Birds of Prey. Collins

Brown LH (1976) Observation on Pel's Fishing Owl (*Scotopelia peli*). Bull Brit Ornith Cl 96(2):49–53

Browning MR (1982) Checklist of Strigiformes. 61 pp, unpublished

Browning MR (1990) Erreneous emmendations of names proposed by Hekstra (Strigidae, *Otus*). Proc Biol Soc Wash 103(2):452

Browning MR, Banks RC (1990) The identity of Pennant's "Wapacuthu Owl" and the subspec. name of the popul. of *Bubo virginianus* from west of the Hudson Bay. J Raptor Res 24(4):80–83

Bruce MD, Dowsett RJ (2004) The correct name of the Afrotropical mainland subspecies of Barn Owl *Tyto alba*. Bull Brit Ornith Cl 124(3):184–187

Brüll H (1984) Greifvögel und Eulen Mitteleuropas. Philler, Minden

Buchanan OM (1964) The Mexican races of the Least Pygmy Owl. Condor 66(2):103–112

Buchanan OM (1971) The Mottled Owl *Ciccaba virgata* in Trinidad. Ibis 103:105, 106

Buckholz PG, Edwards MH, Ong BG, Weir RD (1984) Difference by age and sex in the size of Saw Whet Oel. J Field Ornithol 55(2):204–213

Buckingham DL, Dutson GCL, Newman JL (1995) Birds of Manus. Cambridge

Buller WL (1882) History of the birds of New Zealand. 1. 2nd ed

Buller WL (1888) Birds of New Zealand. Suppl. 2

Buller WL (1904) On a new species of owl from New Zealand. Ibis XV:639 (*Sceloglaux rufifacies*)

Bullock (1990) Birds of the Republic of the Seychelles

Bunn DS, Warburton AB, Wilson RDS (1982) The Barn Owl. Poyser

Burg G (1921) Weidmann 9:6 (*Bubo bubo engadinensis*)

Burleigh TD (1958) Georgia birds. University of Oklahoma Press, 746 pp

Burton E (1836) Proc Zool Soc Lond 1935:152 (*Noctua brodiei*)

Burton JA, et al (1984) Owls of the world. P. Lowe

Burton JA, et al (1992) Owls of the world, 3rd ed. P. Lowe

Busch FD (1986) Brauner Fischuhu im Zoo Hannover gezüchtet. Gefiederte Welt 110(3):81

Busse H (1966) Der Falke 1966(13):250–251 (*Surnia ulula*)

Busse H (1968) Der Falke 1968(15):178–179 (*Bubo omissus*)

Busse H (1969) Der Falke 1969(8):256 (*Pulsatrix perspicillata perspicillata*)

Busse H (1976) Der Falke 1976(11):394–395 (*Bubo poensis*)

Busse H (1977) Der Falke 1977(6):214–215 (*Phodilus badius*)

Busse H (1978) Der Falke 1978(1):34 (*Glaucidium siju*)

Busse H (1984) Der Falke 31(4):142 (*Ketupa ketupu*)

Busse H (1986) Der Falke 76:242, 243 (*Scotopelia bouvieri*)

Busse H (1988) Der Falke 1988:314–315 (*Bubo capensis*)

Busse H (1989) Der Falke 11:324 (*Strix hylophila*)

Butler AL (1899) The birds of the Andaman and Nicobar Islands. J Bombay Nat Hist Soc 12

Butler TY (1979) The birds of Ecuador and the Galapagos Archipelago. The Rhamphastos Agency

Büttikofer J (1889) Notes Leyden Mus 11:34 (*Bubo letti*)

Büttikofer J (1896) Notes Leyden Mus 18:165 (*Ketupa minor*)

Buturlin SA (1907a) J Orn 55:233, 235 (*Syrnium uralense nikolskii*)

Buturlin SA (1907b) Psovaia I Ruzheinaia Ochota 13:87 (*Syrnium cinereum sakhalinense*), (*Nyctala magna*), (*Nyctala caucasica*)

Buturlin SA (1907c) Orn Monatsb 15:100 (*Surnia ulula pallasi*)

Buturlin SA (1908) J Orn 56:287 (*Bubo bubo jakutensis*), (*Nyctala jakutorum*)

Buturlin SA (1910a) Mess Orn 1:119 (*Scops semitorques ussuriensis*), 260 (*Scops scops sibirica*)

Buturlin SA (1910b) Nascha Ochota 4:10 (*Microscops*), 11 (*Cryptoglaux tengmalmi transvolgensis*), 11 (*Cryptoglaux tengmalmi sibirica*)
Buturlin SA (1910c) Mess Orn 187 (*Ninox scutulata ussuriensis*)
Buturlin SA (1911) Mess Orn 26 (*Bubo bubo yenniseensis*)
Buturlin SA (1912) Nascha Ochota 45 (*Scops scops ferghanensis*), 46 (*Scops scops irtyshensis*)
Buturlin SA (1915) Mess Orn 6:133 (*Strix uralensis yenisseensis*)
Buturlin SA (1928) Opredelitel Ptits SSSR 114 (*Bubo bubo tauricus*), (*Bubo bubo tarimensis*)
Butynski JM, Arenonga U, Ndera B, Hart JF (1996) The world rarest owl. Owls Mag Winter NY:2–4 (also front- and backcover)
Butynski JM, Arenonga U, Ndera B, Hart JF (1997) Rediscovery of the Congo Bay (Itombwe) Owl *Phodilus prigoginei*. Bull ABC 4(1):32–35
Byers C (1992) Scops Owl (*Otus scops*) and Striated Owl (*Otus brucei*). Birding World 5(3):107–110
Cabanis JL (1855) Dr. Gundlachs Beiträge zur Ornithologie Cuba's. Strigidae. J Orn 3:465–467 (*Gymnoglaux*)
Cabanis JL (1872) J Orn 20:316 (*Strix amauronota*)
Cabanis JL (1875) J Orn 22:26 (*Scops obsoleta*)
Calburn S, Kemp A (1987) The owls of Southern Africa. Cape Town
Caldwell HR, Caldwell JC (1931) South China Birds. p 232 (*Tyto longimembris albifrons*)
Campbell B (1974) The dictionary of birds in colour. M. Joseph, 352 pp
Campbell B, Lack E (eds) (1985) A dictionary of birds. Poyser
Campbell KL (1977) Observation on the Fishing Owl *Scotopelia peli* on the Tana River. Kenya Bull East Afr Nat Hist Soc 1977:36, 37
Canevari M, Canevari P, Carrizo GR, Harris G, Mata JR, Straneck RJ (1991) Nueva Guia de las Aves Argentinas, vol 1 and 2. Fundacion Acindar
Cardiff SW (1980) Status and distribution of Elf Owl in California. Unpublished
Cardoso da Silva JM, Coelho G, Gonzaga LP (2002) Discovered on the brink of extinction: A new species of Pygmy-Owl (Strigidae: *Glaucidium*) from Atlantic Forest of NE Brazil. Ararajuba 10(2): 123–130
Carlyon J (1985) Rediscovery of the Barred Owl in the Eastern Cape. Afr Wildl 39(1):22, 23
Carlyon J, Meakin P (1985) Barred owls in the Eastern Cape. Bokmakierie 32(2):44, 45
Carriker MA (1910) An annotated list of Birds of Costa Rica, incl. Cocos Islands. Ann Carnegie Mus 6
Carriker MA (1935) P Acad Nat Sci Phila 87:313 (*Ciccaba minima*)
Cassin J (1849) P Acad Nat Sci Phila 4:121 (*Ephialtes sagittatus*), (*Ephialtes Watsonii*), 124 (*Syrnium albo-gularis*), 124 (*Syrnium virgatum*), 157 (*Nyctale Harrisii*)
Cassin J (1853) P Acad Nat Sci Phila 6:185 (*Ephialtes elegans*), 186 (*Ephialtes hendersoni*)
Cassin J (1854) Illustrations of the birds of California, Texas, Oregon, British and Russian America, 6.178 (*Bubo virginianus* var. *pacificus*), 180 (*Scops mccalli*)
Cassin J (1862) Birds of California and Texas. Lippincott
Castro I, Phillips A (1996) A guide to the birds of the Galapagos Islands. C. Helm
Catterall M (1997) Results of the 1996 bird survey of Buton Island, Sulawesi, Indonesia. Ecosurveys, Spilby
Cayley NW (1929) Emu 28:162 (*Ninox ooldeanensis*), (*Ninox yorki*)
Cayley NW (1931) What bird is that? A guide to the birds of Australia. Illustrated by the Author. p 32 and pl 5 (*Tyto novaehollandiae troughtoni*)
Cayley NW (1961) What bird is that? Angus and Robertson (reprint)
Chafer CJ (1992) Observations of the Powerful Owl *Ninox strenua* in the Illawarra and Shoalhaven regions of New South Wales. Aust Bird Watcher 14(8):289–300
Chafer CJ, Anderson M (1994) Sooty Owls in the Hacking River catchment. Aust Birds 27(3):77–84
Chapin JB (1930) Geographic variation on the African Scops Owl. Am Mus Novit 412:1–11 (*Otus scops graueri*)
Chapin JB (1932) Am Mus Novit 570:3 (*Glaucidium tephronotum medje*), (*Glaucidium tephronotum lukolelae*)
Chapman FM (1914) Bull Am Mus Nat Hist 33:318 (*Speotyto cunicularia punensis*)
Chapman FM (1915) Bull Am Mus Nat Hist 34:370 (*Asio flammeus bogotensis*)
Chapman FM (1917) Bull Am Mus Nat Hist 36: 254
Chapman FM (1922) Descriptions of apparently new birds from Colombia, Ecuador and Argentina. Am Mus Novit 31:1–8 (*Ciccaba aequatorialis*), (*Glaucidium brasilianum tucumanum*)
Chapman FM (1923) Am Mus Novit 67:1 (*Ciccaba albigularis meridensis*)
Chapman FM (1926) The distribution of birdlife in Ecuador. Bull Am Mus Nat Hist 55:1–784
Chapman FM (1928) Am Mus Novit 332:3 (*Otus guatemalae napensis*)
Chapman FM (1929) New birds from Mt. Duida, Venezuela. Am Mus Novit 380:1–11 (*Otus choliba duidae*), (*Glaucidium brasilianum duidae*), (*Glaucidium brasilianum ucayalae*)
Chapman FM (1931) The upper bird life of Mts. Roraima and Duida. Bull Am Mus Nat Hist 63:1–135
Chapman FM (1939) The upper zonal birds of Mt. Auyan-Tepui, Venezuela. Am Mus Novit 1051:6 (*Glaucidium brasilianum olivaceum*)
Chasen FN (1935) Bull Raffles Mus 11:84 (*Bubo ketupu büttikoferi*), 86 (*Otus spilocephalus vandewateri*)
Chasen FN (1937) Treubia 16:216 (*Phodilus badius parvus*)
Chasen FN (1939) The birds of the Malay Peninsula, vol 4. Witherby
Chasen FN, Kloss CB (1926) Ibis 279 (*Otus bakkamoena mentawi*)
Cherrie GK, Reichenberger E (1921) Am Mus Novit 27:1 (*Strix chacoensis*)
Chew HH (1999) SCRO Mag 1999(2):53–56
Chubb C (1916a) Birds Brit Guiana I:290 (*Ciccaba superciliaris macconnelli*)
Chubb C (1916b) Bull Brit Ornith Cl 36:46 (*Ciccaba albitarse goodfellowi*), (*Asio galapagoensis aequatorialis*)

Claffey PM (1996) The status of Pel's Fishing Owl *Scotopelia peli* in Togo. Bénin Gap Bull ABC 4(2):135, 136

Clancey PA (1968) Durban Mus Novit 8(11):119 (*Glaucidium perlatum diurnum*)

Clark AH (1907) Proc U S Nat Mus 32:470 (*Bubo tenuipes*), 471 (*Syrnium ma*), (*Syrnium uralense japonicum*), 472 (*Syrnium uralense hondoensis*)

Clark RJ, Mikkola H (1989) A preliminary revision of threatened and near-threatened nocturnial birds of prey in the world. Rapt Mod World 371–388

Clark RJ, Smith DG (1996) Checktable of Strigiformes systematics. (19 pp in letter)

Clark RJ, Smith DG, Kelso LH (1978) Working bibliograpy of owls of the world. Nat Wildlife Fed Raptor Inform Center, Technic Ser 1

Clements J (1981) Birds of the world: A checklist. Croom Helm

Coates BJ (1976) Birds of Papua New Guinea. Port Moresby

Coates BJ (1985) The birds of Papua New Guinea, vol 1, Non-Passerines. Alderly

Coates BJ, Bishop KD, Gardner D (1997) A guide to the birds of Wallacea. Dove

Coats S (1979) Species status and phylogenetic relationship of the Andean Pygmy Owl *Glaucidium jardinii*. Amer Zool 19(3):892

Cole (1906) Bull Mus Comp Zool 50:125 (*Otus choliba thompsoni*)

Collar NJ, Crosby MJ, Stattersfield AJ (1994) Birds to watch, vol 2: The world list of threatened birds. Bird Life International, Cambridge, UK

Collins CT (1963) Notes on the feeding behaviour, metabolism and weight of the Saw Whet Owl. Condor 65:528–530

Connor J (1988) Update Spotted Owl. Living Bird Q Autumn:32–34

Cooper N, Cooper M (1992) Spotted Eagle Owl. Promerops 202:12

Cooper W (1861) Proc Calif Acad Sci 2:118 (*Athene whitneyi*)

Cory CB (1886) Auk 3:471 (*Speotyto dominicensis*)

Cory CB (1915) Field Mus Nat Hist Pub Orn 1:298 (*Otus choliba margaritae*), 299 (*Speotyto cunicularia arubensis*), 299 (*Speotyto cunicularia beckeri*), 300 (*Speotyto cunicularia intermedia*)

Cory CB (1918) Field Mus Nat Hist Pub Zool Ser 13(2):40 (*Speotyto cunicularia minor*)

Coues E (1866) P Acad Nat Sci Phila 51 (*Micrathene*)

Coues E (1889) Auk 4:71 (*Micropallas*)

Coues E (1899) Osprey 3:144 (*Psiloscops flammeola*)

Cracraft J (1983) Species concepts and speciation analysis. Curr Orn 1:159–187

Craighead JJ, Craighead FC (1956) Hawks, owls and wildlife. Stackpole

Cramp S, et al. (1985) Handbook of the birds of Europe, the Middle East and North Africa, vol 6. Oxford University Press

Creaman (ed) (n.y.) D'après Natur de 1450 à 1800. Catalogue, Strasbourg

Cuello J, Gerzenstein E (1962) Las Aves del Uruguay. Comm Zoológ del Museo de Historia Natural de Montevideo 6(93):191

Cusa N (1984) Tunnicliffe's birds. Measured Drawings. V. Gollancz

Cuvier FG (1829) Regne Anim I:52, pl 20 (*Otus mexicanus*)

Czechura G, Debus S (1997) Australian raptor studies II. Birds Australia Monograph 3, Birds Australia

d'Orbigny AD (1839) In La Sagra's Hist. fis., pol. y nat. Isla de Cuba 3, Aves: 33, 40, pl 2, (*Otus siguapa*), 41 (*Noctua siju*)

da Silva JMC, Coelho G, Gonzaga LP (2002) Discovered on the brink of extinction: A new species of Pygmy Owl, NE Brazil. Ararajuba 10(2):123–130 (*Glaucidium mooreorum*)

Dabbene R (1926) Tres aves nuevas para la avifauna uruguaya. Hornero 3:395 (coll. *Aegolius harrisii*)

Dahl F (1899) Mitt Zool Samml Mus Nat Berlin 1:163

Dance SP (1991) Die schönsten Naturgrafiken: Raubvögel. Swan

Daskam T (1977) Aves de Chile. Lord Cochrane S.A.

Dathe (1985) Der Falke 1985(11):394

Daudin FM (1800) Traite d Orn 2:190 (*Strix huhula*), 199 (*Strix nudipes*), 206 (*Strix phalaenoides*), 207 (*Strix cristata*)

Davey SM (1993) Notes on the habitats of few Australian owl species. In: Olsen (ed) Australian raptor studies. Australas Raptor Assos, pp 126–142

David A, Oustalet EM (1877) Les Oiseaux de Chine, atlas. Paris

Davidson PS, Stones T (1993) Birding in the Sula Islands. Bull Orient Bird Cl 18:59–63

Davis LJ (1972) A field guide to the birds of Mexico and Central America. University of Texas Press, Austin

de Groot RS (1983) Origin status and ecology of the owls of Galápagos. Ardea 71:167–182

de Jong J (1995) De Kerkuil en andere in Nederland vorkomende Uilen. Friese Pers Boekerij

de Naurois R (1975) Le «Scops» de l'Ile Sao Thomé *Otus hartlaubi* (Giebel). Bonn Zool Beitr 26:319–355

de Naurois R (1982) Le statut de l'effraie de l'archipel de Cap Vert, *Tyto alba detorta*. Riv Ital Orn 52:154–166

de Roo AEM (1966) Description of the juvenile plumage of *Bubo shelleyi* (Sharpe and Ussher). Rev Zool Bot Afr 73(3–4):385–389

de Vis CW (1887) Proc Linn Soc New S Wales 2(1)(1886):1135 (*Ninox boobook* var. *lurida*)

de Vis CW (1889) Report Science Expedition Bellenden-Ker Range. 84 (*Ninox lurida*)

de Vis CW (1890) Ann Rept Brit New Guinea 1888/89:58 (*Ninox goodenoviensis*)

Dean K (2004) In: Robert Bateman (ed) Vögel. Komet, Köln

Debus SJS (1993) The mainland Masked Owl *Tyto novaehollandiae*, a review. Aust Bird Watcher 15(4):168–191

Debus SJS (1994) The Sooty Owl, *Tyto tenebricosa* in New South Wales. Aust Birds 28(suppl):4–19

Debus SJS, Maciejewski SE, Mc Allen LAW (1998) The Grass Owl in New South Wales. Aust Birds 31:29–45

Dee TJ (1986) The endemic birds of Madagascar. ICBP, Cambridge

Deignan HG (1941) Auk 58:396 (*Athene brama mayri*)

Deignan HG (1945) The birds of Northern Thailand. Smithonian Institute, US Nat Mus Bull 186

Deignan HG (1950) The races of the Collared Scops Owl, *Otus bakkamoena* Pennant. Auk 67:189–201

Deignan HG (1951) Proc Biol Soc Wash 64:41 (*Ninox scutulata randi*)
Deignan HG (1957) Proc Biol Soc Wash 70:43 (*Otus bakkamoena lemurum*)
Deignan HG (1961) Type specimen of birds of the USNM. Smithonian Institute, US Nat Mus Bull 221:138–158
del Hoyo J, Elliott A, Sargatal J (1999) Handbook of the birds of the world, 5. Lynx Ed., Barcelona
Delacour J (1926) Bull Brit Ornith Cl 47:11 (*Ketupa ceylonensis* sic *orientalis*), (*Strix newarensis laotianus*)
Delacour J (1930) Ois 11:654 (*Strix leptogrammica ticehursti*)
Delacour J (1941) On the subspecies of *Otus scops*. Zoologica 26:133–142
Delacour J (1947) Birds of Malaysia. Macmillen, New York
Delacour J, Jabouille P (1930) Ois 11:406 (*Strix leptogrammica orientalis*)
Delacour J, Jabouille P (1931) Les Oiseaux de L'Indochine Francaise, 2. Paris
Delacour J, Mayr E (1945) Notes on the taxonomy of Philippine birds. Zoologica 30:107–109 (*Ninox philippensis centralis*), (*Ninox philippensis proxima*)
Delacour J, Mayr E (1946) Birds of the Philippines. Owls 112–119
Delin H, Svensson L (1988) Der Kosmos Vogelatlas
Dementiev G (1931) Alauda 3:361 (*Bubo bubo eversmanni*), 364 (*Bubo bubo auspicabilis*)
Dementiev G (1933a) Alauda 4(1932):392 (*Bubo bubo omissus*), 394 (*Bubo bubo inexpectatus*)
Dementiev G (1933b) Sur le position systématique de «*Bubo doerriesi*» Seebohm. Alauda 5:383–388, 339 (*Strix aluco siberiae*)
Dementiev G (1951) Ptitsy Sovietskogo Soiuza I:417 (*Strix uralensis buturlini*)
Dementiev G (1952) Bull Mosc Soc Natur Biol Ser 57(2):91 (*Bubo bubo ognevi*)
Dementiev GP, Gladkov NA (1951) The birds of the Sovjet Union, 1. (Engl. ed. transl. Jerusalem 1966), Striges, pp 380–477
des Murs MGPO, Prévost F (1846) Rev Zool 242 (*Bubo dilloni*)
Dickerman RW (1991) On the validity of *Bubo virginianus occidentalis* Stone. Auk 108(4):964, 965
Dickerman RW (1993) The subspecies of the Great Horned Owl of the Great Planes, with notes on adjacent areas. Kansas Ornith Soc Bull 44(2):17–21
Dickerman RW (1997) The era of A. R. Phillips. Mus of SW Biol, University of New Mexico (*Strix occidentalis juanphillipsae*)
Dickerman RW (2004) Notes on the type of *Bubo virginianus scalariventris*. Bull Brit Ornith Cl 124(1):5, 6
Dickey DR, van Rossem AJ (1938) The birds of El Salvador. Zool Soc FMNH 23:1–609
Dickinson EC (1975) The identity of *Ninox scutulata* Raffles. Bull Brit Ornith Cl 95(3):104, 105
Dickinson EC, Kennedy RS, Parkes KC (1991) The birds of the Philippines. BOU Checklist 12, Tring
Dickinson EC, Pearson D, Remsen V, Rooselaar K, Schodde R (2003) The Howard and Moore complete checklist of the birds of the world, 3rd ed. Pica Press
Diggles (1866) Orn Austral pt 7 (*Strix walleri* = *Tyto longimembris longimembris*)
Dodsworth PTL (1913) The Himalayan Wood Owl (*Syrnium nivicola* Hodgs.). J Bombay Nat Hist Soc 22(3):626–629
Domaniewski J (1933) Acta Orn Zool Mus Polonici 1:79 (*Bubo bubo paradoxus*)
Dossenbach HD (1972) Vögel ferner Länder. Lausanne, 174 pp
Downer A, Sutton R (1990) Birds of Jamaica. Cambridge University Press
Dowsett-Lemaire F (1995) A comment on the voice and status of Vermiculated Fishing Owl *Scotopelia bouvieri* and a correction to Dowsett-Lemaire (1992) on the Maned Owl *Jubula letti*. Bull ABC 3(2):134, 135
Dresser HE (1871–1896) A history of the Birds of Europe, vol 6
Dresser HE (1902) Manual palaearctic birds, I
du Pont JE (1971) Philippine birds. D(el) MNH
du Pont JE (1972) Nemouria 6:6 (*Ninox philippensis ticaoensis*)
Dubois (1902) Syn Av 2:900 (*Strix cabrae*)
Duckett JE (1991) Management of the Barn Owl (*Tyto a. javanica*), as a predator of rats in oil palm plantations in Malaysia. Birds Prey Bull 4:11–23
Dugmore P (1997) The milky Eagle Owl (*Bubo lacteus* Temm.) in the wild and in the confinement. Tyto 1(6): 167–199
Duméril AMC (1806) Zool Anal 34 (*Surnia*)
Dunajewski A (1940) Ann Mus Natl Hungar 33, zool.: 99 (*Strix uralensis carpathica*)
Dunajewski A (1948) Bull Brit Ornith Cl 68:130 (*Strix aluco volhyniae*)
Duncan JR (2003) Owls of the world. Their lives, behaviour and survival. K. Porter, Canada, 319 pp
Dunning JB Jr (1993) CRC handbook of avian body masses. CRC Press, pp 94–100
Dunning JS (1982) South American land birds. Harrowood Books, New Providence
Earhart CM, Johnson NK (1970) Size dimorphism and food habits of North American owls. Condor 72:251–264
Eastwood C (1995) Manus – a trip report. Muruk 7(2):53–55
Eates KR (1938) A note on the resident owls of Sindh. J Bombay Nat Hist Soc 40
Eck S (1968) Der Zeichenparallelismus der *Strix varia*. Zool Abh SMTD 29:283–288
Eck S (1970) Über *Ninox rudolfi*. Zool Abh SMTD 30: 137–140
Eck S (1971) Katalog der Eulen des SMTD. Zool Abh SMTD 30(15):173–218
Eck S (1972) Zool Abh SMTD 30(15)(1971):211 (*Strix virgata sneiderni*)
Eck S (1973) Katalog der ornith Sammlung des ZIKMUL, Strigidae. Zool Abh 32(10):155–169
Eck S (1975) Der Falke 1975(10):351
Eck S, Busse H (1973) Eulen. NBB, Wittenberg
Edwards EP (1972) A field guide to the birds of Mexico. E. P. Edwards, Sweet Briar
Eisenmann E (1955) The species of Middle American birds. Trans Linn Soc NY 7:1–28

Elgood JH, Heigham JB, Moore AM, Nason AM, Sharland RE, Skinner NJ (1994) The birds of Nigeria. An annotated checklist. BOU Checklist
Elliot DG (1867) P Acad Nat Sci Phila 99 (*Scops kennicotti*)
Engelmann F (1928) Die Raubvögel Europas, I. Striges 85–144, Neumann-Neudamm
Enriquez-Rocha P, Rangel-Salazar JL, Holt DW (1994) Distribution of Mexican Owls. Rapt Conserv Today 567–574
Epple W, Rogl M (1989) Die Schleiereule. Luzern
Erard C, Roux F (1983) La Chevêchette du Cap *Glaucidium capense* dans l'ouest Africain. Ois Rev France Orn 53:97–104 (*Glaucidium capense etchecopari*)
Etchécopar RD, Hüe F (1967) The birds of North Africa from the Canary Islands to the Red Sea. Oliver and Boyd
Etchécopar RD, Hüe F (1978) Les Oiseaux de Chine, de Mongolie et de Corée. Non Passerif. Boubée
Evans TD (1997) Preliminary estimates of the population density of the Sokoke Scops Owl *Otus irenae* Ripley, in the East Usambara lowlands, Tanzania. Afr J Ecol 35(4):303–311
Evans TD, Watson LG, Hipkiss AJ, Kiure J, Timmins RJ, Perkin AW (1994) New records of Sokoke Scops Owl *Otus irenae*, Usambara Eagle Owl *Bubo vossleri* from Tanzania. Scopus 18:40–47
Everett M (1977) A natural history of owls. Hamlyn
Eversmann EF (1835) Addenda Pallas Zoogr, fasc. I:3 (*Strix turcomana*)
Eyton TC (1845) Ann Mag Nat Hist 16:228 (*Athene malaccensis*)
Falla RA (1948) Australian Barn Owl (*Tyto alba*), in classified summarised notes. N Z Bird Notes 2:171
Falla RA, Sibson RB, Turbott EG (1966) A field guide to the birds of New Zealand and outlaying islands. Collins
Falla RA, Sibson RB, Turbott EG (1978) Birds of New Zealand. Collins Field Guide
Fanchett et al. (2000) Ibis 2000:485–486
Fergusson-Lees J, Faull E (1992) Endangered birds. Philip
Ffrench R (1973) A guide to the birds of Trinidad and Tobago. Livingston Publ., Wynnewood, PA
Finsch O (1898/1899) Über *Scops magicus* und die verwandten Arten. Notes Leyden Mus 20:163–184 and pl 10–11
Finsch O (1906a) A new owl from Java. Ibis 17:2
Finsch O (1906b) Bull Brit Ornith Cl 16:63 (*Syrnium bartelsi*)
Finsch O (1912) Über eine neue Art Zwergohreule von Java. Orn Monatsb 20:156–159 (*Pisorhina angelinae*)
Fischer GA (1812) Mem Soc Imp Nat Moscou 3:276 (*Strix torquata*)
Fisher J (1990) Thorburn's birds. Mermaid Books
Fitter R (Niethammer G, German ed) (1973) Buch der Vogelwelt Mitteleuropas. Das Beste
Fitzpatrick JW, O'Neill JP (1986) *Otus petersoni*, a new Screech Owl from the Eastern Andes, with systematic notes on *O. colombianus and O. ingens*. Wilson Bull 98(1):1–14
Fjeldsa J, Krabbe N (1990) Birds of the High Andes. Apollo Books
Fleay D (1934) Notes for identification of the Winking Owl. Emu 33:174
Fleay D (1940) The Barking Owl mystery. Vict Nature 57: 71–95
Fleay D (1944) Watching the Powerful Owl. Emu 44:97–112
Fleay D (1949) The Tasmanian Masked Owl. Emu 48: 160–176
Fleay D (1955) The Tasmanian Masked Owl. Emu 55: 203–210
Fleming C (1982) G. E. Lodge - Unpublished bird paintings (New Zealand birds painted by Lodge). M. Joseph
Fleming JH (1916) The Saw Whet Owl of the Queen Charlotte Is. Auk 33:420–423 (*Cryptoglaux acadicus brooksi*)
Fleming RL, Traylor MN (1961) Notes on Nepal birds. Fieldiana Zool 35(8):441–487
Fleming RL, Traylor MN (1968) Distributional notes on Nepal birds. Fieldiana Zool 53(3):145–203
Fletcher BS (1998) A breeding record for Minahassa Owl *Tyto inexspectata* from Dumoga-Bone-National Park, Sulawesi. Forktail 14:50–81
Flint VE, Boehme RL, Kostin YV, Kusnetsov AA (1984) A field guide to birds of the USSR. Princeton University Press
Forbush EH, May JB (1955) Natural history of American birds, E and C North America. Bramhall House, NY
Ford NL (1967) A systematic study of the owls, based on comparative osteology. Ph. D. thesis. University Michigan, Ann Arbor
Forster JR (1772) Philos Trans 62:384 (*Strix brachyotis*). 424 (*Strix nebulosa*)
Franklin J (1831) Proc Comm Zool Soc London 1830–1831:15 (*Otus bengalensis*), 115 (*Noctua indica*)
Franson JC, Little SE (1996) Diagnostic findings in 132 Great Horned Owls. J Raptor Res 25(4):163
Fraser L (1843) Proc Zool Soc Lond X:189 (*Strix poensis*)
Fraser L (1854) Proc Zool Soc Lond (1853) 13 (*Bubo poensis*)
Frederiksson R (1993) Die Bestimmung junger Eulen im Ästlingsstadium. Limicola 7/6:285–310
Freethy R (1992) Owls, a guide for ornithologists. Bishopsgate
Fremlin B (1986) A wildlife heritage of Western Australia. St. George Books
Friedmann H (1929) Auk 46:521 (*Otus senegalensis caecus*)
Friedmann H (1949a) A new heron and a new owl from Venezuela. Smiths Misc Coll 111(9):2–3 (*Asio flammeus pallidicaudas*)
Friedmann H (1949b) The birds of North- and Middle America, XI. US Nat Mus Bull 50:(XIII + 793)
Friedmann H, Deignan HG (1939) Notes on some Asiatic owls of the genus *Otus*, with description of a new form. Journ Wash Acad Sci 29:287–291
Friedmann H, Deignan HG (1941) Journ Wash Acad Sci 29:287 (*Otus sunia distans*)
Friedmann H, Keith S (1968) First specimen of *Otus scops turanicus* (Loudon) from Africa. Bull Brit Ornith Cl 88(6):112
Frisch JD (1981) Aves Brasileiros, I. Verona
Frith C, Frith D (1985) Australian tropical birds. National Library Australia
Frith HJ, et al (1983) Readers Didest complete book of Australian birds. Sydney
Fry CN, Keith S, Urban EK (1988) The birds of Africa, 3. Academic Press
Fuertes LA (1922) Birds of Prey. Nat Geog Soc 37:460–467

Gallagher MD, Roger TD (1980) On some birds of Dhofar and other parts of Oman. Muscat Oman Stud Spec Report 2:369–371 (Strigidae: *Otus pamelae*)
Gallagher MD, Woodlock MW (1980) The birds of Oman. London
Gallagher T (1998) Lost and found. Living Bird Q 1998/Spring: 24–28
Gargett V (1977a) Notes on the Cape Eagle Owl *Bubo capensis mackinderi* in the Matopos, Rhodesia. Ostrich 48(1, 2):41, 42
Gargett V (1977b) Searching for Mackinder's Eagle Owl in the Matopos. Honeyguide 91:20–24
Garrido OH (1978) Nuevo récord de la Lucheza Norteamericana *Tyto alba pratincola* (Bp.) en Cuba. Misc Zoo 7:1–4
Garrido OH (2001) Una nueva subespecie del Sijú de Sabana *Speotyto cunicularia* para Cuba. Cotinga 15:75–78 (*Speotyto cunicularia guatanamensis*)
Gatter (1997) Birds of Liberia. Wiesbaden
Gavrilenko NI (1928) Sbirnik Poltawsk Muz 1:279 (*Bubo bubo nativus*)
Geoffroy Saint Hilaire J (1830) Ann Sci Nat Zool 21:199 (*Phodilus*)
Gerber R (1960) Die Sumpfohreule. NBB-Ziemsen, Wittenberg
Géroudet P (1979) Les Rapaces Diurnes et Nocturnes D'Europe, 5th edn (reprinted 1951–1957). Delachaux and Niestlé
Giebel CGA (1872) Thes Ornith 1:448 (*Noctua hartlaubi*)
Giglioli EH (1900) Avicula 4:57 (*Athene chiaradiae*)
Gilbert Pearson T (1936) Birds of America. Garden City Books
Gill LE (1964) A first guide to the South African birds. Miller
Gilliard ET (1940) Am Mus Novit (1071) 5 (*Speotyto cunicularia apurensis*)
Ginn PJ, McIlleron WG, le S Milstein P (1989) The complete book of Southern African birds. Struik
Glauert L (1945) Emu 44:229, 230 (*Tyto maculosa*)
Gloger CWL (1833) Das Abändern der Vögel durch Einfluss des Klimas. Nach zoologischen, zunächst von den europäischen Landvögeln entnommenen Beobachtungen dargestellt, mit den entsprechenden Erfahrungen bei den europäischen Säugthieren verglichen. Breslau, 142 (*Strix sibirica* "Licht.")
Gloger CWL (1842) Gemein Handb und Hilfsb 1841:226 (*Speotyto*)
Glutz v Blotzheim UN (1962) Die Brutvögel der Schweiz. Aargauer Tagblatt
Glutz v Blotzheim UN, Bauer KM (1980) Handbuch der Vögel Mitteleuropas, 9
Gmelin JF (1788) Syst Nat 1, pt 1:287 (*Strix virginiana*), 289 (*Strix naevia*), and (*Strix brasiliana*), 290 (*Strix wapacuthu*), 291 (*Strix cinerea*), 295 (*Strix javanica*), 287 (*Strix zeylonensis*), 288 (*Strix americana*), 296 (*Strix acadica*), 296 (*Strix novae Seelandiae*)
Gmelin SG (1771) N Comm Petrop 15:447, pl 12 (*Noctua minor*)
Godfrey R (1948) Can Field-Nat 61(6)(1947):196 (*Asio otus tuftsi*)
Goodman SM, Creighton GK, Raxworthy CJ (1991) The food habits of the Madagascar Long eared Owl *Asio madagascariensis* in SE Madagascar. Bonn Zool Beitr 42(1):21–26
Goodman SM, Sabry H (1984) A specimen record of Hume's Tawny Owl *Strix butleri* from Egypt. Bull Brit Ornith Cl 104:79–84
Goodman SM, Thorstrom R (1998) The diet of the Madagascar Red Owl (*Tyto Soumagnei*) in the Masuala Peninsula, Madagascar. Wilson Bull 110:417–421
Gosler A (ed) (1991) Die Vögel der Welt. Franckh-Kosmos
Gosse PH (1847) Birds Jamaica 19 (*Ephialtes grammicus*)
Gould GI (1977) Distribution of the Spotted Owl in California. Western Birds 8:131–146
Gould J (1831–1832) A century of birds from the Himalayan Mountains. Hullmandel
Gould J (1832–1837) The birds of Europe. Hullmandel
Gould J (1837a) Proc Zool Soc Lond 10 (*Otus (Brachyotus) galapagoensis*)
Gould J (1837b) Proc Zool Soc Lond (1836) 140 (*Strix delicatula*), (*Strix castanops*)
Gould J (1838a) Syn Birds Austr 3: pl 47 (*Athene ? strenua*)
Gould J (1838b) Proc Zool Soc Lond 15:99 (*Athene leucopsis*)
Gould J (1845) Proc Zool Soc Lond 80 (*Strix tenebricosus*)
Gould J (1846) Proc Zool Soc Lond 14:18 (*Athene rufa*), (*Athene marmorata*)
Gould J (1848–1869) The birds of Australia. Hullmandel
Gould J (1850–1883) The birds of Asia. Hullmandel
Gould J (1862–1873) The birds of Europe. Walter
Gould J (1875–1888) The birds of New Guinea, vol 1
Grandidier A (1867) Rev Mag Zool II(19):84–88 (*Scops madagascariensis*)
Granvik H (1934) The ornithology of NW Kenya Colony with special regard to the Suk and Tarkana districts. Rev Zool Bot Afr 25:41 (*Glaucidium tephronotum elgonense*)
Grassé PP, 1950 (ed) (1950) Traitée de Zoologie. Oiseaux, XV, Masson
Gray GR (1838) Gould's zoological voyage "Beagle", 3, pt 3, pl 4 and 1839, pt 9, p 34 (*Strix punctatissima*)
Gray GR (1844) Zoological voyage, «Erebus» and «Terror», Birds (1844) 2, pl 1 (*Athene albifacies*)
Gray GR (1861) Strigidae. Proc Zool Soc Lond (1860):344–345 (*Ephialtes leucospila*), (*Athene rufostrigata*), (*Athene hypogramma*)
Gray GR (1869) Hand List Birds 1
Gray JE (1829) In: Griffith's Animal Kingdom of Cuvier, 6, p 75 (*Strix tuidara*)
Gray JE (1834) In: Hardwick Ill. Ind. Zool. II, pl 31 (*Strix hardwickii*)
Gray JE (1841) Gould's zoological voyage "Beagle" III
Green RH (1981) Barn Owls in Tasmania. Aust Bird Watcher 9:94, 95
Greenway JC (1958) Extinct and vanishing birds of the world. Am Comm Intern Wildlife Protect Spec Publ 13, NY
Gregory P (1995) More from Manus. Muruk 7(3):112–115
Grewal B (1995) A photographic guide to the birds of India and Nepal. New Holland
Grimmett R, Inskipp C, Inskipp T (1998) Birds of the Indian Subkontinent. Helm

Grinnell J (1913) Two new races of the Pygmy Owl from the Pacific Coast. Auk 30:222–224 (*Glaucidium gnoma vigilante*)
Grinnell J (1915) Auk 32:60 (*Otus asio quercinus*)
Grinnell J (1928) Auk 45:213 (*Otus asio inyoensis*)
Griscom L (1929) Bull Mus Comp Zool 69:159 (*Ciccaba virgata centralis*)
Griscom L (1930a) Am Mus Novit 438:1 (*Cryptoglaux rostrata*)
Griscom L (1930b) Studies from the Dwight Collection of Guatemala Birds, III. Am Mus Novit 1930(438):1–2
Griscom L (1931) Notes on rare and little known neotropical Pygmy Owls. Proc New Engl Zool Cl 12:37–43 (*Glaucidium minutissimum rarum*)
Griscom L (1932) Bull Mus Comp Zool 72:325 (*Pulsatrix perspicillata chapmani*), 326 (*Lophostrix cristata wedeli*)
Griscom L (1935) Notes on Middle American horned owls. Ibis 13th ser 5:546–547
Griscom L (1937) New name for *Otus flammeolus guatemalae* preoccupied. Auk 54:391 (*Otus flammeolus rarus*)
Griscom L, Greenway JC (1937) Bull Mus Comp Zool 81:421 (*Tyto alba hellmayri*)
Groß R (1999) Der Falke 1999(10):299
Grossman ML, Hamlet J (1965) Birds of Prey of the world. Cassell
Grosvenor G, Wetmore A (1937) The book of birds, 2. Nat Geogr Soc
Grote H (1928) Orn Monatsb 36:79 (*Tyto alba hypermetra*)
Gruson ES (1976) Checklist of world's birds. Quadrangle, NY
Guérin-Méneville FE (1843) Rev Zool 321 (*Bubo cinerascens*), (*Otus abyssinicus*)
Gundlach JC (1871) Catálogo da las Aves Cubanas. Madrid
Gundlach JC (1874) J Orn 22:310 (*Gymnoglaux krugii*)
Gunning JWB, Roberts A (1911) Ann Transvaal Mus III:111 (*Pisorhina capensis intermedia*), (*Pisorhina capensis grisea*), (*Pisorhina capensis pusilla*), (*Glaucidium rufum*)
Günther R, Feiler A (1950) Die Vögel von Sao Thomé. Orn Monatsb
Gurney JH (1889) On an apperently undescribed species of owl from Liu-Kiu Isl. Ibis *104 (Scops capnodes)*, 302 (*Scops pryeri*)
Gurney JJ (1882) Ibis (1882) 132 (*Strix aurantia*)
Gurney JJ (1883) Ibis (1883) 171 (*Ninox goldii*)
Hachisuka M (1925) Comparative hand list of the birds of Japan and the British Isles
Hachisuka M (1934) Birds of the Philippine Islands, 3. Witherby, 51 (*Otus rufescens burbidgei*), 52 (*Otus rufescens malayensis*), 50 (*Mimizuku*)
Hallcux D, Goodman SM (1994) The rediscovery of the Malagasy Red Owl *Tyto soumagnei* (Grand.) in NE Madagascar. Bird Conserv Int 4(4):305–311
Haller H (1978) Zur Populationsökologie des Uhus *Bubo bubo* im Hochgebirge. Orn Beob 75:237–65
Haller W (1951) Zur "kiewitt" Frage (Steinkauz oder Waldkauz?). Mess Orn 3:199–201
Hannecart F, Letocart Y (1983) Oiseau de Nouvelle Caledonie et des Loyautes, II. Noumea, New Caled
Harrison C (1975) Jungvögel, Eier und Nester. Parey
Harrison JM (1957) Exhibition of a new race of the Little Owl from the Iberian Peninsula. Bull Brit Ornith Cl 77: 2–3 (*Athene noctua cantabriensis*)
Hartert E (1892) Bull Brit Ornith Cl 1:13 (*Strix flammea bargei*)
Hartert E (1893) Hartert exhibitet a new Scops Owl. *Pisorhina solokensis* sp. nov. Bull Brit Ornith Cl 1:39
Hartert E (1897a) Novit Zool IV:270 (*Strix flammea sumbaensis*)
Hartert E (1897b) Striges (*Pisorhina silvicola and alfredi*). Novit Zool IV:527–528, Tring
Hartert E (1898a) *Pisorhina sulaensis* sp. nov. Novit Zool V:126, Tring
Hartert E (1898b) Novit Zool V:500 (*Strix flammea contempta*)
Hartert E (1900) Novit Zool 7:228 (*Strix cayelii*), 534 (*Strix flammea schmitzi*)
Hartert E (1901) Bull Brit Ornith Cl 12:25 (*Pseudoptynx solomonensis*)
Hartert E (1903) *Pisorhina manadensis kalidupae* ssp. nov. Novit Zool X:21–22, Tring
Hartert E (1904) The birds of SW Is. of Wetter, Rame, Kisser, Letti and Moa. *Pisorhins manadensi tempe-statis* ssp. nov. and *Ninox b. ocellata* ssp. nov. Novit Zool 11:190–191, Tring
Hartert E (1905) Bull Brit Ornith Cl 16:31 (*Strix flammea gracilirostris*)
Hartert E (1906) *Ninox boobook cinnamomina* ssp. nov. Novit Zool 13:293, Tring
Hartert E (1908) Novit Zool 15:288
Hartert E (1910) Novit Zool 17, Tring: 204–206 (*Glaucidium cuculides persimile*)
Hartert E (1912–1921) Die Vögel der paläarktischen Fauna. Striges, Friedländer, pp 955–1040
Hartert E (1913) Bull Brit Ornith Cl 31:38 (*Tyto alba detorta*)
Hartert E (1914a) *Ninox squamipila* and *Otus m. magicus*. Novit Zool 21, Tring
Hartert E (1914b) *Ninox meeki*. Novit Zool 21:289, Tring
Hartert E (1918) *Ninox goldii*. Novit Zool 25:325, Tring
Hartert E (1923) Die Vögel der Pal Fauna. Nachtrag, 1:381–396, Tring
Hartert E (1924) Novit Zool 31:18 (*Athene noctua solitudinis*)
Hartert E (1925a) *Ninox variegata*. Novit Zool 32:121, Tring (*Ninox variegata superior*)
Hartert E (1925b) *Nesaio*. Novit Zool 32:263
Hartert E (1929a) On various forms of the genus *Tyto*. Novit Zool 35(2):93–104, Tring (*Tyto alba stertens*), (*Tyto alba everetti*), (*Tyto alba kuehni*), (*Tyto longimembris chinensis*), (*Tyto longimembris papuensis*)
Hartert E (1929b) Birds Whitney South-Sea Exp. VIII. Am Mus Novit 364:7 (*Spiloglaux jacquinoti eichhorni*), 6 (*Ninox jacquinoti roseoaxillaris*)
Hartert E, Steinbacher F (1932–1938) Die Vögel der Paläarktischen Fauna, 2. Nachtrag. Friedländ, pp 381–396
Hartlaub G (1849) Rev Mag Zool II(1):496 (*Athene leucopsis*)
Hartlaub G (1852) Rev Mag Zool II(4):3 (*Strix thomensis*)
Hartlaub G (1855) J Orn 3:354 (*Bubo leucostictus*)
Hartlaub G (1877) Die Vögel Madagaskars and der benachbarten Inselgruppen. Halle

Hartlaub G (1879) Proc Zool Soc Lond 295 (*Strix oustaleti*)

Hartlaub G (1899) Abh Naturwiss Ver Bremen 16:271 (*Ninox plateni*)

Hartlaub G, Finsch O (1872) On birds from the Pelew and Machenzie Islands. Proc Zool Soc Lond 90 (*Noctua podargina*)

Haverschmidt F (1962) Beobachtungen an der Schleiereule in Surinam. J Orn 103:236–242

Haverschmidt F (1968) Birds of Surinam. Oliver and Boyd

Haverschmidt F (1970) Barn Owls hunting by daylight in Surinam. Wilson Bull 82:101

Haverschmidt F, Mees GF (1994) Birds of Suriname. VACO

Hay A (= of Tweeddale) (1847) Madras J Lit Sci 13(2):147 (*Scops malayanus*)

Hayman P, Burton P (1988) Das goldene Kosmos Vogelbuch. Kosmos-Franckh

Hayman S (1980) Edward Lear's birds. The Wellfleet Press, 96 pp

Hazevoet CJ (1995) The birds of the Cape Verde Islands. BOU Checklist no. 13

Heather BD, Robertson HA (1997) The field guide to the birds of New Zealand. Oxford University Press

Heidrich P, König C, Wink M (1995) Bioakustik, Taxonomie and molekulare Systematik amerikanischer Sperlingskäuze. Stuttgarter Beitr Naturkd Ser A (Biol) 534:1–47

Heidrich P, Wink M (1994) Tawny Owl (*Strix aluco*) and Hume's Owl (*Strix butleri*) are distinct species. Evidence from nucleotide seq. of the cytochrome b gene. Z. Naturforsch. Sect. C, 49, 3, 4:230–234

Heintzelman DS (1984) Guide to owl watching in North America. Piscataway, NY

Hekstra GP (1982a) I don't give a hoot! A revision of the American Screech Owls (*Otus*, Strigidae). Sijtsma 1–147

Hekstra GP (1982b) Description of twentyfour new subspecies of American *Otus*. Bull Zool Mus Univ Amsterdam 9(7):49–63

Hekstra GP (unpubl) Notes to the Eurasien Scops Owls

Hellmayr CE (1932) The birds of Chile. Publ Field Mus (Zool Series) 308(19), 472 pp

Hennicke CR (1905) Naturgeschichte der Vögel Mitteleuropas, 5. (Neue Naumann)

Henry GM (1955) A guide to the birds of Ceylon. Oxford University Press, London

Henry GM (1971) A guide to the birds of Ceylon, 2nd ed. Oxford University Press, London

Henry GM (1998) A guide to the birds of Sri Lanka, 3rd ed. Oxford University Press, Dehli

Herklots GAC (1961) The birds of Trinidad and Tobago. Collins

Herremans M, Louette M, Stevens J (1991) Conservation, status and vocal and morphological description of the Grand Comoro Scops Owl, *Otus pauliani* Benson, 1960. Bird Conserv Int 1:123–133 (82, 83)

Hesse E (1915) J Orn 63:366 (*Bubo bubo borissowi*)

Hicks J, Olsen P, Greenwood D (1990) Saving the last survivor: Boobook owls … *Ninox novaeseelandiae royana*. Birds Internat 2/1:66–73

Higgins (ed) PJ (1999) Handbook of Australian, New Zealand and Antarctic birds, vol 4. Melbourne

Higuchi H, Momose H (1980) On the calls of the Collared Scops Owl *Otus bakkamoena* in Japan. Tori 29(2/3):91–94

Hill FAR, Lill A (1998a) Density and total population estimates for the threatened Christmas Island Hawk Owl *Ninox natalis*. Emu 98:209–220

Hill FAR, Lill A (1998b) Vocalisation of the Christmas Island Hawk Owl *Ninox natalis*: Individual variation in advertisement calls. Emu 98:221–226

Hilty SL (2003) The birds of Venezuela. Pica Press

Hilty SL, Brown WL (1986) Birds of Colombia. Princeton University Press

Hobcroft J (1997) Sooty Owl *Tyto tenebricosa* in Eungella National Park and Kroombit Tops, Queensland. Aust Bird Watcher 17(2):103, 104

Hobcroft J, James DZ (1997) Record of the Grass Owl from Southern New South Wales. Aust Bird Watcher 17(2):91–93

Hodgson BH (1836a) As Res 19 (1836) 172 (*Bubo nipalensis*), 175 (*Scops sunia*), 176 (*Scops lettia*), 168 (*Ulula newarensis*), 175 (*Noctua tubiger*), 175 (*Noctua Tarayensis*)

Hodgson BH (1836b) J As Soc Beng V:364 (*Cultrunguis nigripes*), (*Cultrunguis flavipes*)

Hodgson BH (1837a) J As Soc Beng VI:369 (*Scops sunis*)

Hodgson BH (1837b) Madras J Lit Sci 5:23 (*Ninox*), pl 14 (*Ninox nipalensis* = *lugubris*)

Hoesch NH, Niethammer G (1940) Die Vogelwelt Deutsch-Südwestafrikas. J Orn 88:1–404

Holdaway RN, Worthy TH (1996) Diet and biology of the Laughing Owl *Sceloglaux albifacies* (Aves, Strigidae) on Takaka Hill, Nelson, New Zealand. J Zool (Lond) 239(3): 545–572

Hollands D (1991) Birds of the night. Reed

Hollands D (1995) Silent hunters of the night. Nature Australia Spring:39–45

Holmes D, Nash S (1989) The birds of Java and Bali. Oxford University Press

Holmes D, Phillipps K (1996) The birds of Sulawesi. Oxford University Press

Hombron JB, Jacquinot H (1853) Voyage Póle Sud, III

Hoogerwerf A, de Boer HF (1947) Chronica Nat 103(7):140 (*Strix leptogrammica chaseni*)

Hoppe R (1967) Kichernder Kauz aus Ecuadors Urwäldern. Steckbrief Vogelkosmos 67:310, 311

Hoppe R (1969) Über den Dschungelkauz. Der Falke 1969/9:319

Horne M (1996) Experiences with the Malay Fish Owl (*Ketupa ketupu*). Tyto 1(2):41–43, and 1(4):101–106

Horsfield T (1821) Trans Linn Soc London 13(1):139 (*Strix badia*), 140 (*Otus rufescens*), (*Otus lempiji*), 140 (*Strix castanoptera*), (*Strix selo-puto*), (*Strix orientalis*), 141 (*Strix Ketupu*)

Hose C (1893) On the avifauna of Mt. Dulit on the Baram district in Sarawak. Ibis 416 (*Otus brookii*)

Hosking E, Flegg J (1982) Eric Hosking's Owls. Pelham Books

Howell SNG, Robbins NB (1995) Species limits of the Least Pygmy Owl (*Glaucidium minutissimum*) complex. Wilson Bull 107(1):7–25

Howell SNG, Webb S (1995) A Guide to the birds of Mexico and N Central America. Oxford University Press

Howie RR, Ritcey R (1957) Distribution, habitat selection and densities of the flammulated owls in British Columbia. In: Nero RW, Clark RJ, Knapton RJ, Hamre RH (eds) Biology and conservation of northern forest owls. Symp Proc Gen Techn Raptors RM 142, US Dep Agric Forest Service, Fort Collins, pp 249–254

Hoy PR (1852) P Acad Nat Soc Phila 6:211 (*Bubo subarcticus*)

Hubbard JP, Crossin RS (1974) Notes on northern Mexican birds. Nemouria 14:1–41

Hüe F, Etchécopar RD (1970) Les Oiseaux du Proche et du Moyen Orient. Paris

Huey L (1926) Birds of NW California. Auk XLIII:360–362 (*Otus asio cardonensis*)

Hume A (1870) Rough Notes 1(2):393 (*Ephialtes Huttoni*), 397 (*Ephialtes plumipes*)

Hume A (1873) Str Feath 1:8 (*Ephialtes brucei*), 11 (*Ninox obscurus*), 315 (*Bubo hemalachana*), 407 (*Ephialtes Balli*), 431 (*Syrnium ochrogenys*), 467 (*Heteroglaux*), 468 (*Heteroglaux blewitti*), 469 (*Athene pulchra*)

Hume A (1875) *Strix De-Roepstorffi* sp. nov. Str Feath 3:390–391

Hume A (1876) Str Feath 4:285 (*Ninox burmanica*)

Hume A (1877) Str Feath 5:138 (*Phodilus assimilis*)

Hume A (1878) Str Feath 6:27, 28 (*Syrnium maingayi*), 316 (*Asio butleri*)

Humphrey PS, Bridge D, Reynolds PW, Peterson RT (1970) Birds of Isla Grande (Tierra del Fuego). UKMNH, Lawrence

Hussain SA, Reza Khan MA (1978) J Bombay Nat Hist Soc 74:335 (*Phodilus badius ripleyi*)

Indrawan M (2004) World Birdwatch 26(4):2

Indrawan M, Somadikarta S (2004) A new nawk-owl from the Togian Islands, Gulf of Tomini, central Sulawesi, Indonesia. Bull Brit Ornith Cl 2004, 124(3):160–171

Inglis CM (1945) The Northern Bay Owl. Journ Bombay Nat Hist Soc 19:93–96

Inskipp C, Inskipp T (1991) Birds of Nepal. C. Helm

Iredale T (1956) Birds of New Guinea, vol 2. Georgian House, Melbourne

Irwin MP St (1981) The birds of Zimbabwe. Salisbury

Jackson C (1978) Wood engravings on birds. Witherby

Jackson C (1994) Bird paintings eighteenth century. Anitque Collectors Club

Jaksic FM, Seib RL, Herrera CM (1982) Predation by Barn Owl (*Tyto alba*) in Mediterranean habitats of Chile, Spain and California: A comparative aproach. Am Midl Nat 107:151–162

Jany E (1955) Neue Vogelformen von den Molukken. J Orn 96(1):106. (*Otus magicus obira*)

Jaramillo A, Burke P, Beadle D (2003) Birds of Chile. Pica Press

Jaume D, McMinn M, Alcover IA (1993) Fosil birds from Bujero Del Silo, La Gomera (Canary Is.). Bol Mus Mun Funchal Sup 2:147–165

Jeanson P (1999) R. Reboussin, Les Oiseaux de France. Neuilly sur Seine

Jellicoe M (1954) The Akun Eagle Owl. Sierra Leone Studies 154–167

Jerdon TC (1839) Madras J Lit Sci 10:86 (*Strix longimembris*)

Jerdon TC (1844) Madras J Lit Sci 13:119 (*Scops malabaricus*), (*Scops griseus = bakkamoena*)

Johansen (1907) Orn Jahrb 18:202

Johnsgard P (1991) Photo Essay: Burrowing Owl. Birder's World Feb 91:30–34

Johnsgard P (2002) North American owls. Biology and natural history. Smithonian Institute Press

Johnson AW (1967) The birds of Chile. Buenos Aires

Johnson DN (ed) (1980) Proceeding of the IV Pan African Ornithological Congress. South African Ornithological Society

Johnson NK (1963) The supposed migratory status of the Flammulated Owl. Wilson Bull 75:174–178

Johnson NK, Jones RE (1990) Geographic differantation and distribution of the Peruvian Screech Owl. Wilson Bull 102(2):199–212

Johnson WD, Collins CT (1975) Notes to the metabolism of the Cuckoo Owlet and Hawk Owl. Bull South Calif Acad Sci 74:44–45

Johnstone RE, Darnell JC (1997) Description of a new ssp. of Boobook Owl from Roti Island, Indonesia. W Aust Nat 21:161–173 (*Ninox boobook rotiensis*)

Jonsson L (1992) Die Vögel Europas and des Mittelmeerraumes. Franckh-Kosmos

Kannan R (1993) Rediscovery of the Oriental Bay Owl *Phodilus badius* in peninsular India. Forktail 8:148, 149

Kanowski J (1998) The abundance of the Rufous Owl *Ninox rufa* in upland and highland rainforests of NE Queensland. Emu 98(1):58–61

Karalus KE, Eckert AW (1974) The owls of North America. Doubelday and Co., NY

Kaup JJ (1829) Skizz Entwick-Gesch Europ Thierw 34 (*Aegolius*)

Kaup JJ (1848) Isis v Oken (*Megascops*) col 768 (*Hieracoglaux*), (*Spiloglaux*), (*Sceloglaux*), col. 769 (*Pseudoscops*), col. 771 (*Pulsatrix*)

Kaup JJ (1851) Arch Naturgesch 17:1 (*Ptilopsis*), 107 (*Rhinoptynx*), 110 (*Pseudoptynx philippensis*)

Kaup JJ (1852) Contrib Orn [Jardine] 105 (*Cepheloptynx*), 109 (*Ctenoglaux*), 110 (*Scops latipennis*), 111 (*Scops flammeolus*), 118 (*Strix glaucops*)

Kaup JJ (1862) Monograph of the Strigidae. Trans Zool Soc London, pp 201–260

Kavanagh RP, Jackson R (1997) Homerange, movements and diet of the Sooty Owl *Tyto tenebricosa* near Royal National Park, Sydney. In: Czechura G, Debus S (eds) Australian raptor studies II. Birds Australia Monograph 3, Birds Australia, pp 2–13

Kavanagh RP, Murray M (1996) Home range, habitat and behaviour of the Masked Owl *Tyto novaehollandiae* near Newcastle, New South Wales. Emu 96(4):250–257

Keith S, Twomey A (1968) New distributional records of some East African birds. *Otus senegalensis nivosus*. Ibis 110(4): 538, 539 (*Bubo africanus tanae*)

Kelso EH (1934d) A list of the owls of America. Biol Leaflet 4:43–45

Kelso EH (1936) Auk 53:82 (*Rhinoptynx clamator oberi*)

Kelso L (1932) Synopsis of the American wood owls of the genus *Ciccaba*. Lancaster, Pa, pp 1–47

Kelso L (1933a) Proc Biol Soc Wash 46:151 (*Ciccaba virgata amplonotata*)

Kelso L (1933b) Biol Leaflet 1 (*Novipulsatrix*)

Kelso L (1933c) Biol Leaflet 2:1 (*Pulsatrix perspicillata boliviana*)

Kelso L (1934a) A key to the owls of the genus *Pulsatrix* Kaup. Auk 51:234–236

Kelso L (1934b) *Asio stygius robustus* ssp. n. Auk 51:522, 523

Kelso L (1934c) Notes on habits of the Spect. Owl. Biol Leaflet 4:89–96

Kelso L (1937a) A Costa Rica race of Jardine's Pygmy Owl. Auk 54:304 (*Glaucidium jardinii costaricanum*), 305 (*Strix indranee rileyi*)

Kelso L (1937b) Biol Leaflet 8:1 (*Otus choliba alticola*), (*Otus choliba pintoi*)

Kelso L (1938a) Biol Leaflet 9, unpaged (*Tyto alba subandeana*), (*Tyto alba zottae*)

Kelso L (1938b) Biol Leaflet 10:3 (*Pulsatrix perspicillata austini*)

Kelso L (1939) Biol Leaflet 11, unpaged (*Speotyto cunicularia boliviana*)

Kelso L (1940a) Variation of the external ear-opening in the Strigidae. Wilson Bull 52:24–29

Kelso L (1940b) Biol Leaflet 12:1 (*Otus vermiculatus helleri*), (*Ciccaba virgata minuscula*), (*Lophostrix cristata amazonica*)

Kelso L (1941) Biol Leaflet 13:1 (*Otus choliba suturutus*)

Kelso L (1942) Biol Leaflet 14:2 (*Otus choliba portoricensis*)

Kelso L (1946) A study to the Spectacled Owls, Genus *Pulsatrix*. Biol Leaflet 33:1–13

Kelso L, Kelso EH (1934) A key to the species of the American owls and a list of the owls of America. Biol Leaflet 4

Kelso L, Kelso EH (1935a) Biol Leaflet 5, unpaged (*Otus clarkii*)

Kelso L, Kelso EH (1935b) Auk 52:451 (*Rhinoptynx clamator mogenseni*)

Kelso L, Kelso EH (1936) The relation of feathering of feet of American owls to humidity of environment and to life zones. Auk 53:51–56 and 215 (*Ciccaba virgata eatoni*)

Kelso L, Kelso EH (1943) Auk 60:448 (*Otus vermiculatus huberi*)

Kemp A (1989) Estimation of biological indices for little known African owls. Rapt Mod World 441–449

Kemp A, Calburn S (1987) The owls of Southern Africa. Struik, Winchester

Kemp A, Kemp M (1998) Birds of Prey of Africa and its Islands. Sasol

Kennedy RS, Gonzales PC, Dickinson EC, Miranda HC, Fisher TH (2000) A guide to the birds of the Philippines. Oxford University Press

Kerlinger P, Lein MR, Sevick BJ (1985) Distribution 6 population fluctuation of wintering Snowy Owls *Nyctea scandiaca* in North America. Can J Zool 63(8):1829–1834

Kerr (1792) Anim Kingdom 1:538

Keve A, Kohl S (1961) Bull Brit Ornith Cl 81:51 (*Athene noctua daciae*)

Khakhlov VA (1915) Mess Orn 6:224 (*Bubo bubo zaissanensis*)

King B, Rasmussen PC (1998) The rediscovery of the Forest Owlet, Athene (Heteroglaux) blewitti. Forktail 1998/14: 51–54

King B, Woodcock M, Dickinson EC (1975) Birds of South East Asia. Collins

King B, Yong B (2001) An unknown scops owl, *Otus* spec., from Sumba, Indonesia. Bull Brit Ornith Cl 121:91–93

King PP (1828) Zool J 3:426 (*Strix rufipes*), 427 (*Strix nana*)

Kleinschmidt O (1901) Orn Monatsb 9:168 (*Strix ernesti*)

Kleinschmidt O (1906) *Srix flammea* in Berajah 1–20, Selbstverlag (*Strix flammea rhenana*)

Kleinschmidt O (1907a) *Strix athene* in Berajah 1–6, (I–III), Selbstverlag

Kleinschmidt O (1907b) Falco 3:65 (*Athene sarda*), (*Athene ruficolor*)

Kleinschmidt O (1909) Falco 5:19 (*Strix saharae*)

Kleinschmidt O (1915) Falco 11:18 (*Strix hostilis*)

Kleinschmidt O (1940) Falco 36:60, footnote (*Tyto flammea hauchecorni*)

Kleinschmidt O (1958) Raubvögel und Eulen der Heimat. Ziemsen

Kloss LB (1930) J Siam Soc Nat Hist Suppl 8:81 (*Otus bakkamoena condorensis*)

Knight F (1997) Field guide to the birds of Australia

Knystaustas AJV, Sibnev JB (1987) Die Vogelwelt Ussuriens. Parey

Kobayashi K (1965) Birds of Japan. Osaka

Kobayashi K, Cho H (1981) Birds of Taiwan. Maeda, Kyoto

Koelz W (1939) New birds from Asia, chiefly from India. Proc Biol Soc Wash 52:61–82 (*Tyto alba crypta*), (*Otus bakkamoena stewarti*), (*Aegolius funerea juniperi*)

Koelz W (1950) New subspecies of birds from SW India. Am Mus Novit 1452:1–10 (*Tyto alba microsticta*), (*Strix ocellata grisescens*), (*Strix ocellata grandis*), (*Strix leptogrammica connectens*), (*Glaucidium radiatum principum*), (*Athene brama albida*)

Koelz W (1952) J Zool Soc India 4:45 (*Otus spilocephalus rupchandi*), (*Otus bakkamoena alboniger*), (*Glaucidium brodiei garoense*)

Koelz W (1954) Contrib Inst Reg Expl 1:22 (*Otus sunia khasiensis*), 27 (*Strix aluco obrieni*)

Koenig A (1936) Die Vögel am Nil, 2. Selbstverlag

Koenig L (1973) Das Aktionssystem der Zwergohreule *Otus scops scops*. J Comp Ethol (Suppl) 13:1–124

Koepcke M (1970) The birds of the Depart. Lima, Peru. Livingston Publ.

Kohl S (1977) Über die taxonomische Stellung südosteuropäischer Habichtskäuze, *Strix uralensis macroura* Wolf, 1810. Studii si Communicarie Muz Brukenthal 21:309–344

Kollibay P (1910) On the Ornithology of the Philippines. Orn Monatsb 18:148, 149 (*Pisorhina leucotis granti*)

König C (1967) Europäische Vögel, III. Belser

König C (1970) Europäische Vögel, II. Belser

König (1983) Auf Darwin's Spuren. Parey

König C (1991a) Taxonomische and ökologische Untersuchungen an Kreischeulen (*Otus* spp.) des südlichen Südamerikas. J Orn 132:209–214

König C (1991b) Zur Taxonomie and Ökologie der Sperlingskäuze (*Glaucidium* spp.) des Andenraumes. Ökol Vögel 13:15–76

König C (1994) Lautäußerungen als interspezifischer Isolationsmechanismus bei Eulen der Gattung *Otus*. Stuttgarter Beitr Naturkd Ser A 511:1–35

König C (1999b) Zur Ökologie und zum Lautinventar des Blaßstirnkauzes *Aegolius harrisi* (Cassin 1849) in Nordargentinien. Orn Mitt 51(4):127–138

König C (2001) Eulenforschung in Südamerika. Gefiederte Welt 2001/5:177–181

König C, Ertel R (1979) Vögel Afrikas. Belser

König C, Heidrich P, Wink M (1996) Taxonomie des Uhus (*Bubo* spec.) im südlichen Südamerika. Stuttgarter Beitr Naturkd Ser A 540:1–9

König C, Straneck R (1989) Eine neue Eule aus Nordargentinien. Stuttgarter Beitr Naturkd Ser A 428:1–20

König C, Weick F (2005) Ein neuer Sperlingskauz. Stuttgarter Beitr Naturkd Ser A 688:1–10

König C, Weick F (in prep) Owls. A guide to the owls of the world, 2nd ed. Pica Press

König C, Weick F, Becking JH (1999) Owls. A guide to the owls of the world. Pica Press

König C, Wink M (1995) Eine neue Unterart des Brasil Sperlingskauzes aus Zentralargentinien: *Glaucidium brasilianum stranecki* n. ssp. J Orn 136:461–465

Kotagama SW, Fernando P (1994) A field guide to the birds of Sri Lanka. Wildlife Heritage Trust

Kówan GM (1996) Records of the Amazonian Pygmy Owl *Glaucidium hardyi* from SE Venezuela. Cotinga 5:71, 72

Krechmar AV (2005) Die Sperbereule *Surnia ulula* in der Taiga NO-Sibiriens. Limicola 6/2005:330–337

Kröher O, Weick F (2004) Anmut im Federkleid – Heimische Vögel. Gollenstein, Blieskastel

Krook H (1988) Das Vogelbuch Olof Rudbecks d. J. 1695–1710. Faksimile, Coeckelberg

Kumar TS (1984) The organ: Bodyweight relationship in the Spotted Owl *Athene b. brama* (Temm.). Rapt Res Cent Publ 3

Kumar TS (1985) The life history of the Spotted Owl (*Athene b. brama*) in Andhra Pradesh. Rapt Res Cent Publ 4

Kuroda N (1923) Bull Brit Ornith Cl 43:122 (*Otus japonicus interpositus*)

Kuroda N (1924) On an apparently new form of Ural Owl. Hondo, Japan 15, 16 (*Strix uralensis pacifica*)

Kuroda N (1928) Tori 5(25):26 (*Otus sunia botelensis*)

Kuroda N (1931) A new subspecies of *Bubu blakistoni* from Sakhalin. Tori 31(7):41, 42 (*Bubo blakistoni karafutonis*)

Kuroda N (1936) Birds of the island of Java, 2. Private Publication, Tokyo

La Touche JDD (1919) Bull Brit Ornith Cl 40:50 (*Strix aluco harterti*)

La Touche JDD (1921) New races of *Bubo*. Bull Brit Ornith Cl 42:12–18 (*Bubo bubo jarlandi*) 29–32

Lafontaine RM, Moulaert N (1999) Une nouvelle espèce du petit duc (*Otus*, Aves) aux Comores: taxonomie et statut de conservat. J Afr Zool 112(2):163–169 (*Otus moheliensis*)

Lambert FR, Rasmussen PC (1998) A new Scops Owl from Sangihe Island, Indonesia. Bull Brit Ornith Cl 118(4):204–217 (*Otus collari*)

Lamberton (1928) Bull Acad Malgache 10(1927):40 (*Asio chauvini*)

Lamfuss G (1998) Die Vögel Sri Lankas. M. Kasparek Verlag

Lamothe L (1993) Papuan Hawk Owl *Uroglaux dimorpha* in the Lae–Bulolo area. Muruk 6(1):14

Land HC (1970) Birds of Guatemala. Wynnewood, Livingston Publ. Co.

Lande R (1988) Demographic model of the Northern Spotted Owl *Strix occidentalis caurina*. Oecologia 75:601–607

Landsdowne JF, Livingston JA (1966) Birds of the northern forest. Mc. Clelland and Stewart

Landsdowne JF, Livingston JA (1967) Birds of the Northern Forest. Mc. Clelland and Stewart

Landsdowne JF, Livingston JA (1968) Birds of the Eastern Forest. Mc. Clelland and Stewart

Langrand O (1990) Birds of Madagascar. Yale University Press

Lasley GW, Sexton C, Hillsman D (1988) First record of Mottled Owl (*Ciccaba virgata*) in the United States. Amer Birds 42(1):23, 24

Latham J (1790) Index Orn I:53 (*Strix coromanda*), 58 (*Strix perspicillata*), 62 (*Strix barbata*), 65 (*Strix fulva*)

Latham J (1801a) Index Orn Suppl XII (*Falco connivens*)

Latham J (1801b) Index Orn Suppl XV (*Strix georgica*), (*Srix boobook*)

Latham J (1801c) Index Orn XVII (*Strix undulata*)

Lawrence GN (1860) Ann Lyc Nat Hist NY 7:257 (*Gymnoglaux newtoni*)

Lawrence GN (1878a) On the members of *Gymnoglaux*. Ibis 20:184–187

Lawrence GN (1878b) Proc U S Nat Mus 1:64 (*Strix flammea* var. *nigrescens*), 234 (*Speotyto amaura*)

Legge WV (1878a) A history of the birds of Ceylon, 1. (*Bubo blighi*)

Legge WV (1878b) Ann Mag Nat Hist 5(1):175 (*Scops minutus*)

Lehmann FC (1946) Auk 63:218 (*Bubo virginianus colombianus*)

Lei F, Tsohsin C (1995) The little owl *Athene noctua plumipes* in China. Newsl. World Work, Group of Birds of Prey and Owls 21(22):22, 23

Lekagul B, Round PD (1991) A guide to the birds of Thailand. Saha Karn Bhaet Co.

Lepage D (2005) avibase.jsp. http://www.bsc-eoc.org.avibase

Leshem Y (1974) Hume's Tawny Owl – Lilith of the desert. Teva Va'Aretz 16:66, 67

Leshem Y (1979) Hume's Waldkauz (*Strix butleri*) – die Lilith der Wüste. Nat Mus 109(11):375–7

Leshem Y (1981a) Israel's raptors – The Negev and Judean Desert. Ann Rep Hawk Trust 11:30–35

Leshem Y (1981b) The occurrence of Hume's Tawny Owl in Israel and Sinai. Sandgrouse 2:100–102

Lesson RP (1828) Man. d'Ornith 1:116 (*Strix magellanicus*)

Lesson RP (1830) Traite d Orn livr 2:106 (*Noctua frontata*), 110 (*Otus wilsonianus*), 114 (*Ketupa*)

Lesson RP (1831) Traite d Orn 1:107 (*Scops lophotes*)

Lesson RP (1836) Oeuvres Compl Buffon 7:261 (*Lophostrix*)

Lesson RP (1839) Rev Zool 289 (*Syrnium ocellatum*)

Lewis A (1996) In search of the Badenga. Bull ABC 3(2): 131–133

Lewis A (1998) Mayotte Scops Owl *Otus r. mayottensis*. Bull ABC 5(1):33, 34

Lewis DP (2002–2005) owlpages, photo gallery. http://www.owlpages.com

Lichtenstein MHK (1823) Verzeichnis der Doubletten des Zoologischen Museums der Koeniglichen Universität zu Berlin nebst Beschreibung vieler bisher unbekannter Arten von Säugethieren, Voegeln, Amphibien und Fischen. Verzeichniss Doubletten Zoologischen Museums Universität Berlin i–x + pp 1–118, 59 (*Strix perlata*), (*Strix decussata*)

Lichtenstein MHK (1842) Verzeichniss einer Sammlung von Säugethieren und Vögeln aus dem Kaffernlande, nebst einer Käfersammlung. 11 (*Strix helveola*), 12 (*Strix licua*)

Lichtenstein MHK (1854) Nomenclator avium Musei Zoologici Berolinensis. 7 (*Ephialtes argentina*)

Ligon JD (1968) The biology of the Elf Owl, *Micrathene whitneyi*. Misc Pub Mus Zool U Michigan 136:1–70

Lindroth PG (1788) Museum naturalium Grillianum Söderforssiense institutum anno. 5 (*Strix liturata*)

Linnaeus (von Linné C) (1758) Systema naturae per regna tria naturae, secundum classes, ordines, genera, species, cum characteribus differentiis, synonymis, locis, 10th ed. 1

Lister JJ, 1889. On the natural history of Christmas Is. Proc Zool Soc Lond (1888):525–527 (*Ninox natalis*)

Liversedge TN (1980) A study of Pel's Fishing Owl *Scotopelia peli* in the "pand handle" region of the Okavango Delta, Botswana. In: Johnson DN (ed) Proceeding of the IV Pan African Ornithological Congress. South African Ornithological Society, pp 291–299

Lloyd G, Lloyd D (1971) Greifvögel and Eulen. Delphin, 160 pp

Loche V (1867) Exploration scientifique de l'Algérie pendant les années 1840–1842. Histoire naturelle des Oiseaux, Zoologie 1:99 (*Phasmotynx capensis a tingitanus*)

Louette M (1988) Les Oiseaux des Comores. Annales Serie Sci Tervueren 255

Lowery GH, Dalquest WW (1951) Birds of the state of Veracruz, Mexico. Univ Kansas Publ Mus Nat Hist 3: 533–649 (*Rhinoptynx clamator forbesi*)

Lowery GH, Newman RJ (1949) New birds from state of San Luis Potoso and the Tuxtla Mts. of Vera Cruz, Mexico. Occ Pap Mus Zool LSU 22:1–4 (*Glaucidium minutissimum sanchezi*)

Lowther EHN (1949) A bird photographer in India. Pl. 76 (*Bubo coromandus*)

Maciejewski SE (1997) The Grass Owl *Tyto c.* (*longimembris*) in NE New South Wales. In: Czechura G, Debus S (eds) Australian raptor studies II. Birds Australia Monograph 3, Birds Australia, pp 54–70

MacKay BK (1994) A celebration of owls. Birds of the World B(1):16–26

MacKenzie JP (1986) Birds of Prey. Harrap

Mackworth-Praed CW, Grant CHB (1957) Birds of Eastern and North Eastern Africa. I. Longmans, Green and Co.

Mackworth-Praed CW, Grant CHB (1962) Birds of the South. Third of Africa. I. Longm.ans, Green and Co.

Mackworth-Praed CW, Grant CHB (1970) Birds of the West Central and Western Africa. I. Longm.ans, Green and Co.

MacLean GL (1993) Robert's birds of South Africa, 6th ed. Trustees of Voelckes Bird Book Fund

Magnin G (1991) Notes: A record of the Brown Fish Owl *Kezupa zeylonensis* from Turkey. Sandgrouse 13 (1): 42

Maijer S, Hohnwald S (1997) Bull Brit Ornith Cl 197:223–235

Makatsch W (1989) Wir bestimmen die Vögel Europas. Neumann, Radebeul

Manuel CB, Gilliard ET (1952) Undescribed and newly recorded Philippine Birds. Am Mus Novit 1545:4–5 (*Otus bakkamoena batanensis*)

Marchant S (1948) The West African Wood Owl. Nigerian Field 13:16–20

Marples BJ (1942) A study of the Little Owl (*Athene noctua*) in New Zealand. Trans Proc Roy Soc N Z 72:237–252

Marshall JT Jr (1939) Territorial behaviour of the Flammulated Screech Owl. Condor 41:71–78

Marshall JT Jr (1942) Food and habitat of the Spotted Owl. Condor 44:66, 67

Marshall JT Jr (1949) The endemic avifauna of Saipan, Tinan, Guam and Palau. Condor 51:207, 208

Marshall JT Jr (1957) Birds of pine-oak woodland in southern Arizona and adjacent Mexico. Pacif Coast Avif Cooper Ornith Soc 32:1–125

Marshall JT Jr (1966) Relationship of certain owls around the Pacific. Nat Hist Bull Siam Soc 21:235–242

Marshall JT Jr (1967) Parallel variation in North and Middle American Screech Owls. Mono W Found Vert Zool 1:1–72

Marshall JT Jr (1978) Systematics of small Asian night birds based on voice. Orn Monogr 25:1–58

Marshall JT Jr (1991) Variable Screech Owl and its relatives. Wilson Bull 103:314, 315

Marshall JT Jr, King B (1988) Genus *Otus*. In: Amadom D, Bull J (eds) Hawks and owls of the world. Checklist West Found Vertebr Zool, pp 331–336

Martinez O (1998) Observaciones preliminares sobre la historia naturaldel Mochuelo Andino *Glaucidium bolivianum* (*jardini*) en el bosquede neblina del PN – ANMI Cotapata. Dpto. La Paz. In: Sagot and Guerrero (1998) 120–123

März R (1968) Der Rauhfußkauz. NBB, Ziemsen

Mason IJ (1983) A new subspecies of Masked Owl, *Tyto novaehollandiae* from southern New Guinea. Bull Brit Ornith Cl 103:122–128

Mason IJ, Schodde R (1980) Subspeciation in the Rufous Owl *Ninox rufa* (Gould). Emu 80(3):141–44 (*Ninox rufa meesi*)

Mathews GM (1911) Bull Brit Ornith Cl 27:62 (*Ninox rufa queenslandica*)

Mathews GM (1912a) Aust Avian Rec 1:34 (*Tyto novaehollandiae mackayi*), 35 (*Tyto novaehollandiae kimberli*), (*Tyto novaehollandiae melvillensis*), 34 (*Tyto novaehollandiae whitei*), 35 (*Tyto novaehollandiae riordani*), 75 (*Ninox strenua victoriae*), 120 (*Ninox connivens addenda*), 34 (*Ninox boobook melvillensis*), 120 (*Ninox boobook royana*)

Mathews GM (1912b) A reference-list to the birds of Australia. Novit Zool 18:99 (*Tyto alba alexandrae*): 257 (*Tyto novaehollandiae perplexa*): 257 (*Tyto tenebricosa multipunctata*), 258 (*Tyto tenebricosa magna*), 255 (*Ninox connivens suboccidentalis*), (*Ninox boobook mixta*), 254 (*Ninox boobook halmaturina*)

Mathews GM (1913a) Aust Avian Rec 1:194 (*Ninox boobook macgillivrayi*)

Mathews GM (1913b) Aust Avian Rec 2:74 (*Spiloglaux boweri*), (*Spiloglaux boobook leachi*), (*Spiloglaux boobook trepellasi*), (*Spiloglaux boobook clelandi*)

Mathews GM (1914a) S Austral Orn 1(2):12 (*Tyto novaehollandiae galei*)

Mathews GM (1914b) Austral Av Rec 2:91 (*Tyto longimembris dombraini = novaehollandiae*)

Mathews GM (1916) Birds Austral. 5(1):353–440. Witherby (*Megastrix tenebricosa perconfusa*), 305 (*Berneyornis*), 332 (*Spiloglaux novaeseelandiae everardi*)

Mathews GM (1917) Aust Avian Rec 3:70 (*Spiloglaux novaeseelandiae tasmanica*)

Mathews GM (1920) Check-list of the birds of Australia. Vol 1 and suppl. I–IV and 1–116

Mathews GM (1926) Bull Brit Ornith Cl 46:131 (*Ninox novaebritanniae novaehibernicae*)

Mathews GM (1927) Systema Avium Australasinarum, 1. BOU, London, I–X and 1–426

Mathews GM (1931) A list of the birds of Australasia. Taylor & Francis, London, pp 1–562

Mathews GM (1946) A working list of Australian birds including the Australian Quadrant and New Zealand. G. M. Mathews, Sydney, 184 pp, 55 (*Spiloglaux boobook parocellata*), (*Spiloglaux ocellata carteri*)

Mathews GM, Neumann O (1939) Six new races of Australian birds from North Queensland. Bull Brit Ornith Cl 54:153–154

Maynard CJ (1899) Appendix to catalogue of the birds of the West Indies. Description of new species. Contr to Sci III, pp 33 (*Speotyto bahamensis*)

Mayr E (1929) Birds collected during the Whitney South Sea expedition. Am Mus Novit, 6–7

Mayr E (1931) Whitney South Sea expedition, 17. Am Mus Novit, 14–15 (*Ninox jacquinoti malaitae*)

Mayr E (1933) Am Mus Novit 609:1 (*Asio flammeus ponapensis*)

Mayr E (1935) Whitney South Sea expedition, 30. Am Mus Novit 820:2–3 (*Tyto alba crassirostris*), (*Tyto alba interposita*), (*Ninox jacquinoti mono*), (*Ninox jacquinoti floridae*)

Mayr E (1936) Birds collected during the Whitney South Sea expedition, 31. Am Mus Novit 828:1–19

Mayr E (1937) Birds collected during the Whitney South Sea expedition, 35. Am Mus Novit 939:1–14; 6 (*Uroglaux*)

Mayr E (1938) Bull Raffles Mus 14:14 (*Otus bakkamoena kangeana*), 15 (*Strix leptogrammica vaga*)

Mayr E (1941) A list of New Guinea birds (New Guinea and adjacent islands). American Museum of Natural History, pp 1–260

Mayr E (1943) *Ninox novaeseelandiae*. Revision of Australasian races. Emu 43:12–16 (*Ninox novaeseelandiae moae*), (*Ninox novaeseelandiae arida*)

Mayr E (1944) The birds of Timor and Sumba. Bull Am Mus Nat Hist 83(2):123–194

Mayr E (1945a) The races of *Ninox philippensis*. Zoologica 30:46 and 108 (*Ninox philippensis centralis*), (*Ninox philippensis proxima*)

Mayr E (1945b) Birds of the Southwest Pacific. Macmillan

Mayr E (1975) Grundlagen der zoologischen Systematik. Parey, Hamburg Berlin, 370 pp

Mayr E, Gilliard ET (1954) Birds of Central New Guinea. Bull Am Mus Nat Hist NY 103:341, pl 17, 18

Mayr E, Meyer de Schauensee R (1939) Birds of the island of Biak. P Acad Nat Sci Phila 91:1–37

Mayr E, Phelps WH (1967) The origin of the bird fauna of S Venezuelan highlands. Bull Am Mus Nat Hist NY 136(5): 269–328

Mayr E, Rand AL (1935) Am Mus Novit 814:3 (*Ninox novaeseelandiae pusilla*)

Mayr E, Short LL (1970) Species taxa of North American birds. Publ Nuttall Orn Club 9:1–127

McCann IR (1982) Grampians birds. An illustrated checklist. Hallsgap Tourist Information Centre

McCracken-Peck R (1952) A celebration of Birds. (L.A. Fuertes). Walker

McGillivrey WB (1987) Reversed size dimorphism in 10 species of northern owls. In: Nero RW, Clark RJ, Knapton RJ, Hamre RH (eds) Biology and conservation of northern forest owls. Symp Proc Gen Techn Raptors RM 142, US Dep. Agric. Forest Service. Fort Collins, pp59–66

McGillivrey WB (1989) Geographic variation in size and reversed size dimorphism of the Great Horned Owl in North America. Condor 91(4):77–786

McGregor RC (1904) Bull Philip Mus 4:17: (*Otus cuyensis*), 18 (*Otus calayensis*)

McGregor RC (1905) Birds of the islands of Romblon, Sibuyan and Cresta de Gallo. Bur Govern Labor 25:11–13 (*Otus romblonis*)

McGregor (1907) Philip J Sci 2

McGregor RC (1909) Manual Philippine birds, vol 1. pp 250–252

McGregor RC (1927) New noteworthy Philippine birds, V. Philip J Sci 32(4):517, 518 (*Phodilus riverae*)

McKinnon J (1990) Birds of Java and Bali. Gadjah Mada University Press, Indonesia

McKinnon J, Phillipps K (1993) A field guide to the birds of Borneo, Sumatra, Java and Bali. Oxford University Press

McLachlan GR, Liversidge R (1963) Roberts birds of South Africa. Central News Agency

Mearns EA (1909) Proc U S Nat Mus 26:437 (*Otus steerei*)

Mebs T (1987) Eulen and Käuze. Franckh-Kosmos

Mebs T, Scherzinger W (2000) Die Eulen Europas. Kosmos

Medway L, Wells DR (1976) The birds of the Malay Peninsula. Witherby and University Malaya

Mees GF (1961) An annotated catalogue of a collection of bird skins from W Pilbara, Western Australia. J Roy Soc West Aust 44:97–143; 106 (*Ninox novaeseelandiae rufigaster*)

Mees GF (1963) Status and distribution of some species of owls in Western Australia. W Aust Nat 8:166–9

Mees GF (1964a) A revision of the Australian owls. Zool Verh Leiden 65:1–62

Mees GF (1964b) Geographical variation in *Bubo sumatranus* (Raffles). Zool Meded Rijksmus Nat Hist Leiden 40(13) (4 pages) (*Bubo sumatranus tenuifasciatus*)

Mees GF (1965) The avifauna of Misool. Novae Guinea Zool 31:139–203

Mees GF (1967) Zur Nomenklatur einiger Raubvögel and Eulen. Zool Meded Rijksmus Nat Hist Leiden 42(14): 143–146

Mees GF (1970) Birds from Formosa. Zool Meded Rijksmus Nat Hist Leiden 44:227–229 and 295–297
Mees GF (1971) Birds from Borneo and Java. Zool Meded Rijksmus Nat Hist Leiden 45:231–232
Meinertzhagen R (1920) Bull Brit Ornith Cl 41:21 (*Otus scops powelli*)
Meinertzhagen R (1930) Nicoll's birds of Egypt, 2. London, 700 pp
Meinertzhagen R (1948) On the *Otus scops* (L.) group, and allied groups, with special reference to *Otus brucei* (Hume). Bull Brit Ornith Cl 69:8–11
Meinertzhagen R (1951) On the genera *Athene* Boie, 1822 and *Speotyto* Gloger, 1842. Bull Brit Ornith Cll 70:8–9
Meinertzhagen R (1954) Birds of Arabia. Oliver and Boyd, Edinburgh London
Meinertzhagen R (1959) Pirates and predatores. Oliver and Boyd, Edinburgh London
Meise W (1933) Zur Systematik der Fischeulen. Orn Monatsb 41:169–173 (*Bubo blakistoni piscivorus*)
Melde M (1984) Der Waldkauz. NBB Ziemsen
Mendelssohn H, Yom-Tov Y, Safriel U (1975) Hume's Tawny Owl, *Strix butleri* in Judean, Negev and Sinai deserts. Ibis 110, 111
Menning K (1989) Gefiederte Welt 1989(5):145–146
Menzbier M (1896) Bull Brit Ornith Cl 6:6 (*Syrnium willkonskii*)
Merriam CH (1891) North Am Fauna 5:96 (*Megascops flammeolus idahoensis*)
Merriam CH (1898) Auk 15:39, 40 (*Syrnium occidentale caurina*)
Merriam CH, Fisher AK (1893) The hawks and owls of the United States. US Dep Agricult Bull 3
Meyburg BU, Chancellor RD (1989) Raptors of the modern world. WWGBPO
Meyburg BU, Chancellor RD (1994) Raptor conservation today. Pica Press
Meyer AB (1882) On *Ninox rudolfi* a new species of Hawk Owl in the Malay Archipelago. Ibis 6:232
Meyer de Schauensee R (1964) The birds of Colombia. Livingston Publ.
Meyer de Schauensee R (1966) The species of birds of South America. Livingston Publ.
Meyer de Schauensee R (1971) A guide to the birds of South America. Oliver and Boyd
Meyer de Schauensee R (1984) The birds of China. Oxford University Press
Meyer de Schauensee R, Phelps WH Jr (1978) A guide to the birds of Venezuela. Princeton University Press
Meyer O (1934) J Orn 82(4):575 (*Tyto aurantia*)
Mikhailov K (2000) Am Ostrand Eurasiens: Riesenfischuhu – der aussterbende Fischjäger der Ussuri-Wildnis. Der Falke 2000/47:68–73
Mikkola H (1981) Der Bartkauz. NBB, Ziemsen
Mikkola H (1983) Owls of Europe. Poyser
Miller AH (1955) The avifauna of the Sierra del Carmen of Coahuila, Mexico. Condor 67:154–178
Miller AH (1965) The syringeal structure of the Asiatic owl *Phodilus*. Condor 67:536–538
Miller AH, Miller L (1951) Geographic variation of the screech owls of the deserts of western North America. Condor 53:161–177 (*Otus asio yumanensis*)
Miller W (1915) Bull Am Mus Nat Hist 34:515 (*Strigonax*)
Millsap BA (1988) Elf Owl Natl Wildl Fed Sci Techn Ser 11:140–144
Millsap BA, Johnson RR (1988) Ferruginous Pygmy Owl. Nat Wildl Fed Sci Techn Ser 11:137–9
Milne-Edwards A (1878) Compt Rend Acad Sci Paris 85(1877):66 (*Heliodilus soumagnei*)
Milne-Edwards A, Grandidier A (1876) Histoire, physique, naturelle et politique de Madagascar 12(1) and Atlas, 1876
Milne-Edwards A, Oustalet JFE (1888) Etudes sur le mammifères et le oiseaux des iles Comores. Nouv Arch Mus Hist Nat Ser 2, 219–297 (*Scops humbloti*)
Minnemann (1984) Der Falke 1984(6):214 (*Athene cunicularia*)
Mirza ZB (1985) New record of Dusky Horned Owl nesting on ground. Pak J Zool 17(1):109, 110
Mishima T (1956) Notes on *Ninox scutulata*. Japan Wildl Bull 15:25, 26
Molina GI (1782) Sagg Stor Nat Chili 263 (*Strix cunicularia*)
Momiyama TT (1923) Dobutsu Zasshi 35:400 (*Otus bakkamoena hatchizionis*)
Momiyama TT (1927a) J Chosen Nat Hist Soc 4:1 (*Syrnium uralense coreensis*)
Momiyama TT (1927b) Bull Brit Ornith Cl 48:21 (*Strix uralensis tatibanai*), (*Strix uralensis morii*), (*Strix uralensis nigra*)
Momiyama TT (1928) New and known forms of the Ural Owl (*Strix uralensis*), from southeastern Siberia, Manchuria, Korea, Sakhalin and Japan. Auk 45:177–185 (*Strix uralensis jinkou*), (*Strix uralensis media*)
Momiyama TT (1930) Dobutsu Zasshi 42:329 (*Bubo bubo yamashinai*)
Momiyama TT (1931) Amoeba 3(1–2):68 (*Ninox scutulata totogo*)
Monk KA, de Fretes Y, Reksodiharjo-Lilley G (1997) Periplus. Singapore
Monroe BL (1968) A distributional survey of the birds of Honduras. AOU Orn Monogr 7:1–458
Monroe BL, Sibley C (1993) A world checklist of birds. Yale University Press
Moon G (1992) The Reed field guide to New Zealand birds. Reed
Mooney NJ (1997) Habitat and seasonality of nesting Masked Owl in Tasmania. In: Czechura G, Debus S (eds) Australian raptor studies II. Birds Australia Monograph 3, Birds Australia, pp 34–39
Moore RT (1937) Proc Biol Soc Wash 50:64 (*Otus asio sinaloensis*), 65 (*Otus guatemalae tomlini*), 65 (*Glaucidium minutissimum oberholseri*), 103 (*Asio stygius lambi*)
Moore RT (1941) Three new races in the genus *Otus* from Central Mexico. Proc Biol Soc Wash 54:151–155 (*Otus asio suttoni*), (*Otus asio sortilegus*), 156 (*Otus vinaceus seductus*)
Moore RT (1947) Two new owls, a swift and a poorwill from Mexico. Proc Biol Soc Wash 60:13 (*Otus cooperi chiapensis*), 31 (*Glaucidium minutissimum griscomi*), (*Glaucidium minutissimum occultum*), 141 (*Aegolius ridgwayi tacanensis*)

Moore RT, Marshall JT Jr (1959) *Otus asio lambi,* new subspecies. Condor 61:224–225

Moore RT, Peters JL (1939) The genus *Otus* of Mexico and central America. Auk 56:38–56 (*Otus trichopsis pumilus*), (*Otus guatemalae dacrysistactus*), (*Otus guatemalae fuscus*)

Moreau RE (1964) The rediscovery of an African owl, *Bubo vosseleri.* BOC Mag 84:47–52

Morony JJ Jr, Bock WJ, Farrand J Jr (1975) Reference list of the birds of the world. American Museum of Natural History

Morris P, Hawkins F (1998) The birds of Madagascar. A Photographical Guide. Pica Press

Mosher JA, Henny CJ (1976) Thermal adaptiveness of plumage color in Screech Owls. Auk 93:614–619

Mueller HC (1990) Can Saw whet Owl be sexed by external measurements? J Field Ornith 61(3):339–346

Mueller O (1983) Grass Owl near Broom. W Aust Nat 15:148

Mukherjee AK (1958) Rec Indian Mus 53(1–2)(1955):301 (*Otus brucei exiguus*)

Müller PLS (1776) Natursyst., Suppl. 69 (*Strix caparoch*), 70 (*Strix domingensis*)

Müller S (1841) Verh nat gesch Nederl Land-en Volkenk 4:110 (*Strix magica*)

Müller S (1845) Verh nat gesch Nederl Land-en Volkenk 279 (*Strix (Athene) guteruhi*)

Mulsow D (1964) Die Uhu-Pyramide. Vogelkosmos 11: 246–248

Munn B (1997) Winged Wolverine. Birder's World 11(5):48–51

Murata S (1914) Karafuto Dobutsu-tyosa Hokuku

Murphy RC, Amadon D (1953) Land birds of America. McCraw-Hill Book

Nakamura K (1975) A record of a Brown Hawk Owl in the Northern Pacifik. Tori 23:37–38

Narosky T, Yzurieta D (2003) Birds of Argentina and Uruguay. V. Mazzini (ed)

Naumann JA (1820–1844) Naturgeschichte der Vögel Deutschlands. Color plates by J. F. Naumann, (Nachtrag 1854)

Naumann JF (Hennicke) (1896–1905) Naturgeschichte der Vögel Deutschlands (Der neue Naumann)

Neelakantan KK (1971) Calls of the Malabar Jungle Owlet (*Glaucidium radiatum malabaricum*). J Bombay Nat Hist Soc 68(3):830–832

Nelson AW (1897) Auk 14:49 (*Megascops marmoratus*)

Nelson AW (1901a) Description of five new birds from Mexico. Auk 18:46 (*Glaucidium palmarum*)

Nelson AW (1901b) Description of a new genus and eleven new species and subspecies of birds from Mexico. Proc Biol Soc Wash 14:169, 170 (*Bubo virginianus mayensis*)

Nelson AW (1903) Proc Biol Soc Wash 16:152 (*Syrnium occidentale lucidum*)

Nelson AW (1910) Proc Biol Soc Wash 23:103 (*Glaucidium gnoma pinicola*)

Nelson AW, Palmer TS (1894) Auk 11:39 (*Megascops pinosus*), 40 (*Megascops ridgwayi*), 41 (*Glaucidium fisheri*)

Nero RW, Clark RJ, Knapton RJ, Hamre RH (eds) (1987) Biology and conservation of northern forest owls. Symp. Proc. Gen. Techn. Raptors. RM 142, US Dep. Agric. Forest Service, Fort Collins

Nesterov PV (1912) Annuaire Mus Zool Acad Imp Sci St Petersb 16(1911):378 (*Bubo bubo armeniacus*)

Neumann O (1893) Orn Monatsb 1:62 (*Glaucidium castaneum*)

Neumann O (1899) J Orn 47:56 (*Pisorhina ugandae*)

Neumann O (1911) *Glaucidium capense scheffleri* n. ssp. Orn Monatsb 19:184

Neumann O (1935) Bull Brit Ornith Cl 55:138 (*Bubo ketupu aagaardi*), (*Bubo ketupu pageli*)

Neumann O (1939) A new species and eight new races from Peleng and Taliaboe (*Tyto nigrobrunnea*). Bull Brit Ornith Cl 59:89–90(421):92 (*Tyto rosenbergi pelengensis*), 106 (*Otus manadensis mendeni*)

Niall I (1985) Portrait of a country artist. (Tunnicliffe), V. Gollancz

Nicholson F (1938) Fla Nat 17:99 (*Syrnium nebulosum sablei*)

Nikolski (1889) Sapiski Imper Akad Nauk St Petersburg

Nielsen L (1961) The Sooty Owl in Queensland Aust. Bird Watcher 1:180

Niethammer G (1957) Ein weiterer Beitrag zur Vogelwelt des Ennedi-Gebirges. Bonn Zool Beitr 8:275–284

Norelli MR (1976) American wildlife painting. Phaidon Press

Norman J, Christidis L, Westerman M, Hill FAR (1998b) Molecular data confirm the species status of the Christmas Island Hawk Owl *Ninox natalis.* Emu 98(4):197–208

Norman J, Olsen P, Christidis L (1998a) Molecular genetics confirm taxonomic affinities of the endangered Norfolk I. Boobook *Ninox n. undulata* . Biol Conserv 86(1):33–36

Northern JR (1965) Notes on the owls of the Tres Marias Is., Nayarit, Mexico. Condor 67:358

O'Neill JP, Graves GR (1977) A new genus and species of owl (Aves: Strigidae) from Peru. Auk 94(3):407–416 (*Xenoglaux*), (*Xenoglaux lowery*)

Oates EW (1877) Str Feath 5:247, 248 (*Otus sagittatus*)

Oates EW (1889–1898) (Blanford WT ed): Fauna of British India, birds (Strigidae 264–311)

Oberholser HC (1904) Revision on American Great Horned Owls. Proc U S Nat Mus 27:177–192

Oberholser HC (1905) Proc U S Nat Mus 28:856 (*Asio maculosus amerimnus*)

Oberholser HC (1908) A new Great Horned Owl from Venezuela, with notes on the names of the American forms. Sci Bull Brooklyn Inst Arts Sci 1(14):371–374 (*Bubo virginianus scotinus*)

Oberholser HC (1914) Proc Biol Soc Wash 27:46 (*Bubo virginianus neochorus*)

Oberholser HC (1915) Critical notes on the subspecies of the Spotted Owl, *Strix occidentalis* (Xantus). Proc U S Nat Mus 49(2106):251–257

Oberholser HC (1917) Proc U S Nat Mus 52:184, 190 (*Strix baweana*)

Oberholser HC (1922) A revision on the American Great Horned Owls. Proc U S Nat Mus 177–192

Oberholser HC (1924) Journ Wash Acad Sci 14:302 (*Phodilus badius abbotti*), (*Strix leptogrammica nyctiphasma*)

Oberholser HC (1932) US Nat Mus Bull 159:40 (*Phodilus badius arixuthus*)

Oberholser HC (1937) Journ Wash Acad Sci 27:337 (*Otus asio clazus*), 354 (*Otus asio swenki*), 356 (*Otus asio mychophilus*)

of Tweeddale A Marq (1878) Ornithology of the Philippines. Proc Zool Soc Lond 939–942 (*Scops everetti*), (*Pseudoptynx gurneyi*), (*Ninox spilocephala*)
Ogilvie-Grant VAL (1894) Bull Brit Ornith Cl 3:51 (*Scops longicornis*)
Ogilvie-Grant VAL (1895) Bull Brit Ornith Cl 4:40
Ogilvie-Grant VAL (1896) Ibis 463 (*Ninox mindorensis*)
Ogilvie-Grant VAL (1906a) Bull Brit Ornith Cl 16:99 (*Pseudoptynx mindanensis*)
Ogilvie-Grant VAL (1906b) Bull Brit Ornith Cl 19:11 (*Heteroscops vulpes*)
Ogilvie-Grant VAL (1906c) Ibis 660 (*Scops erlangeri*)
Ogilvie-Grant VAL (1912) Bull Brit Ornith Cl 29:116 (*Scops spurrelli*)
Ogilvie-Grant VAL, Forbes HO (1899) Bull Liverp Mus 2:2, 3 (*Scops socotranus*)
Olney PJ (1984) The rare Nduk Eagle Owl, ... at the London Zoo. Avic Mag 90(3):129–134
Olrog CC (1963) Lista y distribucion de las aves Argentinas. University Nac. de Tucuman, Inst. M. Lillo
Olrog CC (1976) Neotropica 22(68):180 (*Athene cunicularia patridgei*)
Olrog CC (1979a) Nueva Lista de la Avifauna Argentina. Ministerio Cultura y Educac. Fundac. M. Lillo
Olrog CC (1979b) Acta Zool Lilloana 33:5–7 (*Aegolius harrisii dabbenei*)
Olsen J, Wink M, Sauer-Gürth H, Trost S (2002) A new Ninox owl from Sumba, Indonesia. Emu 102:223–231 (*Ninox sumbaensis*)
Olsen PD (ed) (1993) Australian raptor studies. Australas. Raptor Assos.
Olsen PD, Mooney MN, Olsen J (1989) Status and conservation of the Norfolk Island Boobook *Ninox novaeseelandiae undulata*. In: Meyburg E, Chancellor R (eds) Raptors in the modern world. World Working Group on Birds of Prey and Owls, Berlin, pp 415–422
Olsen PD, Stokes T (1989) State of knowledge of Christmas Island Hawk Owl *Ninox squamipila natalis*. In: Meyburg E, Chancellor R (eds) Raptors in the modern world. Berlin, 411–414
Olsen SL (1995) The genera of owls in *Asioninae*. Bull Brit Ornith Cl 115(1):35–39
Olson P (2001) Feather and Brush. CSIRO
Orlando C (1957) Riv Ital Orn 27:54 (*Bubo bubo meridionalis*)
Osgood WH (1901) North Am Fauna 21:43 (*Nyctale acadica scotaea*)
Oustalet EM (1880) Bull Assoc Sci France 2(2):206 (*Ninox reyi*)
Owen DF (1963) Variation in North American Screech Owls and the subspecies concept. Syst Zool 12:8–14
Page WT (1920) The Bengal Eagle Owl. Bird Notes Ashborne 3:40–41
Pakenham RHW (1937) Bull Brit Ornith Cl 37:112 (*Otus pembaensis*)
Pakenham RHW (1979) Birds of Zanzibar and Pemba. BOU Checklist 2, British Ornithological Union, London
Pallas PS (1771) Reise durch verschiedene Provinzen des Russischen Reichs, 1. 455 (*Stryx uralensis*), (*Strix accipitrina*), 456 (*Strix pulchella*)
Pallas PS (1773) Reise durch verschiedene Provinzen des Russischen Reichs, 3. 706 (*Strix deminuta*)
Palmer AH (1895) The life of Josef Wolf. Longmans and Green
Panov EN (1973) The birds of South Ussuriland. Novosibirsk
Parker TA, Parker SA, Plenge MA (1982) An annotated checklist of Peruvian birds. Buteo Books
Parkes KC, Phillips AR (1978) Two new Caribbean subspecies of Barn Owl (*Tyto alba*) with remarks on variations on other populations. Ann Carnegie Mus 47:479–492 (*Tyto alba bondi*), (*Tyto alba niveicauda*)
Parmelee DF (1972) Canadas incredible arctic owls. Beaver 303:30–41
Parmelee DF, Macdonald SD (1960) The birds of the west-central Ellsmere Island and adjacent areas. Bull Nat Mus Can 169, Biol Ser 63:1–103
Parrot J-L (1908) *Athene cuculoides brügeli* n. ssp. Verh Ornith Ges Bay 8:104–107
Patridge WH (1956) Variaciones geográficas en la Lechuza negra, *Ciccaba huhula*. Hornero 10(2):143
Peale TR (1848) United States exploring expedition. During the years 1838, 1839, 1840, 1841, 1842. Under the command of Charles Wilkes. U.S.N. Mammalia and Ornithology, Philadelphia, 8:74 (*Strix lulu*), 75 (*Noctua venatica*)
Pennant T (1769) Indian zoology 3. (*Otus*), (*Otus bakkamoena*)
Penny M (1974) The birds of Seychelles. Collins
Perrins CM (ed) (1992) Die grosse Enzyklopädie der Vögel. Mosaik
Peters JL (1937) Journ Wash Acad Sci 27:82 (*Nesasio*)
Peters JL (1938) Systematic position of the genus *Ciccaba* Wagler. Auk 55:179–186
Peters JL (1940) Check-list of the birds of the world, 4. Cambridge, Mass.
Peters JL (1943) Bull Mus Comp Zool 92:297 (*Ciccaba albitarse opaca*)
Peters JL, Loveridge A (1935) Proc Biol Soc Wash 48:77 (*Tyto capensis libratus*)
Petersen LR (1979) Ecology of Great Horned Owls and Red Tailed Hawks in SE Wisconsin. Dep Nat Res Tech Bull 111, Madison, Wisconsin
Peterson RT (1949) How to know the birds. Mentor Book
Peterson RT (1963) A field guide to the birds of Texas. Mifflin Comp.
Peterson RT (1980) A field guide to the (eastern) birds. Mifflin Comp.
Peterson RT (1990) A field guide to the western birds. Mifflin Comp.
Peterson RT, Chalif EL (1973) A field guide to the Mexican birds and adjacent Central America. Mifflin Comp.
Peterson RT, Mountfort G, Hollom PAD (1976) Die Vögel Europas. Parey
Pettersson G (1984) Ugglor i Europa. Wiehen
Pforr M, Limbrunner A (1980) Ornithologischer Bildatlas der Brutvögel Europas. II. Neumann-Neud.
Phelps WH, Phelps WH Jr (1951) Proc Biol Soc Wash 64: 65–72 (*Glaucidium brasilianum margaritae*)
Phelps WH, Phelps WH Jr (1953) Proc Biol Soc Wash 66:128 (*Otus albogularis obscurus*)

Phelps WH, Phelps WH Jr (1954) Proc Biol Soc Wash 67:103 (*Otus ingens venezuelanus*)
Philippi RA (1940) Notas Ornithologicas. Rev Chil Hist Nat 44:147–152
Phillips AR (1942) Notes on the migration of the Elf Owl and Flammulated Screech Owls. Wilson Bull 54:132–137
Phillips AR, Marshall JT Jr, Monson G (1964) The birds of Arizona, Tucson. Strigiformes: 47–54
Phillips JC (1911) Auk 28:76 (*Strix virgata tamaulipensis*)
Piechocki R (1985) Der Uhu, 5th ed. NBB, Ziemsen
Piechocki R, März R (1985) Der Uhu. NBB, Ziemsen, Wittenberg
Pittermann W (2005) Der Chacokauz. Gefiederte Welt 6/2005: 170–173
Pizzey G, Doyle R (1998) A field guide to thebirds of Australia. Collins
Poliakov (1915) Mess Orn 6:44 (*Bubo bubo ussuriensis*)
Polivanov VM, Shebajev JV, Labshik VJ (1971) On the ecology of *Otus bakkamoena ussuriensis*. In: Ecology and fauna of birds .... Trudy Acad Sc USSR Vladivostok 2:85–91
Pölking F (1986) Naturfotografie. Jahrbuch 1985
Pontoppidan E (1763) Dansk Atlas 1:617, pl 25 (*Strix flammea*)
Poole C (1996) Arround the Oriental Indonesia. Lesser Masked Owl rediscovered. Bull Orient Bird Cl 23:14
Portenko LA (1972) Die Schnee-Eule. NBB, Ziemsen
Pratt HD, Bruner PL, Berrett DG (1987) The birds of Hawaii and the tropical Pacific. Princeton University Press, NY
Priest CD (1939) The Southern White Faced Owl. Ostrich 10:51–53
Prigogine A (1973) Le statut de *Phodilus prigoginei* Schouteden. Gerfaut 63:177–185
Prigogine A (1983) Un nouveau *Glaucidium* de l'Afrique centrale. Rev Zool Afr 97(31.12.1983):886–895 (*Glaucidium albertinum*)
Prigogine A (1985) Statut de quelques chevêchettes africaines et description d'une nouvelle race de *Glaucidium scheffleri*, du Zaire. Gerfaut 75:131–135 (*Glaucidium scheffleri clanceyi*)
Pucheran J (1849) Rev Mag Zool II(1):29 (*Scops rutilus*)
Puget A, Hüe F (1970) La Chevêchette *Glaucidium brodiei* en Afghanistan. Ois Rev France Orn 40:86–7
Pukinski J (1973) To the ecology of the Eagle Owl (*Ketupa blakistoni doerriesi*) in the basin of the river Bekin. Bull Mosc Soc Natur Biol Ser 78:40–47
Pukinski J (1975) In der Ussuri Taiga. Suche nach dem Riesenfischuhu. Brockhaus
Pyle P (1997) Flight-feather molt patterns and age in North American Owls. Am Birding Assoc, Monogr No. 2
Quoy JRC, Gaimard JP (1830) Voyage "Astrolabe". Zool 1:165 (*Noctua punctulata*), 166 (*Noctua variegata*), 168 (*Noctua zelandica*), 170 (*Scops manadensis*)
Raffaele H (1989) A guide to the birds of Puerto Rico and the Virgin Islands. Princeton University Press
Raffaele H, Wiley J, Garrido O, Keith A, Raffaele J (2003) Birds of the West Indies. Helm
Raffles S (1822) Trans Linn Soc London 13(2):279 (*Strix sumatrana*), 280 (*Strix scutulata*)
Ramsey EP (1877) Proc Linn Soc New S Wales 2:105 (*Ninox novae britanniae*)
Ramsey EP (1879) Proc Linn Soc New S Wales 3:249 (*Ninox undulata*), (*Ninox albomaculata*)
Ramsey EP (1880) Proc Linn Soc New S Wales 4:466 (*Ninox terricolor*)
Ramsey EP (1887) Proc Linn Soc New S Wales (2)1(1886):1086 (*Ninox connivens occidentalis*)
Ramsey EP (1888) Tabular list of all the Australian birds. p 36 (*Ninox albaria*)
Ramsey EP (1890) Catalogue of Australian Striges or nocturnal Birds of Prey in the collection of the Australian Museum at Sydney, N.S. Wales. pp 1–26
Ramsey EP (1898) Catalogue of the Australian Birds in the Australian Museum at Sydney, N.S. Wales, part 2, Striges, 2nd ed. pp 1–30
Rand AL (1950) A new race of owl, *Otus bakkamoena*, from Negros, Philippine Islands. Nat Hist Misc Chicago Acad Sci 72:1–5 (*Otus bakkamoena nigrorum*)
Rand AL (1951) Geographical variation on the Pearl Spotted Owlet. Nat Hist Misc [Chicago] 86:1–6
Rand AL, Fleming RL (1958) Birds from Nepal. Fieldiana Zool 41 (1): 1–218
Rand AL, Gilliard ET (1967) Handbook of New Guinea birds. Weidenfeld and Nicolson
Rasmussen PC (1998) Forest Owlet, *Athene blewitti*, rediscovered after 113-year hiatus. Bird Conserv Int 8(1):109
Rasmussen PC (1999) A new species of Hawk Owl, *Ninox*, from N Sulawesi, Indonesia. (*Ninox ios*). Wilson Bull 111(4):457–464
Rasmussen PC (2000) A new Scops-owl from Great Nicobar I. Bull Brit Ornith Cl 118(3):141–152
Rasmussen PC (2005) Birds of South Asia. The Ripley guide. Vol 1: 180 col pl, vol 2: 683 pp, Washington and Barcelona
Rasmussen PC, Abbott (1998) Living Bird 1998(Spring):28
Rasmussen PC, Collar N (1998) Identification, distribution and status of the Forest Owlet *Heteroglaux blewitti*. Forktail 14:41–49
Rasmussen PC, Collar N (1999) Major specimen fraud in the Forest Owlet *Heteroglaux* (*Athene auct.*) *blewitti*. Ibis 141:11–21
Rasmussen PC, Schulenburg TS, Hawkins F, Voninavoko R (2000) Geographic variation in the Malagasy Scops Owl (*Otus rutilus*), and the existence of an unrecognised species. Bull Brit Ornith Cl 120(2):75–102 (*Otus rutilus*), (*Otus madagascariensis*)
Reichenow A (1893) Diagnosen neuer Vogelarten aus Zentral Afrika. Orn Monatsb 1:60–62 and 65 (*Glaucidium sjöstedti*), 178 (*Glaucidium kilimense*)
Reichenow A (1901) Vog Afr 1:660 (*Asio nisuella* ssp. *maroccanus*), 667, 668 (*Otus icterorhynchus*)
Reichenow A (1902) Vog Afr, Atlas (*Otus icterorhynchus*)
Reichenow A (1903) Orn Monatsb 11:40 (*Pisorhina badia*), 85 (*Bubo bubo kiautschensis*), 86 (*Bubo bubo setschuanus*)
Reichenow A (1905) Vog Afr 3:822 (*Athene spilogaster somaliensis*)
Reichenow A (1906) Orn Monatsb 14:10 (*Bubo ascalaphus trotha*)
Reichenow A (1908) J Orn 56:139 (*Bubo vosseleri*)

Reichenow A (1910) J Orn 58:412 (*Bubo bubo norwegicus*), (*Bubo bubo hungaricus*)

Reiser O (1905) Anz Akad Wiss Wien 52(18):324 (*Bubo virginianus deserti*)

Reisner S (1987) Die schönen Lauscher der Nacht. GEO Nov./ 1987:82–98

Rice A (2004) Der verzauberte Blick. Frederking and Thaler

Richmond CW (1896) Proc US Nat Mus 18:663 (*Speotyto brachyptera*)

Richmond CW (1901) Auk 18:193 (*Cryptoglaux*)

Richmond CW (1903) Birds collected by Dr. W. L. Abbott on the coast and Islands of NW Sumatra. Proc U S Nat Mus 26:494 (*Pisorhina umbra*)

Rickett (1900) Bull Brit Ornith Cl 10:56 (*Scops latouchi*)

Ridgely RS, Greenfield PJ (2001) The birds of Ecuador. 2 vols, Pica Press

Ridgely RS, Gwynne JA (1989) A guide to the birds of Panama. Princeton University Press

Ridgway R (1873a) Bull Essex Inst 5:200 (*Strix flammea* var. *guatemalae*), (*Scops asio floridanus*), (*Syrnium nebulosum* var. *sartorii*)

Ridgway R (1873b) Proc Boston Soc Nat Hist 16:93 (*Glaucidium ridgwayi*)

Ridgway R (1874a) Am Sportsman 4:216 (*Speotyto cunicularia* var. *floridana*)

Ridgway R (1874b) In: Baird SF, Brewer TM, Ridgway R (eds) A history of North American birds; Land birds. Little, Brown and Company, Boston, 3, 90 (*Speotyto cunicularia guadeloupensis*)

Ridgway R (1877a) Field Forest 2:213 (*Scops asio maxwelliae*)

Ridgway R (1877b) US Ged Expl 40, Parallel ornith.: 572 (*Bubo virginianus saturatus*)

Ridgway R (1878) Proc US Nat Mus 1:102 (*Scops brasilianus cassini*), 116 (*Scops cooperi*)

Ridgway R (1880) Proc US Nat Mus 3(8):191 (*Strix nebulosa alleni*)

Ridgway R (1882) Proc US Nat Mus 4:366 (*Asio portoricensis*)

Ridgway R (1886) Auk 1886:333

Ridgway R (1887) Proc US Nat Mus 10:267 (*Megascops vermiculatus*), 268 (*Megascops hastatus*)

Ridgway R (1892) Auk 1892:Pl. 2 (*Megascops flammeola idahoensis*)

Ridgway R (1895) On the correct subspecific names of the Texas and Mexican Screech Owls. Auk 12:389, 390 (*Megascops asio cierascens*)

Ridgway R (1914) Birds of North and Middle America. Striges. US Nat Mus Bull 50:594–825

Riley JH (1913) Proc Biol Soc Wash 26:153 (*Tyto perlatus lucayanus*)

Riley JH (1916) Smiths Misc Coll 66(15):1 (*Asio noctipetens*)

Riley JH (1925) Proc Biol Soc Wash 38:10 (*Strix aluco nivipetens*)

Riley JH (1927) Proc Biol Soc Wash 40:93 (*Otus umbra enganensis*)

Ripley SD (1948a) Zoologica 33:200 (*Glaucidium cuculoides austerum*), (*Glaucidium cuculoides deignani*), 201 (*Glaucidium cuculoides delacouri*)

Ripley SD (1948b) Proc Biol Soc Wash 61:100 (*Athene brama ultra*)

Ripley SD (1953) Tori 13:49 (*Ninox scutulata yamashinae*)

Ripley SD (1964) Systematic and ecological study of New Guinea birds. Peabody Mus Bull 19:37–38

Ripley SD (1966) A notable owlet from Kenya (*Otus irenae*). Ibis 108:136–137

Ripley SD (1976) Reconsieration of *Athene blewitti* (Hume). J Bombay Nat Hist Soc 73(1):1–4

Ripley SD (1977) A revision of the subspecies of *Strix leptogrammica* Temm. 1831. Proc Biol Soc Wash 90:993–1001

Ripley SD, Bond J (1966) The Birds of Socotra and Abd-El-Kuri. Smiths Misc Coll 151(7):23–24

Ripley SD, Rabor DS (1958) Notes on a collection of birds from Mindoro I., Philippines. Peabody Mus Bull 13:40–41

Ripley SD, Rabor DS (1962) Postilla 73:4 (*Ninox scutulata palawanensis*)

Ripley SD, Rabor DS (1968) Two new subspecies of birds from the Philippines. Proc Biol Soc Wash 81:31–36 (*Otus scops mirus*)

Ripley SD, Rabor DS (1971) Postilla 50:4

Risdon DHS (1951) The rearing of a hybrid Virginian × European Eagle Owl at Budley Zoo. Avic Mag 57:199–201

Robbins MB, Howell SNG (1995) A new species of Pygmy Owl (Strigidae: *Glaucidium*), from eastern Andes. Wilson Bull 107(1):1–6

Robbins MB, Stiles FG (1999) A new subspecies of pygmy owl (Strigidae, *Glaucidium*) from the Pacific slpoe of the northern Andes. Auk 116(2):305–315

Roberts A (1922) Ann Transvaal Mus 8:212 (*Tyto capensis damarensis*)

Roberts A (1932) Ann Transvaal Mus 15:26 (*Smithiglaux capensis ngamiensis*)

Roberts TJ (1991) The birds of Pakistan. 1. Nonpasseriformes. Oxford University Press

Roberts TJ, King B (1986) Vocalizations of the owls of the genus *Otus* in Pakistan. Ornis Scand 17:299–305

Robertson CJR (ed) (1985) Readers Digest complete book of New Zealand birds. Sydney

Robinson HC (1911) J Fed Malay St Mus 4:246 (*Bubo coromandus klossi*)

Robinson HC (1927a) Exhibition and description of a new owl (*Athenoptera spilocephala stresemanni*) from Sumatra. Bull Brit Ornith Cl 47:126, 127

Robinson HC (1927b) Note on *Phodilus* Less. with proposed new name *Phodilus badius saturatus*, for the birds from Sikkim. Bull Brit Ornith Cl 47:121, 122

Robinson HC, Chasen FN (1939) The birds of the Malay Peninsula, 4. Witherby

Robinson HC, Kloss EB (1916) J Str Branch R As Soc 73:275 (*Pisorhina vandewateri*)

Robinson HC, Kloss EB (1922) J Fed Malay St Mus 10:261 (*Otus luciae siamensis*)

Robinson JW (ed) (1983) A field guide to the birds of Japan. Wild Bird Soc. Japan

Robson C (2000) A field guide to the birds of South East Asia. New Holland

Rochebrune AT de (1883) Bull Sci Soc Philom 7(7):165 (*Scotopelia oustaleti*)

Roowal, Nath (1849) Rec Ind Mus 46:162

Rothschild W (1917) On *Tyto arfaki* Schleg. Bull Brit Ornith Cl 37:17–19

Rothschild W, Hartert E (1902) Novit Zool 9:405 (*Speotyto cunicularia becki*)

Rothschild W, Hartert E (1907) Novit Zool 14:446 (*Strix flammea meeki*)

Rothschild W, Hartert E (1910) Novit Zool 17:110 (*Bubo bubo hispanus*), 111 (*Bubo bubo interpositus*), 112 (*Bubo bubo aharoni*)

Rothschild W, Hartert E (1913/1914) On some Australian forms of *Tyto*. Novit Zool 20:280–284

Rothschild W, Hartert E (1914a) The birds of the Admiralty Islands north of German New Guinea. Novit Zool 21: 291–298 (*Tyto manusi*)

Rothschild W, Hartert E (1914b) Bull Brit Ornith Cl 33:105 (*Ninox meeki*)

Rothschild W, Hartert E (1918) Novit Zool 25 (*Ninox theomacha goldii*)

Rozendaal EG, Dekker RW (1989) Annotated checklist of the birds of the Dumoga-Bone National Park, North Sulawesi. Kukila 4:85–109

Rutgers A (1966–1970a) Ein Vogelparadies in Farben. Europäische Vögel II. Gorssel (Gould reprint)

Rutgers A (1966–1970b) Ein Vogelparadies in Farben. Australische Vogelwelt II. Gorssel (Gould reprint)

Rutgers A (1966–1970c) Ein Vogelparadies in Farben. Vogelwelt Asiens I. Gorssel (Gould reprint)

Rutgers A (1966–1970d) Ein Vogelparadies in Farben. Vogelwelt von Neuguinea I. Gorssel (Gould reprint)

Safford RJ (1993) Rediscovery, taxonomy and conservation of the Anjouan Scops Owl, *Otus capnodes*. Bird Conserv Int 3:57–74

Safford RJ (1996) Appendix description and photos of the trapped *Otus capnodes*. (In letter)

Sagot F, Guerrero J (1998) Actas del IV Encuentro Boliviano para la Conservación de la Aves 25 a 27 de Octobre 1997. Tarija, Bolivia

Salomonsen F (1957) The birds of Greenland. Copenhagen

Salvadori TA (1874) Ann Mus Civ Genova 6:308 (*Athene dimorpha*)

Salvadori TA (1876) Ann Mus Civ Genova 7 (1875) 904 (*Scops beccarii*), 992 (*Ninox peninsularis*)

Salvadori TA (1881) Atti R Accad Sci Torino 16:619 (*Strix aurantia*)

Salvadori TA (1882) On birds, collected in New Britain. Ibis II:132

Salvadori TA (1887) Ann Mus Civ Genova 24:526 (*Syrnium niasense*)

Salvadori TA (1903) Mem R Accad Sci Torino 2(53):95 (*Scops feae*)

Salvadori TA, d'Albertis LM (1875) Ann Mus Civ Genova 7:802, 809 (*Ninox assimilis*)

Salvadori TA, Festa E (1900) Boll Mus Zool Anat Comp Torino 15(368):32 (*Pulsatrix fasciativentris*)

Salvin O (1897) Bull Brit Ornith Cl 6:37 (*Scops ingens*), (*Scops santa-catarinae*), 38 (*Scops roraimae*)

Salvin O, Godman F du C (1897) Biologia Centrali Americana. Pl 61

Sankaran R (1998) An annotated list of the endemic fauna of the Nicobar Islands. Forktail 13:17–22

Sargeant DE (1994) Vermiculated Fishing Owl: How to see one. Bull ABC 1/2:74

Sarker SU (1985) Owls of Bangladesh and their conservation. Birds Prey Bull 2:103–106

Sarudny (Zarudny) NA, Härms M (1902) Orn Jahrb 13:49 (*Scops semenowi*)

Sassi M (1912) Anz K Akad Wiss Wien Math-Naturw Kl 49:122 (*Asio abyssinicus graueri*)

Sauer F (1985) Sauers Naturführer: Afrikanische Vögel. Fauna

Savigny MJCL (1809) Descr Egypte 1(1):105 (*Noctua glaux*), 110 (*Bubo ascalaphus*)

Sayers BC (1976a) Avic Mag 128–136 (The Boobook Owl)

Sayers BC (1976b) Blakiston's Fish Owl. Avic Mag 82(2):61, 62

Sayers BC (1998a) The Laughing Owl (*Sceloglaux albifacies*). Tyto 3(1):134, 135

Sayers BC (1998b) The Congo or Tanzanian Bay Owl (*Phodilus prigoginei*). Tyto 3(4):134, 135

Schaaf R (1997) Eulen, verhaßt-gefürchtet-geliebt-verehrt. Naturschutz im Kleinen, 17. Landesgirokasse

Schalow H (1908) J Orn 56:109 (*Nyctale tengmalmi pallens*)

Scherzinger W (1968) Er lebt nicht im Urwald und heißt trotzdem Dschungelkauz. Vogelkosmos 5:204–6

Scherzinger W (1969a) Eulen – Grimassenschneider unter den Vögeln. Vogelkosmos 7:226–229

Scherzinger W (1986a) Kontrastzeichnungen im Kopfgefieder der Eulen als visuelle Kommunikationsmittel. Ann Naturhist Mus Wien 88/89 B: 37–56

Scherzinger W (1986b) Lebensgeschichte eines Perlkauzes. Gefiederte Welt 1986/11:305, 306

Scherzinger W (2005) Remarks on Sichuan Wood Owl *Strix uralensis davidi*. Bull Brit Orn Cl 2005:275–286

Schiebel G (1910) Orn Jahrb 21:102 (*Scops scops tschusii*)

Schlegel H (1862) Mus Pays-Bas 2, *Oti*: 2, note (*Otus midas*)

Schlegel H (1863) Mus Pays-Bas 2, *Oti:* 13 (*Bubo orientalis minor*)

Schlegel H (1866) Nederl Tijdschr-Dierk 1865(3):99 and 181 (*Strix Rosenbergi*), 183 (*Ninox ochracea*), 329 (*Ninox aruensis*)

Schlegel H (1871) Nederl Tijdschr-Dierk 1871(4):3 (*Noctua hoedtii*)

Schlegel H (1873) Mus Pays-Bas 2, Noctuae Rev: 3 (*Otus capensis major*), 13 (*Scops siaoensis*)

Schlegel H (1879) Notes Leyden Mus 18:50–52, 101. (*Strix inexspectata*), (*Strix tenebricosa arfaki*)

Schluter P (1987) The Brown Fishing Owl in Israel. Land and Nature 13(4):168–176

Schneider W, Eck S (1995) Schleiereulen. Westarp

Schodde R, Mason IJ (1980) Nocturnal birds of Australia. Landsdown, Melbourne

Schönn R, Schönn S (1987) Auf leisen Schwingen. Arnold

Schönn S (1978) Der Sperlingskauz. NBB, Ziemsen

Schönn S, Scherzinger W, Exo KM, Ille R (1991) Der Steinkauz. NBB, Ziemsen

Schouteden H (1952) Un Strigide nouveau d'Afrique noire: *Phodilus prigoginei* nov. sp. Rev Zool Bot Afr XLVI 34: 423–438

Schouteden H (1954) Fauna du Congo Belge et du Ruanda-Urundi, III. Ois non passer Ann Koningl Mus Belg Kongo Wet 29:1–437
Schüz E (1957) Occipital-Gesicht des Sperlingskauzes (*Glaucidium*). Vogelwarte 19:138–140
Sclater PL (1857) Proc Zool Soc Lond 4 (*Glaucidium californicum*)
Sclater PL (1858) Trans Zool Soc London 4:265, pl 61 (*Scops usta*)
Sclater PL (1859) Proc Zool Soc Lond 131 (*Ciccaba nigrolineata*)
Sclater PL (1877) Birds from New Britain. Proc Zool Soc Lond 108 (*Ninox odiosa*)
Sclater PL (1879a) Description of a new species of owl of the genus Ciccaba. Trans Zool Soc London 4
Sclater PL (1879b) On some new or little known species of *Accipitres* in the collection of the Norwich Museum. Trans Zool Soc London 4
Sclater PL (1879c) Remarks on the Nomenclatur of the British Owls, and on the arrangement of the Order Striges. Ibis 3:346–352
Sclater PL (1883) Proc Zool Soc Lond 52 (*Strix sororcula*), 52 and pl 11 (*Ninox forbesi*)
Sclater PL, Salvin O (1859) Ibis 221 (*Ciccaba nigrolineata*)
Sclater PL, Salvin O (1868a) On new birds. Proc Zool Soc Lond 58, 59 (*Scops barbarus*)
Sclater PL, Salvin O (1868b) On new birds. Proc Zool Soc Lond 56, 57 (*Syrnium fulfescens*)
Sclater PL, Salvin O (1868c) On new American birds. Proc Zool Soc Lond 327–329 (*Gymnoglaux lawrencii*)
Sclater WL (1859) Ibis 221 (*Lophostrix stricklandi*)
Sclater WL (1921) Bull Brit Ornith Cl 42:24 (*Tyto alba erlangeri*)
Sclater WL (1922) Bull Brit Ornith Cl 42:24 (*Asio tingitanus andrewsmithi*)
Scopoli GA (1769) Annus I Hist-Nat 19 (*Strix giu*), 21 (*Strix alba*), 22 (*Srix noctua*)
Scott D (1997) The Long-eared Owl. Hawk and Owl Trust, London
Screech C (1999) A profile of the Pel's Fishing Owl. Tyto 3(6):179–185
Scully J (1881) Ornithology of Gilgit. Ibis 14:423–425 (*Syrnium biddulphi*)
Seebohm H (1884) Proc Zool Soc Lond (1883):466 (*Bubo blakistoni*)
Seebohm H (1895) Bull Brit Ornith Cl 5:4 (*Bubo blakistoni doerriesi*)
Semenov (1899) Zap Imp Akad Nauk Classe Sci Phys Math ser 8(6):14 (*Athene glaux kessleri*)
Serle W (1949) New races of a warbler … and an owl, all from British Cameroons. Bull Brit Ornith Cl 69:74–76
Serventy DL, Whittell HM (1948) A handbook of the birds of Western Australia. pp 1–365
Serventy DL, Whittell HM (1962) Birds of Western Australia, 3rd ed
Severinghaus LL (1986) The biology of Lanyu Scops Owl (*Otus elegans botelensis*). Nat Ecol Conserv Soc Taipei, In Sympos Wildlife Cons 1:143–196
Severinghaus LL (1989) The status and conservation of the Lanyu Scops Owl, *Otus elegans botelensis*. Rapt Mod World 423–431
Severtzov NA (1873) Izv Imp O Liub Est Antr Etn 8, 1872(2): 115 (*Athene orientalis*)
Shany N (1995) Juvenile Papuan Owl *Uroglaux dimorpha* near Vanimo. Muruk 7(2):74
Sharpe RB (1870) Ibis 487 (*Syrnium nuchale*)
Sharpe RB (1871) Ibis 101, 417 (*Scotopelia ussheri*)
Sharpe RB (1875a) Catalogue of birds of the British Museum 2:1–314
Sharpe RB (1875b) Contributions to a history of the Accipitres II. The genus *Glaucidium*. Ibis 35–59 and 256 (*Syrnium davidi = Ptynx fulvescens "David"*), 261 (*Scotopelia bouvieri*), 260 (*Glaucidium cobanense*)
Sharpe RB (1876a) A new species of owl. Proc Zool Soc Lond 673 (*Ninox solomonis*)
Sharpe RB (1876b) On the geographical distribution of Barn Owls. Orn Misc [Rowley] 1:269–298. 2:1–21
Sharpe RB (1884) J Linn Soc Lond Zool 17:439 (*Syrnium bohndorffi*)
Sharpe RB (1886) Ibis 163 (*Bubo milesi*)
Sharpe RB (1888a) Suborder Striges, *Heteroscops luciae*. Ibis 77–79
Sharpe RB (1888b) On a collection of birds from the Island of Palawan. Ibis 193–197 (*Scops fuliginosa*), (*Syrnium whiteheadi*), 478 (*Scops luciae*)
Sharpe RB (1888c) Birds from Guadalcanar. Proc Zool Soc Lond 183–184
Sharpe RB (1891) Scientific research of second Yarkand Mission. Aves
Sharpe RB (1892) Bull Brit Ornith Cl 1(1892):4 (*Scops mantananensis*), (*Scops brookii*)
Sharpe RB (1893) Bull Brit Ornith Cl 3(1893):9 (*Scops sibutuensis*), 55 (*Glaucidium borneense*)
Sharpe RB (1897) Bull Brit Ornith Cl 6:47 (*Ninox everetti*), 247 (*Syrnium nigricantius*)
Sharpe RB (1899a) Bull Brit Ornith Cl 8:40 (*Gisella iheringi*)
Sharpe RB (1899b) Bull Brit Ornith Cl 10:28 (*Bubo mackinderi*)
Sharpe RB (1899c) On a new species of owl (*Gisella iheringi*) from Sao Paulo, Brazil. Bull Brit Ornith Cl 12:2–4
Sharpe RB (1899d) A hand-list of genera and species of birds, 1
Sharpe RB (1901) Bull Brit Ornith Cl 12:3 (*Scops holerythra*)
Sharpe RB, Ussher HT (1872) Ibis 182 (*Huhua shelleyi*)
Shaw F (1989) Birds of America. Arch Cape Press
Shaw G (1809) Gen Zool 7(1):253 (*Strix silvatica*), 257 (*Strix orientalis = Strix seloputo?*)
Shelley GE (1873) Ibis 138 (*Scops icterorhynchus*)
Shields J, King G (1990) Spotted Owls and Forrestry in the American Northwest: A conflict of interest. Birds Internat 2(1):34–45
Shirihai H (1996) The birds of Israel. Academic Press and Unipress
Short LL (1975) A zoogeographic analysis of the South American Chaco avifauna. Bull Am Mus Nat Hist 154: 163–352

Sibley CG (1996) Birds of the World. Thayer Birding Software, (Owls: 12 pp)
Sibley CG, Monroe BL Jr (1990) Distribution and taxonomy of birds of the world. Yale University Press
Sibley D (2000) The North American bird guide. Pica Press
Sibnev BK (1963) Observations of the Brown Fish Owl (*Ketupa zeylonensis*) in Ussuri Krai. Ornitologiia 6:486
Sibnev J (2000) Der Falke 2000:69–72 (*Bubo blakistoni*)
Sick H (1984) Ornithologia Brasileira, 1. Rio de Janeiro
Siebold PF (1844) Fauna Japonica. I. Descript. des Oiseaux
Silsby JD (1980) Inland birds of Saudi Arabia. London
Simmons K (1976) Breeding of the Bengal Eagle Owl. Avic Mag 82:135–138
Simpson K, Day N (1986) Field guide to the birds of Australia. Viking O'Neill
Sinclair I, Ryan P (2003) Birds of Africa, south of the Sahara. Struik, 760 pp
Singer D (1992) Greifvögel und Eulen. Franckh-Kosmos
Skemp JR (1955) Size and colour discrepancy in the Tasmanian Masked Owl. Emu 55:210, 211
Skipnoth P (1979) The great bird illustrations, 1730–1930. Hamlyn, 176 pp
Slater P (1970) A field guide to Australian birds. Non-Passeres. Scottish Acad. Press, Perth
Slater P, Slater P, Slater R (1989) The Slater field guide to Australian Birds, 6nd ed. Weldon
Slud P (1960) The birds of Finca "La Selva", Costa Rica: A tropical west forest locallity. Bull Am Mus Nat Hist NY 121, 46 pp
Slud P (1964) The birds of Costa Rica. Bull Am Mus Nat Hist NY 128, 430 pp
Smallbones G (1906) Orn Monatsb 14:27 (*Surnia ulula tianshanica*)
Smith A (1834) S Afr Q J 2:312 (*Noctua woodfordi*), 314 (*Scops capensis*), 313 (*Noctua capensis*), 316 (*Otus madagascariensis*), (*Otus capensis*), 317 (*Strix capensis*), (*Bubo capensis*)
Smith A (1839) Illustrations of the zoology of South Africa. Birds III
Smith WW (1984) *Sceloglaux albifacies* (Laughing Owl). New Zeal J Sci 2:86, 87
Smith-Svenson T (1969) Eulen, die auf Hügeln brüten. Vogelkosmos 2:54–57
Smythies BE (1960) The birds of Borneo. Oliver and Boyd
Smythies BE, Hughes AM (1984) Birds of Burma. Keltern (reprint from 1940)
Sneidern (1955) Noved Colomb 2:35 (*Ciccaba virgata occidentalis*)
Snyder NFR (1961) R Ontario Mus Publ 54:5 (*Bubo virginianus scalariventris*)
Snyder NFR, Wiley JW (1976) Sexual size dimorphism in hawks and owls of North America. Orn Monogr 20, AOU
Sonobe K (ed), et al. (1983) A field guide to the birds of Japan
Sparks J, Soper T (1979) Owls. Their natural and unnatural history. David and Charles
Sparrman A (1788) Mus Carls Fasc 51, pl 51 (*Strix arctica*)
Spix (1824) Av Bras 1:22–23, Pl. 9, lám. 10a
Sprunt A Jr (1955) North American Birds of Prey. Nat Audub Soc, Bonanza Books
Squire J (1987) Notes on Eastern Grass Owl *Tyto longimembris,* breeding in North Queensland. Aust Bird Watcher 12(2):66
Stachanov WS (1931) Kocsag 4:21 (*Surnia ulula orokensis*)
Steadman DW, Kirch PV (1998) Biogeography and prehistoric exploitation of birds in the Mussau Islands, Papua New Guinea. Emu 98:13–22
Stegmann B (1925) Dokl Akad Nauk SSSR A: 61 (*Strix nebulosa elisabethiae*)
Stegmann B (1926) Bull Brit Ornith Cl 47:39 (*Strix aluco obscurata*)
Stegmann B (1929) Annuaire Mus Zool Acad Sci URSS [Leningrad] 29 (1928) 178 (*Bubo bubo dauricus*), 181 (*Strix uralensis dauricus*)
Steinbach G (1980) Die Welt der Eulen.Hamburg
Steinbach G (1981) Vögel unserer Heimat. Habel
Steinbacher J (1962) Beiträge zur Kenntnis der Vögel von Paraguay. Abh Senckenb Nat Gesell 502:1–106
Steinberg R (1997a) Die Queen-Charlotte-Eule (*Aegolius acadicus brooksi*). SCRO-Mag 1:24–30
Steinberg R (1997b) Der Philippinen- o. Streifenuhu (*Bubo philippensis*). SCRO-Mag 1:31–32
Steinberg R (1998b) SCRO Mag 1998(2):22, 28–31
Steinberg R (1999) Der Gran Chaco Rotfußkauz. SCRO-Mag 1999/2:45–52
Stephens JF (1826) In: Shaw G (ed) Gen Zool 13(2)(1825):61 (*Strix novae hollsandiae*), 62 (*Nyctea*)
Stevenson T, Fanshawe J (2002) Field guide to the birds of East Africa. Poyser
Steyn P (1982) Birds of Prey of Southern Africa. David Philip and Croom Helm
Steyn P (1984) A delight of owls. African owls observed. Tanager Books and David Philip, Dover
Steyn P (1988) Partial albino Pel's Fishing Owl. Afr Wildl 42(3):180
Stiles FG, Skutch AF (1991) A guide to the birds of Costa Rica. A. & C. Black
Stolzmann J (1926) Ann Zool Polon Hist Nat 5:124 (*Otus choliba maxima*)
Stone A (1900) P Acad Nat Sci Phila (1899):478 (*Asio accipitrinus Mc Ilhenyi*)
Stone A (1922) A new Burrowing Owl from Colombia. Auk 39:84 (*Speotyto cunicularia carrikeri*)
Stone W (1896) A revision to the North American Horned Owls, with description of a new ssp. Auk XIII:153–6
Stone W (1897) Am Nat 31(363):237 (*Bubo virginianus pallescens*)
Stone W (1899) On a collection of birds from the vicinity of Bogotá, with a review of the South American species of *Speotyto and Troglodytes*. P Acad Nat Sci Phila 899: 302–304 (*Speotyto cunicularia tolimae*)
Stones AS, Davidson PL, Raharjuningtrah W (1997) Notes on the observation of a Taliabu Masked Owl *Tyto nigrobrunnea* in Taliabu Is., Indonesia. Kukila 9:58, 59
Storer PW (1972) The juvenil plumage and relationships of *Lophostrix cristata*. Auk 89:452–455

Storer RW (1994) Avian exotica: The fishing owls. Birder's World 8(3):66, 67

Stotz DF, Fitzpatrick JW, Parker TA, Moskowitz DK (1996) Neotropical birds. Ecology and conservation. University Chicago Press

Stresemann E (1922) Orn Monatsb 30:64 (*Asio helveola hova*)

Stresemann E (1923) Die Ergebnisse der W. Stötznerschen Expeditionen nach Szetschwan, Osttibet and Tschili. 2 (12). Striges bis Ralli. Abhandl Berl Tierk Völkerk Dresden 16:58–70

Stresemann E (1924) Die Gattung *Strix* im Malayischen Archipel. Orn Monatsb 32:110, 111

Stresemann E (1925) Beiträge zur Ornithologie der indo-australischen Region. Mitt Zool Mus Berlin 12(1):179–195

Stresemann E (1927–1934) Sauripoda: Aves. In: Kükenthal W (ed) Handbuch der Zoologie. De Gruyter

Stresemann E (1928) Orn Monatsb 36:41 (*Aegolius tengmalmi beickianus*), 54 (*Ninox scutulata javanensis*)

Stresemann E (1929) Orn Monatsb 37:47 (*Ninox fusca plesseni*)

Stresemann E (1930) Bull Brit Ornith Cl 50:61 (*Ninox novaeseelandiae remigialis*)

Stresemann E (1931) Orn Monatsb 39:105 (*Tyto inexspectata*)

Stresemann E (1933) Ein zweites Exemplar von *Tyto manusi* Rothsch. and Hart. Orn Monatsb 41:153

Stresemann E (1934) Über Vögel, gesammelt von Kopstein. S-Molukken and Tenimber, 1922–24. Zool Meded 17:15–17, 177 (*sororcula*)

Stresemann E (1938) Orn Monatsb 46:149 (*Ninox perversa*)

Stresemann E (1939) Die Vögel von Celebes. I, II. J Orn 87:299–425

Stresemann E (1940) Orn Monatsb 83:432, 433 (*Tyto inexspectata*)

Stresemann E (1940–1941) Orn Monatsb III. Systematik and Biologie. J Orn 88:1–135, 389–487, 89:1–102

Stresemann E (1951) Die Entwicklung der Ornithologie. Peters, (Aula) (reprint of 1996)

Strickland HE (1852) Contrib Orn 1852:60, pl 10 (*Scops cristata* var.)

Studer A, Teixeira DM (1994) Notes to the Buff fronted Owl *Aegolius harrisi* in Brazil. Bull Brit Ornith Cl 114(1):62, 63

Sung W (ed) (1998) China red data book endangered animals, Aves. Sci. Press, Beijing

Sushkin PP (1932) Alauda 4:395 (*Bubo bubo baschkirikus*)

Sutter E, Barruel P (1958) Die Brutvögel Europas. II. Silva

Sutton GM, Burleigh TD (1939) A new screech owl from Nuevo Leon. Auk 56:174, 175 (*Otus asio semplei*)

Svensson L, Grant JP, Mullarney K, Zetterström D (1999) Der neue Kosmos Vogelführer. Kosmos

Swainson W (1832) Fauna Bor Am 2:86 (*Bubo arcticus*)

Swainson (1836) Class Birds I. p 327

Swainson W (1837a) Birds W Afr 1 (*Scops senegalensis*)

Swainson W (1837b) Class Birds II. p 217 (*Scotophilus*)

Swarth HS (1910) Two new owls from Arizona. Univ Calif Pub Zool 7(1):1–8 (*Otus asio gilmani*), (*Strix occidentale huachucae*)

Swinhoe R (1863) Ibis 216 (*Athene pardalota*), 218 (*Bubo caligatus*)

Swinhoe R (1866) Ibis 396 (*Strix pithecops*)

Swinhoe R (1870a) Ibis (new ser) 6:342 (*Ephialtes umbratilis*)

Swinhoe R (1870b) On Chinese zoology. Proc Zool Soc Lond 447–448 (*Athene plumipes*)

Swinhoe R (1870c) Ann Mag Nat Hist Ser 4, 6:152 (*Ephialtes glabripes*), 153 (*Ephialtes Hambroecki*)

Swinhoe R (1874) Ibis (ser 3) 4:269 (*Lempijius erythrocampe*)

Sykes WH (1832) Proc Comm Zool Soc London 82 (*Strix Indranee*)

Taczanowski L (1891) Mem Acad Imp Sci St Petersb 7, Sci Math Phys Nat 39:128 (*Glaucidium passerinum orientale*)

Takano S (1983) A field guide to the birds of Japan. Wild Bird Soc Japan

Takano T, Hamaguchi T, Morioka T, Kanouchi T, Kabaya T (1991) Wild birds of Japan, 2nd ed. Tokyo

Taverner PA (1942) Canadian races of the Great Horned Owl. Auk 59:234–245

Taverner PA (1943) Birds of Canada. Owls. pp 257–265

Taylor PB (1987) Mackinder's Eagle Owl *Bubo c. mackinderi*, breeding in the Kenyan Rift Valley. Scopus 11:13–18

Temminck CJ (1820a) Pl col livr 1: pl 4 (*Strix lactea*)

Temminck CJ (1820b) Pl col livr 3: pl 16 (*Strix leucotis*)

Temminck CJ (1820c) Pl col livr 4: pl 20 (*Strix leschenault*)

Temminck CJ (1821a) Pl col livr 7: pl 39 (*Strix pumila*)

Temminck CJ (1821b) Pl. col livr 8: pl 46 (*Strix maugei*)

Temminck CJ (1821c) Pl col livr 9: pl 50 (*Strix africana*)

Temminck CJ (1821d) Man Orn ed 2, 1:97 (*Strix infuscata*)

Temminck CJ (1821e) Pl col livr 12: pl 68 (*Strix brama*)

Temminck CJ (1822) Pl col livr 25: pl 145 (*Strix atricapilla*), 146 (*Strix grallaria*)

Temminck CJ (1823a) Pl col livr 30: pl 174 (*Strix strepitans*)

Temminck CJ (1823b) Pl col livr 39: pl 230 (*Strix pagodorum*)

Temminck CJ (1824) Pl col livr 49: pl 289 (*Strix hirsuta*)

Temminck CJ (1825) Pl col livr 63: pl 373 (*Strix hylophila*)

Temminck CJ (1827) Pl col livr 73: pl 432 (*Strix furcata*)

Temminck CJ (1831) Pl col livr 88: pl 525 (*Strix leptogrammica*)

Temminck CJ, Schlegel H (1844–1850) In: Siebold's Fauna Japonica. Aves 25

Tengmalm (1793) K Vet Akad Nya Handl 14:267 (*Strix liturata*)

Thorburn A (1990) The bomplete illustrated "Thorburn's Birds". Select edit, Wiltshire, reprint

Thorstrom R, de Roland LAR (1997) First nest record and nesting behaviour of the Madagascar Red Owl *Tyto soumagnei*. Ostrich 68(1):42, 43

Thorstrom R, et al. (1995) Var Fagelvarld. p 95

Thorstrom R, Hart S, Watson RT (1997) New record, ranging behaviour, vocalization and food of the Madagascar Red Owl *Tyto soumagnei*. Ibis 139(3)477–481

Thunberg KP (1798) K Vet Akad Nya Handl 19:184 (*Strix lapponica*)

Ticehurst CB (1922) Description on new races of Indian birds. Bull Brit Ornith Cl 42:57, 122 (*Otus bakkamoena deserticolor*), (*Otus bakkamoena gangeticus*), (*Otus bakkamoena marathae*)

Ticehurst CB (1923) Ibis (1923):242 (*Otus sunia leggei*)

Tickell SR (1833) J As Soc Beng 2:572 (*Strix radiata*), (*Strix lugubris*)

Todd WCE (1916) Proc Biol Soc Wash 29:98
Todd WCE (1917) Proc Biol Soc Wash 30:6 (*Bubo viginianus elutus*)
Todd WCE (1947) Two new owls from Bolovia. Proc Biol Soc Wash 60:90–96 (*Ciccaba albitarse tertia*), (*Pulsatrix melanota philoscia*)
Townsend CH (1890) Proc U S Nat Mus 13:133 (*Speotyto rostrata*)
Traylor MAJ (1952) A new race of *Otus ingens* (Salvin), from Colombia. Nat Hist Misc Chicago Acad Sci 99:1–4 (*Otus ingens colombianus*)
Traylor MAJ (1958) Variation in South American Great Horned Owls. Auk 75:143–149
Trischitta (1939) (pamphlet) Alcune nuove….: 2 (*Athene noctua salentina*)
Tristram HB (1880) Description of a new genus and species of owl from the Seychelles islands. Ibis 4:456–459, 458 (*Gymnoscops insularis*)
Tristram C (1889) On a small collection of birds from the Louisiade and d'Entrecasteaux Islands. Ibis 6(1):557–558 (*Ninox rosseliana*)
Trommer G (1983) Greifvögel. Ulmer
Tsaka-Tsukasa N (1931) Tori 7(31):14 (*Strix uralensis momiyamae*)
Tschchikwischwili (1930) Bull Mus Georgic 5:97 (*Bubo bubo transcaucasicus*)
Tschusi zu Schmidhoffen V (1903) Orn Jahrb 14:139 (*pisorhina scops zarudnyi*), 166 (*Syrnium uralense sibiricum*)
Tschusi zu Schmidhoffen V (1904) Orn Jahrb 15:101 (*Pisorhina scops erlangeri*), 102 (*Pisorhina scops graeca*), (*Pisorhina scops tuneti*), 104 (*Pisorhina scops cycladum*)
Tsohsin C (1987) A synopsis to the birde of China. Sci. Press, Beijing
Turk A (1998) The Rufous Fish Owl. Tyto 3(5):144–148
Turner DA (1974) Cape Grass Owl in Ethiopia. Bull Brit Ornith Cl 94:38, 39
Uchida S (1922a) Handlist of Japanese birds
Uchida S (1922b) Revised handlist of Japanese birds
Uchida S (1926) Revised handlist of Japanese birds
van Balen DJ (1968) Birds of Sumatra. Ardea 56
van der Weyden WJ (1974) Vocal affinities of the Puerto Rican and Vermiculated Screech Owls, (*Otus nudipes* and *Otus guatemalae*). Ibis 116:369–372
van der Weyden WJ (1975) Scops and Screech Owls: Vocal evidence for a basic subdivision in the genus *Otus*. Ardea 63:65–77
van Marle JG, Voous KH (1988) The birds of Sumatra. An annotated checklist. BOU Checklist 10
van Rossem AJ (1932) A southern race of the Spotted Screech Owl. Trans San Diego Soc Nat Hist VII(17):183–186 (*Otus trichopsis mesamericanus*)
van Rossem AJ (1937) The Ferruginous Pygmy Owl of northwestern Mexico. Proc Biol Soc Wash 50:27, 28 (*Glaucidium brasilianum cactorum*)
van Rossem AJ (1938) Condor 40:258 (*Otus trichopsis guerrensis*)
Vaurie C (1960a) Systematic notes on palaearctic birds. 44. Strigidae, *Bubo*. Am Mus Novit 2000:1–13
Vaurie C (1960b) The genus *Athene*. Am Mus Novit 2015:1–21
Vaurie C (1960c) The genera *Otus*, *Aegolius*, *Ninox* and *Tyto*. Am Mus Novit 2021:1–21
Vaurie C (1965) The birds of the palaearctic fauna. 1. Non Passeres. Strigidae, Witherby, pp 578–634
Verheyen RK (1946) Bull Mus R Hist Nat Belg 22:1 (*Glaucidium tephronotum kivuense*)
Vieillot LJP (1807) Ois Am Sept 1:52, pl 20 (*Bubo clamator*), 53, pl 22 (*Bubo nudipes*)
Vieillot LJP (1817) Nouv.Dict.Hist.Nat 7:22 (*Strix ferox*), and (*Strix fusca*), 33 (*Strix superciaris*), 34 (*Strix suinda*), 39 (*Strix choliba*), 44 (*Strix nacurutu*), 45 (*Strix maculata*)
Vieillot LJP (1818) Nouv.Dict.Hist.Nat 7:26 (*Strix perlata*)
Vielliard JME (1989) Rev Bras Zool 6:685–693 (*Glaucidium hardyi*)
Vigors NA (1831) Proc Comm Zool Soc London 8 (*Noctua cuculoides*)
Vigors NA, Horsfield T (1826) Trans Linn Soc London 15:189 (*Noctua maculata*)
Vincent J (1990) Type locality of the Barred Owl (*Glaucidium capense*). Bull Brit Ornith Cl 110(4):170, 171
Volmer B (1999) Der Falke 1999(10):298–300
von Berlepsch H (1901) Bull Brit Ornith Cl 12:6 (*Pulsatrix sharpei*)
von Berlepsch (1908) Novit Zool 15: 288
von Berlepsch H, Stolzmann J (1892) On birds from Peru. Proc Zool Soc Lond 387–388 (*Speotyto cunicularia nanodes*)
von Berlepsch H, Stolzmann J (1902) On the ornithology of Central Peru. Proc Zool Soc Lond 40–41 (*Soeotyto cunicularia juninensis*)
von Berlepsch H, Taczanowski L (1884) On birds collected in western Ecuador. Proc Zool Soc Lond 309, 310 (Strigidae: *Bubo nigrescens*)
von Boetticher H (1929) Eine neue Rasse der Kanincheneule *Speotyto cunicularia* (Mol). Senckenbergiana 11:386–392 (*Speotyto cunicularia pichinchae*)
von Erlanger C (1897) Orn Monatsb 5:192 (*Bubo ascalaphus barbarus*), (*Bubo ascalaphus desertorum*)
von Erlanger C (1904) Beiträge zur Vogelkunde Nordafrikas. J Orn 18:231–233 (*Asio leucotis nigrovertex*), (*Asio abyssinicus*)
von Heuglin T (1863) Beiträge zur Ornithologie Nordostafrikas. J Orn:12, 13 1(*Surniuum sic umbrinum*), 14 (*Athene spilogastra* = Nomen nudum)
von Heuglin T (1869) Ornithologie Nordostafrikas, vol 1, 119 (*Noctua spilogastra*)
von Jordans A (1924) J Orn 72:407 (*Otus scops mallorcae*), 409 (*Tyto alba kleinschmidti*)
von Jordans A (1950) In Syllegomena biologica: 176. Geest and Portig (*Strix aluco clanceyi*)
von Jordans A, Neubaur F (1932) Falco 28:9 (*Otus leucotis margarethae*)
von Jordans A, Steinbacher J (1942) Ann Naturhist Mus Wien 52:234 (*Athene noctua grüni*)
von Lehmann FC (1946) Two new birds from the Andes of Colombia. Auk 63:218–222
von Linné C (Linnaeus) (1758) Systema naturae per regna tria naturae, secundum classes, ordines, genera, species, cum characteribus differentiis, synonymis, locis, 10th ed. 1

von Loudon H (1905) Orn Monatsb 13:129 (*Pisorhina scops turanicus*)
von Madarász J (1900) Magyar Madarai 203 (*Glaucidium setipes*)
von Madarász J (1901a) Termes Fuzet 24:272 (*Scops cypria*)
von Madarász J (1901b) Orn Monatsb 9:54 (*Asio canariensis*)
von Madarász J (1904) Ann Mus Natl Hungar 2:115 (*Strix stictica*)
von Müller JW (1853/1854) Beiträge zur Ornithologie Afrikas
von Pelzeln A (1863) Verh Zool-Bot Ges Wien 15:1125 (*Syrnium superciliare*)
von Pelzeln A (1872) J Orn 20:23 (*Strix insularis*)
von Spix J (1824) Av Bras 1:22, pl 9 (*Strix crucigera*), 23 (*Strix albomarginata*)
von Tschudi JJ (1844) Arch Naturgesch 10(1):266 (*Noctua melanota*)
von Zedlitz A (1908) Orn Monatsb 16:172, 173 (*Scotopelia peli fischeri*), (*Scotopelia peli salvago-raggii*)
Voous KH (1964) Wood Owls of the genera *Strix* and *Ciccaba*. Zool Meded Rijksmus Nat Hist Leiden 471–8
Voous KH (1966) The distribution of owls in Africa in relation to general zoogeographocal problems. Ostrich (Supplem.) 6, pp 499–506
Voous KH (1983) Birds of the Netherlands Antilles. Zutphen
Voous KH (1988) Owls of the Northern Hemisphere. Collins
Wagler JG (1832) Isis v Oken, Col 275 (*Glaucidium gnoma*), 276 (*Scops trichopsis*), 1221 (*Nyctalops stygius*)
Walden (= A of Tweeddale) (1874) Ann Mag Nat Hist 4(13): 123 (*Scops modestus*), 129–131 (*Ninox affinis*)
Walden (= A of Tweeddale) (1875) Trans Zool Soc London 9:145 (Lempijius megalotis)
Wallace AR (1863) On the birds of Bouru. Proc Zool Soc Lond 22 (*Athene hantu*), 487 (*Scops silvicola*), 488 (*Athene florensis*)
Warakagoda DH (2001) The discovery of a new owl. Loris 22:45–47
Warakagoda DH, Rasmussen P (2004) A new species of scops-owl from Sri Lanka. Bull Brit Ornith Cl 124(2):85–105
Watson J (1980) The case of the vanishing owl. Wildlife 22: 39, 39
Webb HP (1992) Field observations of the birds of Santa Isabel, Solomon Is. Emu 92(1):52–57
Webster JD, Orr RT (1958) Variation in Great Horned Owls of Middle America. Auk 75:134–142
Weick F (1985) Check list, Strigidae, distribution, measurements and bodymasses, plumage etc., 58 pp (unpubl.)
Werther (1998): Die mittleren Hochländer des nördlichen Deutsch-Ost-Afrika. 272
Weick F (1999a) Zur Taxonomie der amerikanischen Uhus (*Bubo* spp.). Okol Vogel 21:363–387
Weick F (2003–2005) Neue and wiederentdeckte Eulen der zurückliegenden 15 Jahre. Gefiederte Welt 2003–2005, Fortsetzungen. Ulmer
Weick F (2004) Europas heimliche Jäger. Kalender, Messner
Wells DR (1986) Further parallels between the Asian Bay Owl *Phodilus badius* and *Tyto* species. Bull Brit Ornith Cl 106:12–15
Weske JS, Terborgh JW (1981) *Otus marshalli*, a new species of Screech Owl from Peru. Auk 98(1):1–7
Wetmore A (1922) New forms of neotropical birds. Journ Wash Acad Sci 12(14):323–325 (*Glaucidium nanum vafrum*)
Wetmore A (1935) The type specimen of Newton's Owl. Auk 52:186–187
Wetmore A (1968) The birds of the Republik of Panama, 2. Smiths Misc Coll 150(2):1–603
Wetmore A, Swales BH (1931a) The birds of Haiti and the Dominican Republic. US Nat Mus Bull 150:1–483
Wetmore A, Swales BH (1931b) US Nat Mus Bull 155:41 and 239 (*Speotyto cunicularia troglodytes*)
Wheeler L (1938) A new Wood Owl from Chile. Field Mus Nat Hist Pub Zool Ser 20:471–82 (*Strix rufipes sanborni*)
Whistler H (1963) Popular handbook of Indian birds, 4th ed. Oliver and Boyd (reprint from 1949)
White CMN, Bruce MD (1986) The birds of Wallacea. BOU Checklist 7, Zool Soc London
White GB (1974) Rarest Eagle Owl in trouble. Oryx 12: 484–486
Whitehead J (1899) Birds collected in the Philippine Islands. Ibis 7(5):96–100 (*Scops mindorensis*)
Widodo W, Cox JH, Rasmussen PC (1999) Rediscovery of the Flores Scops Owl *Otus alfredi* on Flores, Indonesia, and reaffirmation of its specific status. Forktail 15:15–23
Wied-Neuwied MAP (1820) Reise nach Brasilien in den Jahren 1815 bis 1817. 1:105 (*Strix ferruginea*), 366 (*Strix pulsatrix*)
Wied-Neuwied MAP (1830) Beiträge zur Naturgeschichte Brasiliens, 3(1), pp 242 (*Strix minutissima*)
Wiley JW (1986) Status and conservation of raptors in the West Indies. Birds Prey Bull 3:57–70
Wilkinson R (1999) The African Fishing Owl with particular reference to historical records from European zoos. Tyto 3(6):166–175
Willard DE, Forster MS, Barroclough GF, Dickerman RW, Cannell PF, Coats S, Carcraft JL, O'Neill JP (1991) The birds of Cerro de la Neblina Territorio Federal Amazonas, Venezuela. Fieldiana Zool new series 65:1–80
Williams GR, Harrison M (1972) The Laughing Owl *Sceloglaux albifacies,* a general survey of a near extinct species. Notornis 19(1):4–19
Williams JG (1963) A field guide to the birds of East and Central Africa. Collins
Williams JG (1971) Säugetiere and seltene Vögel in den Nationalparks Ostafrikas. Parey
Williams JG (1973) Die Vögel Ost- and Zentralafrikas. Parey
Williams RS, Tobias JA (1996) West Peruvian Screech Owl. Cotinga 6:76–77
Wilson A (1975) Bird engravings. Dove Publ. (1766–1813 reprint)
Wink M, Heidrich P (1999) In: König et al. (eds) Molecular evolution and systematics of the owls. pp 39–57
Winkelmann C (1990) Informationen über die Bergwald-Kreischeule *Otus hoyi* nov. spec. and Anmerkungen zur Artzugehörigkeit südamerikanischer Eulen. Trochilus 11:133–135
Witherby H (1905) Bull Brit Ornith Cl 15:36 (*Syrnium aluco mauritanicum*)

Witherby H (1920) Brit Birds 13:283 (*Athene noctua mira*)
Wolf J (1810) In: Meyer, Wolf (eds) Meyer and Wolfs Taschenbuch der deutschen Vogelkunde, I. p 84 (*Strix uralensis macroura*)
Wolfe W, de la Torre J (1990) A photographic study of the North American species owls, their life and behaviour. Crown Publ.
Woods RW (1975) The birds of the Falkland Islands. Oswestry
Wotzkow C (1990) Aspectos reproductivos de *Glaucidium and Gymnasio* en la Ciénaga de Zapata. Pitirre 3(3):10
Wüst W (1970) Die Brutvögel Mitteleuropas. Bayrischer Schulbuch-Verlag
Wylie SR (1976) Breeding the Mottled Owl at the St. Louis Zoo. Avic Mag 82:64–65
Xantus J (1860) P Acad Nat Sci Phila (1859) 193 (*Syrnium occidentale*)
Xianji W, Lan Y (1994) Distribution and conservation of Strigiformes in Yunnan Province, China. Rapt Conserv Today 579–586
Yalden D (1973) Prey of the Abyssinian Long eared Owl, *Asio abyssinicus*. Ibis 115:605, 606
Yamamoto S (1994) Mating behaviour of Blakiston's Fish Owl, *Ketupa blakistoni*. Rapt Conserv Today 587–590
Yamashina Y (1936) Tori 9:220 (*Strix aluco yamadae*)
Yamashina Y (1938) Tori 10:1 (*Pyrroglaux*)
Yang L, Li G (1989) A new subspecies of *Athene brama*, (Spotted Little Owl) – *Athene brama poikila* (Belly mottled Little Owl). Zool Res 10(4):303–308. English summary
Yealland JJ (1968) Breeding of the Javan Fish Owl (*Ketupa ketupu*) at the London Zoo. Avic Mag 74:17–18
Yealland JJ (1969a) Breeding of the West African Wood – Owl (*Ciccaba woodfordi nuchalis*) at the London Zoo. Avic Mag 75:53
Yealland JJ (1969b) Breeding of the Magellan Eagle Owl (*Bubo virginianus nacurutu*) London Zoo. Avic Mag 75:53, 54
Yen KY (1932) Ois Rev France Orn 3:242 (*Tyto longimembris melli*)
Zaletaev VS (1962) Ornitologiia 4:190 (*Bubo bubo gladkovi*)
Zarudny (Sarudny) NA (1905a) Orn Jahrb 16:141 (*Ketupa semenowi*), 142 (*Bubo bubo nikolskii*)
Zarudny (Sarudny) NA (1905b) Orn Monatsb 13:49 (*Syrnium sancti-nicolai*)
Zarudny (Sarudny) NA (1911) Orn Monatsb 19:14 (*Syrnium härmsi*), 34 (*Syrnium blanfordi*)
Zarudny NA, Bilkewitch (1918) Izvestia Zakaspiiskago Muzeia 16 (*Asio otus turcmenica*)
Zarudny NA, von Loudon H (1904) Orn Jahrb 56 (*Carine noctua caucasica*)
Zarudny NA, von Loudon H (1906) Orn Monatsb 14:151 (*Asio accipitrinus pallidus*)
Zarudny NA, von Loudon H (1907) Orn Monatsb 15:2 (*Surnia ulula korejevi*)
Zedlitz O (1908) Kurze Notizen zur Ornis von Nordost-Afrika. Orn Monatsb 16:172–174
Zheng Z, et al. (1980) New records of Chinese birds from Xizang, Tibet. Acta Zool Sinica 26:286–287
Zhitkov BM, Buturlin SA (1906) Zapiski Russkogo Geographicheskogo Obschestva 41:272 (*Bubo bubo ruthenus*)
Zimmerman DA (1972) Avifauna of Kakamega Forest. Bull Am Mus Nat Hist Wash 149:291–292
Zimmerman DA, Turner DA, Pearson DJ (1996) Birds of Kenya and Northern Tanzania. A. & C. Black

Part VII Teil VII

Indices Indices

Index of Scientific Owl Names	Index der wissenschaftlichen Eulennamen
Index of Vernacular English Owl Names	Index der englischen Eulennamen
Index of Vernacular German Owl Names	Index der deutschen Eulennamen
Index of Vernacular French Owl Names	Index der französischen Eulennamen
Index of Vernacular Spanish (Portuguese) Owl Names	Index der spanischen (portugiesischen) Eulennamen
Index of Geographical Terms	Index der geographischen Namen

Index of Scientific Owl Names / Index der wissenschaftlichen Eulennamen

(Names in brackets are synonyms / In Klammern stehende Namen sind Synonyme)

A

abyssinicus, Asio 239, *257*
(*abyssinicus, Otus*) 239
(*acadica, Strix*) 200, 202
acadicus, Aegolius 202, *251*
(*accipitrina, Strix*) 241
(*adspersa, Strix*) 16
Aegolius 200
Aegolius acadicus 202, *251*
Aegolius acadicus acadicus 202, 266
(*Aegolius acadicus brodkorbi*) 202
Aegolius acadicus brooksi 203
(*Aegolius funerea juniperi*) 201
Aegolius funereus 200, *251*
Aegolius funereus caucasicus 201
Aegolius funereus funereus 200
Aegolius funereus magnus 201
Aegolius funereus pallens 201
Aegolius funereus richardsoni 202
Aegolius harrisii 204, *204*
Aegolius harrisii dabbenei 205
Aegolius harrisii harrisii 204
Aegolius harrisii iheringi 204
Aegolius ridgwayi 203
(*Aegolius ridgwayi tacanensis*) 203
(*Aegolius tengmalmi beickianus*) 201
(*aequatorialis, Ciccaba*) 82, 90
affinis, Ninox 219
(*affinis, Strix*) 17
(*africana, Strix*) 111
africanus, Bubo 111, *252*
(*alba, Strix*) 15
alba, Tyto 15, *249*
(*albaria, Ninox*) 214
albertinum, Glaucidium 184, *281*
(*albicollis, Otus*) 237
albifacies, Athene 232
albifacies, Sceloglaux *232*
albigularis, Otus (*Macabra*) *90*
(*albitarse, Syrnium*) 141
albitarsis, Strix 141
(*albiventer, Glaucidium*) 162
(*albiventris, Scops*) 57
(*albo-gularis, Syrnium*) 90
albogularis, Otus 82, 90
(*albomaculata, Ninox*) 209
(*albomarginata, Strix*) 144
alfredi, Otus 39, 261, *272*
(*alfredi, Pisorhina*) 39
alius, Otus 61, 263, *275*
aluco, Strix 127, 133, *254*
(*Aluco, Strix*) 133
(*amaura, Speotyto*) 196
(*amauronota, Strix*) 31
(*americana, Strix*) 235
angelinae, Otus *42*, 261
(*angelinae, Pisorhina*) 42
(*arctica, Strix*) 241
(*arcticus, Bubo*) 100
(*arixuthus, Phodilus*) 35
(*aruensis, Ninox*) 207
ascalaphus, Bubo 109, *110*, *252*
(*Ascalaphus, Bubo*) 109
Asio 234
Asio abyssinicus 239, *257*
Asio abyssinicus abyssinicus 239
Asio abyssinicus graueri 239
(*Asio accipitrinus Mc Ilhenyi*) 241
(*Asio accipitrinus pallidus*) 241
(*Asio butleri*) 136
(*Asio canariensis*) 238
Asio capensis 244, *257*
Asio capensis capensis 244
Asio capensis hova 245
Asio capensis tingitanus 244
(*Asio Chauvini*) 240
Asio clamator 234, *257*
Asio clamator clamator 234
Asio clamator forbesi 234
Asio clamator midas 235
Asio clamator oberi 235
Asio flammeus 240
Asio flammeus bogotensis 243
Asio flammeus domingensis 242
Asio flammeus flammeus *240*, 241, *257*
Asio (*flammeus*) *galapagoensis* 243
Asio flammeus pallidicaudas 242
Asio flammeus ponapensis 242
Asio flammeus sandwichensis 242
Asio flammeus sanfordi 243
Asio flammeus suinda 243
(*Asio flammeus suindus*) 140
(*Asio galapagoensis aequatorialis*) 243
(*Asio helveola hova*) 245
(*Asio leucotis nigrovertex*) 94
(*Asio maculosus amerimnus*) 111
Asio madagascariensis 240, *257*
(*Asio magellanicus algistus*) 99
(*Asio magellanicus heterocnemis*) 100
(*Asio magellanicus icelus*) 99
(*Asio magellanicus lagophonus*) 99
(*Asio magellanicus melancercus*) 102
(*Asio magellanicus mesembrinus*) 102
(*Asio nisuella ssp. maroccanus*) 244
(*Asio noctipetens*) 236
Asio otus 237
Asio otus canariensis 238
Asio otus otus 237, *257*
Asio otus tuftsi 238
(*Asio otus turcmenica*) 237
Asio otus wilsonianus 238
(*Asio portoricensis*) 242
Asio stygius 235, *257*
Asio stygius barberoi 237
(*Asio stygius lambi*) 235
Asio stygius robustus 235, *236*
Asio stygius siguapa 236
Asio stygius stygius 236
(*Asio tingitanus andrewsmithi*) 244
asio, Otus 74, *251*
asio, Strix 70
(*Asio, Strix*) 74
(*aspersus, Megascops*) 76
(*assimilis, Ninox*) 208
assimilis, Phodilus (*badius*) 35
(*assimilis, Phodilus*) 35
Athene 188
(*Athene ? fortis*) 209
Athene albifacies 232
(*Athene bactrianus*) 191
Athene blewitti 193, *282*
Athene brama 192, *251*, *282*
Athene brama albida 192
Athene brama brama 193
Athene brama indica 192
(*Athene brama mayri*) 193
Athene brama pulchra 193
Athene brama ultra 192
(*Athene capense*) 178
(*Athene castanonota*) 179
(*Athene Chiaradiae*) 188
(*Athene cuculoides bruegeli*) 180
Athene cunicularia 194
Athene cunicularia amaura 196
Athene cunicularia boliviana 198
Athene cunicularia brachyptera 196
Athene cunicularia carrikeri 197
Athene cunicularia cunicularia 199
Athene cunicularia floridana 195
Athene cunicularia grallaria 198
Athene cunicularia guadeloupensis 196
Athene cunicularia hypugaea *194*, *251*
Athene cunicularia juninensis 198
Athene cunicularia minor 196
Athene cunicularia nanodes 198
Athene cunicularia partridgei 199
Athene cunicularia pichinchae 197
Athene cunicularia punensis 197
Athene cunicularia rostrata 195
Athene cunicularia tolimae 197
Athene cunicularia troglodytes 195
(*Athene dimorpha*) 231
(*Athene florensis*) 215
(*Athene glaux kessleri*) 189
(*Athene hantu*) 225
(*Athene humeralis*) 207
(*Athene hypogramma*) 224
(*Athene indigena*) 189
(*Athene jacquinoti*) 229
(*Athene leucopsis*) 43, 214
(*Athene malabaricus*) 178
(*Athene malaccensis*) 218
(*Athene marmorata*) 213
Athene noctua 188, *251*
Athene noctua bactriana 191
(*Athene noctua cantabriensis*) 188
(*Athene noctua daciae*) 189
Athene noctua glaux 189
(*Athene noctua Grüni*) 188

(Athene noctua impasta) 191
Athene noctua indigena 189
Athene (noctua) lilith 190
Athene noctua ludlowi 191
(Athene noctua mira) 188
Athene noctua noctua 188, 266
Athene noctua orientalis 191
Athene noctua plumipes 191
(Athene noctua salentina) 189
(Athene noctua solitudinis) 189
Athene noctua somaliensis 190
Athene noctua spilogastra 190
Athene noctua vidalii 188
(Athene ocellata) 212
(Athene orientalis) 191
(Athene pardalotum) 177
(Athene plumipes) 191
(Athene pulchra) 193
(Athene rufa) 206
(Athene rufostrigata) 208
(Athene Sarda) 188
(Athene spilogaster somaliensis) 190
(Athene spilogastra) 190
(Athene squamipila) 225
(Athene strenua) 206, 207
(Athene taeniata) 229
(Athene Vidalii) 188
(Athene Whitelyi) 180
(Athene whitneyi) 186
(Athenoptera spilocephalus stresemanni) 39
(atricapilla, Strix) 84
atricapillus, Otus 84, 251, 265, 278
aurantia, Tyto 25

B

(bactrianus, Athene) 191
(badia, Pisorhina) 38
(badia, Strix) 34, 35
badius, Phodilus 34, 249, 272
(bahamensis, Speotyto) 195
bakkamoena, Otus 36, 65
(Balli, Ephialtes) 38
balli, Otus 38, 261
barbarus, Otus 81, 265
(barbarus, Scops) 81
(barbata, Strix) 152
bartelsi, Strix 131
(bartelsi, Syrnium) 131
(baweana, Strix) 127
beccarii, Otus 58, 263
(beccarii, Scops) 58
bengalensis, Bubo 109
(bengalensis, Otus) 109
(Bergiana, Nyctale) 142
(Berneyornis) 206
(biddulphi, Syrnium) 135
blakistoni, Bubo 120, 253
(Blakistoni, Bubo) 120
(blanfordi, Syrnium) 135
blewitti, Athene 193, 282
(blewitti, Heteroglaux) 188
(Blewitti, Heteroglaux) 193
(Bohndorffi, Syrnium) 137
(boholensis, Otus) 68
bolivianum, Glaucidium 166, 255, 279
boobook, Ninox 210
(Boobook, Strix) 213
(borellianum, Syrnium) 140
(borneense, Glaucidium) 177
(bouruensis, Scops) 57
bouvieri, Scotopelia 126, 253
(boweri, Spiloglaux) 213
(brachyotis, Strix) 241
(brachyptera, Speotyto) 196
brama, Athene 192, 251, 282
(brama, Strix) 193
(brasiliana, Strix) 171
brasilianum, Glaucidium 169, 279
brodiei, Glaucidium 176
(Brodiei, Noctua) 176
brookii, Otus 62
(brookii, Scops) 63
(Brucei, Ephialtes) 49
brucei, Otus 49, 50, 250
Bubo 99
Bubo africanus 111, 252
Bubo africanus africanus 111
(Bubo africanus kollmanspergeri) 112
Bubo (africanus) milesi 112
Bubo africanus tanae 112
(Bubo arcticus) 100
Bubo ascalaphus 109, 110, 252
(Bubo Ascalaphus) 109
(Bubo ascalaphus barbarus) 109
(Bubo ascalaphus desertorum) 109
(Bubo ascalaphus trotha) 111
Bubo bengalensis 109
Bubo blakistoni 120, 253
(Bubo Blakistoni) 120
Bubo blakistoni blakistoni 120
Bubo blakistoni doerriesi 120
(Bubo blakistoni karafutonis) 120
(Bubo blakistoni piscivorus) 120
Bubo bubo 104, 252
(Bubo bubo aharonii) 106
(Bubo bubo armeniacus) 106
(Bubo bubo auspicabilis) 106
(Bubo bubo baschkirikus) 106
(Bubo bubo borissowi) 107
Bubo bubo bubo 104, 105
(Bubo bubo dauricus) 107
(Bubo bubo engadinensis) 104
(Bubo bubo eversmanni) 108
(Bubo bubo gladkovi) 108
Bubo bubo hemachalana 109
Bubo bubo hispanus 104
(Bubo bubo hungaricus) 104
(Bubo bubo inexpectatus) 107
Bubo bubo interpositus 106
Bubo bubo jakutensis 107
(Bubo bubo jarlandi) 107
Bubo bubo kiautschensis 107
(Bubo bubo meridionalis) 104
(Bubo bubo nativus) 106
Bubo bubo nikolskii 108
(Bubo bubo norwegicus) 104
(Bubo bubo ognavi) 104
Bubo bubo omissus 108
(Bubo bubo paradoxus) 108
Bubo bubo ruthenus 105
(Bubo bubo setschuanus) 107
Bubo bubo sibiricus 106
(Bubo bubo swinhoei) 107
(Bubo bubo tarimensis) 108
(Bubo bubo tauricus) 106
(Bubo bubo tenuipes) 107
(Bubo bubo tibetanus) 109
(Bubo bubo transcaucasicus) 106
Bubo bubo turcomanus 108
Bubo bubo ussuriensis 107
(Bubo bubo yamashinai) 107
Bubo bubo yenisseensis 106
(Bubo bubo zaissanensis) 106
(Bubo caligatus) 132
Bubo capensis 110, 252
(Bubo Capensis) 111
Bubo capensis capensis 111
Bubo capensis dilloni 110
Bubo capensis mackinderi 111
Bubo cinerascens 112
(Bubo clamator) 234
Bubo coromandus 117
Bubo coromandus coromandus 118
Bubo coromandus klossi 118
(Bubo Dilloni) 110
(Bubo fasciolatus) 113
(Bubo Hemalachana) 109
(Bubo ketupu aagaardi) 123
(Bubo ketupu büttikoferi) 124
(Bubo ketupu pageli) 124
Bubo lacteus 117, 253
(Bubo lettii) 153
Bubo leucostictus 118, 252
(Bubo mackinderi) 111
Bubo magellanicus 103, 252, 284, 285
Bubo milesi 112
(Bubo nigrescens) 103
Bubo nipalensis 114
Bubo nipalensis blighi 115
Bubo nipalensis nipalensis 114
(Bubo nudipes) 81
(Bubo orientalis minor) 115
Bubo philippensis mindanensis 119, 119
Bubo philippensis philippensis 119
Bubo poensis 113, 252
(Bubo Poensis) 113
Bubo shelleyi 116, 253
Bubo spp. 284, 285
(Bubo subarcticus) 100
Bubo sumatranus 115
Bubo sumatranus strepitans 116
Bubo sumatranus sumatranus 115
Bubo sumatranus tenuifasciatus 116
Bubo virginianus 99, 252, 285
(Bubo virginianus andicolus) 103
(Bubo virginianus colombianus) 103
Bubo virginianus deserti 103, 285
Bubo virginianus elachistus 102, 285
(Bubo virginianus elutus) 102
Bubo virginianus heterocnemis 100, 284, 285
(Bubo virginianus leucomelas) 99
Bubo virginianus mayensis 102, 284, 285
Bubo virginianus nacurutu 102, 285
(Bubo virginianus neochorus) 100
Bubo virginianus nigrescens 103, 285
Bubo virginianus occidentalis 100, 284
Bubo virginianus pacificus 99, 285

(Bubo virginianus pallens) 101
Bubo virginianus pallescens *284*, *285*
Bubo virginianus saturatus 99, *284*, *285*
(Bubo virginianus scalariventris) 100
(Bubo virginianus scotinus) 102
(Bubo virginianus var. *pacificus)* 99
Bubo virginianus virginianus 101, 266, *284*
Bubo virginianus wapacuthu 100, *284*
Bubo vosseleri 114, *252*
bubo, Bubo 104, *252*
(Bubo, Strix) 99, 104
burhani, Ninox 223, 266, *283*
(burmanica, Ninox) 217
(butleri, Asio) 136
butleri, Strix 136, *254*

C

(cabrae, Strix) 29
(calayensis, Otus) 55
californicum, Glaucidium 163, *255*, 266
(caligatus, Bubo) 132
(canariensis, Asio) 238
(caparoch, Strix) 160
(capense, Athene) 178
capense, Glaucidium 183, *281*
capensis, Asio 244, *257*
capensis, Bubo 110, *252*
(Capensis, Bubo) 111
(capensis, Noctua) 183
(capensis, Otus) 244
(Capensis, Scops) 51
(Capensis, Strix) 29
capensis, Tyto 29
capnodes, Otus 45, *250*, 262, *274*
(capnodes, Scops) 45
(Carine brama fryi) 193
(Carine noctua caucasica) 189
(Carine noctua lilith) 190
castaneum, Glaucidium 182
(castanonota, Athene) 179
castanonotum, Glaucidium 179
castanops, Tyto 29
(castanoptera, Strix) 181
castanopterum, Glaucidium 181
(Caucasica, Nyctala) 201
(cayelii, Strix) 26
(Cephelptynx) 206
chacoensis, Strix 141
(Chauvini, Asio) 240
(Chiaradiae, Athene) 188
choliba, Otus 77, *251*, *277*
(choliba, Strix) 78
(Ciccaba) 127
(Ciccaba aequatorialis) 82, 90
(Ciccaba albitarse goodfellowi) 141
(Ciccaba albitarse opaca) 141
(Ciccaba albitarse tertia) 141
(Ciccaba albogularis meridensis) 90
(Ciccaba huhula) 127
(Ciccaba minima) 82
(Ciccaba myrtha) 130
(Ciccaba nigrolineata) 142
(Ciccaba superciliaris macconnelli) 140
(Ciccaba virgata amplonotata) 138
(Ciccaba virgata centralis) 139
(Ciccaba virgata eatoni) 139
(Ciccaba virgata minuscula) 140
(Ciccaba virgata occidentalis) 140
cinerascens, Bubo 112
(cinerea, Strix) 151
cirstata, Lophostrix *154*, 154, *257*, 266
clamator, Asio 234, *257*
(clamator, Bubo) 234
clarkii, Otus 80, 265
cnephaeus, Otus *276*
cnephaeus, Otus (lempiji) 63, 263
cobanense, Glaucidium (gnoma) 165
(cobanense, Glaucidium) 165
collari, Otus 60, 263, *275*
colombianus, Otus 82, 265
(connivens, Falco) 206, 209
connivens, Ninox 208, *256*
cooperi, Otus 73, 264
(coromanda, Strix) 118
coromandus, Bubo 117
costaricanum, Glaucidium *255*, *279*
crassirostris, Tyto 23
(cristata var.*, Scops)* 154
(cristata, Strix) 154, 155
(crucigera, Strix) 78
(Cryptoglaux) 200
(Cryptoglaux acadica brooksi) 203
(Cryptoglaux ridgwayi) 203
(Cryptoglaux rostrata) 203
(Cryptoglaux tengmalmi sibirica) 201
(Cryptoglaux tengmalmi transvolgensis) 200
(Ctenoglaux) 206
(Cubae, Strix) 19
cuculoides, Glaucidium 179
(cuculoides, Noctua) 179
Cultrungius flavipes 122
(Cultrungius Flavipes) 122
(Cultrungius Nigripes) 121
cunicularia, Athene 194
(cunicularia, Strix) 188, 199
(cuyensis, Otus) 58
(cypria, Scops) 49

D

davidi, Strix (uralensis) 150
(davidi, Syrnium) 150
(De-Roepstorffi, Strix) 23
(decussata, Strix) 78
(delicatula, Strix) 24
delicatula, Tyto 24
(deminuta, Strix) 237
deroepstorffi, Tyto 23
detorta, Tyto 22
(Dilloni, Bubo) 110
(dimorpha, Athene) 231
dimorpha, Uroglaux *231*
(domingensis, Strix) 242
(dominicensis, Speotyto) 195
dubiosa, Ninox sp. nov. 223, *283*

E

(elegans, Ephialtes) 55
elegans, Otus 55
enganensis, Otus 62, 263, *275*
(Ephialtes argentina) 85
(Ephialtes Balli) 38
(Ephialtes Brucei) 49
(Ephialtes elegans) 55
(Ephialtes glabripes) 65
(Ephialtes grammicus) 233
(Ephialtes Hambroecki) 40
(Ephialtes hendersoni) 51
(Ephialtes Huttoni) 40
(Ephialtes japonicus interpositus) 55
(Ephialtes leucospila) 56
(Ephialtes nicobaricus) 53
(Ephialtes ocreata) 75
(Ephialtes Plumipes) 65
(Ephialtes sagittatus) 36
(Ephialtes spilocephalus) 39
(Ephialtes umbratilis) 65
(Ephialtes Watsonii) 84
(erlangeri, Scops) 94
(erythrocampe, Lempijius) 65
etchecopari, Glaucidium 182, *255*
(everetti, Ninox) 221
(everetti, Scops) 68

F

(Falco connivens) 206, 209
(fasciata, Nyctale) 142
(fasciaventris, Pulsatrix) 158
(fasciolatus, Bubo) 113
(feae, Scops) 52
(ferox, Strix) 171
(ferruginea, Strix) 171
(fisheri, Glaucidium) 165
(flammea, Strix) 15
(Flammea, Strix) 16, 241
(flammeola, Scops (Megascops)) 46
flammeolus, Otus 36, *46*, *250*
flammeus, Asio 240
flavipes, Cultrungius 122
(Flavipes, Cultrungius) 122
flavipes, Ketupa 122, *122*
(florensis, Athene) 215
(forbesi, Ninox) 225
(fortis, Athene ?) 209
(franseni, Ninox) 207
(frontata, Noctua) 209
(fuliginosa, Scops) 69
fuliginosus, Otus 69, 264
(fulva, Strix) 215
(fulvescens, Ptynx) 150
fulvescens, Strix 144
(fulvescens, Syrnium) 144
(funerea, Strix) 159, 200
funereus, Aegolius 200, *251*
(furcata, Strix) 19
(fusca, Strix) 211
(fuscescens, Strix) 150

G

galapagoensis, Asio (flammeus) 243
(galapagoensis, Otus (Brachyotus)) 243
(Gaucidium (sic) *scheffleri)* 183
(Georgica, Strix) 147
(Gisella) 200

(Gisella iheringi) 204
(giu, Strix) 47
(glabripes, Ephialtes) 65
Glaucidium 161
Glaucidium albertinum 184, 281
(Glaucidium albiventer) 162
Glaucidium bolivianum 166, 255, 279
(Glaucidium borneense) 177
Glaucidium brasilianum 169, 279
Glaucidium brasilianum brasilianum 171
(Glaucidium brasilianum cactorum) 169
Glaucidium brasilianum duidae 170
(Glaucidium brasilianum margaritae) 170
Glaucidium brasilianum medianum 170
Glaucidium brasilianum olivaceum 170
(Glaucidium brasilianum pallens) 171
Glaucidium brasilianum phaloenoides 170
(Glaucidium brasilianum saturatum) 169
Glaucidium brasilianum stranicki 171
Glaucidium brasilianum ucayalae 170
Glaucidium brodiei 176
Glaucidium brodiei borneense 177
Glaucidium brodiei brodiei 176
(Glaucidium brodiei garoense) 177
Glaucidium brodiei pardalotum 177, 255
(Glaucidium brodiei peritum) 177
Glaucidium brodiei sylvaticum 177
Glaucidium californicum 163, 255, 266
Glaucidium californicum californicum 164
Glaucidium californicum grinnelli 163
Glaucidium californicum pinicola 164
Glaucidium californicum swarthi 163
Glaucidium capense 183, 281
Glaucidium capense capense 183
Glaucidium capense etchecopari 182
Glaucidium capense ngamiense 183
(Glaucidium capense robertsi) 183
Glaucidium (capense) scheffleri 183, 255
Glaucidium castaneum 182
Glaucidium castanonotum 179
Glaucidium castanopterum 181
(Glaucidium cobanense) 165
Glaucidium costaricanum 255, 279
Glaucidium cuculoides 179
(Glaucidium cuculoides austerum) 180
Glaucidium cuculoides bruegeli 180
Glaucidium cuculoides cuculoides 179
(Glaucidium cuculoides deignani) 180
(Glaucidium cuculoides delacouri) 180
(Glaucidium cuculoides fulvescens) 180
Glaucidium cuculoides persimile 180
Glaucidium cuculoides rufescens 180
Glaucidium cuculoides whitelyi 180, 255
Glaucidium etchecopari 182, 255
(Glaucidium ferox rufus) 171
(Glaucidium fisheri) 165
Glaucidium gnoma 165
(Glaucidium Gnoma) 165
Glaucidium (gnoma) cobanense 165
Glaucidium (gnoma) costaricanum 165
Glaucidium gnoma gnoma 165
(Glaucidium gnoma grinnelli) 163
(Glaucidium gnoma hoskinsii) 164
(Glaucidium gnoma pinicola) 164
(Glaucidium gnoma vigilante) 164
Glaucidium griseiceps 173, 280
Glaucidium hardyi 174, 280
Glaucidium hoskinsii 164
Glaucidium jardinii 166, 279
(Glaucidium jardinii costaricanum) 165
(Glaucidium kilimense) 163
Glaucidium minutissimum 174, 280
(Glaucidium minutissimum) 174
(Glaucidium minutissimum griscomi) 172
(Glaucidium minutissimum oberholseri) 172
(Glaucidium minutissimum occultum) 173
(Glaucidium minutissimum rarum) 173
(Glaucidium mooreorum) 174
Glaucidium nanum 167, 167
(Glaucidium nanum) 161
(Glaucidium nanum vafrum) 167
Glaucidium nubicola 166, 279
Glaucidium palmarum 172, 280
Glaucidium parkeri 280
Glaucidium passerinum 161, 255, 266
Glaucidium passerinum orientale 162
Glaucidium passerinum passerinum 161
Glaucidium perlatum 162, 255
(Glaucidium perlatum diurnum) 163
Glaucidium perlatum licua 163
Glaucidium perlatum perlatum 162
Glaucidium peruanum 167, 279
(Glaucidium pycrafti) 175
Glaucidium radiatum 178, 255
Glaucidium radiatum malabaricum 178
(Glaucidium radiatum principum) 178
Glaucidium radiatum radiatum 178
Glaucidium ridgwayi 169, 255
Glaucidium ridgwayi cactorum 169
Glaucidium ridgwayi ridgwayi 169
(Glaucidium rufum) 183
Glaucidium sanchezi 172, 280
(Glaucidium scheffleri clanceyi) 183
(Glaucidium setipes) 161
Glaucidium sicki 174
Glaucidium siju 168
Glaucidium siju siju 168
Glaucidium siju vittatum 168
Glaucidium sjoestedti 181
(Glaucidium sjöstedti) 181
Glaucidium tephronotum 175, 255
Glaucidium tephronotum elgonense 176
(Glaucidium tephronotum kivuense) 176
(Glaucidium tephronotum lukolelae) 175
Glaucidium tephronotum medje 175
Glaucidium tephronotum pycrafti 175
Glaucidium tephronotum tephronotum 175
Glaucidium tucumanum 171, 279
(glaucops, Strix) 21
glaucops, Tyto 21, 249
(glaux, Noctua) 189
gnoma, Glaucidium 165
(Gnoma, Glaucidium) 165
(goldii, Ninox) 227
(goodenoviensis, Ninox) 227
(grallaria, Strix) 198
(grammicus, Ephialtes) 233
grammicus, Pseudoscops 233
(granti, Ninox) 230
granti, Ptilopsis 94, 251, 266
(griseata, Lophostrix) 154
griseiceps, Glaucidium 173, 280
(griseus, Scops) 66
guatemalae, Otus 86
(guatemalae, Scops) 86
gurneyi, Mimizuku 96, 266
(gurneyi, Pseudoptynx) 96
(guteruhi, Strix (Athene)) 211
(Gymnasio lawrencei exsul) 93
Gymnoglaux 93
(Gymnoglaux krugii) 89
Gymnoglaux lawrencii 93
Gymnoglaux lawrencii exsul 93, 266
Gymnoglaux lawrencii lawrencii 93, 266
(Gymnoglaux Newtoni) 89
Gymnoscops insularis 61

H

(Hambroecki, Ephialtes) 40
(hantu, Athene) 225
(hardwickii, Strix) 121
hardyi, Glaucidium 174, 280
(härmsi, Syrnium) 135
harrisii, Aegolius 204, 204
(Harrisii, Nyctale) 200, 204
(hartlaubi, Noctua) 43
hartlaubi, Otus 43, 261
(hastatus, Megascops) 86
(Heliodilus Soumagnei) 23
(helveola, Strix (Brachyotus)) 244
(Hemalachana, Bubo) 109
(hendersoni, Ephialtes) 51
(Heteroglaux) 188
(Heteroglaux blewitti) 188
(Heteroglaux Blewitti) 193
(Heteroscops vulpes) 41
(Hieracoglaux) 206
(hirsuta, Strix) 206, 217
(Hoedtii, Noctua) 226
(holerythra, Scops) 38
hoskinsii, Glaucidium 164
(hostilis, Strix) 15
hoyi, Otus 85, 251, 265, 278
(Huhua shelleyi) 116
(huhula, Ciccaba) 127
huhula, Strix 143
(humbloti, Scops) 44
(humeralis, Athene) 207
(Huttoni, Ephialtes) 40
(Hybris nigrescens noctividus) 21
hylophila, Strix 142
(hypogramma, Athene) 224
(hypugaea, Strix) 194

I

icterorhynchus, Otus 37
(icterorhynchus, Scops) 37
(iheringi, Gisella) 204
(Indica, Noctua) 192
(indigena, Athene) 189
(Indranee, Strix) 129
(inexspectata, Strix) 26
inexspectata, Tyto 26
(infuscata, Strix) 171
ingens, Otus 82
(ingens, Scops) 82

insularis, Gymnoscops 61
insularis, Otus 61, *250*, 263
(*insularis, Strix*) 21
insularis, Tyto 21
ios, Ninox 224, 266, *282*
ireneae, Otus 38, *250*, 261
(*italicus, Otus*) 237

J

(*jacquinoti, Athene*) 229
jacquinoti, Ninox 229
(*jakutorum, Nyctala*) 201
jardinii, Glaucidium 166, *279*
(*jardinii, Phalaenopsis*) 166
(*javanensis, Ketupa*) 121
(*javanica, Strix*) 18
Jubula 153
Jubula lettii *153*, *257*, 266

K

kalidupae, Otus (*manadensis*) 60
kalidupae, Otus (*mantananensis*) 263
kennicotti, Otus 70, *251*
(*Kennicotti, Scops*) 70
Ketupa 121
(*Ketupa ceylonensis* (sic) *orientalis*) 122
Ketupa flavipes 122, *122*
(*Ketupa javanensis*) 121
Ketupa ketupu 123, *253*
Ketupa ketupu aagaardi 123
Ketupa ketupu ketupu 123
Ketupa ketupu minor 124
Ketupa ketupu pageli 124
(*Ketupa minor*) 124
(*Ketupa semenowi*) 121
Ketupa zeylonensis 121, *253*
Ketupa zeylonensis leschenault 121
Ketupa zeylonensis orientalis 122
Ketupa zeylonensis semenowi 121
Ketupa zeylonensis zeylonensis 122
ketupu, Ketupa 123, *253*
(*Ketupu, Strix*) 121
(*ketupu, Strix*) 123
(*kilimense, Glaucidium*) 163
(*kirchhoffi, Strix*) 15
koeniswaldiana, Pulsatrix 158, *254*, 266
(*Koeniswaldianum, Syrnium*) 158
koepckeae, Otus 79, 265, *277*
(*krugii, Gymnoglaux*) 89

L

(*labuanensis, Ninox*) 218
(*lactea, Strix*) 117
lacteus, Bubo 117, *253*
lambi, Otus 73, 264
(*lapponica, Strix*) 152
(*latipennis, Scops*) 51
(*latouchi, Scops*) 40
lawrencii, Gymnoglaux *93*
lempiji, Otus 63, *250*, *276*
(*Lempiji, Srix* (sic)) 63
(*Lempijius erythrocampe*) 65
(*Lempijius megalotis*) 68
leptogrammica, Strix *129*, 130, *254*
lettia, Otus 64, *276*
(*lettia, Scops*) 64
(*lettii, Bubo*) 153
lettii, Jubula *153*, *257*, 266
(*leucopsis, Athene*) 43, 214
leucopsis, Ninox 214
(*leucopsis, Otus*) 241
(*leucospila, Ephialtes*) 56
leucostictus, Bubo 118, *252*
leucotis, Ptilopsis 94, 266
(*leucotis, Strix*) 94
(*licua, Strix*) 163
lilith, Athene (*noctua*) 190
(*Liturata, Strix*) 148
longicornis, Otus 42, 261
(*longicornis, Scops*) 42
(*longimembris, Strix*) 30
longimembris, Tyto 30, *249*
Lophostrix 154, 155
Lophostrix cirstata *154*, 154, *257*, 266
(*Lophostrix cristata amazonica*) 155
Lophostrix (*cristata*) *stricklandi* 154, 266
Lophostrix cristata wedeli 154
(*Lophostrix griseata*) 154
(*Lophostrix stricklandi*) 154
(*lophotes, Scops*) 84
loweryi, Xenoglaux *185*, 266, *281*
(*luciae, Scops*) 41
(*lugubris, Strix*) 206, 216
(*lulu, Strix*) 24
lurida, Ninox (*boobook*) 213
(*lurida, Ninox*) 213

M

ma, Strix (*aluco*) 136
(*ma, Syrnium*) 136
Macabra 90
(*macabrum, Syrnium*) 91
(*mackinderi, Bubo*) 111
(*macroptera, Ninox*) 215
(*Maculata, Noctua*) 214
(*maculata, Strix*) 17, 235
(*maculosa, Tyto*) 30
madagascariensis, Asio 240, *257*
madagascariensis, Otus 44, 240, 261, *274*
(*Madagascariensis, Otus*) 240
(*madagascariensis, Scops*) 44
magellanicus, Bubo 103, *252*, *284*, *285*
(*magellanicus, Strix*) 103
(*magica, Strix*) 56
magicus, Otus 55
(*magna, Nyctala*) 201
(*maingayi, Syrnium*) 130
(*malabaricus, Athene*) 178
(*malabaricus, Scops*) 66
(*malaccensis, Athene*) 218
(*malayanus, Scops*) 54
manadensis, Otus 59, *275*
(*manadensis, Scops*) 59
mantananensis, Otus 58
(*mantananensis, Scops*) 58
(*mantis, Strix*) 36
manusi, Tyto 27
(*marmorata, Athene*) 213
(*marmoratus, Megascops*) 86
marshalli, Otus 83, *278*
(*maugei, Strix*) 211
(*McCallii, Scops*) 75
meeki, Ninox 226
(*megalotis, Lempijius*) 68
megalotis, Otus *68*, *250*
Megascops 70
(*Megascops asio aikeni*) 70
(*Megascops asio cineraceus*) 70
(*Megascops asio macfarlanei*) 70
(*Megascops asio saturatus*) 70
(*Megascops asio xantusi*) 71
(*Megascops aspersus*) 76
(*Megascops flammeola idahoensis*) 46
(*Megascops hastatus*) 86
(*Megascops marmoratus*) 86
(*Megascops pinosus*) 76
(*Megascops ridgwayi*) 76
(*Megascops vermiculatus*) 87
(*Megascops vinaceus*) 72
(*Megastrix tenebricosa perconfusa*) 32
(*melanota, Noctua*) 158
melanota, Pulsatrix 158
(*mendeni, Otus*) 60
mentawi, Otus 62, *250*, 263
(*mexicanus, Otus*) 234
Micrathene 186
Micrathene whitneyi *186*, *255*, 266, *281*
Micrathene whitneyi graysoni 187
Micrathene whitneyi idonea 187
Micrathene whitneyi sanfordi 187
Micrathene whitneyi whitneyi 186
(*Micropallas*) 186
(*Micropallas socorroensis*) 187
(*Micropallas whitneyi idonea*) 187
(*Micropallas whitneyi sanfordi*) 187
(*Microscops*) 200
milesi, Bubo 112
milesi, Bubo (*africanus*) 112
Mimizuku 96
Mimizuku gurneyi *96*, 266
(*mindanensis, Pseudoptynx*) 119
mindorensis, Ninox 222, 266
mindorensis, Otus 43, 261
(*mindorensis, Scops*) 43
(*minima, Ciccaba*) 82
(*minor, Ketupa*) 124
(*minor, Noctua*) 241
(*minutissima, Strix*) 174
minutissimum, Glaucidium 174, *280*
(*minutissimum, Glaucidium*) 174
(*minutus, Scops*) 53
mirus, Otus 42, 261
(*modestus, Scops*) 53
moheliensis, Otus 44, *250*, *262*, *276*
(*mooreorum, Glaucidium*) 174
(*morotensis, Scops*) 56
multipunctata, Tyto 32, *249*
(*myrtha, Ciccaba*) 130

N

(*nacurutu, Strix*) 102
(*naevia, Strix*) 74
(*nana, Strix*) 167

nanum, Glaucidium 167, 167
(nanum, Glaucidium) 161
napensis, Otus 88
natalis, Ninox 225, 226
nebulosa, Strix 151
Nesasio 246
Nesasio salomonensis 246
newarensis, Strix 131
(newarensis, Ulula) 131
(Newtoni, Gymnoglaux) 89
(niasense, Syrnium) 130
(nicobaricus, Ephialtes) 53
(nigrescens, Bubo) 103
(nigricantius, Syrnium) 137
(Nigripes, Cultrungius) 121
nigrobrunnea, Tyto 25, 249
(nigrolineata, Ciccaba) 142
nigrolineata, Strix 142, 143
(nigrolineatum, Syrnium) 143
Ninox 206
Ninox affinis 219
Ninox affinis affinis 219
Ninox affinis isolata 219
(Ninox affinis rexpimenti) 219
(Ninox albaria) 214
(Ninox albomaculata) 209
(Ninox aruensis) 207
(Ninox assimilis) 208
Ninox boobook 210
Ninox boobook boobook 213, 256
Ninox boobook cinnamomina 211
Ninox boobook fusca 211
(Ninox boobook halmaturina) 213
Ninox (boobook) lurida 213
(Ninox boobook macgillivrayi) 212
(Ninox boobook melvillensis) 212
(Ninox boobook mixta) 212
Ninox boobook moae 211
Ninox boobook ocellata 212
Ninox boobook plesseni 211
Ninox boobook pusilla 212
Ninox boobook remigialis 212
Ninox boobook rotiensis 210
(Ninox boobook royana) 215
(Ninox boobook var. lurida) 213
Ninox burhani 223, 266, 283
(Ninox burmanica) 217
Ninox connivens 208, 256
(Ninox connivens addenda) 209
Ninox connivens assimilis 208
Ninox connivens connivens 209
Ninox connivens occidentalis 209
Ninox connivens peninsularis 209
Ninox connivens rufostrigata 208
(Ninox connivens suboccidentalis) 209
Ninox dubiosa sp. nov. 223, 283
(Ninox everetti) 221
(Ninox forbesi) 225
(Ninox franseni) 207
(Ninox fusca plesseni) 211
(Ninox goldii) 227
(Ninox goodenoviensis) 227
(Ninox granti) 230
Ninox ios 224, 266, 282
Ninox jacquinoti 229
Ninox jacquinoti eichhorni 229
Ninox jacquinoti floridae 230
Ninox jacquinoti granti 230
Ninox jacquinoti jacquinoti 229
Ninox jacquinoti malaitae 230
Ninox jacquinoti mono 230
Ninox jacquinoti roseoaxillaris 230
(Ninox labuanensis) 218
Ninox leucopsis 214
(Ninox lurida) 213
(Ninox macroptera) 215
Ninox meeki 226
Ninox mindorensis 222, 266
Ninox natalis 225, 226
Ninox nipalensis 206
(Ninox novae britanniae) 228
(Ninox novaebritanniae novaehibernicae) 228
Ninox novaeseelandiae 214
Ninox novaeseelandiae albaria 214
(Ninox novaeseelandiae arida) 212
(Ninox novaeseelandiae moae) 211
Ninox novaeseelandiae novaeseelandiae 215
(Ninox novaeseelandiae pusilla) 212
(Ninox novaeseelandiae remigialis) 212
(Ninox novaeseelandiae rufigaster) 212
Ninox novaeseelandiae undulata 215
(Ninox obscurus) 217
Ninox ochracea 223, 256, 266, 282
Ninox odiosa 228
(Ninox ooldeanensis) 212
(Ninox peninsularis) 209
(Ninox perversa) 224
Ninox philippensis 220
Ninox philippensis centralis 221
Ninox philippensis philippensis 220, 266
Ninox philippensis proxima 222
Ninox philippensis reyi 221
Ninox philippensis spilocephala 221, 266
Ninox philippensis spilonota 221
(Ninox philippensis ticaoensis) 222
(Ninox plateni) 222
Ninox punctulata 227, 283
(Ninox Reyi) 221
(Ninox rosseliana) 227
Ninox rudolfi 210, 283
Ninox rufa 206, 256
Ninox rufa humeralis 207
(Ninox rufa marginata) 206
Ninox rufa meesi 206
Ninox rufa queenslandica 207
Ninox rufa rufa 206
Ninox scutulata 215, 256, 283
Ninox scutulata borneensis 218
Ninox scutulata burmanica 217
Ninox scutulata florensis 215
Ninox scutulata hirsuta 217
(Ninox scutulata isolata) 219
Ninox scutulata japonica 216
Ninox scutulata javanensis 218
Ninox scutulata lugubris 216
Ninox (scutulata) obscura 217
Ninox scutulata palawanensis 217
Ninox scutulata randi 218
Ninox scutulata scutulata 218, 266
(Ninox scutulata totogo) 216
(Ninox scutulata ussuriensis) 215
(Ninox scutulata yamashinae) 216
(Ninox solomonis) 228
Ninox spilocephala 221
(Ninox spilonotus) 221
Ninox squamipila 224
Ninox squamipila forbesi 225
Ninox squamipila hantu 225
Ninox squamipila hypogramma 224
Ninox squamipila squamipila 225
Ninox strenua 207, 208, 256
(Ninox strenua victoriae) 207
Ninox sumbaensis 222, 256, 283
Ninox superciliaris 219, 220, 256
(Ninox terricolor) 227
Ninox theomacha 226, 256
Ninox theomacha goldii 227
Ninox theomacha hoedtii 226
Ninox theomacha rosseliana 227
Ninox theomacha theomacha 227
(Ninox undulata) 207
Ninox variegata 228
(Ninox Variegata superior) 229
Ninox variegata variegata 228
(Ninox yorki) 213
nipalensis, Bubo 114
nipalensis, Ninox 206
nivicola, Strix (aluco) 135
(nivicolum, Syrnium) 135
(noctipetens, Asio) 236
(Noctua Brodiei) 176
(Noctua capensis) 183
(Noctua cuculoides) 179
(Noctua frontata) 209
(Noctua glaux) 189
(Noctua hartlaubi) 43
(Noctua Hoedtii) 226
(Noctua Indica) 192
(Noctua Maculata) 214
(Noctua melanota) 158
(Noctua minor) 241
(Noctua nudipes) 93
(Noctua ochracea) 223
(Noctua podargina) 92
(Noctua punctulata) 206, 227
(Noctua siju) 168
(Noctua spilogastra) 190
(Noctua Tarayensis) 192
(Noctua tubiger) 176
(Noctua variegata) 228
(Noctua venatica) 215
(Noctua Woodfordi) 138
(Noctua zelandica) 215
noctua, Athene 188, 251
(noctua, Strix) 188
(novae-Seelandiae, Strix) 206, 215
novaehollandiae, Tyto 28
novaeseelandiae, Ninox 214
(Novipulsatrix) 156
nubicola, Glaucidium 166, 279
(nuchale, Syrnium) 137
(nudipes, Bubo) 81
(nudipes, Noctua) 93
nudipes, Otus 89
(nudipes, Strix) 89
(Nyctala Caucasica) 201
(Nyctala jakutorum) 201
(Nyctala magna) 201

(Nyctale) 200
(Nyctale acadica scotaea) 202
(Nyctale Bergiana) 142
(Nyctale fasciata) 142
(Nyctale Harrisii) 200, 204
(Nyctale Richardsoni) 202
(Nyctale tengmalmi pallens) 201
(Nyctalops stygius) 236
Nyctea 97
Nyctea scandiaca *97*, 266, *253*

O

(obira, Otus) 56
obscura, Ninox (scutulata) 217
(obscurus, Ninox) 217
(obsoleta, Scops) 50
(occidentale, Syrnium) 145
occidentalis, Strix 144, *145*, *254*
(ocellata, Athene) 212
ocellata, Strix 128
(ocellatum, Syrnium) 128
ochracea, Ninox 223, *256*, 266, *282*
(ochracea, Noctua) 223
(ochrogenys, Syrnium) 129
(ocreata, Ephialtes) 75
odiosa, Ninox 228
(ooldeanensis, Ninox) 212
(orientalis, Athene) 191
(orientalis, Strix) 116
(Orientalis, Strix) 127
Otini 261
Otus 36
(Otus abyssinicus) 239
(Otus albicollis) 237
Otus albogularis 82, 90
Otus albogularis albogularis 90, 265
Otus albogularis macabrum 91, 266
Otus albogularis meridensis 90, 266
(Otus albogularis obscurus) 90
Otus albogularis remotus 91, 266
Otus alfredi 39, 261, *272*
Otus alius 61, 263, *275*
Otus angelinae *42*, 261
Otus asio 74, *251*
Otus asio asio 74, 264
(Otus asio brewsteri) 70
(Otus asio cardonensis) 71
(Otus asio clazus) 71
Otus asio floridanus 75, 264
(Otus asio gilmani) 71
Otus asio hasbroucki 75, 264
(Otus asio inyoensis) 71
Otus asio maxwelliae 74, 264
Otus asio mccallii 75, 264
(Otus asio mychophilus) 71
(Otus asio naevius) 264
(Otus asio quercinus) 70
(Otus asio semplei) 75
(Otus asio sinaloensis) 72
(Otus asio sortilegus) 72
(Otus asio suttoni) 72
(Otus asio swenki) 74, 264
(Otus asio yumanensis) 71
Otus atricapillus 84, *251*, 265, *278*
(Otus atricapillus argentinus) 265
(Otus atricapillus ater) 84
(Otus atricapillus fulvescens) 84
(Otus atricapillus inambarii) 84
(Otus atricapillus morelius) 84
Otus bakkamoena 36, 65
(Otus bakkamoena alboniger) 64
(Otus bakkamoena aurorae) 67
Otus bakkamoena bakkamoena 66, 263
(Otus bakkamoena batanensis) 55
(Otus bakkamoena cnephaeus) 63
(Otus bakkamoena condorensis) 64
Otus bakkamoena deserticolor 66, 263
Otus bakkamoena gangeticus 66, 264
(Otus bakkamoena hatchizionis) 68
(Otus bakkamoena hypnodes) 63
(Otus bakkamoena kangeana) 64
(Otus bakkamoena lemurum) 64
(Otus bakkamoena linae) 67
(Otus bakkamoena manipurensis) 64
Otus bakkamoena marathae 66
(Otus bakkamoena mentawi) 62
(Otus bakkamoena nigrorum) 69
(Otus bakkamoena plumipes) 65, 263
(Otus bakkamoena stewarti) 66
Otus balli 38, 261
Otus barbarus 81, 265
Otus beccarii 58, 263
(Otus bengalensis) 109
(Otus boholensis) 68
(Otus (Brachyotus) galapagoensis) 243
Otus brookii 62
Otus brookii brookii 63, 263
Otus brookii solokensis 62, 263
Otus brucei 49, 50, *250*
Otus brucei brucei 49, 262
Otus brucei exiguus 50, 262
Otus brucei obsoletus 50, 262
Otus (brucei) pamelae 50, 262
Otus brucei semenowi 50, 262
Otus (brucei) socotranus 51, 262
(Otus calayensis) 55
(Otus capensis) 244
(Otus capensis major) 245
Otus capnodes 45, 262, *250*, *274*
Otus choliba 77, *251*, *277*
(Otus choliba alilicuco) 79
(Otus choliba alticola) 78
(Otus choliba catingensis) 78
(Otus choliba caucae) 78
(Otus choliba chapadensis) 78
Otus choliba choliba 78, 265
Otus choliba crucigerus 78, 264
Otus choliba decussatus 78, 265
Otus choliba duidae 77
(Otus choliba guyanensis) 78
(Otus choliba kelsoi) 78
(Otus choliba koepckeae) 79
Otus choliba luctisonus 77, 264
Otus choliba margaritae 77, 264
(Otus choliba maximus) 85
(Otus choliba montanus) 78
(Otus choliba pintoi) 85
(Otus choliba portoricensis) 78
Otus choliba suturutus 78
(Otus choliba thompsoni) 86
(Otus choliba urugaii) 79
Otus choliba uruguaiensis 79, 265
Otus choliba wetmorei 79, 265
Otus clarkii 80, 265
Otus cnephaeus *276*
Otus collari 60, 263, *275*
Otus colombianus 82, 265
Otus cooperi 73, 264
(Otus cooperi chiapensis) 73
(Otus cuyensis) 58
Otus elegans 55
Otus elegans botelensis 55
Otus elegans calayensis 55, 263
Otus elegans elegans 55, 262
(Otus elegans interpositus) 263
Otus enganensis 62, 263, *275*
Otus flammeolus 36, *46*, *250*
(Otus flammeolus borealis) 46
Otus flammeolus flammeolus 262
(Otus flammeolus frontalis) 46
(Otus flammeolus guatemalae) 46
(Otus flammeolus idahoensis) 262
(Otus flammeolus meridionalis) 46
(Otus flammeolus rarus) 46, 262
Otus fuliginosus 69, 264
Otus guatemalae 86
(Otus guatemalae bolivianus) 88
Otus guatemalae cassini 86, 265
(Otus guatemalae centralis) 87
Otus guatemalae dacrysistactus 87
(Otus guatemalae fuscus) 86, 265
Otus guatemalae guatemalae 86, 265
Otus guatemalae hastatus 86, 265
(Otus guatemalae napensis) 88
Otus guatemalae pacificus 80
(Otus guatemalae pallidus) 87
(Otus guatemalae peteni) 87
(Otus guatemalae pettingilli) 86
(Otus guatemalae rufus) 80
(Otus guatemalae thokpsoni) 86, 265
(Otus guatemalae tomlini) 86
Otus hartlaubi 43, 261
Otus hoyi 85, *251*, 265, *278*
Otus icterorhynchus 37
Otus icterorhynchus holerythrus 38, *250*, 261
Otus icterorhynchus icterorhynchus 37, 261
Otus ingens 82
Otus ingens colombianus 82
Otus ingens ingens 82, 265
(Otus ingens minimus) 265
Otus ingens venezuelanus 82
Otus insularis 61, *250*, 263
Otus ireneae 38, *250*, 261
(Otus italicus) 237
Otus kennicotti 70, *251*
Otus kennicotti aikeni 70, 264
Otus kennicotti bendirei 70, 264
(Otus kennicotti brewsteri) 264
Otus kennicotti cardonensis 71
(Otus kennicotti gilmani) 264
Otus kennicotti kennicotti 70, 264
(Otus kennicotti saturatus) 264
Otus kennicotti suttoni 72, 264
Otus kennicotti vinaceus 72
Otus kennicotti xantusi 71
Otus kennicotti yumanensis 71
Otus koepckeae 79, 265, *277*

Otus lambi 73, 264
Otus lempiji 63, *250*, *276*
Otus (*lempiji*) *cnephaeus* 63, 263
Otus lempiji hypnodes 63, 263
Otus lempiji kangeana 64, 263
Otus lempiji lempiji 63, 263
Otus lempiji lemurum 64, 263
Otus lettia 64, *276*
Otus lettia erythrocampe 65, 263
Otus lettia glabripes 65, 263
Otus lettia lettia 64, 263
Otus lettia umbratilis 65, 263
(*Otus leucopsis*) 241
(*Otus leucotis Margarethae*) 94
Otus longicornis 42, 261
(*Otus luciae siamensis*) 41
Otus (*Macabra*) *albigularis* *90*
Otus madagascariensis 44, 240, 261, *274*
(*Otus Madagascariensis*) 240
Otus magicus 55
Otus magicus albiventris 57, 263, *272*
Otus magicus bouruensis 57, 263
Otus magicus leucospilus 56, 263
Otus magicus magicus 56, 263
Otus magicus morotensis 56, 263
Otus magicus obira 56
Otus (*magicus*) *sulaensis* 57, 263
Otus (*magicus*) *tempestatis* 57, 263
Otus manadensis 59, *275*
Otus (*manadensis*) *kalidupae* 60
Otus manadensis manadensis 59
Otus manadensis mendeni 60
(*Otus manadensis obsti*) 59
Otus (*manadensis*) *siaoensis* 59
Otus mantananensis 58
Otus mantananensis cuyensis 58, 263
Otus (*mantananensis*) *kalidupae* 263
Otus mantananensis manadensis 263
Otus mantananensis mantananensis 58, 263
Otus mantananensis mendeni 263
Otus mantananensis romblonis 58, 263
Otus (*mantananensis*) *siaoensis* 59, 263, *275*
Otus mantananensis sibutuensis 59, 263
Otus marshalli 83, *278*
Otus mayottensis 44, 262, *276*
Otus megalotis *68*, *250*
Otus megalotis everetti 68, 264
Otus megalotis megalotis 68, 264
Otus megalotis nigrorum 69, 264
(*Otus mendeni*) 60
Otus mentawi 62, *250*, 263
(*Otus mexicanus*) 234
Otus mindorensis 43, 261
Otus mirus 42, 261
Otus moheliensis 44, *250*, 262, *276*
Otus napensis 88
Otus napensis bolivianus 88
Otus napensis helleri 88, 265
Otus napensis napensis 88, 265
Otus nudipes 89
Otus nudipes newtoni 89, 265
Otus nudipes nudipes 89, 265
(*Otus obira*) 56
Otus pacificus 80, 265, *277*
(*Otus palustris*) 241
Otus pauliani 45, *250*, 262
Otus pembaensis 46, *250*, 262
Otus petersoni 83, *278*
Otus roboratus 80, 265, *277*
(*Otus romblonis*) 58
Otus roraimae 88, 265
Otus rufescens 36
(*Otus rufescens burbidgei*) 36, 261
Otus rufescens malayensis 37, *273*, 261
(*Otus rufescens mantis*) 261
Otus rufescens rufescens 36, 261
Otus rutilus 43, *250*, 261, *274*
(*Otus rutilus mayottensis*) 44
(*Otus rutilus pauliani*) 45
Otus sagittatus 36, *250*, 261
Otus sanctae-catarinae 85, 265
Otus scops *47*, *250*
Otus scops cycladum 48
Otus scops cyprius 49, 262
(*Otus scops erlangeri*) 262
(*Otus scops longipennis*) 262
Otus scops mallorcae 48
Otus scops mirus 42
(*Otus scops powelli*) 48
Otus scops pulchellus 48, 262
Otus scops scops 47, 262
(*Otus scops sibiricus*) 262
Otus scops turanicus 49, 262
Otus seductus 72, 264
(*Otus seductus colimensis*) 72
Otus semitorques 67, *250*
Otus semitorques pryeri 68, 264
Otus semitorques semitorques 67, 264
Otus semitorques ussuriensis 67, 264
Otus senegalensis 51
(*Otus senegalensis caecus*) 51
Otus senegalensis feae 52, 262
(*Otus senegalensis graueri*) 51, 262
(*Otus senegalensis hendersoni*) 262
Otus senegalensis nivosus 52, 262
(*Otus senegalensis pamelae*) 50
(*Otus senegalensis pusillus*) 262
Otus senegalensis senegalensis 51, 262
(*Otus siguapa*) 236
Otus silvicola 69, 264
Otus spilocephalus 39
Otus spilocephalus hambroecki 40, 261
Otus spilocephalus huttoni 40, 261
Otus spilocephalus latouchi 40, 261
Otus spilocephalus luciae 41, 261
(*Otus spilocephalus rupchandi*) 40, 261
Otus spilocephalus siamensis 41, 261
Otus spilocephalus spilocephalus 39, 261
Otus spilocephalus vandewateri 41, 261
Otus spilocephalus vulpes 41
(*Otus steerei*) 59
Otus stresemanni 39, 261, *272*
Otus sunia 52
(*Otus sunia botelensis*) 55
(*Otus sunia distans*) 53, 262
Otus sunia japonicus 54, 262
(*Otus sunia khasiensis*) 53, 262
Otus sunia leggei 53, 262, *273*
Otus sunia malayanus 54, 262
Otus sunia modestus 53, 262
(*Otus sunia nicobaricus*) 262
Otus sunia rufipennis 53, 262
Otus sunia stictonotus 54, 262
Otus sunia sunia 53, 262
Otus thilohoffmanni 37, *250*, 261, *273*
Otus trichopsis 76, *251*
Otus trichopsis aspersus 76, 264
(*Otus trichopsis guerrensis*) 76, 264
(*Otus trichopsis inexpectus*) 76
Otus trichopsis mesamericanus 76, 264
(*Otus trichopsis pumilus*) 76, 264
(*Otus trichopsis ridgwayi*) 264
Otus trichopsis trichopsis 76, 264
Otus umbra 61, 263, *275*
(*Otus umbra enganensis*) 62
Otus vermiculatus 87, 265
(*Otus vermiculatus helleri*) 88
(*Otus vermiculatus huberi*) 87
Otus vinaceus seductus 72
Otus watsonii 83
(*Otus watsonii morelia*) 265
Otus watsonii usta 84, 265
Otus watsonii watsonii 84, 265
(*Otus Wilsonianus*) 238
(*Otus wilsonianus*) 238
otus, *Asio* 237
(*otus*, *Strix*) 234
(*Otus*, *Strix*) 237
(*oustaleti*, *Scotopelia*) 125
(*oustaleti*, *Strix*) 30

P

pacificus, *Otus* 80, 265, *277*
(*pagodorum*, *Strix*) 127
palmarum, *Glaucidium* 172, *280*
(*palustris*, *Otus*) 241
pamelae, *Otus* (*brucei*) 50, 262
(*pardalotum*, *Athene*) 177
parkeri, *Glaucidium* *280*
(*parva*, *Strix*) 18
(*passerina*, *Strix*) 161
passerinum, *Glaucidium* *161*, *255*, 266
pauliani, *Otus* 45, *250*, 262
peli, *Scotopelia* *125*, *253*
(*peli*, *Strix*) 125
pembaensis, *Otus* 46, *250*, 262
(*peninsularis*, *Ninox*) 209
(*perlata*, *Strix*) 20, 162
perlatum, *Glaucidium* 162, *255*
perspicillata, *Pulsatrix* *156*, *254*
(*perspicillata*, *Strix*) 156
peruanum, *Glaucidium* 167, *279*
(*perversa*, *Ninox*) 224
petersoni, *Otus* 83, *278*
(*Phalaenopsis*) 161
(*Phalaenopsis jardinii*) 166
(*phaloenoides*, *Strix*) 170
(*Phasmoptynx Capensis tingitanus*) 244
philippensis, *Ninox* 220
(*philippensis*, *Pseudoptynx*) 119
Phodilus 34
(*Phodilus arixuthus*) 35
(*Phodilus assimilis*) 35
Phodilus badius *34*, *249*, *272*
(*Phodilus badius abbotti*) 35
Phodilus badius arixuthus 35
Phodilus (*badius*) *assimilis* 35

Phodilus badius badius 35
Phodilus badius parvus 35
Phodilus (badius) ripleyi 34
Phodilus badius saturatus 34
(Phodilus prigoginei) 33
(Phodilus riverae) 35
(pinosus, Megascops) 76
(Pisorhina alfredi) 39
(Pisorhina angelinae) 42
(Pisorhina badia) 38
(Pisorhina capensis grisea) 51
(Pisorhina capensis intermedia) 51
(Pisorhina capensis pusilla) 51
(Pisorhina leucotis granti) 94
(Pisorhina manadensis kalidupae) 60
(Pisorhina manadensis tempestatis) 57
(Pisorhina scops bascanica) 48
(Pisorhina scops cycladum) 48
(Pisorhina scops erlangeri) 47
(Pisorhina scops graeca) 47
(Pisorhina scops tuneti) 47
(Pisorhina scops turanicus) 49
(Pisorhina scops zarudnyi) 48
(Pisorhina solokensis) 62
(Pisorhina sulaensis) 57
(Pisorhina ugandae) 51
Pisorhina umbra 61
(Pisorhina vandewateri) 41
(pithecops, Strix) 30
(plateni, Ninox) 222
(plumipes, Athene) 191
(Plumipes, Ephialtes) 65
(podargina, Noctua) 92
podarginus, Pyrroglaux 92, 266
poensis, Bubo 113, 252
(Poensis, Bubo) 113
(poensis, Strix) 17
(portoricensis, Asio) 242
(pratincola, Strix) 19
(prigoginei, Phodilus) 33
prigoginei, Tyto 33, 249, 272
(pryeri, Scops) 68
(Pseudoptynx gurneyi) 96
(Pseudoptynx mindanensis) 119
(Pseudoptynx philippensis) 119
(Pseudoptynx solomonensis) 246
Pseudoscops 233
Pseudoscops grammicus 233
(Psiloscops) 36
Ptilopsis 94, 95, 97
Ptilopsis granti 94, 251, 266
Ptilopsis leucotis 94, 266
(Ptynx fulvescens) 150
(pulchella, Stryx (sic)) 48
(pulchra, Athene) 193
Pulsatrix 156
(Pulsatrix fasciaventris) 158
Pulsatrix koeniswaldiana 158, 254, 266
Pulsatrix melanota 158
(Pulsatrix melanota philoscia) 158
Pulsatrix perspicillata 156, 254
(Pulsatrix perspicillata austini) 156
Pulsatrix perspicillata boliviana 157
Pulsatrix perspicillata chapmani 156
Pulsatrix perspicillata perspicillata 157
Pulsatrix perspicillata saturata 156, 266
(Pulsatrix perspicillata trinitatis) 157
Pulsatrix pulsatrix 157
(Pulsatrix sharpei) 156, 158
pulsatrix, Pulsatrix 157
(pulsatrix, Strix) 157
(pumila, Strix) 174
(punctata, Strix) 29
(punctatissima, Strix) 22
punctatissima, Tyto 22
punctulata, Ninox 227, 283
(punctulata, Noctua) 206, 227
(pusilla, Strix) 18
(pycrafti, Glaucidium) 175
(pygmea, Scops) 51
Pyrroglaux 92
Pyrroglaux podarginus 92, 266

R

(Radiata, Strix) 178
radiatum, Glaucidium 178, 255
(Reyi, Ninox) 221
(Rhabdoglaux) 206
(Rhinoptynx) 234
(Rhinoptynx clamator forbesi) 234
(Rhinoptynx clamator mogenseni) 235
(Rhinoptynx clamator oberi) 235
(Richardsoni, Nyctale) 202
ridgwayi, Aegolius 203
(ridgwayi, Cryptoglaux) 203
ridgwayi, Glaucidium 169, 255
(ridgwayi, Megascops) 76
ripleyi, Phodilus (badius) 34
(riverae, Phodilus) 35
roboratus, Otus 80, 265, 277
(romblonis, Otus) 58
roraimae, Otus 88, 265
(roraimae, Scops) 88
(Rosenbergii, Strix) 27
rosenbergii, Tyto 27
(roseoaxillaris, Spiloglaux) 230
(rosseliana, Ninox) 227
(rostrata, Cryptoglaux) 203
(rostrata, Speotyto) 195
rudolfi, Ninox 210, 283
(rufa, Athene) 206
rufa, Ninox 206, 256
rufescens, Otus 36
(rufescens, Strix) 36, 150
(rufifacies, Sceloglaux) 232
(rufipennis, Scops) 53
rufipes, Strix 141, 254
(rufostrigata, Athene) 208
(rufum, Glaucidium) 183
rutilus, Otus 43, 250, 261, 274
(rutilus, Scops) 43

S

(sagittatus, Ephialtes) 36
sagittatus, Otus 36, 250, 261
(saharae, Strix) 189
salomonensis, Nesasio 246
sanchezi, Glaucidium 172, 280
(sanctae-catarinae, Scops) 85
(sancti-nicolai, Syrnium) 134
(Sandwichensis, Strix) 242
(Sarda, Athene) 188
scandiaca, Nyctea 97, 266, 253
(scandiaca, Strix) 97
Sceloglaux 232
Sceloglaux albifacies 232
(Sceloglaux rufifacies) 232
(scheffleri, Gaucidium (sic)) 183
scheffleri, Glaucidium (capense) 183
(Scops albiventris) 57
(Scops asio bendirei) 70
(Scops asio Floridanus) 75
(Scops asio var. maxwelliae) 74
(Scops barbarus) 81
(Scops beccarii) 58
(Scops bouruensis) 57
(Scops brasilianus var. cassini) 86
(Scops brookii) 63
(Scops Capensis) 51
(Scops capnodes) 45
(Scops cristata var.) 154
(Scops cypria) 49
(Scops erlangeri) 94
(Scops everetti) 68
(Scops feae) 52
(Scops fuliginosa) 69
(Scops griseus) 66
(Scops guatemalae) 86
(Scops holerythra) 38
(Scops humbloti) 44
(Scops icterorhynchus) 37
(Scops ingens) 82
(Scops Kennicotti) 70
(Scops latipennis) 51
(Scops latouchi) 40
(Scops lettia) 64
(Scops longicornis) 42
(Scops lophotes) 84
(Scops luciae) 41
(Scops madagascariensis) 44
(Scops malabaricus) 66
(Scops malayanus) 54
(Scops manadensis) 59
(Scops mantananensis) 58
(Scops McCallii) 75
(Scops (Megascops) flammeola) 46
(Scops mindorensis) 43
(Scops minutus) 53
(Scops modestus) 53
(Scops morotensis) 56
(Scops obsoleta) 50
(Scops pryeri) 68
(Scops pygmea) 51
(Scops roraimae) 88
(Scops rufipennis) 53
(Scops rutilus) 43
(Scops sanctae-catarinae) 85
(Scops scops ferghanensis) 48
(Scops scops irtyshensis) 48
(Scops scops sibirica) 48
(Scops scops tschusii) 47
(Scops Semenowi) 50
(Scops semitorques ussuriensis) 67
(Scops Senegalensis) 51
(Scops siaoensis) 59
(Scops sibutuensis) 59

(Scops silvicola) 69
(Scops socotranus) 51
(Scops spurrelli) 37
(Scops stictonotus) 54
(Scops sunia) 53
(Scops trichopsis) 76
(Scops usta) 84
(Scops whiteheadi) 68
scops, Otus *47, 250*
(Scops, Strix) 47
(Scotiaptex) 127
Scotopelia 125
Scotopelia bouvieri 126, *253*
(Scotopelia oustaleti) 125
Scotopelia peli *125, 253*
(Scotopelia peli fischeri) 125
(Scotopelia peli Salvagoraggii) 125
Scotopelia ussheri 126, *253*
(Scotophilus) 200
scutulata, Ninox 215, *256, 283*
(scutulata, Strix) 218
seductus, Otus 72, 264
(Selo-puto, Strix) 127
seloputo, Strix 127
(semenowi, Ketupa) 121
(Semenowi, Scops) 50
semitorques, Otus 67, *250*
senegalensis, Otus 51
(Senegalensis, Scops) 51
(setipes, Glaucidium) 161
(sharpei, Pulsatrix) 156, 158
shelleyi, Bubo 116, *253*
(shelleyi, Huhua) 116
siaoensis, Otus (manadensis) 59
siaoensis, Otus (mantananensis) 59, 263, 275
(siaoensis, Scops) 59
(sibirica, ? Strix) 106
(sibutuensis, Scops) 59
sicki, Glaucidium 174
(siguapa, Otus) 236
siju, Glaucidium 168
(siju, Noctua) 168
silvicola, Otus 69, 264
(silvicola, Scops) 69
sjoestedti, Glaucidium 181
(sjöstedti, Glaucidium) 181
(Smithiglaux) 178
(Smithiglaux capensis ngamiensis) 183
(socorroensis, Micropallas) 187
socotranus, Otus (brucei) 51, 262
(socotranus, Scops) 51
(solokensis, Pisorhina) 62
(solomonensis, Pseudoptynx) 246
(solomonis, Ninox) 228
(sororcula, Strix) 27
sororcula, Tyto 26
(Soumagnei, Heliodilus) 23
soumagnei, Tyto 23, *249*, 261, *271*
(Speotyto) 188
(Speotyto amaura) 196
(Speotyto bahamensis) 195
(Speotyto brachyptera) 196
(Speotyto cunicularia apurensis) 196
(Speotyto cunicularia arubensis) 196
(Speotyto cunicularia beckeri) 198
(Speotyto cunicularia becki) 194
(Speotyto cunicularia boliviana) 198
(Speotyto cunicularia carrikeri) 197
(Speotyto cunicularia cavicola) 195
(Speotyto cunicularia guadeloupensis) 196
(Speotyto cunicularia guantanamensis) 195
(Speotyto cunicularia intermedia) 198
(Speotyto cunicularia juninensis) 198
(Speotyto cunicularia minor) 196
(Speotyto cunicularia nanodes) 198
(Speotyto cunicularia pichinchae) 197
(Speotyto cunicularia punensis) 197
(Speotyto cunicularia tolimae) 197
(Speotyto cunicularia troglodytes) 195
(Speotyto cunicularia var. *floridana)* 195
(Speotyto dominicensis) 195
(Speotyto rostrata) 195
(spilinotum, Syrnium) 143
spilocephala, Ninox 221
(spilocephalus, Ephialtes) 39
spilocephalus, Otus 39
(spilogastra, Athene) 190
(spilogastra, Noctua) 190
(Spiloglaux) 206
(Spiloglaux boobook clelandi) 214
(Spiloglaux boobook leachi) 213, 214
(Spiloglaux boobook parocellata) 212
(Spiloglaux boobook tregellari) 213
(Spiloglaux boweri) 213
(Spiloglaux jacquinoti eichhorni) 229
(Spiloglaux novaeseelandiae everardi) 212
(Spiloglaux novaeseelandiae tasmanica) 214
(Spiloglaux ocellata carteri) 212
(Spiloglaux roseoaxillaris) 230
(Spiloglaux theomacha) 227
(spilonotus, Ninox) 221
(splendens, Strix) 16
(spurrelli, Scops) 37
(squamipila, Athene) 225
squamipila, Ninox 224
(squamulatum, Syrnium) 138
(Srix (sic) *Lempiji)* 63
(steerei, Otus) 59
(stictica, Strix) 20
(stictonotus, Scops) 54
(strenua, Athene) 206, 207
strenua, Ninox *207, 208, 256*
(strepitans, Strix) 116
stresemanni, Otus 39, 261, *272*
stricklandi, Lophostrix (cristata) 154, 266
(stricklandi, Lophostrix) 154
(stridula, Strix) 127
(Strigonax) 121
Strix 127
(Strix (?) Novae Hollandiae) 28
(Strix acadica) 200, 202
(Strix accipitrina) 241
(Strix adspersa) 16
(Strix affinis) 17
(Strix africana) 111
(Strix alba) 15
Strix albitarsis 141
(Strix albomarginata) 144
Strix aluco 127, 133, *254*
(Strix Aluco) 133
Strix aluco aluco 133
Strix aluco biddulphi 135
(Strix aluco clanceyi) 133
Strix aluco härmsi 135
(Strix aluco harterti) 135
Strix (aluco) ma 136
Strix aluco mauritanica 133
Strix (aluco) nivicola 135
(Strix aluco nivipetens) 135
(Strix aluco obrieni) 135
(Strix aluco obscurata) 134
Strix aluco sanctinicolai 134
Strix aluco siberiae 134
Strix aluco sylvatica 133
(Strix aluco volhyniae) 133
Strix aluco willkonskii 134
Strix (aluco) yamadae 136
(Strix amauronota) 31
(Strix americana) 235
(Strix arctica) 241
Strix asio 70
(Strix Asio) 74
(Strix (Athene) guteruhi) 211
(Strix Athene ruficolor) 189
(Strix atricapilla) 84
(Strix badia) 34, 35
(Strix barbata) 152
Strix bartelsi 131
(Strix baweana) 127
(Strix Boobook) 213
(Strix brachyotis) 241
(Strix (Brachyotus) helveola) 244
(Strix brama) 193
(Strix brasiliana) 171
(Strix Bubo) 99, 104
Strix butleri 136, *254*
(Strix cabrae) 29
(Strix caparoch) 160
(Strix Capensis) 29
(Strix castanoptera) 181
(Strix cayelii) 26
Strix chacoensis 141
(Strix choliba) 78
(Strix cinerea) 151
(Strix coromanda) 118
(Strix cristata) 154, 155
(Strix crucigera) 78
(Strix Cubae) 19
(Strix cunicularia) 188, 199
(Strix De-Roepstorffi) 23
(Strix decussata) 78
(Strix delicatula) 24
(Strix deminuta) 237
(Strix domingensis) 242
(Strix ferox) 171
(Strix ferruginea) 171
(Strix flammea) 15
(Strix Flammea) 16, 241
(Strix flammea bargei) 19
(Strix flammea contempta) 20
(Strix flammea gracilirostris) 15
(Strix flammea meeki) 24
(Strix flammea obscura) 16

(*Strix Flammea rhenana*) 16
(*Strix flammea schmitzi*) 15
(*Strix flammea sumbaensis*) 24
(*Strix flammea* var. *Guatemalae*) 19
(*Strix flammea* var. *nigrescens*) 21
(*Strix fulva*) 215
Strix fulvescens 144
(*Strix funerea*) 159, 200
(*Strix furcata*) 19
(*Strix fusca*) 211
(*Strix fuscescens*) 150
(*Strix Georgica*) 147
(*Strix giu*) 47
(*Strix glaucops*) 21
(*Strix grallaria*) 198
(*Strix hardwickii*) 121
(*Strix hirsuta*) 206, 217
(*Strix hirsuta borneensis*) 218
(*Strix hirsuta japonica*) 216
(*Strix hostilis*) 15
Strix huhula 143
Strix huhula albomarginata 144
Strix huhula huhula 143
Strix hylophila 142
(*Strix hypugaea*) 194
(*Strix Indranee*) 129
(*Strix indranee rileyi*) 130
(*Strix indranee shanensis*) 132
(*Strix inexspectata*) 26
(*Strix infuscata*) 171
(*Strix insularis*) 21
(*Strix javanica*) 18
(*Strix Ketupu*) 121
(*Strix ketupu*) 123
(*Strix kirchhoffi*) 15
(*Strix lactea*) 117
(*Strix lapponica*) 152
Strix leptogrammica *129*, 130, *254*
(*Strix leptogrammica chaseni*) 130
(*Strix leptogrammica connectens*) 129
Strix leptogrammica indranee 129
Strix leptogrammica leptogrammica 130
Strix leptogrammica maingayi 130
Strix leptogrammica myrtha 130
Strix leptogrammica niasensis 130
(*Strix leptogrammica nyctiphasma*) 130
(*Strix leptogrammica orientalis*) 132
(*Strix leptogrammica ticehursti*) 132
Strix leptogrammica vaga 131
(*Strix leucotis*) 94
(*Strix licua*) 163
(*Strix Liturata*) 148
(*Strix longimembris*) 30
(*Strix lugubris*) 206, 216
(*Strix lulu*) 24
(*Strix maculata*) 17, 235
(*Strix magellanicus*) 103
(*Strix magica*) 56
(*Strix mantis*) 36
(*Strix maugei*) 211
(*Strix minutissima*) 174
(*Strix nacurutu*) 102
(*Strix naevia*) 74
(*Strix nana*) 167
Strix nebulosa 151
(*Strix nebulosa alleni*) 147
Strix nebulosa lapponica *152*, *254*
Strix nebulosa nebulosa 151, 266
Strix newarensis 131
Strix newarensis caligata 132
Strix newarensis laotiana 132
(*Strix newarensis laotianus*) 132
Strix newarensis newarensis 131
Strix newarensis ticehursti 132
Strix nigrolineata 142, *143*
(*Strix nivicola yamadae*) 136
(*Strix noctua*) 188
(*Strix novae-Seelandiae*) 206, 215
(*Strix novaeseelandiae maculata*) 215
(*Strix nudipes*) 89
(*Strix occidentale huachucae*) 146
Strix occidentalis 144, *145*, *254*
Strix occidentalis caurina 145
(? *Strix occidentalis juanaphillipsae*) 146
Strix occidentalis lucida 146
Strix occidentalis occidentalis 145
Strix ocellata 128
Strix ocellata grandis 128
Strix ocellata grisescens 128
Strix ocellata ocellata 128
(*Strix orientalis*) 116
(*Strix Orientalis*) 127
(*Strix otus*) 234
(*Strix Otus*) 237
(*Strix oustaleti*) 30
(*Strix pagodorum*) 127
(*Strix parva*) 18
(*Strix passerina*) 161
(*Strix peli*) 125
(*Strix perlata*) 20, 162
(*Strix perspicillata*) 156
(*Strix phaloenoides*) 170
(*Strix pithecops*) 30
(*Strix poensis*) 17
(*Strix pratincola*) 19
(*Strix pulsatrix*) 157
(*Strix pumila*) 174
(*Strix punctata*) 29
(*Strix punctatissima*) 22
(*Strix pusilla*) 18
(*Strix Radiata*) 178
(*Strix Rosenbergii*) 27
(*Strix rufescens*) 36, 150
Strix rufipes 141, *254*
(*Strix rufipes sanborni*) 142
(*Strix saharae*) 189
(*Strix Sandwichensis*) 242
(*Strix scandiaca*) 97
(*Strix Scops*) 47
(*Strix scutulata*) 218
(*Strix Selo-puto*) 127
Strix seloputo 127
Strix seloputo baweana 127
Strix seloputo seloputo 127
Strix seloputo wiebkeni 128
(? *Strix sibirica*) 106
(*Strix sororcula*) 27
(*Strix splendens*) 16
(*Strix stictica*) 20
(*Strix strepitans*) 116
(*Strix stridula*) 127
(*Strix suinda*) 140, 243
(*Strix sumatrana*) 115
(*Strix superciliaris*) 219
(*Strix sylvatica*) 133, 177
(*Strix tenebricosa arfaki*) 32
(*Strix tenebricosus*) 32
(*Strix tengmalmi*) 200, 202
(*Strix thomensis*) 22
(*Strix torquata*) 156, 161
(*Strix Tuidara*) 20
(*Strix turcomana*) 108
(*Strix Ulula*) 159
(*Strix undulata*) 215
Strix uralensis 148
(*Strix uralensis buturlini*) 149
Strix (*uralensis*) *davidi* 150
(*Strix uralensis dauricus*) 149
Strix uralensis fuscescens 150
Strix uralensis hondoensis 150
Strix uralensis japonica 150
(*Strix uralensis jinkou*) 149
Strix uralensis liturata 148, *254*
Strix uralensis macroura 148
(*Strix uralensis media*) 150
(*Strix uralensis momiyamae*) 150
(*Strix uralensis morii*) 149
(*Strix uralensis nigra*) 150
Strix uralensis nikolskii 149
(*Strix uralensis pacifica*) 150
(*Strix uralensis tatibanai*) 149
Strix uralensis uralensis 149
Strix uralensis yenisseensis 149
Strix varia 146
(*Strix varia albescens*) 146
(*Strix varia albogilva*) 147
(*Strix varia brunnescens*) 146
Strix varia georgica 147
Strix varia helveola 147
Strix varia sartorii 147
Strix varia varia 146, *254*, 266
(*Strix varius*) 146
Strix virgata 138, *254*
Strix virgata borelliana 140
Strix virgata centralis 139
Strix virgata macconnelli 140
(*Strix virgata sneiderni*) 140
Strix virgata squamulata 138, *139*, 266
Strix virgata superciliaris 140
Strix virgata tamaulipensis 139
Strix virgata virgata 139
(*Strix virginianus*) 101
(*Strix walleri*) 30
(*Strix wapacuthu*) 100
Strix woodfordi 137, *254*
Strix woodfordi nigricantior 137
Strix woodfordi nuchalis 137
Strix woodfordi umbrina 137
Strix woodfordi woodfordi 138
(*Strix zeylonensis*) 122
(*Strix zorca*) 47
(*Stryx* (sic) *pulchella*) 48
(*Stryx* (sic) *uralensis*) 149
stygius, *Asio* 235, *257*
(*stygius*, *Nyctalops*) 236

(*subarcticus, Bubo*) 100
(*suinda, Strix*) 140, 243
sulaensis, Otus (*magicus*) 57, 263
(*sulaensis, Pisorhina*) 57
(*sumatrana, Strix*) 115
sumatranus, Bubo 115
sumbaensis, Ninox 222, *256*, *283*
sunia, Otus 52
(*sunia, Scops*) 53
(*superciliare, Syrnium*) 140
superciliaris, Ninox 219, *220*, *256*
(*superciliaris, Strix*) 219
Surnia 159
Surnia ulula 159, *256*
Surnia ulula caparoch *160*
(*Surnia ulula korejewi*) 160
(*Surnia ulula orokensis*) 159
(*Surnia ulula pallasi*) 159
Surnia ulula tianschanica 160
Surnia ulula ulula 159
(*Surnium ?* (sic) *umbrinum*) 137
(*sylvatica, Strix*) 133, 177
(*Syrnium albitarse*) 141
(*Syrnium albo-gularis*) 90
(*Syrnium aluco mauritanicum*) 133
(*Syrnium bartelsi*) 131
(*Syrnium biddulphi*) 135
(*Syrnium blanfordi*) 135
(*Syrnium Bohndorffi*) 137
(*Syrnium borellianum*) 140
(*Syrnium cinereum sakhalinense*) 152
(*Syrnium davidi*) 150
(*Syrnium fulvescens*) 144
(*Syrnium härmsi*) 135
(*Syrnium Koeniswaldianum*) 158
(*Syrnium ma*) 136
(*Syrnium macabrum*) 91
(*Syrnium maingayi*) 130
(*Syrnium nebulosum helveolum*) 147
(*Syrnium nebulosum sablei*) 147
(*Syrnium nebulosum* var. *Sartorii*) 147
(*Syrnium niasense*) 130
(*Syrnium nigricantius*) 137
(*Syrnium nigrolineatum*) 143
(*Syrnium nivicolum*) 135
(*Syrnium nuchale*) 137
(*Syrnium occidentale*) 145
(*Syrnium occidentale caurinum*) 145
(*Syrnium occidentale lucidum*) 146
(*Syrnium ocellatum*) 128
(*Syrnium ochrogenys*) 129
(*Syrnium sancti-nicolai*) 134
(*Syrnium spilinotum*) 143
(*Syrnium squamulatum*) 138
(*Syrnium superciliare*) 140
(*Syrnium uralense coreensis*) 149
(*Syrnium uralense hondoensis*) 150
(*Syrnium uralense japonicum*) 150
(*Syrnium uralense nikolskii*) 149
(*Syrnium uralense sibiricum*) 149
(*Syrnium virgatum*) 139
(*Syrnium Whiteheadi*) 128
(*Syrnium Wiebkeni*) 128
(*Syrnium willkonskii*) 134
(*Syrnium woodfordi* var. *sansibaricum*) 137
(*Syrnium woodfordi* var. *suahelicum*) 137

T

(*taeniata, Athene*) 229
Taenioglaux 178
(*Tarayensis, Noctua*) 192
tempestatis, Otus (*magicus*) 57, 263
tenebricosa, Tyto 31
(*tenebricosus, Strix*) 32
(*tengmalmi, Strix*) 200, 202
tephronotum, Glaucidium 175, *255*
(*terricolor, Ninox*) 227
(*theomacha, Spiloglaux*) 227
thilohoffmanni, Otus 37, *250*, 261, *273*
(*thomensis, Strix*) 22
thomensis, Tyto 22
(*torquata, Strix*) 156, 161
trichopsis, Otus 76, *251*
(*trichopsis, Scops*) 76
(*tubiger, Noctua*) 176
tucumanum, Glaucidium 171, *279*
(*Tuidara, Strix*) 20
(*turcomana, Strix*) 108
Tyto 15
Tyto alba 15, *249*
Tyto alba affinis 17
Tyto alba alba 15, *16*
(*Tyto alba alexandrae*) 24
Tyto alba bargei 19
(*Tyto alba bellonae*) 24
(*Tyto alba bondi*) 19
Tyto alba contempta 20
Tyto alba crassirostris 23
(*Tyto alba crypta*) 18
Tyto alba detorta 22
Tyto alba erlangeri 18
Tyto alba ernesti 16
(*Tyto alba everetti*) 24
Tyto alba furcata 19
Tyto alba gracilirostris 15
Tyto alba guttata 16
(*Tyto alba hauchecorni*) 20
Tyto alba hellmayri 20
Tyto alba hypermetra 17, *271*
(*Tyto alba interposita*) 24
Tyto alba javanica 18
(*Tyto alba kleinschmidti*) 15
(*Tyto alba kuehni*) 24
(*Tyto alba lifuensis*) 24
(*Tyto alba microsticta*) 18
(*Tyto alba niveicauda*) 19
Tyto alba pratincola 19, 261
Tyto alba schmitzi 15
Tyto alba stertens 18
(*Tyto alba subandeana*) 20
Tyto alba tuidara 20
(*Tyto alba zottae*) 20
Tyto aurantia 25
Tyto capensis 29
(*Tyto Capensis Damarensis*) 29
(*Tyto capensis libratus*) 29
Tyto castanops 29
Tyto crassirostris 23
Tyto delicatula 24
Tyto delicatula delicatula 24
Tyto delicatula interposita 24
Tyto delicatula meeki 24
Tyto delicatula sumbaensis 24
Tyto deroepstorffi 23
Tyto detorta 22
Tyto glaucops 21, *249*
Tyto inexspectata 26
Tyto insularis 21
Tyto insularis insularis 21
Tyto insularis nigrescens 21
Tyto longimembris 30, *249*
(*Tyto longimembris albifrons*) 31
Tyto longimembris amauronota 31
(*Tyto longimembris baliem*) 31
Tyto longimembris chinensis 31
(*Tyto longimembris dombraini*) 29
(*Tyto longimembris georgiae*) 30
Tyto longimembris longimembris 30
(*Tyto longimembris melli*) 31
Tyto longimembris papuensis 31
(*Tyto maculosa*) 30
Tyto manusi 27
Tyto multipunctata 32, *249*
Tyto nigrobrunnea *25*, *249*
Tyto novaehollandiae 28
Tyto novaehollandiae calabyi 28
(*Tyto novaehollandiae galei*) 28
Tyto novaehollandiae kimberli 28
(*Tyto novaehollandiae mackayi*) 28
(*Tyto novaehollandiae melvillensis*) 28
Tyto novaehollandiae novaehollandiae 28, *249*
(*Tyto novaehollandiae perplexa*) 28
(*Tyto novaehollandiae riordani*) 29
(*Tyto novaehollandiae troughtoni*) 29
(*Tyto novaehollandiae whitei*) 28
(*Tyto perlatus lucayanus*) 19
Tyto prigoginei 33, *249*, *272*
Tyto punctatissima 22
Tyto rosenbergii 27
Tyto rosenbergii pelengensis 28
Tyto rosenbergii rosenbergii 27
Tyto sororcula 26
Tyto sororcula cayelii 26
Tyto sororcula sororcula 27
Tyto soumagnei 23, *249*, 261, *271*
Tyto tenebricosa 31
Tyto tenebricosa arfaki 32
(*Tyto tenebricosa magna*) 32
Tyto tenebricosa multipunctata 32
Tyto tenebricosa tenebricosa *32*
Tyto thomensis 22

U

(*ugandae, Pisorhina*) 51
(*Ulula newarensis*) 131
(*Ulula, Strix*) 159
ulula, Surnia 159, *256*
umbra, Otus 61, 263, *275*
umbra, Pisorhina 61
(*umbratilis, Ephialtes*) 65
(*umbrinum, Surnium ?* (sic)) 137
(*undulata, Ninox*) 207
(*undulata, Strix*) 215
uralensis, Strix 148
(*uralensis, Stryx* (sic)) 149
Uroglaux 231
Uroglaux dimorpha *231*

ussheri, Scotopelia 126, *253*
(*usta, Scops*) 84

V

(*vandewateri, Pisorhina*) 41
varia, Strix 146
(*Variegata superior, Ninox*) 229
variegata, Ninox 228
(*variegata, Noctua*) 228
(*varius, Strix*) 146
(*venatica, Noctua*) 215
(*vermiculatus, Megascops*) 87
vermiculatus, Otus 87, 265
(*Vidalii, Athene*) 188
(*vinaceus, Megascops*) 72
virgata, Strix 138, *254*
(*virgatum, Syrnium*) 139
virginianus, Bubo 99, *252*, *285*
(*virginianus, Strix*) 101
vosseleri, Bubo 114, *252*
(*vulpes, Heteroscops*) 41

W

(*walleri, Strix*) 30
(*wapacuthu, Strix*) 100
(*Watsonii, Ephialtes*) 84
watsonii, Otus 83
(*whiteheadi, Scops*) 68
(*Whiteheadi, Syrnium*) 128
(*Whitelyi, Athene*) 180
(*whitneyi, Athene*) 186
whitneyi, Micrathene *186*, *255*, 266, *281*
(*Wiebkeni, Syrnium*) 128
(*willkonskii, Syrnium*) 134
(*Wilsonianus, Otus*) 238
(*wilsonianus, Otus*) 238
(*Woodfordi, Noctua*) 138
woodfordi, Strix 137, *254*

X

Xenoglaux 185, *281*
Xenoglaux loweryi *185*, 266, *281*

Y

yamadae, Strix (*aluco*) 136
(*yorki, Ninox*) 213

Z

(*zelandica, Noctua*) 215
zeylonensis, Ketupa 121, *253*
(*zeylonensis, Strix*) 122
(*zorca, Strix*) 47

Index of Vernacular English Owl Names / Index der englischen Eulennamen

A

Abyssinian Long-eared Owl 239
African Barred Owlet 183, *281*
African Grass Owl 29
African Scops Owl *51*
African Wood Owl 137
Akun Eagle Owl 118
Albertine Owlet 184, *281*
Amazonian Pygmy Owl 174, *280*
American Great Horned Owls *284, 285*
Andaman Barn Owl 23
Andaman Hawk Owl 219
Andaman Scops Owl 38
Andean Pygmy Owl 166
Anjouan Scops Owl 45, *274*
Ashy-faced Barn Owl 21
Asian Barred Owlet 179
Austral Pygmy Owl *167*
Australian Barn Owl 24
Australian Masked Owl 28

B

Baja (or Cape) Pygmy Owl 164
Balsas Screech Owl 72
Band-bellied (or Rusty-barred) Owl 158
Bare-shanked Screech Owl 80
Barking (or Winking) Owl *208*
Barn Owl *16*
Barred (Malay) Eagle Owl 115
Barred Owl 146
Bartel's Wood Owl 131
Bay Owl *34, 272*
Bearded Screech Owl *81*
Biak Scops Owl 58
Bismarck Hawk Owl 228
Black and White Owl 142, *143*
Black-banded Owl 143
Blakiston's Eagle Owl 120
Blakiston's Fish Owl 120
Boang Barn Owl 23
Boreal Owl 200
Brown Fish Owl 121
Brown Hawk Owl 215, *283*
Brown (or Short-browed) Spectacled Owl 157
Brown Wood Owl *129*
Buff-fronted Owlet *204*
Buffy (Malay) Fish Owl 123
Burrowing Owl *194*

C

Cape Eagle Owl 110
Cape Pygmy Owl (Baja Pygmy Owl) 164
Cape Verde Barn Owl 22
Central American Pygmy Owl 173, *280*
Chaco Owl 141
Chestnut Owlet 182
Chestnut-backed Owlet 179
Chestnut-barred Owlet 182
Christmas Hawk Owl 225, *226*
Cinnabar Hawk Owl 224, *282*
Cinnamon (or Sandy) Scops Owl 37
Cinnamon Screech Owl 83, *278*
Cloudforest Pygmy Owl 166, *279*
Cloudforest Screech Owl 83, *278*
Colima Pygmy Owl 172, *280*
Collared Owlet 176
Collared Scops Owl 63, 64
Colombian Screech Owl 82
Common Barn Owl 15
Common Scops Owl *47*
Congo Bay Owl 33
Crested Owl *154*
Cuban Pygmy Owl 168
Cuban Screech (Bare-legged) Owl *93*

D

Dominica Barn Owl 21
Dubious Hawk Owl 223, 283
Dusky Eagle Owl 117

E

Eastern Grass Owl 30
Eastern Screech Owl *74*
Elf Owl *186, 281*
Enggano Scops Owl 62
Eurasian Eagle Owl 104, *105*
Eurasian Pygmy Owl *161*
Eurasian Scops Owl 47

F

Fearful Owl *246*
Ferruginous Pygmy Owl 169
Flammulated Owl *46*
Flores Scops Owl 39, *272*
Forest Eagle Owl 114, *115*
Forest Owlet 193, *282*
Fraser's Eagle Owl *113*
Fulvous Owl 144

G

Galapagos Barn Owl 22
Giant Scops Owl *96*
Golden Masked Owl 25
Grand Comoro Scops Owl 45
Great Grey Owl 151, *152*
Great Horned Owl 99
Greater Sooty Owl 31, *32*
Guatemalan Screech Owl 86

H

Himalayan Wood Owl 131
Hume's (Tawny) Owl 136

I

Indian (collared) Scops Owl 65
Itombwe Owl 33

J

Jamaican Owl *233*
Japanese Scops Owl 67
Javan Owlet 181
Javan Scops Owl *42*
Jungle Hawk Owl 226
Jungle Owlet *178*

K

Koepcke's Screech Owl 79, *277*

L

Lamb's (or Oaxaca) Screech Owl 73
Laughing Owl *232*
Lesser Antilles Barn Owl 21
Lesser Masked Owl 26
Lesser Sooty Owl 32
Little Owl 188
Little Sumba Hawk Owl 222, *283*
Long-tufted Screech Owl 85
Long-whiskered Owlet *185, 281*
Luzon Scops Owl 42

M

Madagascar Hawk Owl *220*
Madagascar Long-eared Owl 240
Madagascar Red Owl 23, *271*
Madagascar Scops Owl 43, *274*
Magellan Horned Owl 103
Malay Eagle Owl 115
Malay Fish Owl 123
Maned Owl *153*
Mantanani Scops Owl 58
Manus Hawk Owl 226
Manus Masked Owl 27
Marsh Owl 244
Mayotte Scops Owl 44
Mentawai Scops Owl 62
Miles' Spotted Eagle Owl 112
Minahassa Masked Owl 26
Mindanao Scops Owl 42
Mindoro Hawk Owl 222
Mindoro Scops Owl 43
Mohéli Scops Owl 44, *276*
Moluccan Hawk Owl 224
Moluccan Scops Owl 55
Montane Forest-Screech Owl 85, *278*
Morepork 214
Mottled Owl 138, *139*
Mottled Wood Owl 128
Mountain Pygmy Owl 165
Mountain Scops Owl 39

N

Nduk Eagle Owl (Usambara Eagle Owl) 114
New Britain Hawk Owl (Russet Hawk Owl) 228
New Ireland Hawk Owl 228
New Zealand Boobook 214
Nicobar Scops Owl 61, *275*
Northern Hawk Owl 159, *160*
Northern Long-eared Owl 237
Northern Pygmy Owl 163
Northern Saw-whet Owl 202
Northern White-faced Owl 94

O

Oaxaca Screech Owl (Lamb's Screech Owl) 73
Ochre-bellied Hawk Owl 223, *282*
Oriental Scops Owl 52, *273*

P

Pacific Screech Owl *73*
Palau Owl *92*
Palawan Scops Owl 69
Pallid Scops Owl 49
Papuan Hawk Owl *231*
Pearl Spotted Owl 162
Pel's Fishing Owl *125*
Pemba Scops Owl 46
Pernambuco Pygmy Owl 174, *280*
Peruvian Pygmy Owl 167, *279*
Peruvian Screech Owl 80
Pharaoh Eagle Owl 109, *110*
Philippine Eagle Owl *119*
Philippine Hawk Owl 220
Philippine Scops Owl *68*
Powerful Owl *207*
Puerto Rican Screech Owl 89

R

Rajah Scops Owl 62
Red-chested Owlet 175, *176*
Reddish Scops Owl 36, *273*
Ridgway's Pygmy Owl 169
Rio Napo Screech Owl 88
Rock Eagle Owl 109
Roraima (Foothill) Screech Owl 88
Rufescent Screech Owl 82
Rufous (Hawk) Owl 206
Rufous Fishing Owl 126
Rufous-banded Owl 141
Rufous-legged Owl 141
Russet (or New Britain) Hawk Owl 228
Rusty-barred Owl 142
Ryukyu Scops Owl 55

S

Sandy Scops Owl (Cinnamon Scops Owl) 37
Sangihe Scops Owl 60, *275*
Santa Barbara (Bearded) Owl 81
Sâo Tomé Barn Owl 22
Sâo Tomé Scops Owl 43
Serendib Scops Owl 37, *273*
Seychelles Scops Owl 61
Shelley's Eagle Owl 116
Short-browed Spectacled Owl (Brown Spectacled Owl) 157
Short-eared Owl *240*
Sichuan (Wood) Owl 150
Sick's Pygmy Owl 174, *280*
Simeulue Scops Owl 61
Sjoestedt's Owlet *181*
Snowy Owl *97*
Sokoke Scops Owl 38
Solomon Hawk Owl 229
Southern Boobook 210
Southern White-faced Owl *94*
Speckled Hawk Owl 227, *283*
Spectacled Owl *156*
Spotted Eagle Owl 111
Spotted Owl 144, *145*
Spotted Owlet 192, *282*
Spotted Wood Owl 127
St. Vincent Barn Owl 21
Stresemann's Mountain Scops Owl 39
Striated Scops Owl 49
Striped Owl 234
Stygian Owl 235, *236*
Subtropical Pygmy Owl 173, *280*
Sulawesi Masked Owl 27
Sulawesi Scops Owl 59
Sumba Boobook 210, *283*
Sunda Scops Owl 63

T

Taliabu Masked Owl *25*
Tamaulipas Pygmy Owl 172, *280*
Tasmanian Boobook 214
Tasmanian Masked Owl 29
Tawny Fish Owl *122*
Tawny Owl 133
Tawny-bellied Screech Owl 83
Tawny-browed Owl 158
Tengmalm's Owl 200
Togian Hawk Owl 223, *283*
Torotoroka Scops Owl 44, *274*
Tropical Screech Owl 77
Tucuman Pygmy Owl 171, *279*
Tumbes Screech Owl 80, *277*

U

Unspotted Saw-whet Owl 203
Ural Owl 148
Usambara (or Nduk) Eagle Owl 114

V

Variable Screech Owl 84, *278*
Vermiculated Eagle Owl 112
Vermiculated Fishing Owl 126
Vermiculated Screech Owl 87
Verreaux's Eagle Owl 117

W

Wallace's Scops Owl 69
Western Screech Owl 70
Whiskered Screech Owl 76
White-browed Hawk Owl 219
White-fronted Scops Owl 36
White-throated Screech Owl *90*
Winking Owl (Barking Owl) *208*

Y

Yungas Pygmy Owl 166, *279*

Index of Vernacular German Owl Names / Index der deutschen Eulennamen

A

Afrika-Waldkauz (Woodfordkauz) 137
Afrika-Zwergohreule *51*
Albertseekauz 184, *281*
Amazonas-Zwergkauz 174, *280*
Amerikanische Uhus *284, 285*
Andamanen-Schleiereule 23
Andamanen-Zwergohreule 38
Andamanenkauz 219
Anden-Sperlingskauz 166
Angelina Zwergohreule *42*
Anjouan-Zwergohreule 45, *274*
Arabien-Fleckenuhu (Milesuhu) 112
Araukaner-Sperlingskauz (Patagonien-Sperlingskauz) 167
Äthiopienohreule 239
Australien-Schleiereule 24

B

Balsas-Kreischeule 72
Bartelskauz 131
Bartkauz 151, *152*
Beccari-Zwergohreule 58
Bengalenuhu 109
Bergwald-Kreischeule 85, *278*
Bindenfischeule (Pelfischeule) *125*
Bindenhalskauz 142, *143*
Bindenkauz 158
Bindenuhu (Shelley-Uhu) 116, 253
Bismarckkauz 228
Blaßstirnkauz *204*
Blaßuhu (Milchuhu) 117
Blewitt-Kauz 193, *282*
Boang-Schleiereule 23
Boobookkauz (Kuckuckskauz) 210
Brahma-Kauz 192, *282*
Brasil-Sperlingskauz 169
Brasilkauz 142
Brauner Brillenkauz 157
Brillenkauz *156*

C

Chaco-Kauz 141
Chaco-Sperlingskauz (Tucuman-Sperlingskauz) 171, *279*
Cholibaeule 77
Christmas-Kauz 225, *226*
Colima-Zwergkauz 172, *280*
Comoren-Zwergohreule 45
Cooper-Kreischeule (Mangroven-Kreischeule) *73*

D

Davidkauz 150
Dschungelkauz *178*
Dunkle Ohreule (Styx-Eule) 235, *236*

E

Einfarbkauz 226
Elfenkauz *186, 281*
Enggano-Zwergohreule 62
Etchécoparkauz 182

F

Fahlkauz 136
Falkenkauz (Schildkauz) 215, *283*
Fischuhu, (Brauner) 121
Flecken-Kreischeule 76
Fleckenkauz 144, *145*
Fleckenrußeule 32
Fleckenuhu 111
Flores-Zwergohreule 39, *272*
Fuchseule 39

G

Galapagos-Schleiereule 22
Gelbbrauenkauz 158
Gelbfußuhu 118
Gelbschnabelzwergohreule 37
Gilbkauz 144
Gnomen-(Sperlings)kauz 165
Goldeule 25
Graseule 30
Graukopf-Zwergkauz 173, *280*
Grauuhu (Sprenkeluhu) 112
Große Rußeule *32*
Guinea-Uhu *113*

H

Habichtskauz 148
Halsband-Zwergohreule 64
Hartlaub-Zwergohreule 43
Haubenkauz *154*, 155
Himalaja-Braunkauz 131
Himalaja-Fischuhu *122*
Hispaniolaschleiereule 21
Hoskins-Sperlingskauz 164

I

Indien-Halsbandeule 65

J

Jacquinotkauz 229
Jamaikaeule *233*
Japan-Halsbandeule 67
Java-Zwergohreule 42

K

Käferuhu 118
Kaninchenkauz (Präriekauz) *194*
Kap Verde-Schleiereule 22
Kapgraseule 29
Kapkauz 183, *281*
Kapohreule 244
Kapuhu 110
Kastanienkauz 182
Kastanienrückenkauz 179
Kläfferkauz *208*
Kleine Antillen-Schleiereule 21
Kleiner Sumbakauz 222
Koepcke-Kreischeule 79, *277*
Kolumbien-Kreischeule 82
Koromandeluhu 117
Kritzel-Kreischeule (Marmor-Kreischeule) 87
Kuba-Kreischeule *93*
Kuba-Sperlingskauz 168
Kuckuckskauz 179, 210
Kurzbrauen-Brillenkauz 157

L

Lachkauz (Weißwangenkauz) *232*
Lambs Kreischeule 73
Lowery-Zwergkauz *185, 281*
Luzon-Zwergohreule 42

M

Madagaskar-Kauz 219, *220*
Madagaskar-Ohreule 240
Madagaskar-Zwergohreule 43, *274*
Magellan-Uhu 103
Mähneneule *153*
Malaienkauz *129*
Malaienuhu 115
Malegasseneule 23, *271*
Manado-Zwergohreule 59
Mangokauz 128
Mangroven-(Cooper-)Kreischeule *73*
Manus-Schleiereule 27
Manuskauz 226
Marmor-(Kritzel-)Kreischeule 87
Marmorfischeule 126
Maskeneule *34, 272*
Mayotte-Zwergohreule 44
Mentawai-Zwergohreule 62
Milchuhu (Blaßuhu) 117
Milesuhu (Arabien-Fleckenuhu) 112
Minahassaeule 26
Mindanao-Ohreule (Rotohreule) *96*
Mindanao-Zwergohreule 42
Mindoro-Zwergohreule 43
Mindorokauz 222
Moheli-Zwergohreule 44, *276*
Molukken-Zwergohreule 55
Molukkenkauz 224

N

Nacktbein-Kreischeule 80
Nacktfußeule 89

Nebelwald-Kreischeule 83, *278*
Nebelwald-Sperlingskauz 166, *279*
Nepaluhu 114, *115*
Neubritannienkauz 228
Neuhollandeule 28
Neuirlandkauz 228
Neuseeland-Boobook 214
Nicobaren-Zwergohreule 61, *275*
Nordbüscheleule (Weißgesichteule) 94

O

Oaxaca-Kreischeule 73
Ockerbauchkauz 223, *282*
Orient-Zwergohreule 52, *273*
Ost-Kreischeule *74*

P

Pagodenkauz 127
Palaueule *92*
Palawan-Halsbandeule 69
Parker-Zwergkauz 173, *280*
Patagonien-(Araukaner-)Sperlingskauz *167*
Pelfischeule (Bindenfischeule) 125
Pembaeule 46
Perlkauz 162
Pernambuco-Zwergkauz 174, *280*
Peru-Kreischeule 80
Peru-Sperlingskauz 167, *279*
Philippinen-Halsbandeule *68*
Philippinen-Uhu (Streifenuhu) *119*
Philippinen-Zwergohreule 58
Philippinenkauz 220
Ponderosa-Zwergohreule *46*
Prachtkauz *181*
Präriekauz (Kaninchenkauz) *194*
Prigogine-Eule 33
Pünktchenkauz 227, *283*

R

Radscha-Zwergohreule 62
Raufußkauz 200
Ridgway-Sperlingskauz 169
Ridgwaykauz 203
Riesenfischuhu 120
Riesenkauz *207*
Rio Napo-Kreischeule 88
Rocky Mountains-Sperlingskauz 163
Roraima-Kreischeule 88
Rosenbergeule 27
Rostkauz 142, 206
Rotbrust-Sperlingskauz 175, *176*
Rote Fischeule 126
Röteleule 36, *273*
Rötelkauz 141
Roter Buschkauz 206
Rotfußkauz 141
Rotgesicht-Kreischeule 86
Rotohreule (Mindanao-Ohreule) *96*
Rotrücken-Fischeule 126
Rundflügelkauz *231*
Rußeule 31

S

Sägekauz 202
Salomoneneule *246*
Salomonenkauz 229
Salvin-Kreischeule 82
Sangihe-Zwergohreule 60, *275*
Santa Catarina-Kreischeule 85
Sâo Tomé-Schleiereule 22
Schildkauz (Falkenkauz) 215, *283*
Schleiereule 15, *16*, *272*
Schmuck-Zwergohreule 55
Schnee-Eule *97*
Schreieule 234
Schwarzkappen-Kreischeule 84, *278*
Serendib-Zwergohreule 37, *273*
Seychelleneule 61
Shelley-Uhu (Bindenuhu) 116, 253
Sick-Zwergkauz 174, *280*
Simeulue-Zwergohreule 61
Sokoke-Zwergohreule 38
Sperbereule 159, *160*
Sperlingskauz *161*
Sprenkelkauz 138, *139*
Sprenkeluhu (Grauuhu) 112
Steinkauz 188
Streifen-Zwergohreule 49
Streifenkauz 146
Streifenohreule 234
Streifenuhu (Philippinen-Uhu) *119*
Stresemann-Zwergohreule 39
Styx-Eule (Dunkle Ohreule) 235, *236*
Südbüscheleule (Weißgesichteule) *94*
Sulawesi-Schleiereule 27
Sumbakauz 210, *283*
Sumbakauz, Kleiner *272*
Sumpfohreule *240*
Sunda-Fischuhu 123
Sunda-Zwergohreule 63

T

Taliabu-Eule *25*
Tamaulipas-Zwergkauz 172, *280*
Tanimbareule 26
Tasman-Boobookkauz 214
Tasmanien-Schleiereule 29
Togian-Falkenkauz 223, *283*
Torotoroka-Zwergohreule 44, *274*
Trillerkauz 181
Tropfenkreischeule *81*
Tucuman-(Chaco-)Sperlingskauz 171, *279*
Tumbes-Kreischeule 80, *277*

U

Uhu 104, *105*
Usambara-Uhu 114

V

Vielfleck-Falkenkauz 223, *283*
Virginiauhu 99

W

Wachtelkauz 176
Waldkauz 133
Waldohreule 237
Wallace-Zwergohreule 69
Watson-Kreischeule 83
Weißgesichteule (Nordbüscheleule *and* Südbüscheleule) 94
Weißkehl-Kreischeule *90*
Weißstirneule 36
Weißwangenkauz (Lachkauz) *232*
West-Kreischeule 70
Woodfordkauz (Afrika-Waldkauz) 137
Wüstenkauz 136
Wüstenuhu 109, *110*

Y

Yungas-Sperlingskauz 166, *279*

Z

Zebrakauz 143
Zimt-Kreischeule 83, *278*
Zinnoberkauz 224, *282*
Zwergohreule *47*

Index of Vernacular French Owl Names / Index der französischen Eulennamen

C

Chevêche brame (Chouette brame) 192
Chevêche d'Athene 188
Chevêche forestière (Chouette des forêts) 193
Chevêchette à collier 176
Chevêchette à dos marron 179
Chevêchette à pieds jaunes 175
Chevêchette à queue barrée 181
Chevêchette à tête grise 173
Chevêchette australe 167
Chevêchette barrée (ou cuculoide) 179
Chevêchette brune 169
Chevêchette cabouré (ou des montagnes) 165
Chevêchette châtaine (Chevêchette marron) 182
Chevêchette cuculoide (Chevêchette barrée) 179
Chevêchette d'Amazonie 174
Chevêchette de Colima 172
Chevêchette de Cuba 168
Chevêchette de Etchécopar 182
Chevêchette de Hoskins 164
Chevêchette de la jungle 178
Chevêchette de Lowery 185
Chevêchette de Parker 173
Chevêchette de Pernambuco 174
Chevêchette de Ridgway 169
Chevêchette de Rocheuses 163
Chevêchette de Tucuman 171
Chevêchette des Andes 166
Chevêchette des montagnes (Chevêchette cabouré) 165
Chevêchette des nuages 166
Chevêchette des saguaros (Chouette elfe) 186
Chevêchette des Tamaulipas 172
Chevêchette des yungas 166
Chevêchette du Cap 183
Chevêchette du Graben 184
Chevêchette du Pérou 167
Chevêchette marron (ou châtaine) 182
Chevêchette naine 174
Chevêchette perlée 162
Chevêchette spadicée 181
Chouette à collier 158
Chouette à joue blanche (Chouette rieuse) 232
Chouette à lunettes 156
Chouette à pieds rouge 141
Chouette aboyense (Ninoxe aboyense) 208
Chouette africaine 137
Chouette barrée 146
Chouette (ou Chevêche) brame 192
Chouette chevêche ou Chevêche d'Athéne 188
Chouette chevêchette 161
Chouette de (Sitchouan) David 150
Chouette de Bartels 131
Chouette de Brésil (dryade) 142
Chouette de Butler 136
Chouette de Harris (Nyctale de Harris) 204
Chouette de l'Oural 148
Chouette de Ridgway (Nyctale immaculée) 203
Chouette de Sitchouan 150
Chouette de sourcils jaune 158
Chouette de Sumba (Ninoxe de Sumba) 210
Chouette de Tengmalm 200
Chouette des forêts (Chevêche forestière) 193
Chouette des Himalayas 131
Chouette des pagodes (Chouette obscure) 127
Chouette des Solomones 246
Chouette des terriers 194
Chouette dryade 142
Chouette du Chaco 141
Chouette elfe (Chevêchette des saguaros) 186
Chouette épervière 159
Chouette fasciée (striée rouge) 141
Chouette fauve 144
Chouette hirsute (Ninoxe hirsute) 215
Chouette huhul (obscure) 143
Chouette hulotte 133
Chouette lapone 151
Chouette leptogramme 129
Chouette masquée (à pieds rouge) 141
Chouette noire et blanche 142
Chouette obscure (ou des pagodes) 127
Chouette ocellée (indienne) 128
Chouette (ou Ninoxe) papoue 231
Chouette pêcheuse de Bouvier 126
Chouette pêcheuse de Pel 125
Chouette pêcheuse rousse 126
Chouette rayée (barrée) 146
Chouette rieuse (ou à joue blanche) 232
Chouette Scie (Petit Nyctale) 202
Chouette striée 138
Chouette striée rouge 141
Chouette tachetée 144

D

Duc à aigrettes (Hibou à casque) 154
Duc à crinière (Hibou à crinière) 153

E

Effraie d'Hispaniola 21
Effraie de Boang 23
Effraie de clochers 15
Effraie de Madagascar 23
Effraie de Manus 27
Effraie de Minahassa 26
Effraie de prairie 30
Effraie de Prigogine 33
Effraie de Rosenberg 27
Effraie de São Tomé 22
Effraie de Taliabu 25
Effraie de Tanimbar 26
Effraie de Tasmanie 29
Effraie des Andamanes 23
Effraie des Galapagos 22
Effraie dorée 25
Effraie du Cap 29
Effraie masquée 28
Effraie ombrée 31
Effraie piquetée 32

G

Grand-duc à aigrettes 113
Grand-duc africain 111
Grand-duc américain 99
Grand-duc bruyant (de Malaise) 115
Grand-duc Coromandel 117
Grand-duc de Blakiston 120
Grand-duc de Magellanie 103
Grand-duc de Malaise 115
Grand-duc de montagne 110
Grand-duc de Nepal 114
Grand-duc de Shelley 116
Grand-duc de Verreaux 117
Grand-duc des Indes 109
Grand-duc des Philippines 119
Grand-duc des Usambara 114
Grand-duc du Cap (de montagne) 110
Grand-duc du désert 109
Grand-duc sombre (Coromandel) 117
Grand-duc tacheté 118
Grand-duc vermiculé 112

H

Harfang des neiges 97
Hibou à casque (Duc à aigrettes) 154
Hibou à crinière (Duc à crinière) 153
Hibou d'Abyssinie 239
Hibou de Gurney 96
Hibou de la Jamaique 233
Hibou des marais 240
Hibou du Cap (ou Hibou marais) 244
Hibou Grand-duc (d'Europe) 104
Hibou maître bois (Hibou obscur) 235
Hibou malagache 240
Hibou marais 244
Hibou moyen-duc 237
Hibou obscur (Hibou maître bois) 235
Hibou pêcheur brun (Kétoupa brun) 121
Hibou pêcheus malais (Kétoupa malais) 123
Hibou Petit-duc 47
Hibou redoutable 246
Hibou strié 234

K

Kétoupa brun (Hibou pêcheur brun) 121
Kétoupa (Hibou pêcheus) malais 123
Kétoupa roux 122

N

Ninoxe à sourcils blancs 219
Ninoxe (ou Chouette) aboyense 208
Ninoxe brune 226
Ninoxe cinabre 224
Ninoxe coucou 210
Ninoxe de Christmas 225
Ninoxe de Jacquinot 229
Ninoxe de l'Amirauté (ou de Manus) 226
Ninoxe de Mindoro 222
Ninoxe de Nouvelle Irlande 228
Ninoxe de Nouvelle-Seelandia 214
Ninoxe (ou Chouette) de Sumba 210
Ninoxe des Andaman 219
Ninoxe des Moluques 224
Ninoxe des Philippines 220
Ninoxe géante (Ninoxe puissante) 207
Ninoxe (ou Chouette) hirsute 215
Ninoxe ocrée 223
Ninoxe odieuse 228
Ninoxe papoue (Chouette papoue) 231
Ninoxe pointillée 227
Ninoxe puissante (ou géante) 207
Ninoxe rousse 206
Nyctale de Harris (Chouette de Harris) 204
Nyctale immaculée (Chouette de Ridgway) 203

P

Petit Nyctale (Chouette Scie) 202
Petit-duc à bec jaune 37
Petit-duc à collier 64
Petit-duc à face blanche de Grant 94
Petit-duc à face blanche de Temminck 94
Petit-duc à front blanc 36
Petit-duc Africain 51
Petit-duc bridé 81
Petit-duc (Scops) Choliba 77
Petit-duc d'Anjouan 45
Petit-duc d'Enggano 62
Petit-duc d'Irene 38
Petit-duc d'Orient 52
Petit-duc de Beccari 58
Petit-duc de Bruce 49
Petit-duc de Clark 80
Petit-duc de Cooper 73
Petit-duc de Cuba 93
Petit-duc de Florès 39
Petit-duc de Grande Nicobar 61
Petit-duc de Hoy 85
Petit-duc de Java 42
Petit-duc de Karthala 45
Petit-duc de Lamb 73
Petit-duc de Luzon 68
Petit-duc de malagache 43
Petit-duc de Manado 59
Petit-duc de Mantanani 58
Petit-duc de Marshall (Scops de Marshall) 83
Petit-duc de Mayotte 44
Petit-duc de Mentawai 62
Petit-duc de Mindanao 42
Petit-duc de Mindoro 43
Petit-duc de Moheli 44
Petit-duc de Montagne (ou tacheté) 39
Petit-duc de Palau 92
Petit-duc de Palawan 69
Petit-duc de Pemba 46
Petit-duc de Peterson 83
Petit-duc de Puerto Rico 89
Petit-duc de Salvin 82
Petit-duc de Sangihe 60
Petit-duc de Sâo Tomé 43
Petit-duc de Simalur 61
Petit-duc de Stresemann 39
Petit-duc de Sunda 63
Petit-duc de Tumbes 80
Petit-duc de Wallace 69
Petit-duc des Andamanes 38
Petit-duc des montagnes 70
Petit-duc des Philippines (Luzon) 68
Petit-duc du Balsas (Scops du Balsas) 72
Petit-duc du Japon 67
Petit-duc élégant 55
Petit-duc guatémaltèque 86
Petit-duc indien (des Indes) 65
Petit-duc longicorn 42
Petit-duc mystérieux 55
Petit-duc nain 46
Petit-duc radjah 62
Petit-duc roussâtre 36
Petit-duc scieur 61
Petit-duc scops 47
Petit-duc tacheté (Petit-duc de Montagne) 39
Petit-duc Torotoroka 44
Petite Ninoxe de Sumba 222
Phodile calong 34
Phodile de Prigogine 33

S

Scops à gorge blanche 90
Scops á moustaches (Scops tacheté) 76
Scops Choliba 77
Scops d'Amérique 74
Scops d'Elliot 70
Scops de Colombia 82
Scops de Koepcke 79
Scops (Petit-duc) de Marshall 83
Scops de Rio Napo 88
Scops de Roraima 88
Scops de Salvin 85
Scops de Watson 83
Scops (Petit-duc) du Balsas 72
Scops du Pérou 80
Scops tacheté (ou á moustaches) 76
Scops variable 84
Scops vermiculé 87

Index of Vernacular Spanish (Portuguese) Owl Names / Index der spanischen (portugiesischen) Eulennamen

A

Alicuco Común (Autillo Choliba) 77
Alicuco de Santa Catarina 85
Alicuco Fresco (yungueno) 85
Alicuco tropical 84
Alicuco yungueno (Alicuco Fresco) 85
Autillo Africano 51
Autillo Cariblanco Norteno 94
Autillo Cariblanco Soreno 94
Autillo Chino 64
Autillo Choliba (Alicuco Común) 77
Autillo de Andamán 38
Autillo de Anjouan 45
Autillo de Biak 58
Autillo de Célebes 59
Autillo de Colombiano 82
Autillo de Flores 39
Autillo de Java 42
Autillo de la Enggano 62
Autillo de la Mantanani 58
Autillo de la Sangihe 60
Autillo de la Simeulue 61
Autillo de la Sonda 63
Autillo de las Comores 45
Autillo de las Mentawai 62
Autillo de las Palau 92
Autillo de Luzón 42
Autillo de Mayotte 44
Autillo de Mindanao 42
Autillo de Mindoro 43
Autillo de Moheli 44
Autillo de Nicobar 61
Autillo de Palawn 69
Autillo de Pemba 46
Autillo de Santo Tomé 43
Autillo de Seychelles 61
Autillo de Sokoke 38
Autillo de Tumbes 80
Autillo de Wallace 69
Autillo Elegante 55
Autillo Europeo 47
Autillo Filipino 68
Autillo Flameado 46
Autillo Frontiblanco 36
Autillo Indio 65
Autillo Japonés 67
Autillo Malagache 43
Autillo Moluqueno 55
Autillo Montano 39
Autillo Oriental 52
Autillo Persa 49
Autillo Piquigualdo 37
Autillo Rajá 62
Autillo Rojizo 36
Autillo Torotoroka 44

B

Búho Abisinio 239
Búho Africano 111
Búho Americano (nacurutú) 99
Búho Barrado 116 146
Búho Bengali 109
Búho Campestre 240
Búho Ceniciento 112
Búho Chico 237
Búho Corniblanco 154
Búho cornudo Cariblanco 234
Búho de Ákun 118
Búho de Anteojos (Lechuzón de Anteojos) 156
Búho de Coromandel 117
Búho de Crin 153
Búho de El Cabo 110
Búho de Guinea 113
Búho de las Salomón 246
Búho de Mindanao 96
Búho de Usambara 114
Búho del Sáhara (Búho desértico) 109
Búho Filipino 119
Búho Gritón 234
Búho Jamaicano 233
Búho Lechoso 117
Búho Magellanico (Tucúquere) 103
Búho Malayo 115
Búho Malgache 240
Búho Manchado (Cárabo Californiano) 144
Búho Manchú 120
Búho Moro 244
Búho nacurutú (Búho Americano) 99
Búho Negruzco 235
Búho Nepali 114
Búho Nival 97
Búho Pescador Leonado 122
Búho Pescador Malayo 123
Búho Real 104

C

Caburé (Tecolotito) de Ridgway 169
Caburé Común (Mochuelo Caburé) 169
Caburé de Chaco 171
Caburé miudinho de Sick 174
Caburé Patagón 167
Caburé yungueno (Mochuelo yungueno) 166
Cárabo Africano 137
Cárabo Árabe 136
Cárabo Barrado (Búho Barrado) 146
Cárabo Bataraz 141
Cárabo Blanquinegro 142
Cárabo Brasileno 142
Cárabo Café (Lechuza Café) 138
Cárabo Californiano (Búho Manchado) 144
Cárabo Chaqueno 141
Cárabo Común 133
Cárabo de Bartels 131
Cárabo de las Pagodes 127
Cárabo de Sichuán 150
Cárabo fulvo (Cárabo Guatemalteco) 144
Cárabo Gavilán 159
Cárabo Guatemalteco (fulvo) 144
Cárabo Lapón 151
Cárabo Negro 143
Cárabo Ocelado 128
Cárabo Oriental 129
Cárabo Patiblanco 141
Cárabo Pescador Común 125
Cárabo Pescador Marmorata 126
Cárabo Pescador Rojizo 126
Cárabo Uralense 148
Curucucú de Piedemonte 88
Curucucú Orejudo 90

L

Lechuza Australiana 28
Lechuza Café (Cárabo Café) 138
Lechuza Campestre 240
Lechuza Común 15
Lechuza cornuta 34
Lechuza de Célebes 27
Lechuza de El Cabo 29
Lechuza de la Espaniola 21
Lechuza de la Manus 27
Lechuza de la Taliabu 25
Lechuza de la Tanimbar 26
Lechuza de Minahassa 26
Lechuza de Patilarga 30
Lechuza de Tengmalm 200
Lechuza del Congo 33
Lechuza Dorada 25
Lechuza Jamaicano 233
Lechuza Malgache 23
Lechuza Moteada 32
Lechuza Negra 143
Lechuza Tenebrosa 31
Lechuza vizcachera 194
Lechuzita acanelada (Mochuela Canela) 204
Lechuzón Barrado 158
Lechuzón de Anteojos (Búho de Anteojos) 156

M

Mochuelo Acollarado 176
Mochuelo Alpino (Mochuelo Chico) 161
Mochuelo Amazónico 174
Mochuelo Andino (Tecolotito Andino) 166
Mochuelo Brahmán 192
Mochuelo Caburé (Caburé Común) 169
Mochuelo Californiano (Norteamericano) 163
Mochuelo Canela (Lechucita acanelada) 204
Mochuelo Castano 182
Mochuelo Chico (Alpino) 161
Mochuelo Común 188
Mochuelo Cuco 179
Mochuelo de Alberto 184
Mochuelo de Blewitt 193
Mochuelo de Ceilán 179
Mochuelo de El Cabo 183
Mochuelo de Etchécopar 182

Mochuelo de Java 181
Mochuelo de Jungle 178
Mochuelo de Lowery (Mochuelo Peludo) 185
Mochuelo de Pacifico (Mochuelo Peruano) 167
Mochuelo de Pernambuco 174
Mochuelo del Congo 181
Mochuelo Gnomo (Tecolotito serrano) 165
Mochuelo Minimo (eneno) de Sick 174
Mochuelo Norteamericano (Mochuelo Californiano) 163
Mochuelo Pechirrojo 175
Mochuelo Peludo (de Lowery) 185
Mochuelo Perlado 162
Mochuelo Peruano (de Pacifico) 167
Mochuelo Siju 168
Mochuelo (Caburé) yungueno 166

N

Ninox Australiano 210
Ninox de Andamán 219
Ninox de la Christmas 225
Ninox de la Manus 226
Ninox de la Solomón 229
Ninox de las Bismarck 228
Ninox de Nueva Bretana 228
Ninox de Sumba 210
Ninox Filipino 220
Ninox Hálcon 231
Ninox Labrador 208
Ninox Malgache 219
Ninox Maori 214
Ninox Mindoro 222
Ninox Moluqueno 224
Ninox Ocráceo 223
Ninox Papú 226
Ninox Pardo 215
Ninox Punteado 227
Ninox Reidor 232
Ninox Robusto 207
Ninox Rojizo 206

P

Pescador de Ceilán 121

T

Tecolote abetero Norteno (Tecolotito Cabezón) 202
Tecolote abetero Sureno 203
Tecolote Barbudo 81
Tecolote Bigotudo 76
Tecolote de Clark 80
Tecolote de Cooper 73
Tecolote de Oaxaca 73
Tecolote de Salvin 82
Tecolote del Balsas 72
Tecolote del Rio Napo 88
Tecolote Flameado 46
Tecolote Guatemalteco 86
Tecolote Occidental 70
Tecolote Oriental 74
Tecolote vermiculado 87
Tecolotito (Mochuelo) Andino 166
Tecolotito Cabezón (Tecolote abetero Norteno) 202
Tecolotito Centroamericano 173
Tecolotito de Colima 172
Tecolotito de Hoskins 164
Tecolotito de los Saguaros 186
Tecolotito de Parker 173
Tecolotito de Ridgway (Caburé de Ridgway) 169
Tecolotito Ecuatoriano 166
Tecolotito serrano (Mochuelo Gnomo) 165
Tecolotito Tamaulipeco 172
Telocote de Puerto Rico (Tecolote Múcaro) 89
Telocote Múcaro (de Puerto Rico) 89
Tucúquere (Búho Magellanico) 103

U

Urcututú acanelado (de Peterson) 83
Urcututú de Koepcke 79
Urcututú de Marshall 83
Urcututú de Peterson (Urcututú acanelado) 83
Urcututú del Amazonas 83
Urcututú Peruano 80
Urucurcá Chico 158

Index of Geographical Terms / Index der geographischen Namen

A

Aberdeen, Andaman Islands 23
Aberdeen Point, Port Blair, Andaman Islands 219
Abyssinia, Ethiopia 110, 112
Aceh 36
ad Volgam, Samaram, Taicum 48
Adelaide, South Australia 28
Admirality Islands 27, 226
Afghanistan 49, 108, 191
–, eastern 50
–, northern 50
–, southern 50, 192
Africa 51, 94, 95, 137, 153, 162
–, central 116, 125, 162
–, eastern 29, 110
–, north 47, 110
–, northeast 17
–, northwest 48, 110, 133, 237, 241, 244
–, southern 110
–, southwest 94
–, Sub-Saharan 111, 113
–, tropical 17
–, western 116, 162, 244
African woodland, central 183
Aguilas, Spain 104
Ah Chung, Fukien 40
Ahmadnagar, Bombay 49
Ajaccio, Corsica 47
Akrani Range 193
Alaska 97, 99, 159, 160, 200, 201, 202
–, central 151
–, northern 240, 241
–, southeast 99, 146, 163
–, southern 70, 202
–, western 99, 240, 241
Albania 189
Alberta 164
–, central 99, 100, 202
–, northern 99, 100
Albertine Rift 184
Alcudia, Mallorca 15, 48
Aleutian Islands 52, 97
Alexandra, Northern Territory 24
Alexandrowski Mountains 106
Algeria 18, 104, 133, 244
Algoma District, Ontario 100
Alma, New Mexico 164
Alor Island 211
Alpes Uralensis 149
Alps 17, 133, 200
Altai 48, 106
Altenburg, Germany 16
Alto Paraná, Paraguay 158
Älvkarleby, Uppland, Svecia 148
Amalepac Oaxaca 202
Amami Oshima 216
Amani, Tanganyika 114
Amazon 170
–, River 84
–, Valley 20
Amazonia 155
Amazonian Brazil 84
–, southern 84
Ambato, East Ecuador 82, 90
Amboina 56
Ambon 56
America 74, 99
–, boreal 100
Amerique méridionale Columbie 91
Anadyr 97
Anaimalai-Nelliampathy Hills, Kerala 34
Ancash, Quebrada Yanganuco, Peru 79
Andaman Islands 38, 52, 53, 215, 217, 219
Andean region 194
Andes 77, 82, 102, 103, 166, 167, 170, 173, 204, 279
–, central 91, 198
–, de Cumana, northern Venezuela 87
–, eastern 90, 91, 167, 185
–, of Quito, Ecuador 166
–, western 91, 166
Andhra Pradesh 128
Andundi, Zaire 182
Angaur 92
Angola 137, 183
–, northern 29, 137
–, northwest 113, 126, 162
–, southern 138, 183
–, western 118
Anhui 132
Anjouan Island 45
Annam 123, 132
–, central 30
–, south 30
Annobon Island 51, 52
Antigua Island, West Indies 196
Apatzingan 72
Apazote, Campeche, Mexico 139
Apurimae, Peru 167
Arabia 18, 110, 188
–, eastern 49
–, southeast 65
–, southwest 112
Arabian Peninsula 136, 190
Aragua 154
Aral Sea, eastern 49
Arbhem Land 206
Archangelsk 148
Argentina 20, 85, 102, 142, 168, 171, 235
–, central 99, 142, 170, 171
–, eastern 102
–, northeast 79, 85, 138, 140, 144, 171, 199, 204, 235, 236
–, northern 77, 79, 141, 143, 156–158, 167, 234, 235, 237
–, northwest 143, 198, 204, 205
–, Province Salta, Rosario 79
–, southern 77, 103
–, western 103, 142
Argentine Patagonia 168
Arizona 164, 202
–, northern 146
–, southeast 76, 165
–, southern 71, 101, 169, 186
–, southwest 71
Armenia 106
Aroa River, British New Guinea 32
Aroe Island 206, 207
Arroyo de Juan Sánchez, Nayarit 172
Aru Island 56, 206, 207
Aruba Island, Dutch Antilles 196
Aruncha Pradesh 40
Ashanti, Gold Coast 37, 125
Ashkabad, South Transcaspia 108
Asia 241
–, central 47, 159, 160, 188, 240
–, eastern 47, 188
–, Minor 47, 121, 133, 188
–, northern 237
–, southeast 39, 179, 192
–, southern 192
Assam 18, 34, 53, 64, 118, 121, 180
–, eastern 217
–, northeast 192
–, southern 123
–, western 216
Atacama Desert 168
Atelante, Quebec 146
Atlantic Coast 194
Atlas Mountains 104, 110
Atorae Islands 237
Aurowana, northern Somaliland 190
Australia 24, 25, 28, 30, 210, 212, 214, 215
–, central 30
–, eastern 30, 208
–, northeast 32, 208
–, northern 30, 206, 208
–, northwest 208
–, southeast 31, 32, 208
–, southern 209, 212
–, southwest 29, 208, 209, 212
–, western 29, 209
Austria 17, 148, 189
Azara, Paraguay 102
Azerbaijan 191
Azores 237

B

Baba (Babber) Island 211
Babelthuap 92
Babizos, Northeast Sinaloa 235
Bac Tan Tray, North Indochina 180
Bacan 56, 208, 224
Baghdad, Iraq 50
Bahamas 19
Bahia 78, 103, 157, 174
–, southern 174
Bahia de Caraques 197
Baikal Lake 106
Baja, California 71, 144, 164, 186, 187, 194
–, northwestern 99, 239
–, northeast 101
–, northern 145

–, southern 70, 71, 102
Baku, eastern Caucasus 189
Balearic Islands 48
Bali 63, 115, 116, 123, 181, 218
Baliem Valley Netherlands, New Guinea 31
Balkans 133, 148, 200
Baltic countries 16, 148, 188, 200
Baluchistan 50, 136
Balzar Mountains, western Ecuador 80
Ban Mak Khaeng, eastern Thailand 193
Ban Nong kho, southeast Thailand 180
Bang Nara, Peninsular Siam 123
Banggai Islands 27, 28, 59, 60
Bangladesh 30, 53, 64, 118, 122, 128, 178, 180
Bangor, Maine 202
Banka Islands 36, 63, 115, 116, 123, 218
Bankok, Siam 180
Banks Island 24, 209
Banton 58
Banyak Island 129, 130
Bartle Frere 213
Basco, Batan Island 55
Basilan 68, 218, 221
Baskan, northeast Turkestan 48
Bass Strait 214
Batan Island 55
Batjan, Moluccas 56, 224
Batudaka Island 223
Bauro Island, San Cristobal, Solomon Archipelago 230
Bawean Island 127
Bay Island, Honduras 19
Beata Island 195
Begemeder, Ethiopia 137
Belarus 148
Belem, Pará 84
Belgium 188
Belitung Island 35, 63, 123, 130
Belize 173, 235
Bellavista, Peru 80
Bellenden-Ker Range 213
Bellona Island, Solomon Islands 24
Bengal 66
–, lower 128
–, western 65
Benin 125
Benkoka, northern Borneo 131
Benkoker 131
Benteng Village, Togian Island, Indonesia 223
Bequia 21
Bering Sea 240
Bering Strait 240, 241
Bermuda 19
Bhutan 121, 122, 131, 178–180
Biak Island 58
Bibar 128
Biliran 220
Biliton Island 35
Bioko Island 17, 113, 137
Bipindi, Cameroon 38
Bismarck Archipelago 25, 228, 229
Bitye, Cameroon 175
Black Forest 200
Black Sea 133
Blackwater, Pinal County, Arizona 71
Blaya River 134
Blida 133
Blue Mountains, Lushai Hills, Assam 135
Boang Island 23
Bogotá, Colombia 20, 78, 87, 141, 143, 204, 243
Bohemian Forest 148
Bohol 68, 119, 221
Bolivia 78, 102, 103, 141, 143, 154, 156, 167, 198, 204, 234
–, central 79, 82, 90, 91, 158
–, eastern 170, 204, 235, 236
–, northern 82, 84, 88, 155, 157, 170, 173
–, southern 85, 141, 157, 199
–, west-central 243
–, western 79, 82, 158, 198, 205
Bombetok Bay, Madagascar 245
Bonda, Santa Marta, Columbia 170
Bongao 221
Bonin 216
Boracay 221
Borahbum 178
Borneo 35, 36, 39, 41, 58, 62, 63, 115, 116, 123, 129, 130, 176, 177, 218
–, central 130
–, northern 64, 124, 131
–, northwest 63, 124
–, southern 18, 130
Borodinos 216
Boror, Mozambique 183
Bosporus 104
Botal Tobago 55, 216
Botswana 125, 137
–, northern 138, 183
Bougainville, Solomon Archipelago 229, 246
Brahmaputra River 53, 217
Brazil 20, 84, 99, 140, 142, 154, 157, 170, 174, 194, 196, 198, 234, 236
–, Amazonian 156
–, Ceará 171
–, central 78, 140, 198, 204, 234
–, eastern 78, 102, 157, 174, 198, 204
–, northeast 78, 140, 143, 171, 234
–, northern 20, 89, 140, 155, 170, 196, 236
–, southeast 79, 85, 103, 138, 140, 142–144, 174, 204, 236, 243
–, southern 85, 158, 199, 235
Brickaville District, Madagascar 240
Brisbane, Queensland 30
British
–, Columbia 19, 70, 99, 144, 145, 163, 164, 202, 202, 203, 237, 238
–, Guiana 102, 196
–, Isles 16, 237, 240, 241
Buad 220
Buchara, Syria 50
Buenos Aires 77, 79, 141, 170
Buka Island, Solomon Archipelago 24, 229
Bukina Faso 125
Bulgaria 17, 148
Bunguran Island (North Natuna Islands) 35
Burdekin River 207
Buru Island, Moluccas 26, 56, 57, 225
Burundi, northwest 33
Busnah, Phooljan, India 193
Butung Island 224, 227
Bwamba Forest 182

C

Caatinga Forest 174
Cabinda 118
Cabo San Lucas 102
Cachar 18
Caffaria, Cape Province 51
Caicara, Rio Orinoco, Venezuela 102
Cairns, North Queensland 213
Cajamarca 204
Calamian Islands 127, 128
Calayan Island 55, 216
California 71, 99, 102, 163, 164, 202, 240, 241
–, central 145
–, northeast 100
–, northern 70, 99, 146, 151
–, southeast 71, 101, 186
–, southern 71, 145
Cambell Bay, Great Nicobar 61
Cambodia 18, 114, 127, 180, 193
Cameroon 37, 113, 117, 118, 175, 182, 244
–, Highlands 29
–, southern 38, 153, 175
Camiguin Sur 221
Camorta, Nicobar Islands 53, 217, 219
Canada 100, 160, 237
–, central 74
–, northern 97
–, northwest 70
–, south-central 74, 238
–, southeast 238
–, southern 74, 151, 194
–, southwest 46, 144–146
–, western 238
Canary Islands 237, 238, 244
–, eastern 15
Candia District, Crete 48
Cantaura, Anzoategui, northern Venezuela 242
Canton, Kwangtung Province, China 65
Cape 111, 117, 138, 244
–, eastern 183
Cape Colony 244
Cape Horn 103
Cape of Good Hope 111
Cape Province 110, 163
Cape Town, South Africa 17, 29, 30, 111
Cape Verde Islands 22
Cape York, Northern Queensland 28, 206, 209, 212, 213
Capes Melovoy 108
Car Nicobar Island 219
Carabo 221
Caracas Mountain 204
Carapari, Bolivia 157
Cardwell, Queensland 206, 213
Caribbean Islands 194, 234, 240
Carnibia 188
Caroline Island 240, 242
Carpathia 189
Carpathian Mountains 148, 200
Carpentaria, gulf of 206
Carriacou 21
Carteret Harbor, New Ireland 228
Caspian Sea 133, 200, 241
Catagan, Mount Malindang, Mindanao 218

Catamarca 85
Catanduanes 68, 119, 220
Catemaco, Vera Cruz 86
Cauca 204
Caucasus 134, 189, 200, 201, 241
Cayambe, Ecuador 20
Cayenne 143, 157, 234
Cayman Brac 19
Cebu 218, 221
Cechce, western Ecuador 103
Celebes 227
Central Africa 126
Central African Republic, southern 182
Central America 19, 77, 99, 102, 169, 194
Central Asia 49, 191
Central Hondo, Province Idzu 150
Central Mexico 146
Central Siberian Plateau 149
Ceram 225
Cerro Cantoral, Honduras 76
Cerro Coneja 172
Cerro de la Candelaria, Costa Rica 203
Cerro Mali, Darien, Panama 87
Cerro Pechoaima, Rio Negro 82
Ceylon 35, 53, 66, 115, 122, 129, 179, 217
Chaco 141, 171, 237
Chad 110, 244
Chadron, Dawner County, Nebraska 74
Chandranagore 121
Changlo hsien, Hupeh 135
Channel Islands 16
Chapada, Mato Grosso 84
Chelyabinsk 134
Chiang Mai, northern Thailand 53
Chiapas 76, 81, 169, 173, 235
Chichen Itza, Yucatan 86, 102
Chihuahua 70, 76, 164, 165
–, western 72
Chile 20, 103, 168, 199
–, central 142
–, northern 167
–, south-central 142
Chiloe Island 142
Chimandega, Nicaragua 19
China 54, 64, 132, 176, 180, 188
–, central 107, 122, 151, 177, 191
–, east-central 237
–, eastern 34, 118, 133, 135, 177, 216
–, northeast 54, 67, 107, 120, 149, 151, 152, 160, 161, 201
–, northern 97, 136, 159, 161, 191, 216, 240, 241
–, northwest 160, 191
–, southeast 30, 31, 34, 39, 40, 54, 65, 107, 121, 122, 135, 179, 180
–, southern 34, 54, 104, 131, 177, 179, 191, 217
–, southern central 18
–, southwest 18, 114
–, west-central 180
–, western 50, 107, 191, 201
Chinchipa 167
Chirchik Basin, Russian Turkestan 135
Chiriqui 156
Chisec, Guatemala 173
Chiu Lung Shan, Chihli 107
Chivela, Oaxaca, Mexico 139
Choctum, Guatemala 173
Choiseul Island, Solomon Archipelago 229, 246
Christmas Island, Indian Ocean 225
circumpolar 97
Ciremai Mountain 131
Clarence River, New South Wales 32
Clarion Island, Revillagigedo Group 195
Clearwater, southwest Florida 147
Coahuila 70, 165
–, southeast 202
Coban, Alta Vera Paz, Guatemala 165
Cobourg Peninsula 212
Cochabamba, northern Bolivia 88, 90, 91, 204
–, south 85
Cochinchina 30
Cofre de Perote, Las Vigas, Vera Cruz, Mexico 76
Colima, southwest Mexico 72
Collingwood Bay, New Guinea 24
Colombia 20, 90, 91, 102, 103, 140, 141, 143, 156, 166, 204, 234, 235, 243
–, Amazonian 154, 170
–, central 82
–, eastern 78, 83, 88, 143, 157, 170, 197
–, northeast 84, 154
–, northern 82, 166, 170, 173
–, northwest 77, 81, 87, 169
–, southeast 158
–, southwest 155, 173
–, western 19, 82, 140, 143, 154, 197
Colorado 70, 74, 100, 164, 202
–, southern 146
–, Springs 74
Commandeur Island 97
Comoro Islands 17, 44, 45
Comurapa Santa Cruz 204
Concon, Intendencia de Valparaiso, Chile 167
Congo 29
–, Basin 113, 125, 126, 175, 176
–, central 95
–, northern 37, 38, 182
–, northwest 118
–, River 118
–, southern 244
Constanza, Santa Domingo 236
Convención, Cusco, Peru 83
Cooktown 213
Cordillera
–, Central 166
–, de Cóndor 83
–, de Cutucú 83
–, de Rio Chico, Patag., Argentina 20
Córdoba 141, 171
–, northern 171
Coromandel Coast 118
Corpus Christi, Texas 147
Corrientes, northeast Argentina 199
Corsica 16, 189
Costa Rica 77, 81, 87, 156, 165, 173, 203, 234
–, central 165
–, northwest 73
Cozumel Island 86, 87, 235
Crete 48, 189
Crimea 17, 189
Cuba 19, 93, 168, 194, 195, 235, 236, 240, 242
–, central 93
–, eastern 93
–, southern 195
–, western 93
Culebra 89
Curacao 19
Cusco, Peru 88
Cushi 204
Cuyo Island 58
Cyclades 48
Cyprus 49, 190
Czech Republic, southern 189

D

Dabocrom, Gold Coast 118
Dailami, Wadi Bisha, Arabia 50
Daito Island 55
Dakto, Annam 122
Dampara, Dholbhum, Bengal 216
Danao 222
Darjeeling 39
Darmstadt, Germany 16
Daru Islands 28
Davao, Mindanao 119
Dawson River 207
Deir ez Zor, eastern Syria 190
Denmark 188
D'Entrecasteaux Archipelago 227
Departemento
–, Cusco 83
–, Pasco 83
Derby, northwest Australia 209
Desert of Kara Kum, Transcaspia 49
Devuli, Tanzania 163
Dholbhum 178
Dinagat 68, 96, 221
Djaguarasapá, Alto Paraná, Paraguay 142
Djebel Trüe, Arabistan 108
Dochen, Rham Tso Lake, southern Tibet 191
Dogwa, Oriomo River, Territory of Papua 212
Dolsk, Volhynia 133
Dominica, Lesser Antilles 21
Duenas, Guatemala 46
Duida
–, Mountains 77
–, tepui 89
Durango 76, 147, 235
Durasno, Chihuahua, Mexico 72
Durban, Natal, South Africa 111
Dutch Antilles 196
Dzungaria 106

E

East Africa 239
East Beverly, Western Australia 28
East Bismarck Archipelago 23
East Kimberley, Western Australia 28
East Surinam 84
East Tonkin 30
Ecuador 20, 90, 91, 103, 138, 140, 141, 143, 154–156, 166, 170, 173, 235, 243
–, eastern 88, 158
–, northeast 82, 90
–, northwestern 82, 204

–, southeast 83, 167
–, southern 80
–, southwestern 80
–, western 143, 156, 167, 197
Efulen, Cameroon 38
Egypt 18, 189, 190
–, eastern 136
–, northern 50, 110
Eisenberg, Germany 16
El Carmen, Chihuahua 76, 85
El Paso County, Colorado 70
El Rosario, Lower California 71
El Salvador 46, 144, 194
El Taumbo, Cauca, Colombia 82
Elfkarlely 148
Endeavour River 206, 207
Engadin, Switzerland 104
Enggano Island 62, 275
England 15, 133, 188
Ennedi Mountains, Sahara, Chad 112
Entrecasteaux Archipelago 226
Environs d'Ufa, western Urals 106
Eregli, southern Turkey 106
Eritrea 110, 188, 239
Errington, Vancouver Island 163
Escazú, Costa Rica 77
Escorial, Venezuela 90
Espirito Santo 158
Ethiopia 29, 110, 113, 125, 137, 162, 163, 239
–, east 52
–, northeast 110
–, northern 117, 190
–, northwestern 52
–, western 110
Ethiopian
–, coastland 190
–, highlands 110, 244
Eurasia 151, 152, 159, 188, 237
–, northern 159
Europe 47, 104, 133, 159–161, 200, 237, 241
–, central 47, 97, 148, 161, 240
–, eastern 47, 133, 161
–, northern 97, 133, 148, 161
–, southeast 148
–, southern 47
–, southern-central 161
–, western 47, 188
Everard Ranges, central Australia 212

F

Falkland Islands 240, 243
Fanti, Gold Coast 37, 116, 126, 137
Far East, Russian 201
Fars 18
Fatehgarh 66
Faxina, Sao Paulo 198
Félicité Island 61
Fenno-Scandia 151, 152
Fergusson Island 227
Fernando Póo (Bioko) Island 17, 113
Fiji Islands 24, 25, 30
Finland 148, 159
Fitzroy River 212
Five Finger Mountains, Hainan 180
Flinder Island 214
Flores Island, Lesser Sundas 18, 30, 39, 56, 57, 69, 215, 216, 272
Florida 74, 75, 101, 146, 147, 195, 202
Florida Island, Solomon Archipelago 230
Foothills, Pasadena 164
Forêt de Kambatule, Bwanendeke, Kivu 176
Formosa 40, 65, 132, 177, 237
Fort Huachuca, Arizona 71
Fort Mojave, Arizona 186
Fort Tejon, California 145
Fort Walla Walla, Washington 99
Fort Wheeler, Paraguayan Chaco 141
Fox Island River, Newfoundland 100
France 16, 47, 133, 188, 240
Frankston, Victoria 213
Friuli (Friaul), Italy 15
Front Range, Colorado 46
Fuego, Tierra del 199
Fuertoventura, Canary Islands 15
Fuga 218
Fukumoto, North Ryukyu Island 216
Funchal, Madeira 15
Fusan, South Korea 107, 136
Futsing, Fukien 31
Futteghur, northwest India 121
Futuna Island 24

G

Gabon 111, 116, 126, 182
–, northern 37, 38, 153
–, southeast 95
Galapagos Archipelago 22, 240, 243
Gambia 17, 51
Gambo, Ethiopia 94
Gamdalkal, Bachtiari region, Iran 134
Ganges 109
Gansu Province 151
Gayana 242
Gede Mountain 131
Geelvink Bay 58
Georgia 74, 147
–, south 147, 195
German East Africa, Ukami, Tanganyika 137
Germany 16
–, western 16, 188
Ghana 37, 116, 118, 126, 153, 175
Ghats, eastern, Madras 53, 66
Gilgit 135
Gilolo, eastern 208
Goa 178
Goiás 85
Goldie River, New Guinea 227
Gonave Island 195, 236
Goodenough Island, D'Entrecasteaux Archipelago 227
Goto 67
Goudenough 227
Graham Island, Queen Charlotte Islands 203
Gran Canaria 238
Grand Canyon Village, Arizona 71
Grand Cayman Islands 19
Grand Comoro 45
Great Barrier Island 215
Great Britain 133, 188
Great Khinghan 120
Great Lakes 74, 146
Great Nicobar Island 61, 219
Great Plains, western 194
Greater Antilles 81, 233, 242
Greater Sunda Islands 34, 35, 215
Greece 133, 189, 200
–, northern 47
–, south 47, 48
Greenland 97, 240
Grenada 21
Guadalcanal, Solomon Archipelago 230
Guadalupe Island 194
Guadeloupe Island, West Indies 196
Guamuchil, Sinaloa, Mexico 72
Guanajuato 138, 146, 187
Guangdong 54, 122
Guangsi 122
Guapongo, Guerrero, Mexico 46
Guatemala 46, 86, 87, 144, 154, 165, 169, 173, 203, 235
–, central 81
–, northern 81
Guerrero 101, 138, 165, 235
–, northeast 172
–, western 72
Guiana 155, 196
Guianas 20, 78, 143
Guimaras 221
Guinea 118, 126
Gujarat 128
Gulf of Bengal 38
Gulf of Guinea 43, 52
Gulf of Mexico 74
Gulf of St. Lawrence 202
Gulf of Tomin 283
Gunong Semanggol, Perak, Malay States 118
Gunong Tahan, Malay Peninsula 41

H

Hachijo Island 68
Hacienda Mirador 86
Haidarabad, Sind 66
Haifa 189
Hainan Island 40, 64, 65, 122, 131, 132, 177, 179, 180
Haiti 195
Hakodate 120
Hall Island 97
Halmahera 56, 208, 224
Hari-rud, Transcaspia 108
Harrach-Bache 244
Hattam, Mount Arfak, New Guinea 32
Hawaiian Islands 240, 242
Heilongjiang 201
Hen Island 215
Hessen 241
Hidalgo, Texas 172, 187, 202
Hilong, Hilong Peak, Province of Agusau, Mindanao 42
Himachal Pradesh 178
Himalayas 39, 40, 65, 104, 109, 114, 121, 122, 128, 131, 135, 176–179, 200, 237
–, northern 191
–, northwest 65, 135, 201
–, Simla-Almora district 179

–, southern 191
–, western 65, 179
Hispaniola Island 19, 21, 194, 195, 235, 236, 240, 242
Hmachal Pradesh 178
Hokkaido 67, 107, 120, 150, 237
Holarctic 200
Holland 188
Hondo 150
–, central 150
–, northern 150
Honduras 86, 87, 102, 144, 154, 165, 173, 194, 203, 235
Hong Kong 52
Honshu, southern 150
Huachuca Mountains, Arizona 146
Huancabamba 204
Hudson Bay 100, 151, 160
Hudson Strait 151
Huiyaumba, Inambari, Cuzco, Peru 84
Humboldt Bay, California 163
Hungary 17, 104, 161, 189

I

Iberian Peninsula 16, 104, 133, 188, 237, 240
–, central 48
–, northern 48
Iceland 97, 241
Idaho 151, 164, 202
–, northern 70, 151
Incachaca, Bolivia 91, 141
Indepedence, Inyo County, California 71
India 18, 30, 48, 52, 65, 121, 127, 128, 129, 178, 193
–, central 66, 118, 216
–, east-central 193
–, eastern 39, 64
–, northeast 34, 122, 180
–, northern 39, 53, 104, 118, 135, 216
–, northwest 49, 66, 121, 122, 135, 201
–, southeast 66
–, southern 53, 114, 217
–, southwest 34, 66, 178
–, west-central 193
–, western 128
Indian Ocean 225
Indian River, Florida 75
Indian Subcontinent 109, 192, 215
–, central 192
–, northern 192
Indochina 53, 64, 122
–, northern 177, 180
–, southern 122, 180
Indonesia 18, 210, 217, 222, 224
Indramaju, Cheribn, Java 218
Indrawan 283
Inner Mongolia 191
Insula Norfolk 215
Iran 18, 49, 108, 121, 136, 191, 192, 237
–, northwest 106, 134
–, southeast 65, 66
–, southern 18, 50, 192
–, western 134
Iraq 18, 49, 121
–, eastern 50, 108, 110, 191
–, northeast 134
–, northern 50
–, western 110
Ireland 188
Irian Jaya 231
Irtysh River 134
Ishim 134
Isla de Juvendad 195
Isla de la Juventud 168
Isla de Vàsques 89
Isle of Pines, Cuba 19, 168, 195, 236
Israel 49, 50, 110, 121
–, central 48, 106
–, eastern 136
–, southern 136
Issyk-Kul, Tian-Shan 191
Isthmus 144
Italy 16, 47, 104, 133, 189, 237
Itombwe
–, Massive 33
–, Forest 272
Ivory Coast 37, 118, 126, 153, 175, 182
Izu Island 67, 68, 216

J

Jaco Island 24
Jahrum 18
Jakschi 120
Jalapa, Nicaragua 87
Jalapa, Vera Cruz, Mexico 86
Jalisco, Mexico 72, 169
Jamaica 19, 233
Jambi 127
James Bay 100
James Island, Galapagos Archipelago 22
Janáubu, North Minas Geraës 78
Japan 54, 67, 97, 107, 148, 180, 215, 216, 237
–, northern 67, 104
–, southern 55
Japen Island 32
Java 18, 35, 36, 59, 63, 115, 116, 123, 127, 181
–, central 131
–, northern 127
–, western 42, 131, 218
Jerripani, Mussorie 40
Jiangsu 31
Jilin 136
Jima, Ecuador 82
Joazeira, Bahia, Brazil 103
Johnston River, Queensland 32
Jolo Island, Philippines 36, 216, 221
Jordan 48, 106, 136
Juan Fernandez Islands 240
Juguy 167
Jujuy 85, 204, 205
Junin 204
Juxta flumen Amazonum 78

K

Kabaena 227
Kagamega 176
Kai Nong, Siam (Thailand) 41
Kakamega 175
Kalao 18
Kalaotoa 18
Kalidupa Island 30, 59, 60
Kamchatka 159, 201, 240, 241
Kangaroo Island 213
Kangean Island 18, 63, 64
Kanowit, Sarawak, northern Borneo 64
Kansas 100
–, eastern 101
Kansu 237
Kapiti Island 215
Kara Aktachi, Crimea 106
Karkas Island 24, 209
Kashka Su, Tian Shan 201
Kashmir 65, 179
Kasiruta 56
Kathiawar Peninsula 128
Kau, Kenya 125
Kazakhstan 106, 108
Kazan 200
Keetmanskoop 111
Kei Islands 212
Kenya 29, 30, 38, 110, 125, 137
–, central 94, 111
–, eastern 183
–, northern 94
–, southeast 52, 111, 112
–, southern 95
–, west-central 111
–, western 175, 176
Kerala 34, 128, 178
Ketchum, Idaho 46
Khanka Lake, Ussuriland 67, 201
Khasi Mountains, Assam 53
Khaw Nok Ram, Trong, Lower Siam 130
Kiaochow, Shantung, China 107
Kibwezi, southern Kenya 183
Kilimanjaro, East Africa 163
Kimberleys 206
Kirgizia 49
Kirrima Range 213
Kisar Island 24
Kislovodsk, northern Caucasus 201
Kiukiang, North Kiangsi 107
Kohima, Nagaland 40
Kolidoo, Tenasserim 180
Kolyma 201
Kommandeur Islands 241
Konkan, southern 178
Korea 67, 120, 136, 148, 149
–, central 216
–, eastern 107
–, northern 54
–, southern 216
Korinchi 39
–, Peak, Sumatra 41
Koror 92
Koryakland 97
–, western 151, 152
Kotmalie, Ceylon 53
Krain, Slowenia 47, 188
Krasnoyarsk, Siberia 48, 106, 149
Kulu, India 109
Kuril Islands 67, 120
Kwa Mtessa, Uganda 51
Kwobia, Gold Coast 162
Kyushu 67, 150

L

La Caldera 85
La Cornisa 85
La Guasimas, Sinaloa, Mexico 86
La Morelia, Rio Capueta, Colombia 84
La Palma 238
La Pampa 170
La Peca 83
Labobo 59, 60
Labrador 97, 100, 159, 160, 202, 240, 241
Labuan Island 218
Lagos, southern Portugal 188
Laguna Perida, Peten, Guatemala 87
Lajeh, Arabia 18
Lake Baikal 48
Lake Balkash 191
Lake Chad 244
Lake Enerdekh 149
Lake Khanka, southern Ussuriland 215
Lake Kokonor, eastern Tsinghai 191
Lake of Woods County, Minnesota 146
Lake Tanganyika 183, 239
Lakhimpur District, northeast Assam 192
Lali Mountains, southeast Kenya 52, 112
Laloki, New Guinea 209
Lambayeque 80
Lamia, Greece 47
Lance au Loupe, Labrador 100
Lano de Garzas 172
Lanyu Island 55, 216
Lanzarote 15
Laos 18, 40, 131, 217
–, northern 132, 180
–, southern 132, 180, 193
Lapland 152
Larat Island 26, 27
Laredo, Santander, northwest Spain 188
Last Mountain Lake, Saskatchewan, Canada 238
Latvia 188
Lenkoran, Talych 134
Lepanto, North Luzon 68
Les Glacières 133
Lesser Antilles 21, 196
Lesser Sunda Islands 25, 39, 56, 57, 69, 210, 211, 215, 222, 225
Leti Island 211
Levant 188
Leyte 68, 119, 220
Lianhuashan Nature Reserve 151
Liberia 37, 113, 116, 118, 126, 153, 162, 175, 182
Lichiang Mountains, Yunnan 135
Lifu, Loyalty Island 24
Lima, Peru 79, 198
Limburg, The Netherlands 188
Linares, Salamanca, Spain 133
Liqua River, Northeast Capeland 163
Lithuania 151, 152
Little Barrier Island 215
Livadia, Cyprus 49
Loceri, Sardinia 16
Lomblen 56, 57
Lombok 18, 56, 57
Lonauli, western Ghats 129
Londa, Bombay Presid 18
Long Island 24
Lop Nor, Tarim Basin, Sinkiang 108
Lopé, Ogowe River, Gaboon 126
Loquilocom, Samar 35
Lord Howe Island 29, 214
Lorica, Bolivar, Colombia 102
Los Esemiles, Chalatenango, El Salvador 76
Louisiade Archipelago 226, 227
Louisiana 75
Lower Congo 29
Lower Guinea 116
Loyalty Islands 24, 25
Lukolela, Middle Congo River 175
Lushai Hills, Bangladesh 64
Lutu, Timor Laut, Tanimbar Islands 225
Luzon 31, 43, 68, 119, 216, 218, 220

M

Mackay, Queensland 28
Mackenzie Valley 100
Madagascar 17, 23, 43, 44, 219, 240, 244, 245, 274
–, central 17
–, eastern 23, 43
–, northeast 23, 219
–, northern 43
–, southern 219
–, southwest 44, 219
–, western 44
Madeira 15
Madhya Pradesh 128
–, eastern 193
Madopolo auf Bisa (Kep Obi) 56
Mafia Island 183
Magdalena Valley 197
Maharashtra 128
–, northwest 193
Mahé Island, Seychelles 61
Maine, northern 100
Makran Coast 136
Malabar Coast 66, 178
Malacca 36, 52, 54, 130, 218
Malaita Island, Solomon Archipelago 230
Malay Peninsula 18, 34–37, 39, 41, 52, 54, 63, 122, 123, 127, 129, 130, 215
–, northern 217
–, southern 63, 218
Malaysia, Borneo 176, 177, 218
Malenge Island 223
Mali 110
–, central 117
–, southern 125
Mallorca 133
Manabi 167
Manam Island 24, 209
Manchuria 107, 188
–, central 162
–, south 67
–, southeast 67, 149, 216
–, west 120
Mandano, Celebes 59
Mangole 57
Manila 68
Manipur, Assam 34, 64
Manitoba 238
–, central 100
–, southern 100
Mantanani Island 58
Manus Island, Admirality Islands 27, 226
Maore Island 44
Marakech, southern Morocco 189
Maranón 167
Maraynioc, Peru 141
Margarita Island, Venezuela 20, 77, 170, 196
Maria Island 29
Mariduque 68, 218
Marinduque 220
Markha River 149
Marudo Bay, Bengkoke River 124
Masbate 222
Massana, Eritrea 190
Masset, Graham Island, Queen Charlotte Islands 202
Mato Grosso 102
–, northern 84, 155
–, southern 78
Mau Forest 175, 176
Maun, Ngamiland 183
Mauretania 18, 110, 190
Mayotte (Maore) Island 44, 276
Mazatán, Chiapas, southeast Mexico 73
Mazatlan, Sinaloa, Mexico 86, 138
Mediterranean 104
–, Islands 237
–, region 47
Medje, Ituri Forest, Congo 175
Mekran Coast 136
Melbourne 214
Melville Island, Northern Territory 28, 212
Mendoza 79
Mengtsz, South Yunnan 107
Mentawai Island 62, 129, 130
Merauke 28
–, River 209
Mérida, Montana Sierra Valle, Venezuela 78, 204
Mexican Plateau 72
Mexico 46, 72, 75, 76, 86, 87, 101, 102, 138, 147, 164, 165, 169, 186, 194, 202, 234, 235, 239
–, central 46, 70, 76, 143, 144, 146, 147, 172, 186, 187, 202
–, eastern 86, 139
–, northeast 75, 139, 172
–, northern 71, 75, 76, 101, 144, 146, 237
–, northwest 71, 74, 86, 163, 164, 186
–, southeast 73, 76, 173
–, southern 46, 73, 81, 139, 144, 146, 154, 156, 165, 173, 203, 234, 235
–, southwest 72
–, Tehuantepec City, Oaxaca 138
–, western 86, 138, 169, 172, 235
Miass River 134
Michoacan, southwest Mexico 76, 146
–, central 202
Micronesia 242
Mid West Australia 212
Middle America 138, 154, 156, 234
Middle East 49, 106, 133, 189, 237
–, interior 190
Minahassa, Sulawesi 26
Minami-Daitojima 55
Minas Geraës 142, 236

Mindanao Island 42, 68, 96, 119, 216, 218, 221, 275
Mindoro Island 43, 215, 218, 221, 222
–, Highlands 43
Minnesota 151
–, northeast 151
–, western 100
Mirador, Vera Cruz, Mexico 147, 235
Mirafloras, Lower California 187
Mirgorod, Poltava, Ukraine 106
Misiones, Arroyo, River Uruguai 79, 85, 140, 142, 144, 157, 158, 174, 237
Misool Island 226
Misori Island 58
Mitchell, Iowa 100
Mitchell River 206
Mitumba Mountains 239
Moa Island 209, 211
Mohéli Island 44, 45, 276
Moluccas 26, 56, 215, 224, 225
–, northern 56, 208
Monelido, Boni and Gorontalo, Sulawesi 27
Mongolia 162
–, northeast 149, 152, 241
–, northern 106, 151, 152, 160
Mono Island, Solomon Archipelago 230
Montana 164, 202
–, eastern 74
–, western 70, 151
Monte Sociedad, Paraguayan Chaco 237
Monterey, Northern California 99
Monterrey, Nuevo Leon, Mexico 75
Montevideo, Uruguay 85, 235
Morelos 101, 138
–, southwest 172
Morocco 47, 104, 133, 190, 237, 241, 244
–, central 48, 244
–, north 48, 244
Morotai Island, Moluccas 56, 208
Morropon, northwest Peru 80
Motzorongo, Vera Cruz 86
Mouleina, Biskra, southern Algeria 189
Mount Anderson 212
Mount Auyan-Tepui, southeast Venezuela 170
Mount Baco 43
Mount Cameroon 181
Mount Duida, Venezuela 77, 170
Mount Dulaugan 43
Mount Dulit, Sarawak, Borneo 63
Mount Elgon 175, 176
Mount Epa, New Guinea 208
Mount Hálcon 43
Mount Kabobo 239
Mount Kalulong, Sarawak, Borneo 177
Mount Karthala, Grand Comoro Island 45
Mount Kenya 111, 239
Mount Kina Balu, Borneo 41
Mount Ophir, Malacca 37
Mount Pichincha, Ecuador 197, 243
Mount Roraima, Guyana 78
Mount Tacitaro, Michoacan, Mexico 146
Mount Todera, Air, southern Sahara 189
Mountains
–, nguru 114
–, ulguru 114
Mozambique 110, 125, 137, 183
–, south-central 183
–, western 30, 111
Mpwapwa, Tanganyika 137
Muna Island 227
Munroe County, Florida 147
Mupin, Szechwan 150
Murgab 237
Murray River 213
Murree, Kotegurh and Garlawal 65
Musangakye, Zaire 184
Muscat (Masqat), Arabia 112
Muusi 33
Mwali Island 44, 45
Myanmar 18, 30, 34, 39, 40, 53, 64, 121, 179, 193, 217
–, central 132
–, northeast 122
–, northern 131, 132, 135, 180
–, northwest 180
–, southern 36, 115, 116, 118, 123, 127, 129, 130, 131
–, western 109, 118, 128, 178
Mysore 128
–, western 178

N

Nadu, tamil 128
Nagaland 34
Namableda, Boror, East Africa 51
Namibia 95, 162, 183
–, southern 110, 111
Nandi 175, 176
Napo, Ecuador 84
Narino, Colombia 140, 204
Nasei 55
Natal 95
Natuna Islands, northern 218
Naxos Island, Cyclades 48
Nayarit 169
Ndzuani Island 45
Near East 18
Neblina tepui (Neblina Mountain) 89, 204
Negri-lama, Gulf of Tomini, Celebes 223
Negros Island 69, 218, 221
Neilgherri Mountains, India 30
Nejapa, Oaxaca, Mexico 73
Nepal 34, 39, 52, 53, 64–66, 114, 121, 122, 131, 135, 176, 179
–, central 40, 65
–, eastern 64
–, southern 30, 118
–, Terai 192
Nevada 100, 164
–, central 99
–, southern 186
–, western 99
Nevis Island 196
New Britain 25, 228
–, northern 24
New Brunswick 100, 101, 202
New Caledonia 24, 25, 30
New Guinea 31, 32, 206, 207, 208, 210, 226, 227
–, central 209, 231
–, eastern 31, 209
–, northwest 231
–, southeast 24, 30, 227, 231
–, southern 28, 212
–, western 31
New Hanover, Bismarck Archipelago 228, 229
New Holland 213
New Ireland 24, 228
New Mexico 144, 164, 200, 202
–, southern 101
–, southwest 186
New Providence, The Bahamas 19, 195
New South Wales 24, 28, 29, 207, 209, 213, 214
–, eastern 207
New York 74
New Zealand 188, 214, 215, 232
–, North Island 215
–, South Island 215
Newfoundland 160, 202
Ngadzidja Island 45
Nias Island 35, 123, 124, 129, 130
Nicaragua 235
–, central 76
–, northern 76, 86, 87
Nicasio, California 70
Nichlaul, United Provinces, northern India 128
Nicobar Islands 52, 217, 219
–, central 53
–, southernmost 61
Niger 110, 125
Nigeria 118, 125
–, southeast 126
Nikolsk-Ussuriysk, South Ussuriland 107
Nile Valley 18
Nissan Island 24
Nive 24
Nligiris 128
Noong-zai-ban, Manipur 180
Norfolk Island 215
Normanby Island 227
North Africa 18, 104, 133, 188, 190
North America 19, 144, 146, 151, 159, 194, 200, 202, 237, 240, 241
–, eastern 74, 100
–, western 163
North Cape Province 95
North Carolina 146, 147, 202, 240
North Eurasia 97
North Island 215, 232
North Korea 216
North Luzon 42
North Mato Grosso 84
North Misiones 85
North Natuna Islands 35, 63
North Pakistan 108
North Queensland 28
North Territory, northern 206
North Tienshan 109
North Vietnam 131, 132, 180
Northern Territory 28, 209, 212
Norway 104, 148, 159, 240
Nova Hollandia 209
Nova Scotia 100, 101, 146, 202, 238
Nova Zeelandia 215
Novo Redondo, Angola 51
Nuevo León 139, 165, 239
–, southwest 202
Nyungwe forest 33

O

Oaxaca 73, 76, 147, 156, 165, 172, 202
–, central 202
–, northern 173
Ob, river 106
Obi Islands 56, 208
Obihiro, Hokkaido 107
Oceania 207, 229
Ogwarra, New Guinea 31
Ohsumi, Kyushu 150
Okanagan Valley, British Columbia, Canada 46
Okhotsk 162, 237
–, Coast 149
–, Sea 148
Okinawa, Ryukyu Islands 67, 68
Oklahoma 74, 101
–, central 75
–, northern 238
–, western 70, 71
Old Kandahar, Afghanistan 191
Olekminsk 162
Ollantaytambo, Peru 103
Oman 50, 66, 112, 136
Omara (Ormara) 136
Omilteme, Guerrero, Mexico 76
Ontario 100
–, northern 100, 202
–, southern 74, 101
Ooldea, South Australia 212
Orange Free State, South Africa 51
Oregon 163, 164, 200
–, northeast 99
–, southern 99
–, western 99, 145
Orenburg 200
–, eastern 241
Oria, Siriohi/Raiputane, western India 178
Orinoco River, Venezuela 84
Orissa 128
Otini 36
Outer Mongolia 191

P

Pacasmayo, western Peru 198
Pacific Islands, south-west 25
Pacific Slope 156
Pagai Islands
–, north 62
–, south 62
Pagalu 52
Pakha, Tonkin 132
Pakistan 18, 39, 49, 128, 131, 135
–, eastern 49, 118
–, northeast 179
–, northern 40, 50, 52, 53, 65, 176, 177, 237
–, northwestern 49, 121
–, southern 65, 66, 136, 192
–, western 50
Palau Islands 92
Palawan Island 69, 127, 128, 217
–, southern 58
Palestine 110
Palmas, Boyaca, Columbia 197
Palo Pinto County, Texas 75
Paluma 213
Pamir 109
Panama 19, 81, 87, 165, 169, 173, 194, 234
–, eastern 140, 154, 156
–, northern 87
–, western 102, 139, 154, 156, 165, 203
Panay 221
Pangerango Mountain 42, 131
Papua New Guinea 231
Pará, west 155
Parador de Montana 171
Paraguay 78, 102, 143, 157, 171, 174, 199, 204, 234, 235, 237, 243
–, eastern 78, 140, 142, 144, 158, 170, 171
–, northern 141, 143
–, southeast 85
–, southern 142
–, western 79
Paramaribo, Surinam 20
Paraná 85
Paros 48
Parry's Creek 212
Pasadena, California 70
Pasco 204
Pasis Datar, Mount Pangerango, Java 131
Patagonia 194
Patrie ignorée 209
Pearl Island 77
Pegu, Burma 193, 217
Peleliu 92
Peleng Island 27, 28, 59, 60
Pemba Island 17, 46
Penablanca, Popayan, Cauca, Colombia 103
Pennsylvania 19, 146, 238
Permé, Darien, Panama 154, 156, 173
Pernambuco 174
Perth 212
Peru 20, 82, 88, 90, 91, 154, 158, 167, 173, 204
–, central 83, 103, 166, 198, 204
–, eastern 78, 83, 84, 143, 155, 156, 174, 234
–, northeast 83, 84
–, northern 79, 91, 141, 158, 166, 185, 204
–, northwestern 79, 80, 83, 103, 141, 143, 156, 197, 243
–, southeast 158
–, southern 83, 143, 170, 243
–, southwest 198
–, western 167, 198
Philippines 30, 31, 34, 35, 43, 55, 68, 69, 119, 128, 215–218, 220–222
–, northeast 220
–, northern 43
–, southeast 221
–, southern 42, 96
–, southwest 58, 69
–, west-central 58
–, western 127
Pichincha 204
Pines Island 93
Plains of Tolima, Colombia 197
Platte River 194
Pohnpei Island 242
Point Barrow, Alaska 241
Poland 16, 188
–, northern 148
Polillo 220
Pomeriana 241
Ponapé Island 240, 242
Pondicherry 128, 193
Port Essington 206
Port Famine, Straits de Magellan 141, 167
Port Keats, Northern Territory 209
Port Moresby 24
Portezuelo, Hidalgo, Mexico 72
Porto Jimenez, Costa Rica 76
Porto Santo 15
Portugal 244
Prasli north Island 61
Presidente Medici Rondônia, Brazil 174
Presidio, Vera Cruz 234
Pretoria 51
Principe Islands 22
Provincia Carchi, Ecuador 166
Puebla 187
Pueblo 202
Puerto Bertoni 158
Puerto Casado, Paraguay 79, 171
Puerto Princesa, Palawan 128
Puerto Rico 89, 242
Pulau Padang Island, Sumatra 63
Pulo, Condor 64
Puna Island, Ecuador 197
Punjab 52, 66, 131
–, northern 201
Pyrenees 200

Q

Quebec 101
–, central 202
–, southern 202
–, southwest 151
Qued Kasserine, central Tunisia 109
Queen Charlotte Islands 202, 203
Queensland 206, 207, 208, 209, 212
–, northeast 29, 32, 213
–, northwest 209
–, southeast 32, 207
Quellon, Chiloe Island, Chile 142
Quelpart 67
Quinghai, southeast 151
Quito, Ecuador 141, 197

R

Radjasthan 128
Raffles Bay 212
Rahuri, Ahmedanggar 49
Raipur 66
Rameswaram, Madras, India 193
Rantan, southeastern Borneo 116
Red Sea Coast 188, 190
Red Sea Mountains 136
Reghin, eastern Transylvania, Romania 189
Repok Mountains, Flores Islands 39
Revillagigedo Islands 195
Riau Archipelago 123, 218
Rif Mountains 110

Rio Cauca, El Tambo, Colombia 78
Rio Comberciato 88
Rio de Janeiro 144
Rio de la Plata 243
Rio Grande 74
Rio Grande do Belmonte, Bahia 157
Rio Grande do Sul 79, 85, 142, 157
Rio Martinez, Tamaulipas 139
Rio Mayo Valley, northern Peru 185
Rio Mondaih, Paraguay 142
Rio Negro 79, 168
Rio Preto, Bahia, Brazil 198
Rio Suturutu, Buenavista, Santa Cruz 78
Rio Tehuantepec 73
Rioja 185
River Ob 106
Roba-Shalo, Ethiopia 94
Rocky Mountains 70, 146, 151, 200
Romang Island 211
Romania 106, 189
Romblon Island 58
Roraima, British Guiana 88, 196
Roraima tepui 89
Rosario de Lerma, Salta, Argentina 171
Rossel Island, Louisiade Archipelago 227
Roti Island, Indonesia 24, 210
Rotuma Island 24
Roumania
  –, eastern 189
  –, southern 189
Rügen, Germany 16
Rumpin River, Pahang State, Malaysia 63
Russia 188, 200
  –, central 97, 161
  –, central European 105
  –, eastern 105
  –, European 149
  –, northern 97, 104, 148
  –, northwest 148, 161
  –, southern 189
  –, western 16, 133
Russian Altai 191
Ruteng Mountains 39
Ruwenzori Mountains 239
Rwanda
  –, northern 184
  –, southwest 33
Ryukyu Island 55, 67, 216
  –, south 68

S

Saadatabad, Kerman, Iran 192
Sabah 131
Sabi Valley, southeast Tanzania 163
Sabtang Island 55
Sacapulas, Rio Negro Valley, Guatemala 203
Sachalin 54
Sacramento, California 99
Sadi Malka, Ethiopia 51
Sado 67
Sahara 110, 190
Sahel Zone 137
Sakhalin 64, 67, 97, 104, 107, 120, 148, 149, 159, 161, 162, 201, 240
  –, northern 151, 152
Salak Mountain 131
Salem, Oregon 70
Salitres 103
Salta, Argentina 85, 166, 167, 205
Salvador, eastern 203
Samar 34, 35, 68, 119, 220
Samoa 24, 25
San Antonio, Texas 101
San Cristobal Island 230
San Jacinto Mountains, California 71
San Jose, Costa Rica 102
San José de Lourdes, Peru 83
San José, Yungas de Cochabamba, Bolivia 158
San Luis Obisco, California 99
San Luis Potosi 172
  –, northern 202
San Martin, northwest 185
Sanana 57
Sandakan 131
Sanga Sanga 221
Sangihe Island 27, 60, 275
Santa Ana, Costa Rica 73
Santa Ana, Bolivia 82
Santa Anita, Lower California 71
Santa Barbara, Vera Paz, Guatemala 81
Santa Catarina 79, 85, 158, 174
Santa Cruz Island 24
Santa Isabel 246
Santa Rosa de la Calamuchita 171
Santa Sevilla, Island of Pines 93
Santarem 103
Santiago 22
Santiago del Estero 171, 237
Santo Domingo, Oaxaca, Mexico 156
Sao Jacinto 222
Sao Marcello 198
Sao Paulo, Brazil 78, 204
Sao Tiago, Cape Verde Islands 22
Sâo Tomé Island 22, 43
Sao Vincente 22
Sarasota Bay, Florida 195
Sarawak 124
Sarayacu, Rio Ucayali, Peru 170
Sardinia 16, 47, 133, 188, 189
Sarepta 48, 241
Sasan, Junagadh, western India 128
Saskatchewan 100, 238
  –, southern 100
Saudi Arabia 136
  –, southern 49, 50
Saur Range 106
Saurashtra 128
Sawu Island 24, 212
Sayans 162
Scandinavia 104, 133, 161
  –, northern 97, 104, 200
  –, western 97
Schalil River 134
Scho 57
Scolah Dras, Sumatra 39
Scotland 188
Scutari, Albania 104
Sea Lion Island, Falkland Islands 243
Sea of Okhotsk 237
Semau Island 211
Semirara 58, 221
Semliki Valley 182
Semmio, Niam-Niam Country, Ubangi-Shari 137
Senegal 51, 52, 94, 117, 162, 244
Senegambia 94, 113, 125, 137, 162
Sennar 51
Sepik River 24, 209
Seram Island 26, 56, 225
Seven Sisters 216
Severn River 241
Sevilla, Bohol 68
Seychelles 61
Shato, Nankou, China 191
Shikoku 67
Shinano, Hondo 150
Shoto 55
Shuikow, Fukien 31
Siao Island 59
Siargao 96, 221
Siasi, Sulu Archipelago 221
Siau Island 59, 275
Siberia 104, 148, 159, 161, 191
  –, central 106, 159, 162
  –, eastern 104, 152, 161, 162, 215
  –, northeast 107, 201, 240
  –, northern 97
  –, southeast 54, 67, 107, 120, 149, 216
  –, southern 201
  –, southwest 189
  –, western 133, 134, 201, 241
Siberut 62
Sibutu Island, Philippines 59, 221
Sibuyan 58, 221
Sichuan 237
  –, central 151
  –, western 151
Sicily 104, 133
Sidemi, lower Ussuri 120
Sidi Ali bin Aouin, southern Tunisia 109
Sierra de Chapada, Mato Grosso 78
Sierra de la Gigante 164
Sierra de la Laguna, Baja California, Mexico 102, 164
Sierra de Perijá, Northwest Venezuela 82, 90
Sierra Leone 52, 118, 126
Sierra Nevada 145, 151
Sikkim 34, 39, 121, 131
  –, eastern 180
  –, western 179
Simalur Island 61
Simeulue Island 275
Simla 176
Sinai Peninsula 136
  –, southern 190
Sinaloa
  –, northern 70, 72
  –, southern 172
Sindh 48
Sintaung, Shan States 132
Sipora Island 62
Siquijor 218, 221
Sitio Calpi 222
Sitka, Alaska 70, 99
Skalistoy, East Caspian Sea 108
Socorro Island 186, 187

Socotra Island 49, 51
Sokoke-Arabuku Forest, Kenya 38
Sokolowa, Siberia 134
Sokotui 107
Solok Mountains, Sumatra 62
Solomon Archipelago 24, 25, 228–230, 246
Somadikarta 283
Somalia 94, 117, 188
–, southern 125, 137, 183
–, western 162
Somaliland 52
Sonora, Mexico 76, 138, 164, 169, 172, 186
–, north-central 71
–, northeast 202
–, northwest 71
–, southern 72, 172
Sorong, New Guinea 231
South Africa 17, 29, 51, 94, 111, 137, 138, 183, 236, 244
–, eastern 30, 163
–, northeastern 125
–, northern 162, 163
–, southeastern 52
–, southern 52, 162
South America 77, 84, 90, 99, 138, 139, 141, 143, 170, 175, 204, 235, 240
–, eastern 194
–, northern 235
–, northwest 173, 194
South Andaman Islands 23, 38
South Asia 48
South Australia 29, 213
–, eastern 213
–, southeast 207
South Bahia 85
South Calamian Island 58
South Cameroon 126
South Carolina 74, 241
South Island 215, 232
South Kuril Islands 67, 107
South Mexico 169
–, Oaxaca 142
South Moluccas 57
South Queensland 29, 213
South Spain 15
South Vietnam 18, 127, 180
Spain 188, 241, 244
–, northern 104
–, northwest 240
–, southeast 188
Spencer Gulf 213
Spitzbergen 97
Sri Lanka 18, 34, 53, 65, 66, 114, 115, 121, 122, 129, 178, 179, 217, 273
–, central 35
–, southern 35
–, southwest 37
St. Christopher Island 196
St. Croix, Virgin Island 89
St. George Island, Solomon Islands 229
St. Lawrence River 202
St. Michael, Alaska 99
St. Vincent, Lesser Antilles 21
State Bahia, Brazil 174
Stato of Espirito Santo, Brazil 158
Stewart Island 215, 232
Strasbourg, France 16
Sturibisi River, British Guiana 140
Sudan 94, 110
–, central 117
–, eastern 163, 190
–, southeast 125, 137
–, southern 137, 244
–, western 162
Sugnur River 152
Sula Island 25, 57
Sula Mangoli 57
Sulawesi 27, 30, 59, 224, 227, 275, 282, 283
–, north-central 26
–, northern 26, 59, 60, 224
Sulu Islands, southwestern 59, 221
Sumatra 18, 36, 39, 41, 62, 115, 116, 123, 127, 129, 130, 176, 177, 218
–, central 63, 64
–, eastern 64
–, northern 36, 63, 64
–, southeast 35
–, southern 63
–, western 124
Sumba Island, Indonesia 24, 30, 210, 222, 283
Sumbawa 56, 57, 69
Sundas 216
Surinam 20, 83, 154, 196
Sweden 104, 133, 148, 159, 161, 200, 237, 241
–, northern 241
–, southern 16, 188
Syria 110
–, northern 49, 50, 121
Szechuan 107

T

Tablas 58, 221
Tacazze River, northern Ethiopia 125
Tadjikistan 49, 50
Tafira, Gran Canaria, Canary Islands 238
Tagula Island 227
Taianzan, northeast Korea 149
Tait-Shafton, Jamaica 233
Taiwan 30, 39, 40, 54, 64, 65, 122, 131, 132, 133, 136, 176, 177, 215, 216
Takata, Tainan district, Taiwan 136
Taliabu, Sula Islands 25, 57
Tallah, Tunisia 47
Tamaulipas, Mexico 86, 139, 165, 172
–, southern 86
Tamil Nadu 114
Tana River 52
Tanae River 112
Tanga Islands 23
Tanganjika Sea 51
Tanglapoi, western Alor 211
Tanimbar Islands 24, 26, 27, 224, 225
Tanzania 30, 46, 110, 114, 125, 137
–, central 183
–, northeastern 38, 114, 183
–, southeast 163
–, southwest 138
Tara, western Siberia 48
Taranga 215
Tarara 28
Tarbagatay 159
Tarim Basin, central 50
Tasmania 29, 210, 214
Tawitawi 221
Tedzhen 237
Teffé, Solimoes, Brazil 84
Tehuantepec City, Oaxaca, Mexico 102
Tenasserim 180, 217
Tenerife 238
Tepa, Barbar Island 211
Ternate Island 56, 224
Terres Magellaniques 103
Texas 75, 146, 147
–, eastern 101
–, northern 146
–, southern 75, 101, 186, 187
–, southwestern 72, 144, 186
–, western 70, 239
Tezu Hills, Mishmi 180
Thailand 18, 34, 63, 64, 115, 116, 121, 123, 180, 193, 217
–, northern 40, 131
–, northwestern 53, 132
–, southeast 180
–, southern 18, 36, 37, 41, 123, 127, 129, 130
–, western 118
The Bahamas 195
The Ghauts 129
The Guianas 140, 143, 154–157, 170, 234
The Hollows, Mackay, Queens Park Creek 207
Three Kings Island 215
Thursday Island 209
Tian Shan 160, 201
Tibet
–, eastern 191
–, southeast 177, 180
–, southern 191
Ticao 222
Tien Shan 48, 135, 159, 200
Tierra del Fuego 20, 103, 142, 168, 194, 243
Timor Island 24, 210, 211
Tobago 20, 235
Todo Mountains 39
Togian Archipelago 223, 283
Tomsk, Siberia 149
Tonga Island 24, 25
Torres Strait 209, 212
Tortuga Island 21
Tourangly, Aral-Sea 108
Transbaikalia 108, 149, 162, 191
Transcaspia 108, 134
Transcaucasia 47, 106, 201
Transvaal 138
–, eastern 183
Travancore 178
Tres Marias Island 186
Tres Reyes 58
Trinidad Island 20, 77, 78, 140, 156, 157, 170
–, northeast 235
Trinkat 219
Triton Bay, New Guinea 207, 227
Tsinghai, western China 201
Tsushima 67
Tucumán 85, 102, 167, 171, 198, 204, 205, 237
Tukangbesi Islands 30, 59, 60
Tumindao Island 59
Tunis, Tunisia 47

Tunisia 47, 48, 133, 241
–, northwest 237
Tura Mountains, Garo Mountains, Assam 177
Turgui 241
Turkestan 133, 135, 191, 237, 241
–, Chinese 108
Turkey 189
–, central 49, 133
–, norhteast 134
–, northern 47
–, southeast 49, 190
–, southern 49, 50, 121
–, western 133
Turkmenia 50, 108, 134
Tymi River, East Sakhalin 107

U

Uganda 29, 137, 163, 175
–, eastern 176
–, northern 94
–, southern 95, 111
–, southwest 176, 182
–, western 113, 239
Ukraine 104, 188
–, southern 106, 189
Ulan Bator 152
Ulyanovsk (Simbirsk) 105
Ungave Peninsula 100
Union Island 21
United Provinces, India 192
United States 46, 97, 238
–, central 74, 146
–, coastal western 163
–, eastern 146
–, northeastern 74
–, northern 74, 151, 159, 160
–, southeastern 147
–, southern 165, 169, 202, 237, 239
–, southwestern 71, 99, 101, 163, 186
–, western 145, 164, 202
Upolu Island, Samoa Group 24
Upper Apure Valley, Venezuela 196
Upper Egypt 109
Upper Guinea 116
Upper Yangtse River, Southeast Tibet 109
Ural Mountains 105, 106, 133, 134, 237
Ural River 108
Ursula Island 58
Urubamba Valley 88
Uruguay 77, 79, 85, 102, 170, 199, 234, 235
–, northeast 204
–, northern 171
–, southern 171
Usambara Mountains 38, 114
Ussuriland 64, 67, 162
Utah, southeast 146
Uttar Pradesh, northern 114
Uzbekistan 50

V

Valle del Zamora, Ecuador 158
Vancouver Island, Victoria, British Columbia 99, 163
Vanuatu Island 24
–, northern 24
Venezuela 20, 77, 78, 102, 138, 140, 141, 143, 154, 156, 157, 170, 204, 234
–, central 196
–, eastern 20
–, northern 77, 82, 87, 154, 196, 242
–, northwest 82, 84, 90, 143, 154, 204, 235
–, southeast 170
–, southern 89, 143, 155
–, western 90, 166
Veracruz 76, 87, 101, 156, 235
–, northern 86
–, southeast 173
Vera Paz, Guatemala 154
Veragua, Panama 173
Vermelho, Paraná, Brazil 85
Vernon, Skagit Valley, Washington 145
Victoria, Australia 29, 32, 207, 209, 213, 214
–, eastern 32
–, southeast 207
Victoria River, Northern Territory 30
Victoria, Vancouver Island 70
Vieques Island 89
Vietnam 18, 30, 31, 34, 114, 121, 123, 192, 217
–, central 132
–, northeastern 180
–, southern 41
Villa Bella de Mato Grosso, River Guepore 140
Vindhyan Hills 109
Virgin Island 89
Virginia 101, 238
Viti Levu, Fiji Islands 30
Volcano Island 216
Volga
–, Basin 105
–, Delta 106
–, River 47, 48, 108, 148
Vulcan Tacaná, Chiapas, southern Mexico 203

W

Wadi Suenut, Jordan Valley, Palestine 106
Waigeo Island 206, 207, 226
Waikouaiti, South Island, New Zealand 232
Waingapo, Sumba Island 24
Wairarapa District, North Island, New Zealand 232
Walea Bahi 223
Walea Kodi 223
Wales 188
Walla Walla, Washington 70
Wallacea 216
Wallis Island 24
Warnambool, Victoria 29
Warnes, Santa Cruz, Bolivia 198
Washington 99, 163, 202
–, western 145
Water 207
Watson Ranch 101
Wellesley, Fed. Malay States 35
West Africa 113, 117, 118, 175, 244
West Asia 49
West Kimberley District 212
West Pakistan 108
West Sumatra 275
Western Australia 28, 30
–, northeast 206
Wetar Island, Lesser Sundas 24, 56, 57
Wyoming 151, 202

X

Xieng-Khouang, Laos 132

Y

Yagachi-shima 67
Yakoshih, northwest Manchuria 120
Yaku-shima 67
Yakutia 201
–, western central 149
Yakutsk 107
Yamato 216
Yamdena Island 26, 27
Yao-Shan, Kwangtung 31
Yapen Island 231
Yemen 112, 136
Ypanema, Sao Paulo 142
Yquitos, upper Amazonian River 155
Ysabel Island, Solomon Islands 229, 246
Yucatán 86, 87, 102
Yugoslavia
–, northwest 189
–, southeast 189
Yukon, southern 238
Yunnan 18
–, southeast 31
–, western 54
Yurinagui Alto 204

Z

Zagros, Southwest Iran 108, 121
Zaire 111, 118, 137, 183
–, central 113, 182
–, eastern 33, 37, 38, 116, 137, 153, 176, 184, 239
–, northeastern 126
–, northern 37, 38, 116, 182
–, northwest 182
–, southeast 183
–, southern 95, 138
Zaisan Nor 106
Zambesi Valley 183
Zambia 30
Zambiza 204
Zamboanga, Mindanao 68, 96, 221
Zamora 167
Zamora-Chinchipe, Ecuador 173
Zankab, Bahr el Abjad 94
Zanzibar Island 17, 137
Zimbabwe 110, 111, 137
Zulia 82